AF873505

GEORGE ALBERT BOULENGER

CATALOGUE
OF THE
LIZARDS
IN THE
BRITISH MUSEUM
(NATURAL HISTORY)

VOLUME II

IGUANIDÆ, XENOSAURIDÆ, ZONURIDÆ
ANGUIDÆ, ANNIELLIDÆ, HELODERMATIDÆ
VARANIDÆ, XANTUSIIDÆ, TEIIDÆ
AMPHISBÆNIDÆ

Elibron Classics
www.elibron.com

Elibron Classics series.

© 2005 Adamant Media Corporation.

ISBN 1-4021-0203-8 (paperback)
ISBN 1-4212-8342-5 (hardcover)

This Elibron Classics Replica Edition is an unabridged facsimile of the edition published in 1885 by Taylor and Francis, London.

Elibron and Elibron Classics are trademarks of Adamant Media Corporation. All rights reserved.

This book is an accurate reproduction of the original. Any marks, names, colophons, imprints, logos or other symbols or identifiers that appear on or in this book, except for those of Adamant Media Corporation and BookSurge, LLC, are used only for historical reference and accuracy and are not meant to designate origin or imply any sponsorship by or license from any third party.

CATALOGUE

OF THE

LIZARDS

IN THE

BRITISH MUSEUM

(NATURAL HISTORY).

SECOND EDITION.

BY

GEORGE ALBERT BOULENGER.

VOLUME II.

IGUANIDÆ, XENOSAURIDÆ, ZONURIDÆ, ANGUIDÆ,
ANNIELLIDÆ, HELODERMATIDÆ, VARANIDÆ,
XANTUSIIDÆ, TEIIDÆ, AMPHISBÆNIDÆ.

LONDON:
PRINTED BY ORDER OF THE TRUSTEES.
1885.

ALERE FLAMMAM.

PRINTED BY TAYLOR AND FRANCIS.

RED LION COURT, FLEET STREET.

PREFACE.

The manuscript of this second volume of the 'Catalogue of Lizards' was so far advanced at the time of publication of the first, that it has been possible to issue both in the same year.

The present volume contains an account chiefly of Lizards of the New World; and among the contributions to our collection, the donations made by Messrs. Salvin and Godman and by the U.S. National Museum are the most extensive and important.

Our best thanks for communications and the loan of specimens are due to Dr. Strauch, Director of the Museum of the Imperial Academy of St. Petersburg, Professor Vaillant, of the Muséum d'Histoire Naturelle de Paris, Dr. Boettger, of the Senckenberg Museum, and M. Lataste.

ALBERT GÜNTHER,
Keeper of the Department of Zoology.

British Museum, N. H.,
November 15, 1885.

INTRODUCTION.

THIS second volume contains an account of the families *Iguanidæ*, *Xenosauridæ*, *Zonuridæ*, *Anguidæ*, *Anniellidæ*, *Helodermatidæ*, *Varanidæ*, *Xantusiidæ*, *Teiidæ*, and *Amphisbænidæ*; it is therefore chiefly devoted to American Lizards.

The increase in the number of species known, and of species and specimens represented in the British Museum, since the publication of the general works by Duméril and Bibron and by Gray is shown in the following tables:—

	Number of Species characterized		
Families.	by Dum. & Bibr.	by Gray.	in present volume.
Iguanidæ	94	126	293
Xenosauridæ	—	—	1
Zonuridæ	6	8	14
Anguidæ	17	25	44
Anniellidæ	—	—	2
Helodermatidæ	1	1	3
Varanidæ	12	23	27
Xantusiidæ	—	—	4
Teiidæ	29	44	108
Amphisbænidæ	15	15	65
Total	174	242	561

Number of Species and Specimens in the British Museum in

	1845.		1885.	
	Species.	Specimens.	Species.	Specimens.
Iguanidæ	83	240	211	1358
Xenosauridæ	—	—	1	4
Zonuridæ	6	17	10	53
Anguidæ	16	38	26	147
Anniellidæ	—	—	1	1
Helodermatidæ ..	1	2	2	8
Varanidæ........	21	87	24	256
Xantusiidæ	—	—	1	7
Teiidæ	21	57	69	356
Amphisbænidæ....	9	21	30	145
Total..	157	462	375	2335

G. A. BOULENGER.

Department of Zoology,
November 13, 1885.

SYSTEMATIC INDEX.

LACERTILIA.

Subord. I. LACERTILIA VERA (continued).

Fam. 6. IGUANIDÆ.

Fam. 7. Xenosauridæ.

Fam. 8. Zonuridæ.

Fam. 9. Anguidæ.

CATALOGUE

OF

LIZARDS.

Order LACERTILIA

(continued).

Suborder I. LACERTILIA VERA

(continued).

Fam. 6. IGUANIDÆ.

Iguaniens, part., *Cuvier, Règne Anim.* ii. 1817.
Ascalabotæ, part., *Merrem, Tent. Syst. Amph.* 1820.
Pneustoidea, Agamoidea, part., *Fitzinger, Neue Classif. Rept.* 1826.
Iguanidæ, *Gray, Phil. Mag.* (2) ii. 1827.
Pachyglossæ, part., *Wagler, Syst. Amph.* 1830.
Dendrobatæ prosphyodontes and Humivagæ prosphyodontes, *Wiegmann, Herp. Mex.* 1834.
Iguaniens pleurodontes, *Duméril & Bibron, Erp. Gén.* iv. 1837.
Corythophanæ, Hypsilophi, Ptychosauri, Polychri, Dactyloæ, Draconturæ, Heterotropides, Steirolepides, Doryphori, Phrynosomata, *Fitzinger, Syst. Rept.* 1843.
Iguanidæ, *Gray, Cat. Liz.* 1845.
Anolidæ, Iguanidæ, *Cope, Proc. Ac. Philad.* 1864.
Iguanidæ, *Boulenger, Ann. & Mag. N. H.* (5) xiv. 1884.

The Lizards of this family resemble very closely, in external as well as internal characters, those of the preceding, or *Agamidæ*, from which they are distinguished by the pleurodont dentition. The distinct heterodontism, so frequent in the latter family, is exhibited to a slight degree by one genus only (*Uraniscodon*); in all others the teeth are subequal in size, and if, as is usually the case, the lateral ones differ from the front ones, the change is gradual. In all genera save *Amblyrhynchus*, *Conolophus*, and *Phymaturus*, the anterior teeth are conical; the lateral ones are also sometimes simply conical or with obtuse crowns, but more frequently compressed and tricuspid; in *Iguana* and *Metopoceros* the crowns are finely denticulated on the edge. In the above-named highly remarkable genera *Amblyrhynchus*, *Conolophus*, and *Phymaturus*, all the teeth are deeply trilobate, or flower-de-luce shaped. The shafts of the teeth are constantly long and cylindrical, and hollowed out at the base. Pterygoid teeth are present in many species; I have not employed this character for the distinction of genera unless accompanied by others. *Chamæleolis* is one of the few Lizards in which teeth are inserted on the palatine bones.

The skull does not differ in any important point from that of the *Agamidæ*. However, dermal cranial ossifications, as, for instance, the horn-like tubercles of *Phrynosoma*, may be present. A supra-orbital arch, such as has been noticed in the Agamoid genus *Lyriocephalus*, occurs in *Corythophanes* and *Phrynosoma*. The clavicle is slender proximally, except in the genera *Basiliscus* and *Læmanctus*. In describing the *Iguania* as having the "clavicle with simple proximal ends," Cope * remarks in a footnote that "the transverse limb of the mesosternum [sternum, *nobis*] extending to the angle of the clavicle, gives an appearance in some of the *Basiliscinæ* of a proximal foramen." I had hitherto held this opinion to be correct, but now find, on careful examination of the pectoral arch of the types alluded to, that the clavicle is really loop-shaped, and that those two genera constitute an exception in the family *Iguanidæ*, which is characterized by a non-dilated clavicle. The interclavicle is T- or anchor-shaped, except in *Phrynosoma*, in which the longitudinal limb is absent. The sternum is frequently perforated by a fontanelle. A more or less complete system of ossified or tendinous abdominal ribs is developed in many genera, in some (*Anolis*, *Polychrus*, *Liosaurus*, &c.) to the same extent as in the *Geckonidæ*.

The tongue is thick and villose, entirely fixed to the floor of the mouth, or slightly free anteriorly, and not or but very feebly nicked. The pupil of the eye is round, and the eyelids well developed. The tympanum is distinct, except in *Holbrookia*.

Femoral pores exist in all North-American genera, and are absent in the great majority of the South-American. In some of the latter the males have a series of pores on the anterior border of the vent.

The scaling of the head and body varies extremely, and gular appendages, crests, and other ornaments are not unfrequent. The upper head-scales are usually small, but in some genera assume a

---

* Proc. Ac. Philad. 1864, p. 227.

shield-like disposition. An enlarged scale is usually present, representing the interparietal of other Lizards, and, through fusion with adjacent scales, sometimes acquires a remarkably large size. Following the example of previous authors, and for convenience, I have used the term "occipital" for this shield, although it is not homologous with that so-called; but I have had to depart from this rule in dealing with the genus *Liolæmus*, in which, owing to the greater development of the parietals and the presence of a true occipital, that terminology might have led to confusion.

The habits of the numerous members of this family are as varied as their physiognomy. All the forms which we have observed in the Agamoids are repeated here, save the parachute-bearing Dragons, which have no pleurodont analogues. On the other hand, such types as the *Anoles*, with their digital expansions, and the semi-marine algivorous *Amblyrhynchus*, are unrepresented in the acrodont series. However, this apparent parallelism between the Agamoid and Iguanoid series of genera is very superficial, and there is, it appears to me, not one form so exactly repeated in both as to deserve to be united into the same genus, were the character of the dentition, on which the family distinction is based, not to be considered.

The great majority of Iguanoids are insectivorous; *Iguana*, *Brachylophus*, *Amblyrhynchus*, *Conolophus*, *Phymaturus*, *Sauromalus*, *Basiliscus*, are herbivorous; and the *Ctenosaura* are reported to be omnivorous. Some species of *Sceloporus* and *Phrynosoma* are at present the only Iguanoids known to be ovoviviparous.

With the exception of two genera occurring in Madagascar, and one in the Friendly and Fiji Islands, the range of this family is restricted to the New World.

Synopsis of the Genera.

I. Femoral pores absent (or a few present, in the male sex only, in the genus *Enyalioides*).

A. Digits more or less dilated or depressed, with smooth transverse lamellæ inferiorly, the distal joint narrower, cylindrical or compressed.

1. Distal joint raised above the penultimate.

Lateral teeth with subspherical crowns; palatine and pterygoid bones toothed . . . . 1. **Chamæleolis**, p. 7.

Lateral teeth tricuspid; tail prehensile.
2. **Xiphocercus**, p. 8.

Lateral teeth tricuspid; tail not prehensile.
3. **Anolis**, p. 11.

2. Distal joint not raised.

Digits scarcely dilated . . . . 4. **Norops**, p. 95.

B. Digits not dilated.

1. Præmaxillary teeth conical.

a. Infradigital lamellæ distinctly keeled.

α. Posterior part of head more or less produced.

A gular sac; head with two ridges uniting posteriorly into an elevated bony crest .... 7. **Corythophanes**, p. 101.

No gular sac; occipital region flat.. 8. **Læmanctus**, p. 104.

No gular sac; toes lobate .... 9. **Basiliscus**, p. 106.

β. Head not produced posteriorly.

* One or more transverse gular folds, or a very large[1] occipital (or interparietal) shield.

† Sternum without fontanelle.

‡ Body more or less compressed; a dorsal crest.

No abdominal ribs; no gular sac; no femoral pores.
10. **Ophryoessa**, p. 111.

Abdominal ribs; male with a more or less developed gular sac and usually a few femoral pores .... 11. **Enyalioides**, p. 112.

‡‡ Body depressed.

A dorsal crest; toes denticulated laterally.
19. **Chalarodon**, p. 128.

No dorsal crest; toes compressed .. 20. **Hoplurus**, p. 129.

No dorsal crest; toes depressed, denticulated laterally.
17. **Pristidactylus**, p. 127.

†† A sternal fontanelle; occipital shield very large.

‡ Caudal scales small or moderately large.

Digits straight; no canine-like teeth 27. **Tropidurus**, p. 171.

Digits strongly bent at the articulations; anterior maxillary teeth longest .... 28. **Uraniscodon**, p. 178.

‡‡ Caudal scales very large and spinose.

No fold across the throat; tail subcylindrical.
29. **Strobilurus**, p. 181.

Two folds across the throat; tail flat.
30. **Urocentron**, p. 182.

---

[1] *I. e.* much larger than the eye-opening.

** No transverse gular fold; occipital shield small.

† A gular appendage.

Body compressed .... 5. **Tropidodactylus**, p. 97.

†† No gular appendage.

‡ Males with anal pores.

Dorsal scales smooth and juxtaposed 23. **Helocephalus**, p. 137.

Dorsal scales keeled and imbricate.. 24. **Liolæmus**, p. 138.

‡‡ No anal pores.

§ Caudal scales forming verticils.

Abdominal ribs; dorsal and ventral scales large and keeled.
18. **Scartiscus**, p. 127.

No abdominal ribs .... 21. **Stenocercus**, p. 131.

§§ Caudal scales not forming verticils.

No abdominal ribs; body slightly depressed; head-scales small an keeled .... 25. **Saccodeira**, p. 158.

Abdominal ribs; body more or less compressed.
26. **Liocephalus**, p. 160.

b. Infradigital lamellæ smooth or indistinctly keeled.

α. A transverse gular fold; no anal pores.

* Body cylindrical or feebly compressed.

A slight dorsal crest or denticulation.
12. **Enyalius**, p. 117.

No crest; dorsal lepidosis heterogeneous.
13. **Anisolepis**, p. 122.

No crest; dorsal scales uniform, granular.
14. **Urostrophus**, p. 123.

** Body depressed; no dorsal crest.

Lateral teeth tricuspid............ 15. **Liosaurus**, p. 124.

Lateral teeth with obtuse or subspherical crowns.
16. **Diplolæmus**, p. 125.

β. No gular fold; male with anal pores.
22. **Ctenoblepharis**, p. 136.

2. Præmaxillary teeth tricuspid.

Body much depressed .... 31. **Phymaturus**, p. 184.

II. Femoral pores present.

A. Third and fourth toes equal.

Body strongly compressed; no dorsal crest.
6. **Polychrus**, p. 98.

B. Fourth toe longer than third.

1. Anterior teeth, as well as the laterals, trilobate.

Tail strongly compressed, crested; a rudiment of web between the toes .... 32. **Amblyrhynchus**, p. 185.

Tail roundish, slightly compressed; toes quite free.
33. **Conolophus**, p. 186.

2. Lateral teeth with denticulated crowns.

A small gular pouch .... 34. **Metopoceros**, p. 187.

A large non-dilatable gular appendage.
35. **Iguana**, p. 189.

3. Lateral teeth tricuspid (sometimes with one or two additional cusps) or subconical.

a Head without spines.

α. Series of femoral pores short; a transverse gular fold.

Body compressed; a dorsal crest; male with a gular sac.
36. **Brachylophus**, p. 192.

Body scarcely compressed; a dorsal crest; no gular sac.
38. **Ctenosaura**, p. 195.

Body depressed; no dorsal crest; dorsal scales equal.
39. **Cachryx**, p. 199.

Body depressed; no dorsal crest; dorsal scales heterogeneous.
40. **Hoplocercus**, p. 199.

β. Series of femoral pores extending on nearly the whole length of the thigh, or no transverse gular fold.

* A vertebral crest or denticulation.

A small gular sac and a feeble transverse gular fold.
37. **Cyclura**, p. 193.

No gular pouch; a strong transverse gular fold.
41. **Dipsosaurus**, p. 201.

** No dorsal crest; body depressed.

† Occipital scale small; a transverse gular fold.

Tail scarcely longer than head and body.
42. **Sauromalus**, p. 202.

Tail much longer than head and body.
43. **Crotaphytus**, p. 203.

†† Occipital very large.

‡ Two transverse gular folds; dorsal scales uniform, minute; caudal scales small.

§ Edge of the gular folds not denticulated.
44. **Petrosaurus**, p. 205.

§§ Edge of the second gular fold denticulated.

Tympanum distinct; no spines on the outer margin of the foot.
45. **Callisaurus**, p. 205.

Tympanum distinct; a series of spines on the outer margin of the foot . . . . 46. **Uma**, p. 206.

Tympanum concealed . . . . 47. **Holbrookia**, p. 207.

‡‡ One or two transverse gular folds; caudal scales large and strongly keeled.
48. **Uta**, p. 210.

‡‡‡ No gular fold .. 49. **Sceloporus**, p. 216.

b. Head bordered posteriorly by bony spines.
50. **Phrynosoma**, p. 239.

1. CHAMÆLEOLIS.

Anolis, part., *Dum. & Bibr.* iv. p. 85.
Chamæleolis, *Cocteau, in R. de la Sagra, Hist. Cuba, Rept.* p. 145; *Gray, Cat. Liz.* p. 196.
Pseudochamæleon, *Fitzing. Syst. Rept.* p. 63.

Tympanum distinct. Parietal bone expanded into a large shield, covered, as well as the rest of the upper surface of the head, with rugose tubercular dermal ossifications. Body compressed, covered above with unequal-sized, flat, soft scales, inferiorly with minute granules. A dorso-nuchal dermal fold or crest. No transverse gular fold. Male with a large, non-inflatable, folding gular appendage. Digits depressed, strongly dilated, inferiorly with smooth transverse lamellæ, the distal joints slender and compressed. No femoral or præanal pores. Tail moderately long, strongly compressed. Lateral teeth with subspherical crowns; pterygoid and palatine bones toothed. No sternal fontanelle. Abdominal ribs.

Cuba.

1. Chamæleolis chamæleontides.

Chamæleolis fernandina, *Gray, Cat.* p. 196.
Anolis chamæleonides, *Dum. & Bibr.* p. 168.
Chamæleolis fernandina, *Coct. l. c.* pl. xv.

Pseudochamæleon cocteaui, *Fitz. l. c.*
Anolis fernandina, *Reinh. & Lütk. Vidensk. Meddel.* 1862, p. 170.
? Chamæleolis porcus, *Cope, Proc. Ac. Philad.* 1864, p. 168.

Head very large, chamæleon-like, twice as long as broad, raised posteriorly; upper surface concave, covered with irregular, rough, bony tubercles, largest on the prominent canthus rostralis; nostril near the tip of the snout; eye-opening very small; ear-opening small, vertically oval, a small dermal process above it; labials numerous; rostral and mental small, latter bisected; several rows of enlarged scales near the infralabials. Gular scales minute, granular; a double row of small triangular lobes on the chin; gular appendage large, extending to the posterior extremity of the thorax. Body strongly compressed. Dorso-nuchal crest attached to the posterior border of the casque, decreasing gradually in height, its edge scarcely denticulated; it is covered with longitudinal rows of roundish flat tubercles, separated by wrinkled subgranular skin. Body covered above and on the sides with large, round, flat scales, separated by intervals occupied by much smaller ones; ventral region minutely granulate. Limbs rather short, the adpressed hind limb reaching axilla; digits strongly dilated. Tail very strongly compressed, as long as head and body, covered with large, juxtaposed, irregular smooth scales, inferiorly with smaller keeled ones. Brownish, variegated with yellowish; gular appendage purplish.

Total length	325	millim.
Head (to extremity of casque)	66	„
Width of head	35	„
Body	110	„
Fore limb	59	„
Hind limb	80	„
Tail	160	„

Cuba.

a. ♂. Cuba. W. S. MacLeay, Esq. [P.].

2. XIPHOCERCUS.

Anolis, part., *Dum. & Bibr.* iv. p. 85; *Gray, Cat. Liz.* p. 199.
Xiphocercus, *Fitzing. Syst. Rept.* p. 67; *Cope, Proc. Ac. Philad.* 1861, p. 215.
Placopsis, *Gosse, Ann. & Mag. N. H.* (2) vi. 1850, p. 346.

Tympanum distinct. Body compressed, covered above with unequal-sized flat scales. Dorso-nuchal crest very slightly developed or absent. Head with large shields. No transverse gular fold. Male with a non-inflatable folding gular appendage. Digits strongly dilated, inferiorly with smooth transverse lamellæ, the distal joints slender and compressed. No femoral or præanal pores. Tail moderately long, strongly compressed, prehensile. Lateral teeth tricuspid; pterygoid teeth. No sternal fontanelle. Abdominal ribs.

Tropical America.

1. Xiphocercus valenciennesii.

Anolius valenciennii, *Gray, Cat.* p. 206.
Anolis valencienni, *Dum. & Bibr.* p. 131; *A. Dum. Arch. Mus.* viii. pl. xix. fig. 2; *Bocourt, Miss. Sc. Mex., Rept.* pl. xiv. fig. 1.
Dactyloa (Xiphocercus) valenciennii, *Fitzing. l. c.* p. 67.
Placopsis ocellata, *Gosse, l. c.*
Anolis leucocephalus, *Hallow. Proc. Ac. Philad.* 1856, p. 226.
Xiphocercus valenciennii, *Cope, l. c.*

Head large, nearly twice as long as broad, nearly twice as long as the tibia; forehead plane, covered with very large symmetrical smooth shields; shields of the supraorbital semicircles in contact between the orbits or separated by one row of scales; occipital much longer than broad, much larger than the ear-opening, in contact with the supraorbital shields or separated by one series of scales; five to eight large smooth supraocular scales, separated from the supraorbitals by one row of granules; canthus rostralis sharp, canthal scales four or five; nostril just above the canthus rostralis; eight or nine upper labials to below the centre of the eye; ear-opening very small. Gular appendage very large in the male, extending beyond the thorax, smaller, but nevertheless well developed, in the female; the greater portion of this appendage perfectly naked; gular scales smooth. Body strongly compressed, not crested, covered with flat smooth roundish granules, largest on the flanks, where they are separated by minute grains, subimbricate on the belly. Limbs rather short, the adpressed hind limb reaches the axilla or the shoulder; twenty-three to twenty-six lamellæ under phalanges II. and III. of the fourth toe. Tail strongly compressed, with a low crest or denticulation above; its length does not much exceed that of head and body. Male with enlarged postanal scales. Cream-coloured or pale brown above, with more or less distinct brown markings; lower surfaces white, the belly sometimes vermiculated with brown; the skin of the gular appendage grey (in spirits).

Total length	191	millim.
Head	27	„
Width of head	15	„
Body	57	„
Fore limb	28	„
Hind limb	44	„
Tibia	13	„
Tail	107	„

Jamaica.

a–b, c–e, f. ♂, ♀, & hgr. Jamaica. P. H. Gosse, Esq. [P.]. (Types of *Placopsis ocellata.*)

g. ♂. Jamaica?

2. Xiphocercus heterodermus.

Anolis heterodermus, *A. Dum. Cat. Méth. Rept.* p. 59, and *Arch. Mus.* viii. p. 516, pl. xix. fig. 4.

Head large, nearly twice as long as broad, nearly twice as long as the tibia; forehead concave, covered with large, irregular, smooth or slightly tectiform shields; shields of the supraorbital semicircles in contact between the orbits, or separated by two rows of scales; parietal and squamosal bones forming slight ridges, with series of enlarged shields; occipital scale much longer than broad, much larger than the ear-opening, in contact with the supraorbital shields or separated by one series of scales; one or two large smooth supra-ocular scales, separated from the supraorbitals by one row of granules; canthus rostralis angular, canthal scales extremely large, four; nine upper labials to below the centre of the eye; ear-opening very small. Gular appendage rather small, with longitudinal series of scales; gular scales smooth; a series of large square shields on each side, close to and much larger than the infralabials. Body strongly compressed, with a slight dorso-nuchal denticulation or crest; dorsal and lateral scales flat, smooth, roundish, widely separated by smaller round scales of irregular size; ventral scales hexagonal or squarish, smooth, strongly imbricate. Limbs rather short; the adpressed hind limb reaches the axilla; twenty lamellæ under phalanges II. and III. of the fourth toe. Tail strongly compressed, with a slight crest or denticulation above, covered with rather large keeled scales, the keels forming straight longitudinal continuous lines; its length does not much exceed that of head and body. Male with strongly en-larged postanal scales. Olive above, some of the larger scales and a band from the loreal region to the fore limb yellowish white; belly pale olive, brown-spotted.

Total length	160	millim.
Head	21	,,
Width of head	11	,,
Body	44	,,
Fore limb	25	,,
Hind limb	33	,,
Tibia	10	,,
Tail	95	,,

Colombia.

a–b. ♂.	Bogota.
? *c*. Skin, not good state, from snout to vent 100 millim.	—— ?

3. ANOLIS *.

Anolis, *Daud. Rept.* iv. p. 50; *Merrem, Tent.* p. 44; *Fitzing. N. Classif. Rept.* p. 17; *Dum. & Bibr.* iv. p. 85; *Bocourt, Miss. Sc. Mex., Rept.* p. 57.
Anolius, *Cuv. R. A.* ii. p. 41; *Gray, Cat. Liz.* p. 199.
Xiphosurus, *Fitzing. l. c., and Syst. Rept.* p. 67; *Gray, l. c.* p. 197.
Dactyloa, *Wagler, Syst. Amph.* p. 148; *Wiegm. Herp. Mex.* p. 16; *Fitzing. Syst. Rept.* p. 66; *Gray, l. c.* p. 198.
Draconura, *Wagl. l. c.*; *Wiegm. l. c.*; *Fitzing, l. c.* p. 69; *Gray, l. c.* p. 207.
Acantholis, *Cocteau, C. R. Ac. Paris,* iii. 1836, p. 226; *Gray, l. c.* p. 206.
Ctenonotus, *Fitzing. l. c.* p. 64.
Semiurus, p. 64.
Microctenus, p. 64.
Ptychonotus, p. 65.
Istiocercus, p. 65.
Eunotus, p. 65.
Deiroptyx, p. 66.
Trachycœlia, p. 66.
Ctenodeira, p. 66.
Tropidopilus, p. 66.
Eudactylus, p. 67.
Heterolepis, p. 67.
Trachypilus, p. 67.
Pristicercus, p. 67.
Ctenocercus, p. 68.
Gastrotropis, p. 68.
Heteroderma, p. 68.
Dracontopsis, p. 69.
Rhinosaurus, *Gray, l. c.* p. 199.
Scytomycterus, *Cope, Journ. Ac. Philad.* (2) viii. 1876, p. 165.

Tympanum distinct. Body more or less compressed, cylindrical, or feebly depressed, covered with juxtaposed or imbricate scales or granules. Dorso-nuchal crest slightly developed or absent. Male with a gular appendage. Digits more or less dilated, inferiorly with smooth transverse lamellæ, the distal joints slender and raised above the basal ones. No femoral or præanal pores. Tail long, cylindrical or compressed, not prehensile. Lateral teeth tricuspid; pterygoid teeth generally present. No sternal fontanelle. Abdominal ribs.

Tropical and Subtropical America.

---

* 1. *Anolius flavescens*, Gray, Cat. p. 202.—*Hab.* —— ?
The specimens are missing in the collection.
2. *Anolis (Draconura) tropidogaster*, Hallow. Proc. Ac. Philad. 1856, p. 224. —Colombia.
3. *A. sericeus*, Hallow. l. c. p. 227.—Jalapa, Mexico.
4. *A. angusticeps*, Hallow. l. c. p. 228.—Cuba.
5. *A. cooperi*, Baird, Proc. Ac. Philad. 1858, p. 254; Cope, eod. loc. 1862, p. 181.—California.
6. *A. longicauda*, Hallow. Proc. Ac. Philad. 1860, p. 481; Cope, l. c.—Nicaragua.
7. *A. rubiginosus*, Bocourt, Ann. Sc. Nat. (5) xvii. 1872, art. 2, and Miss. Sc. Mex., Rept. pl. xvii. bis, fig. 2.—Oaxaca.

Synopsis of the Species.

I. Tail strongly compressed and crested; body strongly compressed, with a dorsal crest of small triangular lobes.

Ventral scales smaller than dorso-laterals; latter flat and smooth .... 1. *equestris*, p. 21.
Ventrals slightly smaller than dorso-laterals; latter flat and keeled; scales on upper surface of snout tubercular................. 2. *ricordii*, p. 22.
Ventrals nearly as large as dorso-laterals; latter flat and keeled; scales on upper surface of snout flat, hexagonal .... 3. *cuvieri*, p. 23.
Ventrals larger than dorso-laterals; latter granular and keeled .... 4. *edwardsii*, p. 24.

II. Tail more or less strongly compressed in the male, with a crested or serrated upper edge, which is at least indicated in the female by a series of strongly enlarged keeled scales. Dorsal scales small, as large as or a little larger than the laterals, and smaller than the ventrals; those on the vertebral line frequently more enlarged. Occipital scale considerably larger than the surrounding ones. No fold across the throat.

A. Ventral scales smooth or very feebly keeled.

1. The scales on the middle of the back larger than those on the flanks.

a. Tibia as long as the distance between the end of the snout and the ear-opening.

Supraorbital semicircles separated from each other by two or three series of scales, and from the occipital by four or five series .. 5. *gundlachii*, p. 25.

b. Tibia shorter than the distance between the end of the snout and the ear-opening.

a. Scales on upper surface of snout not keeled; supraorbital semicircles formed of very large scales, in contact or narrowly separated.

* The width of the dilated part of the middle toes equals at least two thirds the length of the slender distal part.

† Occipital scale separated from the supraorbitals.

Supraocular scales keeled; scales on the vertebral line very feebly enlarged; 18 to 25 lamellæ under phalanges II. and III. of the fourth toe......................... 6. *cristatellus*, p. 26.
Supraocular scales smooth; scales on the vertebral line scarcely larger than those on the flanks; 18 to 21 lamellæ under phalanges II. and III. .... 7. *stratulus*, p. 27.

Supraocular scales smooth or very feebly keeled; scales on the vertebral line abruptly enlarged, forming two longitudinal series; ventral scales nearly as large as antefemorals; 18 to 21 lamellæ under phalanges II. and III. .... 9. *gingivinus*, p. 28.
Supraocular scales smooth or very feebly keeled; scales on the vertebral line enlarged; ventrals much smaller than antefemorals; 23 to 32 lamellæ under phalanges II. and III. .... 10. *leachii*, p. 29.

†† Occipital in contact with supraorbitals.

Supraocular scales smooth .... 11. *alligator*, p. 31.

** The width of the dilated part of the middle toes only half the length of the slender distal part.

Head not twice as long as broad; ventral scales strongly imbricate, nearly as large as the enlarged supraoculars .... 14. *cybotes*, p. 34.
Head twice as long as broad; ventrals much smaller than the largest supraoculars .... 15. *citrinellus*, p. 35.

β. Scales on upper surface of snout keeled; supraorbital semicircles separated by one or two series of scales.

Occipital scale smaller than the ear-opening .. 8. *homolechis*, p. 28.
Occipital larger than the ear-opening ...... 12. *grahami*, p. 32.

2. The scales on the middle of the back not at all larger than those on the flanks.

Supraorbital semicircles in contact; supraocular scales smooth .... 13. *distichus*, p. 33.

B. Ventral scales keeled.

1. Occipital scale in contact with the supraorbitals.

Supraorbital semicircles in contact; supraocular scales keeled .... 16. *acutus*, p. 36.

2. Occipital separated from supraorbitals.

a. Scales on upper surface of snout not keeled; supraorbital semicircles in contact or narrowly separated by one or two series of scales.

α. Tibia shorter than the distance between the end of the snout and the ear-opening.

Supraocular scales keeled .... 17. *krugi*, p. 37.

β. The length of the tibia equals or exceeds the distance between the end of the snout and the ear-opening.

Supraocular scales keeled .... 18. *richardii*, p. 37.
Supraocular scales smooth .... 19. *lineatus*, p. 38.

b. Scales on upper surface of snout keeled; supraorbital semicircles separated by one or two series of scales.

Ventral scales small and feebly imbricate, the keels not forming continuous lines ...... 20. *lineatopus*, p. 39.
Ventral scales large and strongly imbricate, the keels sharp and forming continuous lines .... 21. *sagræ*, p. 40.

III. Tail strongly compressed, with serrated upper edge; a strong fold across the throat.

All the scales very small and keeled........ 22. *vermiculatus*, [p. 41.

IV. Tail cylindrical; dorsal and lateral scales minutely granular, intermixed with scattered enlarged ones.

Ventral scales smooth; supraorbital semicircles in contact .... 23. *loysiana*, p. 42.

V. Tail subcylindrical or slightly compressed, seldom more compressed and with a slightly serrated trenchant upper edge[1]. Dorsal scales, except in some cases those on the vertebral line, not or but indistinctly larger than those on the flanks. Digital expansions strongly or moderately developed, *i. e.* the third phalange of the fourth toe is considerably broader than the basal.

A. Occipital scale at least as large as the ear-opening; scales of the supraorbital semicircles considerably enlarged.

1. Tibia not measuring two thirds the length of the head (to occiput).

a. Occipital separated from supraorbitals.

α. 22 to 27 lamellæ under phalanges II. and III. of the fourth toe.

Scales of the supraorbital semicircles separated from each other by one or two series of scales, and from the occipital by one series. 24. *carolinensis*, [p. 43.
Scales of the supraorbital semicircles separated from each other by two or three series of scales, and from the occipital by three series .... 25. *chlorocyanus*, [p. 44.

β. 18 lamellæ under phalanges II. and III. of the fourth toe.

Occipital separated from supraorbitals by one series of scales; two or three large supraoculars .... 27. *isolepis*, p. 46.

---

[1] The species presenting this exception will at once be recognized as not belonging to division II. of this synopsis, by the very small size of the occipital.

Occipital separated from supraorbitals by two or three series of scales; four loreal scales in a vertical series in front of the orbit .. 28. *beckeri*, p. 46.
Occipital separated from supraorbitals by one series of scales; two loreal rows .... 44. *impetigosus*, [p. 55.

b. Occipital in contact with supraorbitals.

Rostral plate not produced .... 45. *tigrinus*, p. 55.
Rostral produced into a flexible appendage .. 46. *lævis*, p. 56.

2. Tibia measuring at least two thirds the length of the head.

a. Occipital separated from supraorbitals.

α. Tibia quite as long as the distance between the end of the snout and the ear.

* Supraorbital semicircles in contact.

Supraocular scales smooth .... 26. *lucius*, p. 45.

** Supraorbital semicircles separated.

Supraocular scales keeled; supraorbital semicircles separated from each other by three series of scales, and from the occipital by four or five series .... 33. *limifrons*, p. 49.
Supraocular scales keeled; supraorbital semicircles separated from each other by one series of scales, and from the occipital by two series .... 34. *bransfordii*, p. 50.
Supraocular scales smooth................ 35. *trochilus*, p. 50.

β. Tibia shorter than the distance between the end of the snout and the ear.

* Scales on upper surface of snout keeled; supraocular scales keeled.

† Scales on the vertebral region not keeled, perfectly similar to those on the flanks.

Occipital not larger than the ear-opening.... 29. *damulus*, p. 47.

†† Scales on the vertebral region keeled.

‡ Tibia slightly shorter than the distance between the end of the snout and the ear-opening.

The adpressed hind limb does not reach beyond the eye; tail once and two thirds to twice as long as head and body .... 30. *fusco-auratus*, [p. 48.
The adpressed hind limb does not reach beyond the eye; tail not more than once and one third as long as head and body ...... 31. *rodriguezii*, p. 49.

The adpressed hind limb reaches the end of the snout .... 32. *aureolus*, p. 49.

‡‡ Tibia considerably shorter than the distance between the end of the snout and the ear-opening.

Occipital scarcely larger than the ear-opening, separated from the supraorbitals by two series of scales .... 42. *baccatus*, p. 54.
Occipital considerably larger than the ear-opening, separated from the supraorbitals by three or four series of scales .... 43. *guentheri*, p. 54.

** Scales on upper surface of snout smooth.

† Occipital larger than the ear-opening.

‡ Canthus rostralis weak; snout not produced.

Supraorbital semicircles separated by two or three rows of scales; the adpressed hind limb reaches the front of the orbit ...... 36. *bocourtii*, p. 50.
Supraorbital semicircles in contact or separated by one row of scales; the adpressed hind limb does not reach the eye .... 37. *ortonii*, p. 51.

‡‡ Canthus rostralis sharp; end of snout truncate in the female, compressed and produced in the male.

Ventral scales smooth .... 47. *punctatus*, p. 57.
Ventral scales keeled .... 48. *boulengeri*, p. 58.

†† Occipital not larger than the ear-opening.

Supraorbital semicircles in contact .... 50. *buckleyi*, p. 58.

b. Occipital in contact with supraorbitals.

Supraorbital semicircles in contact .... 49. *transversalis*, p. 58.

B. Occipital very small or quite indistinct, at any rate smaller than the ear-opening.

a. Tibia as long or nearly as long as the distance between the end of the snout and the ear-opening.

a. Median dorsal scales keeled.

* Head (to occiput) a little longer than the tibia.

† Caudal scales subequal.

16 to 18 lamellæ under phalanges II. and III. of the fourth toe; ventral scales smooth or indistinctly keeled .... 38. *schiedii*, p. 52.
13 or 14 lamellæ under phalanges II. and III.; ventrals keeled .... 40. *tropidolepis*, p. 53.

†† Tail with a dorsal series of enlarged scales.

Supraorbital semicircles separated by three or four series of scales .... 39. *polylepis*, p. 52.
Supraorbital semicircles separated by one series of scales .... 41. *pulchripes*, p. 53

** Head not longer than the tibia.

Supraorbital semicircles separated by five series of scales .... 58. *pachypus*, p. 63.

β. All dorsal scales perfectly smooth.

Head as long as the tibia; upper head-scales smooth .... 51. *fasciatus*, p. 59.
Head a little longer than the tibia; upper head-scales bi- or tri-carinate .... 52. *gemmosus*, p. 60.

b. Tibia not longer than the distance between the end of the snout and the posterior border of the orbit.

a. Ventral scales smooth or indistinctly keeled.

* Enlarged supraocular scales keeled.

Dorsal scales smooth .... 53. *andianus*, p. 60.
Dorsal scales keeled .... 54. *squamulatus*, [p. 61.

** Supraocular scales not keeled.

† Scales on upper surface of arm pluricarinate.

Ear-opening very small .... 56. *microtus*, p. 62.

†† Scales on upper surface of arm smooth or unicarinate.

Dorsal scales flat and smooth; tail without upper ridge, nearly twice as long as head and body .... 57. *insignis*, p. 63.
Dorsal scales convex, granular; tail little longer than head and body, subtriangular in section, with trenchant, serrate dorsal angle .... 59. *pentaprion*, p. 64.
Dorsal scales granular and keeled; tail at least twice as long as head and body .... 60. *fraseri*, p. 65.
All the scales of the body equal in size .... 55. *latifrons*, p. 62.

β. Ventral scales distinctly keeled.

Ventral scales large and sharply keeled .... 61. *copii*, p. 65.
Ventral scales small and rather feebly keeled 62. *petersii*, p. 66.

VI. Tail strongly compressed; dorsal scales much larger than the laterals.

Dorsal and ventral scales rhomboidal, imbricate, keeled .... 63. *pulchellus*, p. 67.

VII. Tail cylindrical or feebly compressed; dorsal scales larger than the laterals whenever the digital expansions are distinctly developed; dorsal and lateral scales equal or nearly equal in a few species in which the third phalanx of the fourth toe is not distinctly wider than the basal.

A. Dorsal scales as large as or larger than the ventrals.

1. Head (to occiput) longer than the tibia.

a. Dorsal scales not larger than the ventrals.

Dorsal and ventral scales large; occipital enlarged.............................. 64. *semilineatus*, [p. 68.

Scales smooth; occipital not larger than the surrounding scales..................... 91. *oxylophus*, p. 84.

b. Dorsal scales larger than the ventrals.

α. Scales on upper surface of snout smooth.

Dorsal and ventral scales keeled .... 65. *spectrum*, p. 69.

Dorsal and ventral scales smooth .... 67. *lionotus*, p. 70.

β. Scales on upper surface of snout keeled or striated.

Dorsal scales feebly keeled or striated ...... 66. *cyanopleurus*, [p. 69.

Dorsal scales strongly keeled, in about ten longitudinal series .... 86. *humilis*, p. 82.

Dorsal scales strongly keeled, in seven series. 87. *quaggulus*, p. 83.

2. Head not, or scarcely, longer than the tibia; dorsal and ventral scales sharply keeled.

Dorsal scales larger than ventrals; tibia as long as the head.......................... 88. *tropidonotus*, [p. 83.

Dorsal scales larger than ventrals; tibia shorter than the head .... 89. *uniformis*, p. 84.

Dorsal scales not larger than ventrals ...... 90. *metallicus*, p. 84.

B. Dorsal scales smaller than ventrals[1].

1. The length of the tibia is less than the distance between the end of the snout and the ear.

a. Ventral scales smooth.

Head elongate .... 68. *alutaceus*, p. 71.

b. Ventral scales keeled.

α. Occipital considerably longer than the greatest diameter of the ear-opening.

[1] In a few species very slightly.

* Head nearly twice as long as broad.

The adpressed hind limb reaches the eye.... 70. *heliactin*, p. 72.
The hind limb reaches the neck or the ear .. 71. *ustus*, p. 73.

** Head about once and two thirds as long as broad.

† Dorsal scales considerably smaller than ventrals, two or four median rows abruptly enlarged.

Two median dorsal rows enlarged; supraocular scales keeled .... 73. *concolor*, p. 74.
Two median dorsal rows enlarged; supraoculars smooth .... 74. *bourgæi*, p. 74.
Four median dorsal rows enlarged .... 76. *læviventris*, p. 76.

†† Dorsal scales slightly smaller than ventrals, graduating into the laterals.

Enlarged supraocular scales much dilated transversely, in a single longitudinal series, smooth .... 77. *nebulosus*, p. 77.

††† Dorsal scales considerably smaller than ventrals, graduating into the laterals.

Dorsal and ventral scales small; male without a large blue or purplish spot on the gular appendage .... 80. *intermedius*, [p. 78.
Ear-opening not or hardly half the length of the occipital; scales rather large; male with a large spot on the gular appendage . 81. *sallæi*, p. 79.
Ear-opening measuring more than half the length of the occipital; male with a large spot on the gular appendage .... 82. *binotatus*, p. 80.

β. Occipital not, or but slightly, longer than the ear-opening.

* Dorsal scales rhomboidal, keeled.

Dorsal scales nearly as large as ventrals, small; occipital not enlarged .... 75. *salvini*, p. 75.
Dorsal scales nearly as large as ventrals, large; occipital enlarged .... 78. *nebuloides*, p. 77.
Dorsal scales considerably smaller than ventrals, graduating into the laterals........ 83. *cumingii*, p. 80.
Dorsal scales considerably smaller than ventrals, abruptly passing into the laterals.. 85. *crassulus*, p. 81.

** Dorsal scales granular, smooth, very small.

Head very short........................ 106. *obtusirostris*, [p. 95.

2. Tibia at least as long as the distance between the end of the snout and the ear.

a. Dorsal scales keeled.

α. The third phalanx of the fourth toe distinctly wider than the basal.

* Ventral scales smooth or indistinctly keeled.

Supraorbital semicircles separated by one row of scales of nearly equal size............ 72. *cymbops*, p. 73.

** Ventral scales keeled.

† Scales of the supraorbital semicircles in contact medially or separated by one row of very small scales.

Scales on the forehead not keeled; tail with a dorsal series of enlarged scales .... 79. *palpebrosus*, [p. 77.
Scales on the forehead keeled .... 95. *lemurinus*, p. 86.

†† Scales of the supraorbital semicircles separated by one to four series of scales.

‡ Occipital as large as or larger than the ear-opening.

Dorsal scales much larger than laterals, smaller than ventrals; supraorbital semicircles separated from each other by one or two series of scales, and from the occipital by two or three series; 13 or 14 lamellæ under phalanges II. and III. of the fourth toe .. 84. *cupreus*, p. 80.
Dorsal scales very small; supraorbital semicircles separated from each other by two to four series of scales, and from the occipital by three to five series; 13 or 14 lamellæ.. 93. *godmani*, p. 85.
Scales of the three or four median dorsal series nearly as large as the ventrals .... 96. *trachyderma*, [p. 87.
Dorsal scales very small; 17 or 18 lamellæ 98. *biporcatus*, p. 88.

‡‡ Occipital smaller than the ear-opening.

Occipital minute or wanting.............. 91. *pœcilopus*, p. 84.
Occipital enlarged, little smaller than the ear-opening .... 97. *stigmosus*, p. 87.

β. The three basal phalanges of the fourth toe of nearly equal width.

* Head nearly twice as long as broad.

Eleven longitudinal series of large dorsal scales 69. *bitectus*, p. 71.

Two median series of dorsal scales abruptly enlarged .... 94. *radulinus*, p. 86.

** Head short.

† Occipital considerably larger than the ear-opening.

Ventral scales large and sharply keeled .... 99. *chrysolepis*, p. 89.

†† Occipital nearly as large as, or smaller than, the ear-opening.

Dorsal scales very small, slightly larger than the laterals; ventrals large and strongly keeled.............................. 100. *scypheus*, p. 90.
Dorsal and ventral scales small, feebly keeled 102. *leptoscelis*, p. 92.
Scales on the vertebral region nearly as large as the ventrals; latter strongly keeled.... 103. *lentiginosus*, [p. 93.

b. Dorsal scales smooth.

Occipital larger than the ear-opening; upper head-scales keeled .... 101. *nitens*, p. 91.
Occipital larger than the ear-opening; upper head-scales smooth.................... 104. *bombiceps*, [p. 94.
Occipital minute........................ 105. *capito*, p. 94.

1. **Anolis equestris.**

Dactyloa equestris, *Gray, Cat.* p. 198*.
Sloane, Nat. Hist. Jam. ii. p. 333, pl. 273. fig. 2.
Anolis equestris, *Merr. Tent.* p. 45; *Dum. & Bibr.* iv. p. 157; *Coct. in R. de la Sagra, Hist. Cuba, Rept.* p. 114, pl. ix.
Anolius rhodolæmus, *Bell, Zool. Journ.* iii. 1828, p. 235, suppl. pl. xx.
Ctenonotus (Eupristis) equestris, *Fitzing. Syst. Rept.* p. 64.

Head very large, once and three fourths as long as broad, much longer than the tibia; upper surface slightly concave, covered with irregular, rough, bony tubercles, largest on the snout, and especially on the canthus rostralis; loreal rows four or five †; supraocular scales very small; occipital indistinct; squamosal bone forming a strong tubercular ridge; eye-opening small; ear-opening very small; nine or ten upper labials to below the centre of the eye. Gular appendage very large, extending posteriorly to beyond the thorax, a little smaller in the female; its greater portion perfectly naked. Body strongly compressed, with a small dorso-nuchal crest

* O'Shaughnessy remarks, Ann. & Mag. N. H. (4) xv. p. 271, that "the second specimen referred in Dr. Gray's Catalogue to this species is a *Urostrophus vautieri*." As I find in another bottle labelled *Urostrophus vautieri* a specimen of *Anolis equestris*, it is clear that Gray never committed the error which has been attributed to him, but that a confusion of bottles has taken place.

† The loreal rows are counted in a vertical series in front of the orbit.

of triangular erect scales. Dorsal scales large, flat, smooth, suboval, separated by fine granules, slightly larger on the flanks; ventral scales much smaller than dorsals, subhexagonal, imbricate, smooth. Limbs moderately long; the adpressed hind limb reaches the temple; digital expansions well developed; forty-four to forty-six lamellæ under phalanges II. and III. of the fourth toe. Tail strongly compressed, with serrated upper edge, its length nearly twice that of head and body. Male with enlarged postanal scales. Bright blue-green above, pale green inferiorly: upper surface of head yellow or brown; a whitish band above the fore limb; gular appendage pink.

Total length	460	millim.
Head	55	,,
Width of head	31	,,
Body	105	,,
Fore limb	67	,,
Hind limb............	106	,,
Tibia	33	,,
Tail	300	,,

Cuba, Jamaica.

a. ♂.	Cuba.	W. S. MacLeay, Esq. [P.].
b. ♂.	Cuba.	
c. ♀.	—— ?	

2. Anolis ricordii.

Xiphosurus ricordii, *Gray, Cat.* p. 197.
Anolis ricordii, *Dum. & Bibr*. iv. p. 167.
Ctenonotus (Semiurus) ricordii, *Fitzing. Syst. Rept.* p. 64.
Eupristis baleatus, *Cope, Proc. Ac. Philad.* 1864, p. 168.
Xiphosurus ricordii, *O'Shaughn. Ann. & Mag. N. H.* (4) xv. 1875, p. 271.

Head large, once and two thirds to once and three fourths as long as broad, much longer than the tibia; forehead slightly concave, a deeper concavity behind the interorbital region; anterior part of head covered with large, irregular, rough, bony tubercles; scales of the supraorbital semicircles also rough and tubercular, separated by three longitudinal rows of small tubercles; supraocular scales small, the larger ones keeled; occipital indistinct; canthus rostralis prominent, canthal scales four, tubercular; loreal rows six or seven; ten upper labials to below the centre of the eye; ear-opening small, vertically oval. Gular appendage very large in both sexes, extending posteriorly beyond the thorax, entirely covered with granules; gular scales obtusely keeled. Body strongly compressed, with a crest of small erect spines, highest on the nape; this crest more developed in the male than in the female; dorsal and lateral scales small, roundish or oval, keeled, juxtaposed or separated by minute granules; ventral scales slightly smaller, smooth or obtusely keeled, subimbricate. Limbs moderately elongate; the adpressed hind limb reaches the temple or the eye; digital expansions well developed; thirty to thirty-two lamellæ under phalanges II. and III. of the fourth toe. Tail strongly compressed, with feebly serrated upper edge, in the adult male with a high fin-like crest on its anterior half, supported by the much produced neural spines; the length of the tail equals nearly twice that of head and body. Male with enlarged postanal scales. Brown or bluish above, with more or less distinct bluish-white transverse bands, most regular on the limbs; head above yellowish brown; belly white, uniform or brown-spotted; gular appendage yellow.

Total length	416	millim.
Head	46	„
Width of head	28	„
Body	100	„
Fore limb	63	„
Hind limb	108	„
Tibia	33	„
Tail	270	„

San Domingo.

a. ♂.	San Domingo.	
b. ♀.	San Domingo.	M. Sallé [C.]. (Type of *Eupristis baleatus.*)

3. Anolis cuvieri.

Xiphosurus velifer, *Gray, Cat.* p. 197.
Cuv. R. A. ii. p. 42, pl. v. fig. 1.
Anolis cuvieri, *Merr. Tent.* p. 45.
Xiphosurus cuvieri, *Fitzing. N. Class. Rept.* p. 48.
Dactyloa cuvieri, *Wagl. Syst. Amph.* p. 148.
Anolis velifer, *Guér. Icon. R. A., Rept.* pl. xii. fig. 1; *Dum. & Bibr.* iv. p. 164; *Reinh. & Lütk. Vidensk. Meddel.* 1862, p. 260.
Ctenonotus (Semiurus) cuvieri, *Fitzing. Syst. Rept.* p. 64.

Head large, not quite twice as long as broad, much longer than the tibia; forehead and occiput concave; anterior part of head covered with hexagonal, flat, rugose scales; scales of the supraorbital semicircles not enlarged; supraocular scales small, with two series of hexagonal, keeled, enlarged ones; supraciliary edge very prominent, forming a small crest; canthus rostralis prominent; loreal rows five; ten upper labials; ear-opening vertically oval. Gular appendage very large, extending posteriorly beyond the thorax, nearly entirely naked. Body strongly compressed, with a small serrated crest. Dorsal scales ovate-hexagonal, feebly keeled, juxtaposed, separated by minute granules; ventral scales as large as dorsals, squarish, smooth, subimbricate. Limbs moderately elongate; the scales under the thighs as large as the ventrals; digital expansions well developed. Tail strongly compressed, in the male with a high fin-like crest on its anterior half, supported by the much-produced neural spines. Slaty blue above, flanks blackish:

a series of brown spots on the vertebral line; lower surfaces bluish white, the gular appendage white.

Total length	387	millim.
Head	45	,,
Body	97	,,
Fore limb	60	,,
Hind limb	105	,,
Tail	245	,,

Porto Rico and Virgin Islands.

4. Anolis edwardsii.

Dactyloa edwardsii, *Gray, Cat.* p. 198.
Edwards, Glean. Nat. Hist. p. 73, pl. —.
Anolis edwardsii, *Merr. Tent.* p. 45; *Griff. A. K.* ix. p. 228, pl. —.; *Dum. & Bibr.* iv. p. 161.
Ctenonotus (Microctenus) edwardsii, *Fitzing. Syst. Rept.* p. 64.
Dactyloa edwardsii, *Gosse, Nat. Soj. Jamaica*, pl. iv.

Head rather large, once and three fifths to once and three fourths as long as broad, much longer than the tibia; forehead plane or scarcely concave; a concavity behind the interorbital region in the adult male; upper head-scales hexagonal, strongly keeled, largest on the forehead; scales of the supraorbital semicircles scarcely larger than the latter, separated by two or three rows of scales; twelve to seventeen enlarged, keeled supraocular scales, separated from the supraorbitals by one row of granules; occipital as large as or a little larger than the ear-opening, separated from the supraorbital semicircles by three or four series of scales; canthus rostralis sharp, canthal scales five or six; loreal rows five or six; seven or eight labials to below the centre of the eye; ear-opening moderately large, subtriangular. Gular appendage of moderate size, scarcely developed in the female, with rows of distant feebly-keeled scales; gular scales very small, obtusely keeled. Body strongly compressed, with a dorso-nuchal crest of triangular scales; the crest scarcely distinct in the female. Dorsal and lateral scales granular, keeled, the latter somewhat smaller; ventral scales slightly larger than dorsals, subhexagonal, smooth. Limbs moderately long; the adpressed hind limb reaches the temple or the eye; digital expansions well developed; twenty-eight to thirty-two lamellæ under phalanges II. and III. of the fourth toe. Tail strongly compressed, with a crest of triangular scales much more developed than on the back; the caudal scales with annuli of strongly enlarged ones, forming very distinct segments; length of the tail at least twice that of head and body. Male with slightly enlarged postanal scales. Green above, with more or less distinct reddish or whitish oblique transverse bands on the body; female with a vertebral series of round light spots; lower surfaces greenish white.

Total length	438	millim.
Head	44	,,
Width of head	28	,,
Body	94	,,
Fore limb	60	,,
Hind limb	103	,,
Tibia	32	,,
Tail	300	,,

Jamaica.

a–b, c–d, e. ♂.	Jamaica.	P. H. Gosse, Esq. [P.].
f–k. ♂ ♀.	Jamaica.	R. Heward, Esq. [P.].
l. ♂.	Jamaica.	Capt. Parry [P.].
m. ♂.	——?	Sir W. Hooker [P.].
n. ♀.	——?	Zoological Society.
o, p. ♂.	——?	
q. Skeleton.	——?	

5. Anolis gundlachii.

Anolis gundlachi, *Peters, Mon. Berl. Ac.* 1876, p. 705.

Head, to ear-opening, as long as the tibia; the distance between the eye and the ear-opening slightly less than the length of the snout; two sharp frontal ridges; suprarostral scales polygonal, mostly keeled; scales of the supraorbital semicircles separated by two or three rows of smaller scales; fifteen to seventeen enlarged keeled supraocular scales; occipital large, a little smaller than the ear-opening, separated from the supraorbitals by four or five rows of scales. Gular appendage of moderate size. Body compressed, with a low vertebral fold beginning on the nape. Dorsal and lateral scales small, granular, a little enlarged, and forming a double keel on the vertebral fold; ventral scales larger, convex, or feebly keeled. Limbs anteriorly with keeled scales; lower and posterior faces of the arm and thigh finely granulate. Tail strongly compressed, in its basal half with an elevated fin-like crest supported by the much-produced spinous processes. Greyish green, with small black spots forming three or four irregular cross-bands on the body; a few whitish, dark-edged, roundish spots on the sides; snout and caudal crest blackish green, each scale with a yellowish-green spot; skin of gular appendage blackish, the scales lemon-yellow; lower surfaces greenish yellow.

Total length	165	millim.
Head	16·5	,,
Body	38·5	,,
Fore limb	30	,,
Hind limb	50	,,
Tail	110	,,

Porto Rico.

6. Anolis cristatellus.

Cuv. R. A. ii. 2nd ed. p. 49.
Anolis cristatellus, *Dum. & Bibr.* iv. p. 143; *Reinh. & Lütk. Vidensk. Medd.* 1862, p. 249; *Bocourt, Miss. Sc. Mex., Rept.* pl. xiv. fig. 12; *Peters, Mon. Berl. Ac.* 1876, p. 706.
Ptychonotus (Istiocercus) cristatellus, *Fitzing. Syst. Rept.* p. 65.
Xiphosurus cristatellus, *O'Shaughn. Ann. & Mag. N. H.* (4) xv. 1875, p. 271.

Head moderate, once and a half to once and two thirds as long as broad, longer than the tibia; forehead more or less distinctly concave, with two diverging ridges; a concavity behind the interorbital region in the male; upper head-scales smooth; scales of the supraorbital semicircles very large, in contact between the orbits (exceptionally separated by one row of granules), produced forwards as frontal series; six to twelve enlarged, keeled supraocular scales, separated from the supraorbitals by one or two rows of granules; occipital nearly as large as the ear-opening, separated from the supraorbital semicircles by two or three series of scales; canthus rostralis sharp, canthal scales three; five or six loreal rows; six or seven labials to below the centre of the eye; ear-opening moderately large, vertically suboval. Gular appendage of moderate size, scarcely developed in the female, with rows of distant small scales; gular scales very small, feebly keeled. Body rather short, feebly compressed, in the male with a dorso-nuchal fold. Dorsal scales small, granular, smooth, smaller still on the flanks; ventral scales larger, cycloid, imbricate, smooth or feebly keeled. Limbs moderately long; the adpressed hind limb reaches the eye or a little beyond; digital expansions well developed; eighteen to twenty-five lamellæ under phalanges II. and III. of the fourth toe. Tail compressed, strongly in the male; the latter have usually, but not always, a much developed fin-like crest on the basal half of the tail, supported by the much-produced spinous processes of the vertebræ; length of the tail not twice that of head and body. Male without distinctly enlarged postanal scales. Greyish or brownish above, uniform or marbled with brown; females and young with a more or less distinct light vertebral band; lower surfaces whitish, uniform or dotted with brown; gular appendage blackish.

Total length	205	millim.
Head	21	,,
Width of head	13	,,
Body	49	,,
Fore limb	33	,,
Hind limb	56	,,
Tibia	17	,,
Tail	135	,,

West Indies.

a–g, h–n, o–p. ♂, ♀, & hgr.	St. Thomas.	Hr. Riise [C.].

q–t. ♂ ♀.	St. Thomas.	M. Sallé [C.].
u–z. ♂, ♀, & yg.	St. Thomas.	H.M.S. 'Challenger.'
α–γ. ♂.	Dominica.	
δ, ε. ♂ ♀.	Martinique.	
ζ–η. ♂.	W. Indies.	— Bland, Esq. [P.].

7. Anolis stratulus.

Anolis stratulus, *Cope, Proc. Ac. Philad.* 1861, p. 209; *Reinh. & Lütk. Vidensk. Meddel.* 1862, p. 255; *Bocourt, Miss. Sc. Mex., Rept.* pl. xiv. fig. 11.

Head moderate, once and three fourths to twice as long as broad, longer than the tibia; snout rather strongly depressed; forehead concave, with two feeble ridges; upper head-scales smooth; scales of the supraorbital semicircles very large, in contact between the orbits or separated by one series of scales, extending forwards on each side as frontal series; seven to ten enlarged, smooth supra-ocular scales, separated from the supraorbitals by one row of granules; occipital a little larger than the ear-opening, separated from the supraorbital semicircles by one or two series of scales; canthus rostralis angular, canthal scales four or five; loreal rows four or five; six or seven labials to below the centre of the eye; ear-opening moderately large, vertically suboval. Gular appendage large, extending posteriorly beyond the thorax, not developed in the female; gular scales smooth. Body rather short, feebly compressed, without dorso-nuchal fold. Dorsal scales small, granular, scarcely enlarged on the vertebral line; ventral scales much larger, roundish-hexagonal, subimbricate, smooth. Limbs moderately long; the adpressed hind limb reaches the eye; digital expansions well developed; eighteen to twenty-one lamellæ under phalanges II. and III. of the fourth toe. Tail compressed, with unequally serrated upper edge, its length not twice that of head and body. Male with enlarged postanal scales. Pale greyish or brownish above, with rather indistinct darker marblings, and four large dark-brown, sometimes white-edged, spots on the vertebral line; lower surfaces whitish, throat spotted or marbled with brown.

Total length	105	millim.
Head	15	„
Width of head	8	„
Body	30	„
Fore limb	19	„
Hind limb	31	„
Tibia	10	„
Tail	60	„

Porto Rico and Virgin Islands.

a. Several spec., ♂ ♀.	St. Thomas.	Hr. Riise [C.].
b. ♂.	Danish W. Indies.	Copenhagen Museum.

8. **Anolis homolechis.** (Plate I. fig. 4.)

Xiphosurus cristatellus (*non D. & B.*), *Gray, Cat.* p. 197.
Xiphosurus homolechis, *Cope, Proc. Ac. Philad.* 1864, p. 169; *O'Shaughn. Ann. & Mag. N. H.* (4) xv. 1875, p. 271.

Head moderate, nearly once and two thirds as long as broad, longer than the tibia; forehead concave, with two strong ridges; all the upper head-scales keeled; scales of the supraorbital semicircles feebly enlarged, separated by one or two rows of scales; eleven or twelve enlarged supraocular scales, separated from the supraorbitals by one row of smaller keeled scales; occipital smaller than the ear-opening, separated from the supraorbital semicircles by three series of scales; canthus rostralis sharp, canthal scales two; five loreal rows; five labials to below the centre of the eye; ear-opening large, vertically oval. Gular appendage large; gular scales small, smooth. Body rather short, feebly compressed, without fold or crest. Dorso-lateral scales very small, granular, those on the vertebral region a little enlarged, hexagonal, keeled; ventral scales large, cycloid, smooth, strongly imbricate. The adpressed hind limb reaches the eye; digital expansions well developed; eighteen lamellæ under phalanges ii. and iii. of the fourth toe. Tail strongly compressed, with a high fin-like crest supported by the spinous processes. No distinctly enlarged postanal scales. Brown above, whitish inferiorly.

Head	15	millim.
Width of head	8·5	,,
Body	32	,,
Fore limb	22	,,
Hind limb	37	,,
Tibia	12	,,

West Indies.

a. ♂. W. Indies. (Type.)

9. **Anolis gingivinus.** (Plate II. fig. 1.)

Anolis gingivinus, *Cope, Proc. Ac. Philad.* 1864, p. 170, and 1871, p. 220.

Head large in the male, about once and three fourths as long as broad, longer than the tibia; forehead and occiput concave in the adult male; two distinct frontal ridges; upper head-scales not keeled; scales of the supraorbital semicircles large, extending forwards as frontal series, in contact on a small point medially; seven to nine enlarged, smooth or feebly keeled supraocular scales, separated from the supraorbitals by one row of granules; occipital a little larger than the ear-opening, separated from the supraorbital semicircles by two or three series of scales; canthus rostralis sharp, canthal scales three; loreal rows four or five; six or seven labials to below the centre of the eye; ear-opening moderately large, vertically suboval. Gular appendage moderately large; gular scales smooth. Body short, not measuring twice the length of the head in the male, compressed; male with a very slight dorso-nuchal fold. Dorsal scales very small, granular, smaller still on the flanks; a double series of enlarged obtusely keeled scales on the vertebral line; ventral scales much larger, nearly as large as the scales on the anterior face of the femur, roundish-hexagonal, imbricate, perfectly smooth. Limbs moderately long; the adpressed hind limb reaches the eye; digital expansions well developed; eighteen to twenty-one lamellæ under phalanges II. and III. of the fourth toe. Tail strongly compressed, covered with unequal-sized scales, forming distinct segments, with a strong unequally serrated upper ridge; its length not twice that of head and body. Male with enlarged postanal scales. Greyish or pale buff above, with more or less distinct darker spots or variegations; a light, dark-edged band from axil to groin; lower surfaces whitish.

Total length	140	millim.
Head	21	„
Width of head	12	„
Body	39	„
Fore limb	28	„
Hind limb	46	„
Tibia	14	„
Tail	80	„

Anguilla Island, W. Indies.

a–c. ♂ ♀. Anguilla Island. W. J. Cooper, Esq. [P.]. (Types.)

10. **Anolis leachii.**

Anolius leachii, *Gray, Cat.* p. 200.
—— reticulatus, *Gray, l. c.* p. 204.
? Lacerta bimaculata, *Sparrm. Vetensk. Acad. Handl.* v. 1784, p. 179, pl. iv. fig. 4.
Anolis leachii, *Dum. & Bibr.* iv. p. 152; *Cope, Proc. Ac. Philad.* 1871, p. 221; *Bocourt, Miss. Sc. Mex., Rept.* pl. xiv. fig. 13.
—— maculatus, *Gray, Ann. & Mag. N. H.* v. 1840, p. 112.
—— reticulatus, *Gray, l. c.* p. 114.
Ctenonotus bimaculatus, *Fitzing. Syst. Rept.* p. 64.
Xiphosurus ferreus, *Cope, Proc. Ac. Philad.* 1864, p. 168.
Anolis alliaceus, *Cope, l. c.* p. 175.
—— bimaculatus, *O'Shaughn. Ann. & Mag. N. H.* (4) xv. 1875, p. 272.
Xiphosurus oculatus, *Cope, Proc. Am. Phil. Soc.* xviii. 1879, p. 274.

Head moderate, once and a half to once and three fourths as long as broad, longer than the tibia; cheeks strongly swollen in the full-grown male; forehead and occiput concave in the adult male, the former with two distinct ridges; upper head-scales not keeled; scales of the supraorbital semicircles large, in contact between the orbits, or separated by one row of scales, these enlarged scales extending forwards as frontal series; five to twelve enlarged, smooth or feebly keeled supraocular scales, separated from the supraorbitals by one row of granules; occipital as large as or a little smaller than the ear-opening, separated from the supraorbital semicircles by one or two series of scales; canthus rostralis sharp, canthal scales three or four; loreal rows four to six; six or seven labials to below the centre of the eye; ear-opening moderately large, vertically suboval. Gular appendage rather small, absent in the female; gular scales smooth or obtusely keeled. Body compressed in the male, with a dorso-nuchal fold. Dorsal scales small, granular, obtusely keeled, somewhat smaller still on the flanks; a more or less marked double series of enlarged scales on the dorso-nuchal fold, forming a rather strong denticulation on the nape in the adult male; ventral scales larger than dorsals, much smaller than antefemorals, hexagonal, imbricate, smooth or feebly keeled. Limbs moderately long; the adpressed hind limb reaches the eye or the posterior border of the orbit; digital expansions well developed; twenty-three to thirty-two lamellæ under phalanges II. and III. of the fourth toe. Tail compressed, strongly in the male, covered with unequal-sized keeled scales forming more or less distinct segments; a caudal crest is scarcely indicated by a slight denticulation in the female; in the half-grown male it is represented by a strong unequally-serrated ridge, and in the full-grown male it assumes a fin-like appearance, being supported by the much produced neural spines; length of the tail not twice that of head and body. Male with enlarged postanal scales. Brownish or bluish grey above, uniform or spotted, or vermiculated with black; frequently a dark band, edged with white anteriorly, on each side of the head, extending to beyond the insertion of the fore limb: sometimes a series of black ocelli with white centre on the side of the back; full-grown male uniform dark-brown above; lower surfaces whitish, with metallic gloss.

Total length	292 millim.
Head	36 ,,
Width of head	24 ,,
Body	76 ,,
Fore limb	52 ,,
Hind limb	85 ,,
Tibia	26 ,,
Tail	180 ,,

West Indies (Dominica, Guadeloupe, Nevis, S. Eustacius).

a. ♂.	Guadeloupe.	(Type of *Xiphosurus ferreus*.)
b. ♀.	Guadeloupe.	
c–d. ♂ & yg.	Dominica.	G. F. Angas, Esq. [P.].
e–h. ♂.	Dominica.	
i. ♂.	Nevis.	T. Cottle, Esq. [P.].
k. ♀.	——?	(Type of *A. reticulatus*.)
l. Hgr. ♂.	——?	(Type of *A. alliaceus*.)
m. Hgr. ♂.	——?	Bristol Museum [P.].
n, o, p. ♂ & hgr.	——?	

Anolis marmoratus.

Anolius marmoratus, *Gray, Cat.* p. 201.
Anolis marmoratus, *Dum. & Bibr.* iv. p. 139; *Reinh. & Lütk. Vidensk. Meddel.* 1862, p. 258.
Ptychonotus dumerilii, *Fitzing. Syst. Rept.* p. 65.

I am not able to detect in the description a single character of sufficient importance to justify a separation of this form from *A. leachii*. A careful comparison of the types alone can show whether it is to rank as a species, or merely as a variety of *A. leachii*.
Martinique.

11. Anolis alligator.

Anolius cepedii, *Gray, Cat.* p. 201.
—— æneus, *Gray, l. c.* p. 205.
Anolis bimaculatus, part., *Daud. Rept.* iv. p. 55.
—— cepedii, part., *Merr. Tent.* p. 44.
—— alligator, *Dum. & Bibr.* iv. p. 134; *Duvern. R. A., Rept.* pl. xviii. fig. 2; *Cope, Proc. Am. Phil. Soc.* xi. 1869, p. 164; *Bocourt, Miss. Sc. Mex., Rept.* p. 59, pl. xiv. fig. 4.
—— goudotii, *Dum. & Bibr.* p. 108.
Ptychonotus fasciatus, *Fitzing. Syst. Rept.* p. 65.
Dactyloa (Eudactylus) goudotii, *Fitzing. l. c.* p. 67.
Anolis trinitatis, *Reinh. & Lütk. Vidensk. Meddel.* 1862, p. 269.
—— cepedii, *O'Shaughn. Ann. & Mag. N. H.* (4) xv. 1875, p. 272.

Head moderate, once and two thirds to once and three fourths as long as broad, longer than the tibia; snout rather strongly depressed, forehead concave; no frontal ridges; upper head-scales smooth; scales of the supraorbital semicircles strongly dilated, in contact with each other and with the occipital, which is larger than the ear-opening; six to ten enlarged, smooth or indistinctly keeled supraocular scales, separated from the supraorbitals by one row of granules; canthus rostralis sharp in front of the eye, otherwise rather obtuse; canthal scales four or five; loreal rows four or five; five or six labials to below the centre of the eye; eye-opening moderately large, vertically suboval. Gular appendage rather large, not developed in the female; gular scales smooth. Body compressed, in the male with a feeble dorso-nuchal fold. Dorsal scales small, granular, keeled, larger and more distinctly keeled on the vertebral region; ventral scales larger than dorsals, roundish hexagonal, feebly imbricate, smooth or indistinctly keeled. Limbs moderately long; the adpressed hind limb reaches the temple or the posterior border of the orbit; digital expansions well developed; twenty-two to twenty-six lamellæ under phalanges II. and III. of the fourth toe. Tail compressed, with feebly and equally serrated upper edge; its length not twice that of head and body. Male without enlarged postanal scales. Brownish or greenish above, frequently with a darker network enclosing small round whitish spots; lower surfaces whitish or greenish, with metallic gloss.

Total length	200	millim.
Head	22	,,
Width of head	13	,,
Body	58	,,
Fore limb	34	,,
Hind limb	52	,,
Tibia	16	,,
Tail	120	,,

West Indies (South Caribbee Isles and Trinidad).

a. ♂.	Martinique.	
b–c. ♂.	Trinidad.	
d. Hgr. ♂.	—— ?	T. Bell, Esq. [P.]. (Type of *A. æneus*.)
e. ♂.	—— ?	Sir R. Schomburgk [P.].
f, g–h, i–n. ♂ ♀.	—— ?	

12. Anolis grahami.

Anolius punctatus (*non Daud.*), *Gray, Cat.* p. 203.
—— grahami, *Gray, l. c.* p. 274.
Anolis punctatus, *Gray, Ann. & Mag. N. H.* v. 1840, p. 113.
—— iodurus, *Gosse, Ann. & Mag. N. H.* (2) vi. 1850, p. 344; *Cope, Proc. Ac. Philad.* 1861, p. 210; *Bocourt, Miss. Sc. Mex., Rept.* pl. xiv. fig. 17.
—— opalinus, *Gosse, l. c.*; *Cope, l. c.*
—— punctatissimus, *Hallow. Proc. Ac. Philad.* 1856, p. 225.
—— heterolepis, *Hallow. l. c.* p. 230.
—— grahami, *Cope, Proc. Am. Philos. Soc.* xi. 1869, p. 164; *O'Shaughn. Ann. & Mag. N. H.* (4) xv. 1875, p. 273.

Head rather large, once and two thirds to once and three fourths as long as broad, longer than the tibia; forehead very slightly concave, no regular frontal ridges; occiput concave in the male; upper head-scales keeled; scales of the supraorbital semicircles moderately enlarged, separated mesially by one or two series of scales; twelve to seventeen enlarged keeled supraocular scales, separated from the supraorbitals by two or three (more rarely one) series of granules; occipital much larger than the ear-opening, separated from the supraorbitals by three to five series of scales; canthus rostralis angular, canthal scales four; loreal rows four or five; seven or eight (in one specimen six) labials to below the centre of the eye; ear-opening small, suboval. Gular appendage moderately large, absent or merely indicated in the female; gular scales obtusely keeled. Body not compressed; a slight nuchal fold in the male, no dorsal fold. Dorsal scales small, granular, smooth or obtusely keeled, very slightly enlarged on the vertebral region; ventral scales larger, roundish hexagonal, subimbricate, smooth or feebly keeled. Limbs moderately long; the adpressed hind limb reaches the eye or the posterior border of the orbit; digital expansions well developed; twenty-four to twenty-eight lamellæ under phalanges II. and III. of the fourth toe. Tail compressed, more feebly in the female than in the male, with a series of enlarged strongly keeled scales superiorly; its length not twice that of head and body. Male with slightly enlarged postanal scales. Green, purplish, or brownish above, usually with darker reticulation enclosing round light spots; females sometimes with a few angular cross bars on the back, with or without a light vertebral band; throat with dark reticulation.

	♂.	♀.	
Total length	166	119	millim.
Head	21	14	„
Width of head	13	8	„
Body	45	32	„
Fore limb	30	21	„
Hind limb	50	33	„
Tibia	15	10	„
Tail	100	73	„

Jamaica.

a. ♀.	——?	Sir J. MacGregor [P.]. (Types.)
b–c, *d–f*. ♂ ♀.	——?	
g–h. ♂.	——?	
i. ♂.	——?	Bristol Museum [P.].
k. ♂.	Jamaica.	Capt. Parry [P.].
l–m, *n–o*, *p*. ♂.	Jamaica.	P. H. Gosse, Esq. [P.]. (Types of *A. iodurus*.)
q. ♀.	Jamaica.	P. H. Gosse, Esq. [P.]. (Type of *A. opalinus*.)
r–s. ♂.	Jamaica.	P. H. Gosse, Esq. [P.].

13. Anolis distichus.

Anolis distichus, *Cope, Proc. Ac. Philad.* 1861, p. 208, *and Proc. Am. Philos. Soc.* xi. 1869, p. 164.
—— dominicensis, *Reinh. & Lütk. Vidensk. Meddel.* 1862, p. 261.
—— brevirostris, *Bocourt, Nouv. Arch. Mus.* vi. 1870, *Bull.* p. 11, *and Miss. Sc. Mex., Rept.* pl. xiv. fig. 6.

Head rather short, once and a half as long as broad, a little longer than the tibia; forehead concave, frontal ridges distinct; upper head-scales smooth or a little rugose; scales of the supraorbital semicircles very large, in contact between the orbits, extending forwards as frontal series; six to eight large, smooth supraocular scales, separated from the supraorbitals by one or two rows of granules; occipital much larger than the ear-opening, in contact with the supraorbitals, or separated by one series of scales; canthus rostralis angular, canthal scales three; loreal rows four; five labials to below the centre of the eye; ear-opening rather small, suboval. Gular appendage rather small, absent in the female; gular scales smooth. Body scarcely compressed, without dorso-nuchal fold. Dorsal scales minute, granular, all equal; ventral scales larger, roundish, juxtaposed or subimbricate, smooth. Limbs moderately long; the adpressed hind limb reaches the eye or the anterior border of the orbit; digital expansions moderately developed; eighteen lamellæ under phalanges II. and III. of the fourth toe. Tail compressed, rather strongly in the male, feebly in the female, with feebly serrated upper edge; its length not once and a half that of head and body. Male with enlarged postanal scales. Brownish above, with more or less distinct darker dots or vermiculations.

Total length	113 millim.
Head	15 ,,
Width of head	10 ,,
Body	38 ,,
Fore limb	28 ,,
Hind limb	42 ,,
Tibia	13 ,,
Tail	60 ,,

Bahamas, San Domingo.

a. ♂.	S. Domingo.	Copenhagen Museum [P.]. (As typical of *A. dominicensis*.)
b–c. ♀.	——?	

14. **Anolis cybotes.** (Plate I. fig. 5.)

Anolis cybotes, *Cope, Proc. Ac. Philad.* 1862, p. 177, *and Proc. Am. Philos. Soc.* xi. 1869, p. 164.
—— riisei, *Reinh. & Lütk. Vidensk. Meddel.* 1862, p. 264.

Head large in the male, with slightly swollen cheeks, moderate in the female; its width about once and two thirds in its length, which exceeds that of the tibia; forehead concave, occiput concave in the male; two feeble frontal ridges; upper head-scales smooth or a little rugose; scales of the supraorbital semicircles very large, in contact between the orbits; eleven to thirteen enlarged, feebly keeled supraocular scales, separated from the supraorbitals by one or two rows of granules; occipital as long as, but narrower than, the ear-opening, separated from the supraorbitals by two series of scales; canthus rostralis sharp, canthal scales three; six to eight loreal rows; six or seven labials to below the centre of the eye; ear-opening rather large, suboval. Gular appendage of moderate size, scarcely developed in the female; gular scales obtusely keeled. Body short, compressed, in the male with a well-marked dorso-nuchal fold. Dorsal scales minute, granular, considerably enlarged and feebly keeled on the vertebral line; ventral scales large, cycloid, strongly imbricate, smooth. Limbs moderately long; the adpressed hind limb reaches between the orbit and the nostril; digital expansions rather narrow; eighteen to twenty lamellæ under phalanges II. and III. of the fourth toe. Tail compressed, strongly in the male, with feebly serrated upper edge; its length not twice that of head and body. Male with enlarged postanal scales. Grey-brown above, with darker markings; female with a pale vertebral line, and a regular series of large, sharply defined, brown spots on each side of the back; limbs with more or less distinct brown cross bars; lower surfaces whitish.

Total length	193	millim.
Head	25	"
Width of head	15	"
Body	43	"
Fore limb	35	"
Hind limb	62	"
Tibia	18	"
Tail	125	"

San Domingo.

a–b. ♂ ♀.	S. Domingo.	Copenhagen Museum [P.]. (As typical of *A. riisei.*)
c. ♂.	S. Domingo.	

15. **Anolis citrinellus.** (Plate I. fig. 3.)

Anolis citrinellus, *Cope, Proc. Ac. Philad.* 1864, p. 170.

Head moderate, twice as long as broad, longer than the tibia; forehead concave, with diverging frontal ridges; upper head-scales smooth; scales of the supraorbital semicircles large, extending forwards as frontal series, in contact on a small point between the orbits, otherwise separated by one series of scales; six or seven enlarged, feebly keeled supraocular scales, one of which is very much larger than the others, separated from the supraorbitals by one series of granules; occipital as large as the ear-opening, separated from the supraorbitals by two series of scales; canthus rostralis angular, canthal scales three; loreal rows five; six labials to below the centre of the eye; ear-opening moderately large, suboval. A small gular pouch (♀); gular scales smooth. Body slightly compressed, no dorso-nuchal fold. Dorsal scales small, granular, a little enlarged on the vertebral line; ventral scales larger, smooth or indistinctly keeled, roundish-hexagonal, feebly imbricate. Limbs moderately long; the adpressed hind limb reaches the eye; digital expansions rather narrow; twenty lamellæ under phalanges II. and III. of the fourth toe. Tail compressed, with slightly serrated upper edge, not quite twice as long as head and body. Brown above, with darker and lighter angular cross bars on the back and cross bars on the limbs; lower surfaces yellow.

Total length	115	millim.
Head	13	"
Width of head	6·5	"
Body	27	"
Fore limb	19	"
Hind limb	30	"
Tibia	9	"
Tail	75	"

San Domingo.

a. ♀.	S. Domingo.	C. Taylor, Esq. [P.]. (Type.)

16. Anolis acutus.

Anolis acutus, *Hallow. Proc. Ac. Philad.* 1856, p. 228; *Cope, Proc. Ac. Philad.* 1861, p. 209; *Reinh. & Lütk. Vidensk. Meddel.* 1862, p. 252; *Bocourt, Miss. Sc. Mex., Rept.* pl. xv. fig. 2.
—— newtoni, *Günth. Ann. & Mag. N. H.* (3) iv. 1859, p. 212, pl. iv. fig. A.

Head moderate, nearly once and three fourths as long as broad, longer than the tibia; forehead and occiput concave in the male; frontal ridges distinct; upper head-scales smooth; scales of the supraorbital semicircles very large, in contact with each other and with the occipital, the size of which equals or exceeds that of the ear-opening; the supraorbital series produced forwards as frontal series; six to ten enlarged, keeled supraocular scales, separated from the supraorbitals by one row of granules; canthus rostralis angular, canthal scales four; loreal rows four to six; six labials to below the centre of the eye; ear-opening moderate, vertically oval. Gular appendage moderately large, indicated in the female; gular scales obtusely keeled. Body compressed, scarcely in the female; a slight dorso-nuchal fold. Dorsal scales small, granular, much enlarged, rhomboidal and strongly keeled on the vertebral region; ventral scales nearly as large as vertebrals, subhexagonal, imbricate, distinctly keeled. Limbs moderate; the adpressed hind limb reaches the eye or slightly beyond; digital expansions well developed; twenty-one to twenty-four lamellæ under phalanges II. and III. of the fourth toe. Tail strongly compressed, covered with unequal-sized keeled scales forming distinct segments, with a strong unequally serrated upper ridge; its length not twice that of head and body. Male with enlarged postanal scales. Greenish-olive or brownish above, uniform or dotted with brown; a brown band along the posterior side of the thighs in the female; latter sometimes with a light vertebral band crossed by narrow black bars; lower surfaces whitish, with metallic gloss.

Total length	165	millim.
Head	18	,,
Width of head	10	,,
Body	42	,,
Fore limb	26	,,
Hind limb	46	,,
Tibia	15	,,
Tail	105	,,

St. Croix, West Indies.

a–b, c–d, e. ♂ ♀.	St. Croix.	A. and E. Newton, Esqrs. [P.]. (Types of *A. newtoni.*)
f–i, k–l. ♂.	St. Croix.	Hr. Riise [C.].

17. Anolis krugi.

Anolis krugi, *Peters, Mon. Berl. Ac.* 1876, p. 707.

Head longer than the tibia, the snout resembling that of *A. cristatellus*; supraorbital semicircles nearly entirely separated by one series of scales medially, produced forwards as frontal series bearing two ridges; four to six enlarged, keeled, supraocular scales; occipital larger than the ear-opening, separated from the supraorbitals by two or three series of scales; loreal rows five. A dorso-nuchal fold. Dorsal and lateral scales granular, except four median series along the vertebral line which are enlarged and distinctly keeled; ventrals strongly keeled. Tail more than twice as long as head and body, with a dorsal series of enlarged hexagonal keeled scales. Olive-green, back and sides dotted and vermiculated with black; nape with whitish dots; a lateral yellow band; belly greenish yellow.

Total length	170	millim.
Length to vent	46	„
Head	18	„
Fore limb	20	„
Hind limb	38	„

Porto Rico.

18. Anolis richardii.

Anolius occipitalis, *Gray, Cat.* p. 200.
—— richardii, *Gray, l. c.* p. 202.
—— stenodactylus, *Gray, l. c.* p. 204.
Anolis richardii, *Dum. & Bibr.* iv. p. 141; *Bocourt, Miss. Sc. Mex., Rept.* pl. xv. fig. 6; *O'Shaughn. Ann. & Mag. N. H.* (4) xv. 1875, p. 275.
—— occipitalis, *Gray, Ann. & Mag. N. H.* v. 1840, p. 112.
—— stenodactylus, *Gray, l. c.* p. 114.
Ptychonotus (Ctenodeira) richardii, *Fitzing. Syst. Rept.* p. 66.

Head moderate, once and a half as long as broad, as long as the tibia; forehead and occiput deeply concave in the adult male, frontal ridges distinct, diverging; upper head-scales not keeled; scales of the supraorbital semicircles large, separated by one or two series of scales (exceptionally in contact); eight to sixteen enlarged, keeled supraocular scales, separated from the supraorbitals by one or two series of granules; occipital a little larger than the ear-opening, separated from the supraorbitals by one or two series of scales; canthus rostralis distinct, canthal scales three or four; loreal rows five; five or six labials to below the centre of the eye; ear-opening large, vertically oval. Gular appendage large, smaller but well developed in the female; gular scales obtusely keeled. Body compressed; a dorso-nuchal fold, in the adult with a small denticulation formed by a double series of enlarged conical scales. Dorsal scales small, oval, keeled, smallest on the flanks, distinctly enlarged on the dorsal fold; ventral scales nearly as large as the vertebrals, roundish-hexagonal, feebly imbricate, keeled. Limbs long; the adpressed hind limb reaches between the eye and the tip of the snout; digital expansions well developed; twenty-three to twenty-five lamellæ under phalanges II. and III. of the fourth toe. Tail strongly compressed, with serrated upper edge; its length nearly twice that of head and body. No enlarged postanal scales. Brownish or purplish above, uniform or with darker transverse bars on the back; frequently a white spot on the upper lip, below the eye; lower surfaces greenish white.

Total length	393	millim.
Head	33	,,
Width of head	22	,,
Body	90	,,
Fore limb	56	,,
Hind limb	103	,,
Tibia	32	,,
Tail	270	,,

West Indies (Jamaica, Tortola, Dominica, St. Vincent).

a, b. ♂.	W. Indies.	T. Bell, Esq. [P.]. (Types of *A. occipitalis.*)
c. ♀.	W. Indies.	T. Bell, Esq. [P.].
d. ♂.	Jamaica.	J. Winterbottom, Esq. [P.]. (Type of *A. stenodactylus.*)
e–f. ♂ ♀.	Dominica.	

19. **Anolis lineatus.**

Anolius lineatus, *Gray, Cat.* p. 200.
Anolis lineatus, *Daud. Rept.* iv. p. 66, pl. xlviii. fig. 1; *Dum. & Bibr.* iv. p. 146; *Cope, Proc. Am. Phil. Soc.* xxii. 1885, p. 181.
Ptychonotus (Trachycœlia) lineatus, *Fitzing. Syst. Rept.* p. 66.

Head moderate, once and two thirds as long as broad, slightly longer than the tibia; forehead slightly concave, with distinct frontal ridges; upper head-scales not keeled; scales of the supraorbital semicircles large, in contact medially, produced forwards as frontal series; seven enlarged, smooth supraocular scales, separated from the supraorbitals by one row of granules; occipital nearly as large as the ear-opening, separated from the supraorbitals by two series of scales; canthus rostralis angular, canthal scales three; loreal rows three or four; seven labials to below the centre of the eye; ear-opening rather small, oval, oblique. Gular appendage large, extending posteriorly beyond the thorax; gular scales feebly keeled. Body compressed; a slight dorso-nuchal fold. Dorsal scales small, rhomboidal, juxtaposed, feebly keeled; lateral scales smaller, granular; ventral scales larger than dorsals, hexagonal, feebly imbricate, keeled. The adpressed hind limb reaches the eye; digits moderately dilated; twenty-one lamellæ under phalanges II. and III. of the fourth toe. Tail strongly compressed, with feebly serrated upper edge; its length not twice that of head and body. Male with enlarged postanal scales. Sandy-grey above, with indistinct lighter spots forming irregular vertical bars on the sides; a white, black-edged band along each side, from limb to limb; lower surfaces white; a large black spot on the gular appendage.

Total length	202	millim.
Head	21	"
Width of head	12	"
Body	51	"
Fore limb	36	"
Hind limb	62	"
Tibia	19	"
Tail	130	"

Leeward Islands; Martinique?

a. ♂. Curaçoa.

20. **Anolis lineatopus.** (PLATE I. figs. 1, 2.)

Anolius lineatopus, *Gray, Cat.* p. 203.
—— maculatus, *Gray, l. c.*
Anolis lineatopus, *Gray, Ann. & Mag. N. H.* v. 1840, p. 113.
—— maculatus, (*non Gray, suprà*) *Gray, l. c.*; *Cope, Proc. Ac. Philad.* 1861, p. 209; *Reinh. & Lütk. Vidensk. Meddel.* 1862, p. 268; *Bocourt, Miss. Sc. Mex., Rept.* pl. xiv. fig. 18; *O'Shaughn. Ann. & Mag. N. H.* (4) xv. 1875, p. 275.

Head moderate, once and two thirds to once and three fourths as long as broad, longer than the tibia; forehead concave, with strong frontal ridges; occiput concave in the male; upper head-scales keeled; scales of the supraorbital semicircles large, separated medially by one or two series of scales; nine to sixteen enlarged, keeled supraocular scales, in contact with the supraorbitals or separated by one row of granules; occipital nearly as large as the ear-opening, separated from the supraorbitals by three to five series of scales; canthus rostralis angular, canthal scales three or four; loreal rows five to seven; six to eight labials to below the centre of the eye; ear-opening moderately large, suboval. Gular appendage large, extending backwards beyond the thorax, scarcely indicated in the female; gular scales keeled. Body rather short, compressed; male with a slight dorso-nuchal fold. Dorsal scales small, granular, keeled, smallest on the flanks, distinctly enlarged on the vertebral region; ventral scales larger, subhexagonal, slightly imbricate, distinctly keeled. Limbs moderate; the adpressed hind limb reaches the eye or between the latter and the nostril; digital expansions moderately developed; eighteen to twenty-one lamellæ under phalanges II. and III. of the fourth toe. Tail compressed, with very slightly serrated upper edge, its length not twice that of head and body. Male with slightly enlarged postanal scales. Brownish above, with darker markings or reticulations, and with lighter spots confluent into more or less regular transverse bands on the back and limbs; female with five light longitudinal bands, the widest and most distinct extending from the shoulder to the hind limb; lower surfaces whitish.

Total length	197	millim.
Head	23	,,
Width of head	14	,,
Body	49	,,
Fore limb	34	,,
Hind limb	57	,,
Tibia	17	,,
Tail	125	,,

Jamaica.

a. ♀.	——?	Sir J. MacGregor [P.]. (Type.)
b, *c*. ♂.	——?	Sir J. MacGregor [P.]. (Types of *A. maculatus*.)
d. ♂.	——?	
e–i. ♂.	Jamaica.	P. H. Gosse, Esq. [P.].
k. ♂.	Jamaica.	

21. Anolis sagræ.

Anolius nebulosus, (*non Wiegm.*) *Gray, Cat.* p. 203.
Anolis sagræi, *Dum. & Bibr.* iv. p. 149; *Coct. in R. de la Sagra, Hist. Cuba, Rept.* p. 131, pl. xiii.; *Hallow. Proc. Ac. Philad.* 1856, p. 229; *Bocourt, Miss. Sc. Mex., Rept.* p. 80, pl. xv. fig. 14; *O'Shaughn. Ann. & Mag. N. H.* (4) xv. 1875, p. 275.
Dactyloa (Trachypilus) sagræi, *Fitzing. Syst. Rept.* p. 67.
Draconura catenata, *Gosse, Ann. & Mag. N. H.* (2) vi. 1850, p. 346.
Anolis ordinatus, *Cope, Proc. Ac. Philad.* 1864, p. 175.

Head once and two fifths to once and two thirds as long as broad, a little longer than the tibia; forehead very slightly concave; upper head-scales sharply keeled; scales of the supraorbital semicircles a little enlarged, separated by one series of scales; five to eleven enlarged, keeled supraocular scales, in contact with the supraorbitals or separated by one row of granules; occipital nearly as large as the ear-opening, separated from the supraorbitals by two or three series of scales; canthus rostralis distinct, canthal scales two or three; loreal rows four or five; five or six labials to below the centre of the eye; ear-opening moderate, oval. Gular appendage moderately large, indicated in the female; gular scales feebly keeled. Body compressed; dorso-nuchal fold absent or feebly developed. Dorsal scales small, feebly keeled, smallest on the flanks, a little enlarged on the vertebral line; ventral scales larger, rhomboidal, strongly imbricate, strongly keeled. The adpressed hind limb reaches the eye; digits moderately dilated; eighteen to twenty-one lamellæ under phalanges II. and III. of the fourth toe. Tail compressed, strongly in the male, with slightly serrated upper edge; its length not twice that of head and body. Male with enlarged postanal scales. Brown above, with golden gloss, uniform or spotted with darker; sometimes three paler longitudinal bands run along the vertebral line and from axilla to groin; frequently a series of large oval or rhomboidal darker spots on each side of the back; lower surfaces whitish, with metallic gloss.

Total length	160	millim.
Head	17	„
Width of head	11	„
Body	43	„
Fore limb	27	„
Hind limb	44	„
Tibia	14	„
Tail	100	„

Bahamas, Cuba, Jamaica, eastern coast of Central America, Venezuela.

a–b. ♂.	Cuba.	W. S. MacLeay, Esq. [P.].
c. ♂.	Cuba.	
d. ♂.	Bluefields, Jamaica.	P. H. Gosse, Esq. [P.]. (Type of *Draconura catenata.*)
e–g. ♂ ♀.	W. Indies.	(Types of *A. ordinatus.*)
h. ♂.	Yzabal.	
i, k. ♂.	Honduras.	
l–p. ♂ ♀.	Half Moon Cay, Belize coast.	
q, r. ♂.	Caracas.	

22. **Anolis vermiculatus.**

Anolius vermiculatus, *Gray, Cat.* p. 204.
Anolis vermiculatus, *Dum. & Bibr.* iv. p. 128; *Coct. in R. de la Sagra, Hist. Cuba, Rept.* p. 120, pl. x.; *Bocourt, Miss. Sc. Mex., Rept.* pl. xv. fig. 7.
Ptychonotus (Deiroptyx) vermiculatus, *Fitzing. Syst. Rept.* p. 66.

Head rather large, twice as long as broad, longer than the tibia; forehead not concave, occiput concave in the male; upper head-scales small, rugose; scales of the supraorbital semicircles enlarged, separated by one or two series of scales; median supraocular scales very feebly enlarged; occipital smaller than the ear-opening, separated from the supraorbitals by four series of scales; canthus rostralis prominent, canthal scales five or six; seven or eight labials to below the centre of the eye; ear-opening large, vertically oval. Gular appendage scarcely indicated; a strong, angular, transverse gular fold; gular scales keeled. Body compressed; a slight nuchal fold. Dorsal scales small, oval, granular, keeled, smaller on the flanks, a little enlarged on the vertebral region; ventral scales small, larger than dorsals, feebly imbricate, strongly keeled. Limbs long; the adpressed hind limb reaches the anterior border of the orbit; digits rather feebly dilated; twenty-eight to thirty lamellæ under phalanges II. and III. of the fourth toe. Tail strongly compressed, with feebly serrated upper edge, at least twice as long as head and body. Male with enlarged postanal scales. Olive above, the back and neck handsomely reticulated with black; lower surfaces pale olive-brown.

Total length	357	millim.
Head	37	,,
Width of head	19	,,
Body	80	,,
Fore limb	51	,,
Hind limb	92	,,
Tibia	27	,,
Tail	240	,,

Cuba.

a–c. ♂. Cuba.

23. **Anolis loysiana.**

Acantholis loysiana, *Gray, Cat.* p. 206.
Acantholis loysiana, *Cocteau, C. R. Ac. Paris,* iii. 1836, p. 226, *and in R. de la Sagra, Hist. Cuba, Rept.* p. 141, pl. xiv.
Anolis loysiana, *Dum. & Bibr.* iv. p. 100; *Bocourt, Miss. Sc. Mex., Rept.* pl. xiv. fig. 9.
Heteroderma loysianum, *Fitzing. Syst. Rept.* p. 68.
Anolis (Acantholis) argillaceus, *Cope, Proc. Ac. Philad.* 1862, p. 176.

Head nearly twice as long as broad in the male, shorter in the female, much longer than the tibia; forehead concave, frontal ridges indistinct; upper head-scales smooth; scales of the supraorbital semicircles large, in contact medially; three or four large, smooth supraocular scales, in a single longitudinal series, separated from the supraorbitals by granules; occipital larger than the ear-opening, separated from the supraorbitals by two series of scales; canthus rostralis sharp; six upper labials; ear-opening small, roundish. Gular appendage small; gular scales smooth. Body not compressed; no dorso-nuchal fold. Dorsal and lateral scales small, granular, intermixed with scattered enlarged ones, which in the male are conical; ventral scales small, larger than dorsals, roundish, smooth, subimbricate. The adpressed hind limb reaches the neck; digital expansions well developed. Tail cylindrical, a little longer than head and body, with a dorsal series of enlarged flat scales. Greenish or brownish-white above, with a series of brown spots on each side; a dark temporal band; lower surfaces white.

Total length	77	millim.
Head	13	,,
Body	24	,,
Fore limb	15	,,
Hind limb	38	,,
Tail	40	,,

Cuba.

24. Anolis carolinensis.

Anolius porcatus, *Gray, Cat.* p. 201.
—— principalis, *Gray, l. c.* p. 202.
Catesby, Nat. Hist. Carol. ii. p. 55, pl. lxv.
? Lacerta principalis, *Linn. S. N.* i. p. 360.
Anolis bullaris, part., *Daud. Rept.* iv. p. 69.
—— carolinensis, *Dum. & Bibr.* iv. p. 121; *Holbr. N. Am. Herp.* ii. p. 67, pl. viii.; *Cocteau, in R. de la Sagra, Hist. Cuba, Rept.* p. 125, pl. xi.; *Hallow. Proc. Ac. Philad.* 1856, p. 232.
—— porcatus, *Gray, Ann. & Mag. N. H.* v. 1840, p. 112.
Dactyloa (Ctenocercus) carolinensis, *Fitzing. Syst. Rept.* p. 68.
Anolis principalis, *Bocourt, Miss. Sc. Mex., Rept.* pl. xv. fig. 1; *Shufeldt, Am. Nat.* 1883, p. 919.

Head once and two thirds to once and three fourths as long as broad, much longer than the tibia; in the adult male the head very large, with very strong and prominent frontal ridges, converging anteriorly, and canthi rostrales, and slightly concave occiput, the head-scales large and rough, and the nostrils turned entirely upwards; the head moderate in the female, with strongly striated scales; scales of the supraorbital semicircles separated by one or two series of scales; six to eleven enlarged supraocular scales, separated from the supraorbitals by one series of smaller scales; occipital as large as the ear-opening, separated from the supraorbitals by one or two series of scales; canthal scales four; loreal rows three to five; six to nine labials to below the centre of the eye; ear-opening small, roundish subtriangular. Gular appendage moderately large, scarcely indicated in the female; gular scales keeled or striated. Body not compressed; no dorso-nuchal fold. Dorsal and ventral scales equal, keeled, roundish-hexagonal, juxtaposed or subimbricate; lateral scales a little smaller, round. Limbs short; the adpressed hind limb reaches the shoulder or the ear; digital expansions well developed; twenty-two to twenty-seven lamellæ under phalanges II. and III. of the fourth toe. Tail round, with a dorsal series of slightly enlarged scales; its length nearly twice that of head and body. Male with enlarged postanal scales. Green above, head brownish; body and limbs frequently dotted or reticulated with black; sometimes a light vertebral band, and a large blue ocellus above the axil; lower surfaces greenish white, uniform or speckled with blackish.

	♂.	♀.	
Total length	116	140	millim.
Head	26	16	,,
Width of head	15	9	,,
Body	45	34	,,
Fore limb	26	20	,,
Hind limb	43	31	,,
Tibia	13	10	,,
Tail	145	90	,,

S.E. United States; Cuba.

a–f. ♂ & hgr.	N. America.	E. Doubleday, Esq. [P.].	
g. ♀.	N. America.	Sir R. Murchison [P.].	
h. ♂.	N. America.		
i–l. ♂ ♀.	N. Orleans.		
m. ♂.	Texas.		(Types of
n. ♂.	Cuba.	W. S. MacLeay, Esq. [P.].	*A. porcatus.*)
o–r. ♂ & yg.	Cuba.		
s. ♀.	——?	J. O. W. Fabert, Esq. [P.].	
t–u. ♂.	——?	M. Sallé [C.].	

25. Anolis chlorocyanus.

Anolius bullaris, *Gray, Cat.* p. 206.
? Lacerta bullaris, *Linn. S. N.* i. p. 368.
Anolis bullaris, part., *Daud. Rept.* iv. p. 69.
—— chlorocyanus, *Dum. & Bibr.* iv. p. 117; *Reinh. & Lütk. Vidensk. Meddel.* 1862, p. 266; *O'Shaughn. Ann. & Mag. N. H.* (4) xv. 1875, p. 272.
Dactyloa (Xiphosurus) chlorocyana, *Fitzing. Syst. Rept.* p. 67.
Anolis læviceps, *Lichtenst. Nomencl. Rept. Mus. Berol.* p. 7.
—— (Ctenocercus) cœlestinus, *Cope, Proc. Ac. Philad.* 1862, p. 177.

Head moderate, nearly twice as long as broad, much longer than the tibia; forehead very slightly concave, no frontal ridges; upper head-scales rugose, not keeled; scales of the supraorbital semicircles feebly enlarged, separated medially by one or two series of scales; nine to twelve enlarged, feebly striated supraocular scales, separated from the supraorbitals by two series of smaller scales; occipitals small, as large as the ear-opening, separated from the supraorbitals by three series of scales; canthus rostralis angular, canthal scales four; loreal rows three or four; seven or eight labials to below the centre of the eye; ear-opening small, oval. Gular appendage indicated in the female [large in the male]; gular scales obtusely keeled. Body not compressed; no dorso-nuchal fold. Dorsal scales small, granular, smooth or obtusely keeled, very slightly enlarged on the vertebral region; ventral scales larger, subhexagonal, subimbricate, smooth or obtusely keeled. Limbs not very long; the adpressed hind limb reaches the neck; digital expansions well developed; twenty-five lamellæ under phalanges II. and III. of the fourth toe. Tail roundish, with a series of enlarged flat scales superiorly; its length not twice that of head and body. Green, lighter inferiorly; in one of the specimens the head and a large spot on each side of the body, extending to the axil, dark brown.

Total length	151 millim.
Head	16 ,,
Width of head	8·5 ,,
Body	38 ,,
Fore limb	21 ,,
Hind limb	33 ,,

Tibia	10 millim.
Tail	97 „

San Domingo.

a, *b*. ♀. S. Domingo.

26. Anolis lucius.

Anolius lucius, *Gray, Cat.* p. 205.
Anolis lucius, *Dum. & Bibr.* iv. p. 105; *Coct. in R. de la Sagra, Hist. Cuba, Rept.* p. 136, pl. xii.; *Bocourt, Miss. Sc. Mex., Rept.* pl. xiv. fig. 5; *O'Shaughn. Ann. & Mag. N. H.* (4) xv. 1875, p. 272.
Dactyloa lucius, *Fitzing. Syst. Rept.* p. 67.
Anolis (Gastrotropis) argenteolus, *Cope, Proc. Ac. Philad,* 1861, p. 213; *Bocourt, l. c.* fig. 10.

Head moderate in the male, smaller and with much more elongate snout in the female; in the former the width of the head is contained once and two thirds to once and three fourths in the length, in the latter it is contained at least twice; in both sexes the length of the head slightly exceeds that of the tibia; snout much depressed, forehead slightly concave; no frontal ridges; upper head-scales smooth; scales of the supraorbital semicircles large, in contact medially; six to ten enlarged, smooth supraocular scales, separated from the supraorbitals by one or two series of granules; occipital nearly as large as the ear-opening, separated from the supraorbitals by one or two series of scales; canthus rostralis rather obtuse, canthal scales four or five; loreal rows five; six or seven labials to below the centre of the eye; ear-opening rather large, suboval. Gular appendage rather small, not developed in the female; gular scales smooth. Body not compressed; no dorso-nuchal fold. Dorsal scales very small, granular, smooth, uniform in size; ventral scales larger, subhexagonal, not or but slightly imbricate, smooth. Limbs elongate and slender; the hind limb reaches the front border of the orbit in the male, between the latter point and the end of the snout in the female; digital expansions well developed; twenty-one lamellæ under phalanges II. and III. of the fourth toe. Tail very slightly compressed, nearly twice as long as head and body. Male with enlarged postanal scales. Pale grey above, variegated with brown; a light streak from the eye to the tympanum, and another above the temple; four more or less distinct brown chevron-shaped markings across the back; limbs with brown cross bars; lower surfaces whitish; the throat with brown longitudinal lines or reticulations in the female.

	♂	♀
Total length	—*	146 millim.
Head	14	14 „
Width of head	8	6·5 „
Body	31	32 „

* Tail lost.

	♂.	♀.
Fore limb	22	24 millim.
Hind limb	38	41 „
Tibia	12	13 „
Tail	—	100 „

Cuba.

a. ♀.	Cuba.	Prof. Peters [P.].
b–d. ♂ ♀.	——?	

27. **Anolis isolepis.**

Anolis (Ctenocercus) isolepis, *Cope, Proc. Ac. Philad.* 1861, p. 214.

Head large, nearly twice as long as broad, much longer than the tibia; forehead slightly concave; frontal ridges distinct, short; upper head-scales smooth; scales of the supraorbital semicircles large, separated medially by one series of scales; only two or three large, smooth supraocular scales, separated from the supraorbitals by one series of granules; occipital larger than the ear-opening, separated from the supraorbitals by one series of scales; canthus rostralis sharp, projecting, canthal scales five; loreal rows four; seven upper labials to below the centre of the eye; ear-opening small, roundish-subtriangular. Gular appendage rather small; gular scales smooth. Body not compressed; no dorso-nuchal fold. Dorsal and lateral scales equal in size, minute, granular, smooth; ventrals larger, roundish, flat, juxtaposed, smooth. Limbs very short; the adpressed hind limb reaches the neck; digital expansions well developed; eighteen lamellæ under phalanges II. and III. of the fourth toe. Tail cylindrical, with a dorsal series of slightly enlarged flat scales. No enlarged postanal scales. Whitish olive above, with irregular brown markings; upper surface of head obscured by brown marblings; lower surfaces whitish, with scattered brown dots on the throat.

Head	14 millim.
Width of head	7·5 „
Body	24 „
Fore limb	13 „
Hind limb	21 „
Tibia	6·5 „

Cuba.

a. ♂.	Cuba.	Prof. Peters [P.].

28. **Anolis beckeri.**

Anolis beckeri, *Bouleng. Proc. Zool. Soc.* 1881, p. 921.

Head once and three fourths as long as broad, much longer than the tibia; frontal concavity slightly marked, no frontal ridges; upper head-scales smooth; scales of the supraorbital semicircles enlarged, in contact medially, or separated by one series of scales; enlarged supraocular scales numerous, smooth, separated from the supraorbitals by two rows of granules; occipital much larger than the ear-opening, separated from the supraorbitals by two or three series of scales; canthus rostralis rather obtuse, canthal scales four; loreal rows four; eight upper labials to below the the centre of the eye; ear-opening very small, roundish. Gular appendage small, but well developed (♀); gular scales smooth. Body not compressed. Dorsal and lateral scales equal in size, minute, granular, smooth; ventrals similar, but larger. Limbs very short; the adpressed hind limb reaches the shoulder; digital expansions well developed; eighteen lamellæ under phalanges II. and III. of the fourth toe. Tail cylindrical, not much longer than head and body, with a dorsal series of enlarged flat scales. Light grey above, with pure white spots and a dark-brown network; tail encircled by complete dark-brown annuli; beneath white, with a few dark spots.

Total length	113	millim.
Head	15	"
Width of head	8·5	"
Body	40	"
Fore limb	19·5	"
Hind limb	29	"
Tibia	9	"
Tail	58	"

Yucatan.

29. **Anolis damulus.** (PLATE II. fig. 2.)

Anolis damulus, *Cope, Proc. Ac. Philad.* 1864, p. 169.

Head moderate, nearly once and two thirds as long as broad, longer than the tibia; forehead not concave, with slight, strongly diverging frontal ridges; upper head-scales feebly keeled; scales of the supraorbital semicircles enlarged, produced forwards as frontal series, separated by one or two rows of scales; fourteen or fifteen enlarged, keeled supraocular scales, separated from the supraorbitals by one or two series of smaller scales; occipital as large as the ear-opening, separated from the supraorbitals by four series of scales; canthus rostralis angular, canthal scales four; loreal rows five; six labials to below the centre of the eye; ear-opening moderately large, oval. Gular appendage moderately large; gular scales smooth. Body not compressed; no dorso-nuchal fold. Dorsal scales minute, granular, all perfectly equal, there being no vertebral series of enlarged ones; ventral scales larger, hexagonal, subimbricate, smooth. The adpressed hind limb reaches the anterior border of the orbit; digital expansions well developed; twenty-one lamellæ under phalanges II. and III. of the fourth toe. Tail scarcely compressed, with strongly keeled scales, and a series of feebly enlarged scales along the dorsal line; its length not twice that of head and body. Male with enlarged postanal scales. Above metallic light reddish brown, with minute whitish dots on the sides and limbs; lips varied with darker; a whitish streak from below eye to ear; tail with blackish cross bars near the end; belly pale golden, gular region faintly greenish.

Total length	117	millim.
Head	13	,,
Width of head	8	,,
Body	27	,,
Fore limb	20	,,
Hind limb	34	,,
Tibia	11	,,
Tail	77	,,

Hab. —— ?

a. ♂. —— ? Dr. A. Günther [P.]. (Type.)

30. Anolis fusco-auratus.

Anolius fusco-auratus, *Gray, Cat.* p. 205.
Anolis fusco-auratus, *D'Orbigny, Voy. Amér. Mér., Rept.* pl. iii. fig. 2; *Dum. & Bibr.* iv. p. 110; *Guichen. in Gay, Hist. Chile, Rept.* p. 21; *Bocourt, Nouv. Arch. Mus.* vi. 1870, *Bull.* p. 15, *and Miss. Sc. Mex., Rept.* pl. xiv. fig. 16.
Dactyloa (Tropidopilus) fusco-aurata, *Fitzing. Syst. Rept.* p. 67.
—— fusco-aurata, *Tschudi, Faun. Per., Rept.* p. 24.
Anolis viridiæneus, *Peters, Mon. Berl. Ac.* 1863, p. 147.

Head rather small, twice as long as broad, a little longer than the tibia; forehead concave, frontal ridges very short; upper head-scales very small, feebly keeled; scales of the supraorbital semicircles enlarged, separated by two or three (exceptionally one) series of scales; eight to twelve enlarged, keeled supraocular scales, separated from the supraorbitals by one or two series of granules; occipital larger than the ear-opening, separated from the supraorbitals by two or three series of scales; canthus rostralis feeble, canthal scales three or four; loreal rows five or six; six to eight labials to below the centre of the eye; ear-opening moderately large, vertically oval. Gular appendage moderately large, indicated in the female; gular scales smooth. Body not or but very slightly compressed; no dorso-nuchal fold. Dorsal and lateral scales very small, granular, slightly enlarged, and distinctly keeled on the vertebral region; ventral scales small, larger than dorsals, round, juxtaposed, smooth. Limbs slender; the adpressed hind limb reaches the eye, or between the latter and the tympanum; digital expansions not much developed; fourteen to sixteen lamellæ under phalanges II. and III. of the fourth toe. Tail slender, roundish, covered with strongly keeled scales, the median dorsal series a little enlarged; length of the tail once and two thirds to twice that of head and body. No enlarged postanal scales. Grey-brown or olive above, with metallic green or coppery gloss, with ill-defined darker spots; frequently a light, dark-edged cross band between the eyes; females sometimes with a broad pale coppery vertebral band broadly edged with blackish; lower surfaces whitish, uniform, or more or less closely dotted with brown.

Total length	124	millim.
Head	13	"
Width of head	7	"
Body	33	"
Fore limb	20	"
Hind limb	37	"
Tibia	11	"
Tail	78	"

Brazil, Chili, Peru, Ecuador, Venezuela.

a. ♂.	Para.	
b–e. ♂ ♀.	Yurimaguas, Huallaga River.	Dr. Hahnel [C.].
f. ♀.	Pallatanga, Ecuador.	Mr. Buckley [C.].
g. ♀.	Guayaquil.	Mr. Fraser [C.].

31. Anolis rodriguezii.

Anolis rodriguezii, *Bocourt, Miss. Sc. Mex., Rept.* p. 62, pl. xiii. fig. 1.

It is doubtful whether this form deserves to rank as a species. The only important difference I can find, on comparison with the preceding, is the shorter tail, its length being not more than once and one third that of head and body. The coloration is quite similar to that of the banded variety (spec. *g*) of *A. fusco-auratus*.

The specimen described by Bocourt measures 80 millim., in which the tail enters for 42. The specimen which I must refer to this form measures 44 millim. from snout to vent, and the tail 59.

Guatemala.

a. ♀. ——?

32. Anolis aureolus.

Anolis aureolus, *Cope, Proc. Am. Phil. Soc.* xxii. 1885, p. 390.

Apparently very closely related to the two preceding species, from which it is distinguished by longer hind limbs, reaching the end of the snout when extended.

Yucatan, Guatemala.

33. Anolis limifrons.

Anolis (Dracontura) limifrons, *Cope, Proc. Ac. Philad.* 1862, p. 178.
—— limifrons, *Bocourt, Miss. Sc. Mex., Rept.* p. 65, pl. xiv. fig. 20.
? Anolis rivieri, *Thominot, Bull. Soc. Philom.* (7) vi. 1882, p. 251.

Near *A. fusco-auratus*. Head as long as the tibia; upper head-scales very small, keeled; supraorbital semicircles separated from each other by three, and from the occipital by four or five series of scales; twelve enlarged, keeled supraocular scales; occipital nearly as large as the ear-opening, which is large and vertical; loreal rows seven. Dorsal and lateral scales granular, rugulose, all equal. The adpressed hind limb reaches the anterior border of the orbit. Tail cylindrical, without a larger median series of scales. Above bronze brown, beneath rusty white, separation between the two abrupt anteriorly; lips and femora beneath varied with brown.

Size of *A. fusco-auratus*.

Veragua.

34. Anolis bransfordii.

Anolis bransfordii, *Cope, Proc. Ac. Philad.* 1874, p. 67.

Near *A. fusco-auratus.* Upper head-scales smooth; supraorbital semicircles separated from each other by one, and from the occipital by two series of scales; twelve supraocular scales keeled. Gular appendage little developed. Dorsal and lateral scales subequal and smooth. The adpressed hind limb reaches the end of the snout. Tail long and slender, covered with keeled scales. Above golden-lead colour, beneath silvery, the line of separation defined from the orbit to the groin; no cross bands on head or body; front brown-speckled; feet blackish.

Total length 119 millim., length to groin 35.

Nicaragua.

35. Anolis trochilus.

Anolis trochilus, *Cope, Proc. Ac. Philad.* 1871, p. 215, *and Journ. Ac. Philad.* (2) viii. 1876, p. 121.

Near *A. fusco-auratus,* with which it appears to agree except in the following points:—Supraocular scales smooth, six enlarged. The adpressed hind limb reaches the end of the snout. Caudal scales smooth. Gular appendage little developed. Above and below brilliant metallic green, with a few black dots along the vertebral line; head and anterior part of sides brown; a black **V** extending from the auricular openings, which are connected by a broad black band with the orbits; another **V** extends towards the occiput from the limbs to the nuchal **V**, enclosing a narrow brown area with it; top of snout and limbs blackish.

Total length 108 millim., length to groin 37.

Costa Rica and Nicaragua.

36. Anolis bocourtii.

Anolis bocourtii, *Cope, Journ. Ac. Philad.* (2) viii. 1876, p. 167.

Allied to *A. trochilus, bransfordii,* and *ortonii.* Upper head-scales smooth; supraorbital semicircles separated by two or three rows of scales; six enlarged, smooth supraocular scales. Gular appendage large. Dorsal scales thickened but not keeled, the laterals a little smaller still. The adpressed hind limb reaches the front of the orbit. Tail slightly compressed, without median keel, twice as long as head and body. Coppery brown above; below light coppery, frequently dusted and speckled with brown, especially in females.

Total length	135	millim.
Length to ear-opening ..	11	,,
Width of head	6	,,
Length to vent	45	,,
Hind limb	33	,,

Nauta, Upper Amazons.

37. **Anolis ortonii.**

Anolis ortonii, *Cope, Proc. Ac. Philad.* 1868, p. 97.
—— cynocephalus, *Bocourt, Nouv. Arch. Mus.* vi. 1870, *Bull.* p. 13, *and Miss. Sc. Mex., Rept.* pl. xiv. fig. 7.
—— bouvieri, *Bocourt, Miss. Sc. Mex., Rept.* p. 58, pl. xiv. fig. 8; *O'Shaughn. Ann. & Mag. N. H.* (4) xv. 1875, p. 274, *and Proc. Zool. Soc.* 1881, p. 243.

Head moderate, once and three fourths as long as broad, longer than the tibia; forehead concave, no frontal ridges; upper head-scales small, smooth; scales of the supraorbital semicircles enlarged, in contact medially, or separated by one series of scales; six to ten enlarged, smooth supraocular scales, separated from the supraorbitals by one series of granules; occipital much larger than the ear-opening, separated from the supraorbitals by one or two series of scales; canthus rostralis feebly marked, canthal scales four or five; loreal rows four or five; six to eight labials to below the centre of the eye; ear-opening small, oval. Gular appendage large, with rows of widely distant scales, indicated in the female; gular scales smooth. Body not or but slightly compressed; no dorso-nuchal fold. Dorsal and lateral scales very small, granular, scarcely larger on the vertebral region; ventral scales larger, small, round, juxtaposed, smooth. The adpressed hind limb reaches the ear or slightly beyond; digital expansions moderate; eighteen lamellæ under phalanges II. and III. of the fourth toe. Tail slightly compressed, covered with strongly keeled scales, in the male with a very slightly serrated upper edge; its length not twice that of head and body. No enlarged postanal scales. Brown or grey above, with coppery or greenish metallic gloss, with rather ill-defined darker spots or variegations, sometimes forming chevron-shaped marks on the back, the most distinct of which is on the sacrum; lips spotted; throat sometimes black; belly whitish, with metallic gloss.

Total length	137	millim.
Head	14	,,
Width of head	8	,,
Body	36	,,
Fore limb	23	,,
Hind limb	35	,,
Tibia	11	,,
Tail	87	,,

Central and South America.

a. ♂.	Surinam?	
b. ♂.	Pebas.	H. W. Bates, Esq. [C.].
c. ♂.	Canelos, Ecuador.	Mr. Buckley [C.].
d. ♂.	Pallatanga, Ecuador.	Mr. Buckley [C.].
e. ♀.	Guayaquil.	Mr. Fraser [C.].
f. ♀.	Pozuzo, N. Peru.	W. J. Veitch, Esq. [P.].

38. Anolis schiedii.

Dactyloa schiedii, *Wiegm. Herp. Mex.* p. 48.
Anolis gibbiceps, *Cope, Proc. Ac. Philad.* 1864, p. 174.
—— schiedii, *Bocourt, Miss. Sc. Mex., Rept.* p. 64, pl. xiv. fig. 19.

Head once and two thirds to once and three fourths as long as broad, slightly longer than the tibia; forehead concave, frontal ridges obsolete; upper head-scales keeled; supraorbital semicircles separated from each other by two or three, and from the occipital by three to five series of scales; ten to fourteen enlarged, keeled supraocular scales; occipital small, smaller than the ear-opening; canthus rostralis short, canthal scales two; loreal rows seven to nine; seven labials to below the centre of the eye; ear-opening rather large, vertically oval. Gular appendage large, present also, but less developed, in the female; gular scales obtusely keeled. Body short, not compressed; no dorso-nuchal fold. Dorsal scales very small, rhomboidal, keeled, hardly larger than the laterals, which are granular and rugulose; ventrals small, larger than dorsals, rounded, imbricate, smooth. The adpressed hind limb reaches nearly the end of the snout; digital dilatations not much developed; sixteen to eighteen lamellæ under phalanges II. and III. of the fourth toe. Tail subcylindrical, nearly twice as long as head and body, covered with equal keeled scales. No enlarged postanal scales. Coppery above, uniform; sometimes with darker spots, or a light vertebral line in the female; lower surfaces whitish with metallic gloss.

From snout to vent	49	millim.
Head	15	,,
Width of head	9	,,
Fore limb	24	,,
Hind limb	46	,,
Tibia	13·5	,,

Central America; Venezuela.

a. ♀. Caraccas. (Type of *A. gibbiceps.*)

39. Anolis polylepis.

Anolis polylepis, *Peters, Mon. Berl. Ac.* 1873, p. 738.

Near *A. schiedii* and *A. limifrons.* Supraorbital semicircles separated by three or four series of scales; loreal rows nine. Median dorsal scales keeled, smaller than the ventrals, and a little larger than the granular laterals; ventrals smooth. Tail compressed, especially in the male, with a vertebral series of enlarged scales. Otherwise as in *A. schiedii.* A brown band between the eyes; females with a light festooned vertebral band, widening from the scapular region and extending to the eyes; males with a light nuchal marking and small dorsal spots or transverse markings.

Chiriqui.

40. Anolis tropidolepis.

Head once and two thirds to once and three fourths as long as broad, slightly longer than the tibia; forehead concave, no frontal ridges; upper head-scales very small, strongly uni- or tricarinate; scales of the supraorbital semicircles scarcely enlarged, separated by three to five series of scales; six to twelve enlarged, strongly keeled supraocular scales; occipital small, smaller than the ear-opening, separated from the supraorbitals by three to five series of scales; canthus rostralis angular, canthal scales four; loreal rows seven to nine; seven or eight labials to below the centre of the eye; ear-opening rather large, vertically oval. Gular appendage very small, absent in the female; gular scales keeled. Body feebly compressed in the male; no dorso-nuchal fold. Dorsal and lateral scales very small, granular, keeled, those on the vertebral region slightly larger, strongly keeled; ventrals small, larger than dorsals, rounded, imbricate, keeled. The adpressed hind limb reaches nearly the end of the snout; digital dilatations feebly developed; thirteen or fourteen lamellæ under phalanges II. and III. of the fourth toe. Tail cylindrical, slightly compressed in the male, covered with subequal keeled scales. No enlarged postanal scales. Brownish or greenish above, with metallic gloss, with or without small darker spots; an angular dark cross band between the eyes, and frequently a V-shaped dark band from the orbits to the nape; males with a very broad dark band from the eye to the side of the body; female frequently with a light, dark-edged vertebral band; lower surfaces yellowish or greenish, with metallic gloss.

Total length	164	millim.
Head	15	"
Width of head	9	"
Body	39	"
Fore limb	23	"
Hind limb	42	"
Tibia	13	"
Tail	110	"

Costa Rica.

a. Many specs., ♂ ♀. Irazu. F. D. Godman, Esq. [P.].

41. Anolis pulchripes.

Anolis pulchripes, *Peters, Mon. Berl. Ac.* 1873, p. 739.

Head a little longer than the tibia; supraorbital semicircles separated by one series of scales; twelve to fourteen enlarged supraoculars, arranged in three or four longitudinal rows, the two anterior scales of the inner row most enlarged; occipital smaller than the ear-opening, separated from the supraorbitals by two or three series of scales; loreal rows five; nine upper labials; ear-opening vertically oval. Median dorsal scales very small, keeled, hardly larger than the granular laterals; ventrals small, with a feeble keel extending only on the basal half of the scale. Digital expansions well-developed. Tail rounded, with a vertebral series of enlarged scales. Grey-brown above, whitish inferiorly, the throat yellow; loreal region and lips dark-spotted; a narrow dark cross band between the eyes; a black nuchal spot behind the occipital; in front of the shoulder a light, dark-edged line ascending backwards; limbs with broad dark and narrow white cross bands and lines.

Total length	80 millim.
Length to vent	37 „
Fore limb	16·5 „
Hind limb	30 „

Chiriqui.

42. Anolis baccatus.

Anolis baccatus, *Bocourt, Miss. Sc. Mex., Rept.* p. 59, pl. xiv. fig. 14.

Head longer than the tibia; forehead concave; upper head-scales tricarinate; scales of the supraorbital semicircles enlarged, separated by one series of scales; about ten enlarged, feebly keeled supraocular scales; occipital larger than the ear-opening, separated from the supraorbitals by two series of scales; canthus rostralis angular; loreal rows four; ten or eleven upper labials; ear-opening small. Body not compressed. Dorsal scales granular, pearl-like, feebly keeled, scarcely larger than the laterals; ventrals smooth, larger than dorsals, granular. Limbs very short. Tail rounded. Head above reddish-brown; body and limbs greenish-grey, black-marbled; a yellowish zigzag vertebral band; upper lip and lower surfaces yellowish-white; throat finely marbled with light reddish.

Total length	109 millim.
Head, to ear	12 „
Width of head	7·5 „
Length to vent	40 „
Tibia	8·5 „

Mexico.

43. Anolis guentheri.

Anolis guntherii, *Bocourt, Miss. Sc. Mex., Rept.* p. 61, pl. xiv. fig 15.

Head longer than the tibia; forehead concave; upper head-scales unicarinate; scales of the supraorbital semicircles enlarged, separated by two series of scales; about fifteen keeled supraoculars; occipital larger than the ear-opening, separated from the supraorbitals by four series of scales; canthus rostralis angular; loreal rows four; eight or nine upper labials. Body not compressed. Dorsal scales extremely small, almost as small as the granules on the flanks, juxtaposed, very slightly keeled; ventrals larger, smooth. Limb short. Tail short, round, with a median dorsal series of enlarged flat scales. Yellowish above, minutely dotted with whitish; a large light brown, white-edged spot on each shoulder, behind the arm; lower surfaces yellowish-white, throat marbled with grey.

Total length	95 millim.
Head, to ear	13·5 ,,
Width of head	8·5 ,,
Length to vent	50 ,,
Tibia	11 ,,

Mexico.

44. **Anolis impetigosus.** (Plate II. fig. 3.)

Anolis impetigosus, *Cope, Proc. Ac. Philad.* 1864, p. 174.

Head nearly twice as long as broad, much longer than the tibia; a deep rhomboidal frontal concavity; no frontal ridges; upper head-scales smooth; scales of the supraorbital semicircles large, in contact medially; three large, smooth supraocular scales, separated from the supraorbitals by a series of granules, occipital larger than the ear-opening, separated from the supraorbitals by two series of scales; canthus rostralis indistinct; loreal rows two; seven upper labials to below the centre of the eye; ear-opening small, roundish-subtriangular. Female with a large, folded, *inflatable gular pouch*; gular scales smooth; a series of very large chin-shields on each side. Body not compressed. Scales all flat and granular, small and equal on the back and sides, larger on the belly. Limbs very short; the adpressed hind limb reaches hardly the shoulder; digital expansions well developed; eighteen lamellæ under phalanges II. and III. of the fourth toe. Tail cylindrical, a little longer than head and body, without a dorsal series of enlarged scales. Pale brown above, with small, indistinct darker spots; lower surfaces white, gular pouch with large black spots.

Total length	107 millim.
Head	14 ,,
Width of head	7·5 ,,
Body	35 ,,
Fore limb	16 ,,
Hind limb	25 ,,
Tibia	8 ,,
Tail	58 ,,

Hab. ——?

a. ♀. ——? (Type.)

45. **Anolis tigrinus.**

Anolis tigrinus, *Peters, Mon. Berl. Ac.* 1863, p. 143; *Bocourt, Miss. Sc. Mex., Rept.* pl. xiv. fig. 2.

Head twice as long as broad, much longer than the tibia; forehead slightly concave, no frontal ridges; upper head-scales rather large, smooth, only five longitudinal series between the canthals; scales of the supraorbital semicircles large, in contact medially; five to ten enlarged, smooth supraocular scales, in contact with the supraorbitals; occipital much larger than the ear-opening, in contact with the supraorbitals; canthus rostralis short; loreal rows two; eight or nine upper labials; ear-opening very small, subtriangular. Scales small and smooth, the ventrals largest, rounded, the dorsals a little larger than the laterals, with two median rows slightly prominent. The adpressed hind limb reaches the axilla; digital expansions well developed. Tail round, with equal smooth or feebly keeled scales. Brownish above, the body with indistinct dark brown cross bands and yellowish round spots; outer side of limbs with regular small light round spots and a dark network; lower surfaces brownish yellow.

Total length (tail injured)	110	millim.
Head	16	,,
Width of head	8	,,
Fore limb	18	,,
Hind limb	27	,,

Chili?

46. Anolis lævis.

Scytomycterus lævis, *Cope, Journ. Ac. Philad.* (2) viii. 1876, p. 165.

Rostral plate produced into a flexible appendage; a shallow frontal concavity; lateral occipital crests prominent; upper head-scales large and smooth, only three rows between those of the canthus rostralis at the middle; supraorbital semicircles in contact along the entire median line and with the occipital, which is large; loreal rows two; ear-opening small; a series of large chin-shields in contact with the lower labials throughout. Scales of the body smooth, of nearly equal size; those of the tail larger, the median superior series not different from the others. Limbs short, the adpressed hind limb falling considerably short of the axilla; digital expansions rather narrow. Colour above dark grey, below pigmented white; the two colours are abruptly defined below the orbit and the scapula, and there are brown spots behind the axilla; tail distantly annulate.

Total length	139	millim.
Head, to ear	17	,,
Width of head	9	,,
Length to vent	60	,,
Fore limb	18	,,
Hind limb	27	,,
Tibia	7	,,

Huallaga River, Eastern Peru.

47. Anolis punctatus.

Rhinosaurus gracilis, *Gray, Cat.* p. 199.
Anolius punctatus, *Gray, l. c.* p. 205.
Anolis punctatus, *Daud. Rept.* iv. p. 84, pl. xlviii. fig. 2; *Dum. & Bibr.* iv. p. 112; *Guichen. in Casteln. Amér. Mér., Rept.* p. 16; *O'Shaughn. Proc. Zool. Soc.* 1881, p. 241.
—— gracilis, *Wied, Reise Bras.* ii. p. 131, *and Abbild.*
—— viridis, *Wied, l. c.* p. 132, *and Abbild.*
—— violaceus, *Spix, Lac. Bras.* p. 15, pl. xvii. fig. 2.
—— nasicus, *Dum. & Bibr.* p. 115; *Guichen. l. c.* p. 17, pl. iv. fig. 2.
Ptychonotus (Eunotus) gracilis, *Fitzing. Syst. Rept.* p. 65.
Dactyloa (Eudactylus) punctata, *Fitzing. l. c.* p. 67.

Head once and three fourths to twice as long as broad, longer than the tibia; end of snout truncate in the female, pointed, compressed, and projecting in the male; forehead and interorbital space concave; no frontal ridges; upper head-scales smooth; scales of the supraorbital semicircles large, in contact or separated by one or two series of scales; a few slightly enlarged, more or less distinctly keeled, supraocular scales; occipital as large as or a little larger than the ear-opening, separated from the supraorbitals by two or three series of scales; canthus rostalis sharp, canthal scales four; loreal rows four to six; seven to nine upper labials to below the centre of the eye; ear-opening moderately large, suboval. Gular appendage well developed in the male, very small in the female; gular scales smooth or obtusely keeled. Body compressed in the male, with a slight dorso-nuchal fold. Scales small, granular, smooth, scarcely larger on the back than on the sides; ventrals larger, roundish, subimbricate, smooth. The adpressed hind limb reaches the posterior border of the tympanum or between the latter and the ear-opening; digital expansions well developed; twenty-six to twenty-eight lamellæ under phalanges II. and III. of the fourth toe. Tail feebly compressed, without dorsal series of enlarged scales; its length about twice that of head and body. Male with or without enlarged postanal scales. Purplish above, with scattered round white dots; female sometimes with a vertebral series of elongate lighter spots; lower surfaces greenish- or purplish-white.

Total length	262	millim.
Head	24	"
Width of head	12	"
Body	58	"
Fore limb	33	"
Hind limb	59	"
Tibia	18	"
Tail	180	"

Brazil and Upper Amazons.

a. ♂.	Brazil.	
b–c, d. ♂ ♀.	Rio Janeiro.	G. Busk, Esq. [P.].
e. ♂.	Peruvian Amazons.	Mr. E. Bartlett [C.].
f. ♂.	Cayaria, Peru.	Messrs. Veitch [P.].

48. Anolis boulengeri.

Anolis boulengeri, *O'Shaughn. Proc. Zool. Soc.* 1881, p. 242, pl. xxiv. fig. 1.

Doubtfully distinct from the preceding, from which it differs only in the distinctly keeled ventral scales.

a. ♀. Canelos, Ecuador. Mr. Buckley. [C.]. (Type.)

49. Anolis transversalis.

Anolis transversalis, *A. Dum. Cat. Méth. Rept.* p. 57, *and Arch. Mus.* viii. p. 515, pl. xix. fig. 3; *Guichen. in Casteln. Amér. Mér., Rept.* p. 17; *Bocourt, Miss. Sc. Mex., Rept.* pl. xiv. fig. 3.

Head once and three fourths as long as broad, longer than the tibia; upper head-scales smooth; scales of the supraorbital semicircles large, in contact medially; a few enlarged, smooth supraoculars; occipital slightly larger than the ear-opening, in contact with the supraorbitals; ear-opening moderate. Gular appendage small. Dorsal and lateral scales equal, minute, granular; ventrals larger, imbricate, smooth. Limbs moderate, digital expansions well developed. Tail slightly depressed at the base, once and a half as long as head and body. Green above, with broad transverse brown bands, the anterior, or scapular, sending a perpendicular branch along the nape; head vermiculated with brown.

Sarayacu, Peru.

50. Anolis buckleyi.

Anolis buckleyi, *O'Shaughn. Proc. Zool. Soc.* 1880, p. 492, pl. xlix.

Head nearly twice as long as broad, a little longer than the tibia; forehead and interorbital space concave; no frontal ridges; upper head-scales smooth; scales of the supraorbital semicircles large, in contact medially; a few enlarged, smooth supraocular scales; occipital not larger than the ear-opening, separated from the supraorbitals by one series of scales; canthus rostralis angular, canthal scales three; loreal rows four; six or seven labials to below the centre of the eye; ear-opening moderate, oval. Gular appendage very large, extending posteriorly beyond the thorax; gular scales smooth; a series of very large chin-shields on each side, in contact with the lower labials. Body compressed. Scales minute, granular, smooth, scarcely larger on the back than on the flanks; ventrals larger, roundish, subimbricate, smooth. The adpressed hind limb reaches the posterior border of the orbit; digital expansions well developed; twenty-five lamellæ under phalanges II. and III. of the fourth toe. Tail slightly compressed, without dorsal series of enlarged scales; its length nearly twice that of head and body. Male with enlarged postanal scales. Brownish or greenish, with black spots or punctulations on the back and sides, and a black network on the upper surface of the head on the sutures of the scales; fine oblique white or bluish lines down the sides of the body, with rows of black dots between; ventral surface pale bluish, dotted with black; gular appendage bright saffron-yellow, with black longitudinal streaks and small spots.

Total length	263	millim.
Head	33	„
Width of head	12	„
Body	65	„
Fore limb	37	„
Hind limb	62	„
Tibia	19	„
Tail	165	„

Ecuador.

a–b. ♂.	Canelos.	Mr. Buckley [C.]. (Types.)

51. **Anolis fasciatus.** (Plate III. fig. 1.)

Head rather small, once and two thirds as long as broad, as long as the tibia; forehead and interorbital space slightly concave; no frontal ridges; upper head-scales small, smooth; scales of the supraorbital semicircles enlarged, separated by one or two series of scales; a few enlarged, indistinctly keeled, supraocular scales; occipital indistinct, smaller than the ear-opening; canthus rostralis distinct, canthal scales five; loreal rows five; seven upper labials to below the centre of the eye; ear-opening rather large, oval. Gular appendage very large, extending posteriorly beyond the thorax; gular scales smooth. Body compressed, no dorso-nuchal fold. Scales very small, granular, smooth, scarcely larger on the back than on the flanks; ventrals larger, roundish, imbricate, smooth. The adpressed hind limb reaches the eye; digital expansions rather narrow; twenty lamellæ under phalanges II. and III. of the fourth toe. Tail very slightly compressed, without dorsal series of enlarged scales; its length more than twice that of head and body. Male with enlarged postanal scales. Purplish-brown above; body with blackish cross bands forming chevrons turned forwards on the sides, these blackish bands enclosing small lighter spots; belly and lower surface of limbs variegated brown and white; gular appendage yellow (?), with series of small black dots.

Total length	221	millim.
Head	16	„
Width of head	9·5	„
Body	50	„
Fore limb	33	„
Hind limb	53	„
Tibia	16	„
Tail	155	„

Ecuador.

a. ♂.	Guayaquil.	Mr. Fraser [C.].

52. **Anolis gemmosus.** (Plate III. fig. 2.)

Anolis gemmosus, *O'Shaughn. Ann. & Mag. N. H.* (4) xv. 1875, p. 280.

Head nearly twice as long as broad, a little longer than the tibia; forehead concave, no frontal ridges; upper head-scales very small, bi- or tri-carinate; scales of the supraorbital semicircles not enlarged, separated by four or five series of scales; supraocular scales small, tricarinate; occipital not enlarged; canthus rostralis short, angular, canthal scales three; loreal rows six; eight labials to below the centre of the eye; ear-opening rather large, oval. Gular appendage large, extending posteriorly beyond the thorax, absent in the female. Body slightly compressed in the male; no dorso-nuchal fold. Scales very small, granular, smooth, equal on the back and sides; ventrals very small, larger than dorsals, irregular, subimbricate, smooth. The adpressed hind limb reaches the anterior border of the orbit or a little beyond; digital expansions rather narrow; seventeen lamellæ under phalanges II. and III. of the fourth toe. Tail roundish, covered above with equal keeled scales; its length almost three times that of head and body. Male with enlarged postanal scales. Purplish above, with black and whitish dots; sides of belly with round light spots; female with a dorsal series of light, black-edged, arrowhead-shaped markings pointing forwards; lower surfaces whitish, chin variegated with brown; gular appendage pale blue.

Total length	217	millim.
Head	16	"
Width of head	8	"
Body	41	"
Fore limb	28	"
Hind limb	51	"
Tibia	15	"
Tail	160	"

Ecuador.

a. ♀.	——?	J. Brenchley, Esq. [P.].	(Type.)
b. ♂.	Intac, Ecuador.	Mr. Buckley [C.].	

53. **Anolis andianus.**

Anolis squamulatus? (*non Ptrs.*), *Bouleng. Ann. & Mag. N. H.* (5) ix. 1882, p. 458.

Head twice as long as broad, much longer than the tibia; forehead and interorbital region concave, no frontal ridges; upper head-scales very small, rugose; scales of the supraorbital semicircles scarcely enlarged, separated by five series of scales; ten to twelve enlarged, tricarinate supraocular scales, separated from the supraorbitals by a series of granules; occipital not enlarged; canthus rostralis feebly marked, no enlarged canthal scales; loreal rows six; seven labials to below the centre of the eye; ear-opening moderate, oval. Gular appendage absent in the female. Body slightly compressed. Scales very small, granular, smooth, equal on the back and sides; ventrals very small, larger than dorsals, granular, smooth. The adpressed hind limb reaches to between the ear and the orbit; digits moderately dilated; eighteen lamellæ under phalanges II. and III. of the fourth toe. Tail cylindrical, covered above with small, equal, feebly keeled scales, its length almost thrice that of head and body. Dull lilac above, minutely and indistinctly speckled with blackish; lower surfaces whitish.

Total length	230	millim.
Head	18	„
Width of head	9	„
Body	47	„
Fore limb	25	„
Hind limb	42	„
Tibia	12	„
Tail	175	„

Ecuador.

a. ♀. Milligalli, 6200 ft. E. Whymper, Esq. [C.].

54. Anolis squamulatus.

Anolis squamulatus, *Peters, Mon. Berl. Ac.* 1863, p. 145; *Bocourt, Miss. Sc. Mex., Rept.* pl. xiv. fig. 21.

Head nearly twice as long as broad; forehead and interorbital space concave; no frontal ridges; a large rhomboidal occipital concavity; upper head-scales very small, keeled; scales of the supraorbital semicircles scarcely enlarged, separated by five or six series of scales; slightly enlarged, strongly keeled supraocular scales; occipital very small, much smaller than the ear-opening, separated from the supraorbitals by six series of scales; canthus rostralis sharp; loreal rows nine in the middle; eleven or twelve upper labials; ear-opening rather large, oval. Gular appendage long, but not very prominent. Dorsal scales very small, keeled; lateral scales minutely granular; ventrals much larger than dorsals, round, smooth (or keeled?). No dorso-nuchal fold. The adpressed hind limb reaches a little beyond the eye; digital expansions well developed. Tail round, covered with small keeled scales. Head, and broad more or less confluent bands on the body, limbs, and tail dark brown; the interspaces between these bands, as well as the lower surfaces, brownish-grey.

Total length	398	millim.
Head	34	„
Width of head	17·5	„
Body	84	„
Fore limb	54	„
Hind limb	93	„
Tail	280	„

Puerto Cabello and Panama.

55. Anolis latifrons.

Anolis latifrons, *Berthold, Abh. Ges. Götting.* iii. 1847, p. 6, pl. i. fig. 2.

Head as in the preceding, but all the scales smooth. Gular appendage very large. A nuchal fold. All the scales on the body equal, small, smooth, imbricate, the ventrals not larger than the others. Olive-green, dotted and spotted all over with black; a quadrangular large black spot in front of the shoulder; a yellowish-white cross band between the eyes.

Total length	400	millim.
Head	33	„
Width of head	17	„
Body	83	„
Fore limb	61	„
Hind limb	105	„
Tail	284	„

Province Popayan, Colombia.

56. Anolis microtus.

Anolis microtus, *Cope, Proc. Ac. Philad.* 1871, p. 214, *and Journ. Ac. Philad.* (2) viii. 1876, p. 119, pl. xxiv. fig. 2.

Head once and three fourths as long as broad, much longer than the tibia; forehead slightly concave, with a slight median ridge; upper head-scales moderately large, not keeled; supraorbital semicircles separated by two series of scales; occipital indistinct; canthus rostralis distinct, canthal scales three; loreal rows four; seven or eight upper labials to below the centre of the eye; ear-opening small or very small, suboval. Gular appendage large, present in the female; gular scales granular, smooth. Body compressed; a strong nuchal fold. Dorsal and lateral scales small, scarcely larger than ventrals, juxtaposed, flat, smooth or rugose; ventral scales smooth, roundish, subimbricate. Scales on upper surface of limbs pluricarinate. The adpressed hind limb reaches the neck; digital expansions well developed; twenty lamellæ under phalanges II. and III. of the fourth toe. Tail feebly compressed, without upper ridge, nearly twice as long as head and body. Male with slightly enlarged postanal scales. Brownish above, uniform or with yellow spots confluent into oblique vertical bands on the sides of the body.

Total length	341	millim.
Head	33	„
Width of head	18	„
Body	83	„
Fore limb	52	„
Hind limb	73	„
Tibia	21	„
Tail	225	„

Costa Rica.

a–b. ♂ ♀.	**Irazu.**	**F. D. Godman, Esq. [P.].**

57. Anolis insignis.

Anolis insignis, *Cope, Proc. Ac. Philad.* 1871, p. 213, *and Journ. Ac. Philad.* (2) viii. 1876, p. 120, pl. xxiv. fig. 1.

Head nearly once and three fourths as long as broad, much longer than the tibia; forehead slightly concave, no frontal ridges; upper head-scales small, not keeled; supraorbitals and occipital scarcely enlarged; supraoculars small, smooth; canthus rostralis distinct; loreal rows five; seven upper labials to below the centre of the eye; ear-opening moderate, oval. Gular appendage large; gular scales granular, smooth. Body compressed; a distinct nuchal fold. Dorsal and lateral scales very small, juxtaposed, flat, and perfectly smooth, those on the vertebral line slightly enlarged; ventral scales larger, small, rounded, subimbricate, flat, and perfectly smooth. The adpressed hind limb reaches the neck; digital expansions well developed; twenty-seven lamellæ under phalanges II. and III. of the fourth toe. Tail compressed, without upper ridge, nearly twice as long as head and body. Male with enlarged postanal scales. Brownish above, with a black network forming four broad bands across the body; neck with dark lines; a large black-edged ocellus in front of the shoulder; tail with regular broad blackish annuli; limbs with blackish cross bars; lower surfaces brownish white.

Total length	415	millim
Head	41	„
Width of head	22	„
Body	114	„
Fore limb	54	„
Hind limb	88	„
Tibia	27	„
Tail (injured)	260	„

Costa Rica.

a. ♂.	Costa Rica.	O. Salvin, Esq. [C.].

58. Anolis pachypus.

Anolis pachypus, *Cope, Journ. Ac. Philad.* (2) viii. 1876, p. 122, pl. xxiv. fig. 3.

Head once and a half as long as broad, as long as the tibia; forehead concave, no frontal ridges; upper head-scales keeled; scales of the supraorbital semicircles not enlarged, separated by five series of scales; a single row of four small keeled supraoculars near the supraorbital border; four rows of scales between the supraorbitals and the occipital; six canthal scales, continued on the edge of the eyebrow to near its middle in a manner unusual in the *Anoles*; loreal rows eight; ear-opening larger than occipital. Gular appendage well developed. Scales of the abdomen smooth, those of the sides very small, on the back gradually enlarging to two or three larger and faintly keeled rows on the median line, which are much smaller than the ventrals. Limbs stout and clumsy; the adpressed hind limb reaches the end of the snout; digits feebly dilated. Tail slender, cylindrical, with equal scales. Colour emerald-green, with a broad brown band from the orbit to the middle of the side, and a narrower one on each side of the sacrum; a dark brown cross band between the supraciliary borders; a deep brown spot behind the occiput, and a smaller one on the nape; femur with a longitudinal black band behind, and an oblique one across the superior face; tibia and forearm with a broad cross band.

Total length	135	millim.
Head, to tympanum	11	,,
Length to vent	47	,,
Fore limb	22	,,
Hind limb	43	,,
Tibia	12	,,

Pico Blanco, Costa Rica.

59. Anolis pentaprion.

Anolis (Coccoëssus) pentaprion, *Cope, Proc. Ac. Philad.* 1862, p. 178.

Head nearly once and two thirds as long as broad, much longer than the tibia; forehead slightly concave, no frontal ridges; upper head-scales small, rough; supraorbitals and occipital scarcely enlarged; supraoculars small, smooth; canthus rostralis obtuse; loreal rows three; eight upper labials to below the centre of the eye; ear-opening moderate, suboval. Gular appendage large; gular scales minutely granular. Body not compressed; a slight nuchal fold. Dorsal and lateral scales uniformly granular; ventrals a little larger, also granular and smooth. The adpressed hind limb reaches the neck; digital expansions well developed; twenty-two lamellæ under phalanges II. and III. of the fourth toe. Tail little longer than head and body, compressed, subtriangular in section, the vertebral angle trenchant, serrate; four basal angles formed by the continued keels of as many series of large scales. No enlarged postanal scales. Reddish-brown above, marbled with blackish; lower surfaces whitish, with brown variegations.

Total length	170	millim.
Head	21	,,
Width of head	13	,,
Body	54	,,
Fore limb	29	,,
Hind limb	45	,,
Tibia	14	,,
Tail	95	,,

Colombia; Ecuador.

a. ♂.	Guayaquil.	Mr. Fraser [C.].

60. Anolis fraseri. (PLATE IV.)

Anolis fraseri, part., *Günth. Proc. Zool. Soc.* 1859, p. 407.
—— devillei, *Bouleng. Bull. Soc. Zool. France,* 1880, p. 42.

Head not quite once and two thirds as long as broad, much longer than the tibia; forehead slightly concave, no frontal ridges; upper head-scales small, rough, not keeled; supraorbitals and occipital scarcely enlarged; supraoculars small, smooth; canthus rostralis short, prominent, rather obtuse; loreal rows six or seven; eight or nine upper labials to below the centre of the eye; ear-opening moderate, oval. Gular appendage large; gular scales minutely granular, obtusely keeled. Body compressed; a distinct dorso-nuchal fold, strongly developed on the nape. Dorsal and lateral scales granular and keeled, the former scarcely larger than the latter; ventral scales small, larger than dorsals, rounded, sub-imbricate, smooth or obtusely keeled. The adpressed hind limb reaches the ear-opening, or not quite so far; digital expansions well developed; twenty-one or twenty-two lamellæ under phalanges II. and III. of the fourth toe. Tail slightly compressed, with slightly prominent upper edge; its length more than twice that of head and body. No enlarged postanal scales. Purplish brown above, with three darker broad cross bands on the body; a few black spots on scapular region and on nuchal fold; lower surfaces pale brown, with more or less distinct darker dots; gular appendage yellow.

Total length	380	millim.
Head	32	,,
Width of head	20	,,
Body	78	,,
Fore limb	45	,,
Hind limb	75	,,
Tibia	22	,,
Tail	270	,,

Ecuador.

a. ♂.	W. Ecuador.	Mr. Fraser [C.]. (Type.)
b. ♂.	Quito.	
c. ♂.	Nanegal, 3000 ft.	Edw. Whymper, Esq. [C.].

61. Anolis copii.

Anolis fraseri, part., *Günth. Proc. Zool. Soc.* 1859, p. 407; *Bocourt, Miss. Sc. Mex., Rept.* pl. xv. fig. 12.
—— copei, *Bocourt, l. c.* p. 77, pl. xv. fig. 10.

Head once and two thirds to once and three fourths as long as broad, much longer than the tibia; forehead concave, no frontal ridges; upper head-scales small, keeled; scales of the supraorbital semicircles feebly enlarged, separated by two or three series of scales; supraocular scales numerous, keeled; occipital slightly enlarged; canthus rostralis short, angular, canthal scales four or five; loreal rows six or seven; nine to eleven labials to below the centre of the eye; ear-opening moderate, oval. Gular appendage small in the half-grown male, slightly indicated in the female; gular scales keeled. Body slightly compressed; no distinct dorso-nuchal fold. Dorsal and lateral scales granular and feebly keeled, the former not larger, or even somewhat smaller, than the latter; ventrals much larger, rhomboidal, imbricate, strongly keeled. The adpressed hind limb reaches between the ear and the posterior border of the orbit; digital expansions well developed; twenty-two or twenty-three lamellæ under phalanges II. and III. of the fourth toe. Tail round or slightly compressed, with slightly prominent upper edge; its length more than twice that of head and body. No enlarged postanal scales. Brown above, marbled with darker, and with scattered small black spots and occasionally also with yellowish ones; lower surfaces whitish or brownish.

Total length	330	millim.
Head	29	,,
Width of head	17	,,
Body	69	,,
Fore limb	41	,,
Hind limb	73	,,
Tibia	21	,,
Tail	232	,,

Central America, Ecuador.

a–b. ♂ ♀.	Vera Paz, Low forest.	O. Salvin, Esq. [C.].
c. ♂.	Alta Vera Paz.	Mr. F. C. Sarg [C.].
d. ♀.	Costa Rica.	O. Salvin, Esq. [C.].
e. ♀.	W. Ecuador.	Mr. Fraser [C.]. (One of the types of *A. fraseri.*)

62. Anolis petersii.

Anolis biporcatus, (*non Wiegm.*) *Cope, Proc. Ac. Philad.* 1871, p. 215.
—— petersii, *Bocourt, Miss. Sc. Mex., Rept.* p. 79, pls. xiii. fig. 2, & xv. fig. 11.

Head once and a half or once and three fifths as long as broad, considerably longer than the tibia; forehead concave, no frontal ridges; upper head-scales small, rugulose or slightly keeled; scales of the supraorbital semicircles a little enlarged, separated by one or two series of scales; supraocular scales small, feebly keeled; occipital very small; canthus rostralis angular, canthal scales five; six or seven loreal series; nine to eleven upper labials to below the centre of the eye; ear-opening small, oval. Gular appendage moderate; gular scales granular, obtusely keeled. Body compressed; no dorso-nuchal fold. Dorsal and lateral scales equal, small, granular, keeled; ventrals larger, small, subquadrilateral, imbricate, keeled. The adpressed hind limb reaches the ear-opening or slightly beyond; digital expansions well developed; twenty-seven lamellæ under phalanges II. and III. of the fourth toe. Tail roundish, with a vertebral series of enlarged scales, but no prominent ridge; its length more than twice that of head and body. No enlarged postanal scales. Brownish or pale olive above, with dark-brown or reddish-brown marblings; four more or less distinct dark cross bands on the body—the anterior, on the scapula, and the posterior, on the sacrum, best defined; a dark band from the eye to above the tympanum; lips with dark vertical bars; lower surfaces greenish white, dotted with brown; throat with dark-brown marblings.

Total length	302	millim.
Head	26	"
Width of head	18	"
Body	66	"
Fore limb	35	"
Hind limb	62	"
Tibia	18	"
Tail	210	"

Mexico.

a–b. ♂ ♀.	Mexico.	
c. Yg.	Mexico.	M. Sallé [C.].

63. Anolis pulchellus.

Anolius pulchellus, *Gray, Cat.* p. 204.
Anolis pulchellus, *Dum. & Bibr.* iv. p. 97; *Reinh. & Lütk. Vidensk. Meddel.* 1862, p. 257; *Bocourt, Miss. Sc. Mex., Rept.* pl. xvi. fig. 28; *Peters, Mon. Berl. Ac.* 1876, p. 706.
Dactyloa (Heterolepis) pulchella, *Fitzing. Syst. Rept.* p. 66.

Head rather large, twice as long as broad, much longer than the tibia; forehead slightly concave, with two slightly diverging frontal ridges; upper head-scales faintly keeled or striated; scales of the supraorbital semicircles large, extending forwards as frontal series, in contact between the orbits or separated by one series of scales; five to ten enlarged, keeled supraocular scales, separated from the supraorbitals by one or two series of granules; occipital a little larger than the ear-opening, separated from the supraorbitals by two to four series of scales; canthus rostralis angular, canthal scales three or four; loreal rows four or five; five or six labials to below the centre of the eye; ear-opening rather small, roundish-subtriangular. Gular appendage moderately large, with widely separated scales; gular scales keeled. Body compressed; no dorso-nuchal fold. Dorsal scales nearly as large as ventrals, rhomboidal, imbricate, keeled; lateral scales very small, granular; ventral scales rhomboidal, strongly imbricate, keeled. Limbs moderate; the adpressed hind limb reaches the eye or the posterior border of the orbit; digits rather feebly dilated; eighteen or nineteen lamellæ under phalanges II. and III. of the fourth toe. Tail strongly compressed, with sharp, not distinctly serrated upper edge, its length more than twice that of head and body. No enlarged postanal scales. Pale grey-brown above; a white band on the upper lip, extending along each side of the neck and body, bordered above with blackish to the origin of the fore limb; usually two series of white, black-edged ocelli along each side of the body; limbs with more or less distinct dark cross bars; lower surfaces pure white.

Total length	165	millim.
Head	17	,,
Width of head	8	,,
Body	33	,,
Fore limb	18	,,
Hind limb	43	,,
Tibia	11	,,
Tail	115	,,

West Indies (Porto Rico, Virgin Islands, and Martinique).

a. Several spec., ♂ & yg.	S. Thomas.	Hr. Riise [C.].
b. ♂.	S. Thomas.	M. Sallé [C.].

64. **Anolis semilineatus.** (Plate V. fig. 1.)

Anolis semilineatus, *Cope, Proc. Ac. Philad.* 1864, p. 171.

Head moderately large, a little more than twice as long as broad, longer than the tibia; forehead concave, no distinct frontal ridges: upper head-scales feebly keeled; scales of the supraorbital semicircles feebly enlarged, separated by two series of scales; five enlarged, keeled supraocular scales, separated from the supraorbitals by one series of smaller scales; occipital as large as or a little larger than the ear-opening, separated from the supraorbitals by two or three series of scales; canthus rostralis angular, canthal scales four; loreal rows five or six; five or six labials to below the centre of the eye; ear-opening moderate, roundish-subtriangular. Gular appendage moderately large, absent in the female; gular scales feebly keeled. Body compressed; no dorso-nuchal fold. Dorsal scales as large as ventrals, hexagonal, subimbricate, keeled, forming twelve longitudinal series; lateral scales minute, granular; ventral scales rhomboidal, strongly imbricate, keeled. Limbs rather long; the adpressed hind limb reaches the anterior border of the orbit; digits feebly dilated; eighteen lamellæ under phalanges II. and III. of the fourth toe. Tail feebly compressed, more than twice as long as head and body. Male with enlarged postanal scales. Pale brown above; a white band on the upper lip, extending on each side of the neck and body, where it vanishes gradually; female with a pale, dark-edged vertebral band; lower surfaces white.

Total length	141	millim.
Head	12	,,
Width of head	5	,,
Body	26	,,

Fore limb	15	millim.
Hind limb	30	,,
Tibia	9	,,
Tail	103	,,

San Domingo.

a. ♂.	S. Domingo.	(Type.)
b. ♀.	S. Domingo.	

65. **Anolis spectrum.**

Anolis spectrum, *Peters, Mon. Berl. Ac.* 1863, p. 136; *Bocourt, Miss. Sc. Mex., Rept.* pl. xvi. fig. 24.

Head nearly three times as long as broad, longer than the tibia; forehead concave; upper head-scales smooth; scales of the supraorbital semicircles large, separated by one series of scales; three or four enlarged, keeled supraocular scales, in contact with the supraorbitals; occipital nearly as large as the ear-opening, separated from the supraorbitals by one or two series of scales; canthus rostralis feebly marked; loreal rows four; eight upper labials; ear-opening moderately large, roundish. Gular appendage small; gular scales keeled. Body compressed; no dorso-nuchal fold. Six or eight longitudinal series of large, hexagonal, keeled dorsal scales, which are broader than long, and at least twice as large, as the ventrals; lateral scales very small, granular; ventral scales keeled. The adpressed hind limb reaches between the ear and the eye. Tail round or slightly compressed, covered with large strongly keeled scales, the keels forming continuous lines; its length more than twice that of head and body. Yellowish-grey or metallic green above, the sides of the body brown; a sharply defined white spot at the elbow.

Total length	150	millim.
Head	12	,,
Width of head	4·5	,,
Body	28	,,
Fore limb	13	,,
Hind limb	28	,,
Tail	110	,,

Cuba.

66. **Anolis cyanopleurus.**

Anolis (Dracontura) cyanopleurus, *Cope, Proc. Ac. Philad.* 1861, p. 211; *Bocourt, Miss. Sc. Mex., Rept.* pl. xvi. fig. 29.

Head moderately large, twice as long as broad, longer than the tibia; forehead concave, frontal ridges distinct; upper head-scales keeled or striated; scales of the supraorbital semicircles slightly enlarged, separated by one series of scales; three or four enlarged, keeled supraocular scales, separated from the supraorbitals by one row of granules; occipital smaller than the ear-opening, separated from the supraorbitals by two or three series of scales; canthus rostralis rather feebly marked, canthal scales three or four; loreal rows four or five; five or six labials to below the centre of the eye; ear-opening moderate, oval. Gular appendage small; gular scales feebly keeled. Body compressed; no dorsal-nuchal fold. Eight longitudinal series of large, hexagonal, feebly keeled or striated dorsal scales, larger than the ventrals; lateral scales very small, granular; ventral scales hexagonal, imbricate, keeled. Limbs long and slender; the adpressed hind limb reaches between the eye and the nostril; digits rather feebly dilated; fourteen or fifteen lamellæ under phalanges II. and III. of the fourth toe. Tail slightly compressed, covered with large strongly keeled scales, the keels forming continuous lines; its length more than twice that of head and body. No distinctly enlarged postanal scales. Pale metallic green above; sides purplish blue, sometimes with a series of small metallic spots; a white streak on the upper lip; limbs with metallic transverse spots; lower surfaces whitish.

Total length	140	millim.
Head	12	,,
Width of head	6	,,
Body	28	,,
Fore limb	14	,,
Hind limb	31	,,
Tibia	9	,,
Tail	100	,,

Cuba.

a–c. ♂. Cuba. Prof. Peters [P.].

67. **Anolis lionotus.**

Anolis (Dracontura) lionotus, *Cope, Proc. Ac. Philad.* 1861, p. 210.

Head short, broad; frontal ridges almost obsolete, enclosing an ovoid concavity; interorbital space concave; upper head-scales small, smooth; scales of the supraorbital semicircles separated medially by a series of very small scales; three or four supraoculars of large size, surrounded by several smaller scales, some of them weakly keeled; occipital elongate-oval, not in contact with the supraorbitals; canthus rostralis short; loreal rows eight; superior labials eight. Gular appendage moderate. Scales of the sides small, not keeled; of the belly larger, smooth; of the back still larger, in about twelve rows, smooth. The adpressed hind limb reaches the nostril; digital expansions narrow. Tail compressed, the median superior scales flat, offering no rudiment of a crest. General colour above bronze-brown, with a few darker shades on the dorsal region; tail and hinder extremities paler, femora light spotted behind; a light band extends from the angle of the mouth to the middle of the side; beneath it there are brown vermiculations, which become darker near the groin; lips dark; belly immaculate.

Cocuyas de Veraguas, Colombia.

68. Anolis alutaceus.

Anolis (Dracontura) alutaceus, *Cope, Proc. Ac. Philad.* 1861, p. 212.

Head elongate, less so than in *A. spectrum*. Frontal ridges distinct, not elevated, convergent anteriorly, and enclosing a very elongate depression; anterior upper head-scales weakly many-keeled or smooth, those of the supraorbital semicircles separated medially by small scales; five to seven enlarged supraocular scales, separated by granules from the supraorbitals; occipital small, irregular, widely separated from the supraorbitals; canthus rostralis not prominent; upper labials seven; auricular opening moderate. Gular appendage elongate, of female rudimentary. Lateral scales small, gradually blending into the larger dorsals, of which there are about ten rows, weakly keeled; ventral scales smooth, transversely oblong, in some specimens having a carinate appearance upon the thoracic region. The adpressed hind limb reaches the orbit. Tail cylindrical, a little more than twice the length of head and body. Upper parts dark cinnamon, lighter in the female, beneath paler, thickly punctulated with brown on the throat, sides, and extremities; on the head, a light band crosses the anterior part of the supraorbital region, and another extends longitudinally back from the rostral plate.

Total length 134 millim., head to ear 15, tail 91.

Cuba.

69. Anolis bitectus. (Plate V. fig. 2.)

Anolis bitectus, *Cope, Proc. Ac. Philad.* 1864, p. 171.

Head nearly twice as long as broad, as long as or slightly longer than the tibia; snout acuminate; forehead concave, frontal ridges obsolete; upper head-scales strongly bi- or tricarinate; scales of the supraorbital semicircles slightly enlarged, separated by two or three series of scales; six or seven strongly enlarged, pluricarinate supraocular scales, separated from the supraorbitals by one series of smaller scales; occipital as large as the ear-opening, separated from the supraorbitals by three or four series of scales; canthus rostralis distinct, canthal scales four; loreal rows seven; eight labials to below the centre of the eye; ear-opening rather large, vertically oval. Gular appendage large, extending posteriorly beyond the thorax, indicated in the female; gular scales keeled. Body a little compressed; no dorso-nuchal fold. Dorsal scales large, hexagonal, juxtaposed, strongly keeled, in eleven longitudinal series; lateral scales minutely granular; ventral scales larger than dorsals, rounded, imbricate, strongly keeled. Limbs slender; the adpressed hind limb reaches the anterior border of the orbit or the nostril; digits very feebly dilated; thirteen or fourteen lamellæ under phalanges II. and III. of the fourth toe. Tail roundish, slightly compressed, covered with equal, strongly keeled scales; its length not twice that of head and body. No enlarged postanal scales. Pale brown above; a broad dark brown band from the eye to the side of the body; limbs with irregular dark brown transverse bars; a dark transverse band between the eyes; upper lip and lower surfaces whitish.

Total length	129	millim.
Head	15	,,
Width of head	7·5	,,
Body	37	,,
Fore limb	26	,,
Hind limb	46	,,
Tibia	14	,,
Tail	77	,,

Ecuador.

a, b. ♂ ♀. W. Ecuador. Mr. Fraser [C.]. (Types.)

70. Anolis heliactin.

Anolis heliactin, *Cope, Proc. Ac. Philad.* 1864, p. 172; *Bocourt, Miss. Sc. Mex., Rept.* p. 106, pls. xiii. fig. 4, & xvi. fig. 32.

Head narrow and elongate, twice as long as broad, longer than the tibia; forehead concave, frontal ridges obsolete; upper head-scales smooth or feebly keeled; supraorbital semicircles separated by one series of scales; enlarged supraoculars in two longitudinal series, the inner composed of three large scales separated from the supraorbitals by one series of granules; occipital large, much larger than the ear-opening, separated from the supraorbitals by one series of scales; canthus rostralis angular, canthal scales three; loreal rows four; six upper labials to below the centre of the eye; ear-opening moderate, oval, oblique. Gular appendage small; gular scales keeled. Body not compressed; no dorso-nuchal fold. Dorsal scales rather large, hexagonal, subimbricate, keeled, forming about fourteen longitudinal series, and graduating into the small subgranular scales of the sides; ventrals larger than dorsals, rhomboidal, imbricate, strongly keeled. Limbs slender; the adpressed hind limb reaches the eye; digital expansions narrow; fourteen lamellæ under phalanges II. and III. of the fourth toe. Tail roundish. No enlarged postanal scales. Pale bronzy above, with indistinct darker spots; a purple spot on each side of the gular appendage.

Head	12	millim.
Width of head	6	,,
Body	28	,,
Fore limb	17	,,
Hind limb	30	,,
Tibia	10	,,

Mexico.

a. ♂. ——?

71. Anolis ustus.

Anolis ustus, *Cope, Proc. Ac. Philad.* 1864, p. 172.

Head much longer than the tibia, twice as long as broad in the female, a little shorter in the male; forehead very slightly concave, frontal ridges short and divergent; upper head-scales more or less distinctly tricarinate; scales of the supraorbital semicircles large, in contact or separated by one series of scales; four or five large, feebly keeled supraocular scales; occipital large, larger than the eye-opening, separated from the supraorbitals by one series of scales; canthus rostralis sharp, canthal scales four or five; loreal rows four or five; six to eight upper labials to below the centre of the eye; ear-opening small, oval, oblique. Gular appendage small in both sexes; gular scales keeled. Body not compressed; no dorso-nuchal fold. Dorsal scales subhexagonal, juxtaposed, keeled, passing gradually into the much smaller, granular scales of the flanks; ventrals larger than dorsals, rhomboidal, imbricate, strongly keeled. The adpressed hind limb reaches the neck or the ear; digits rather feebly dilated; fourteen or fifteen lamellæ under phalanges II. and III. of the fourth toe. Tail cylindrical, nearly twice as long as head and body. Yellowish brown or greyish above, with rather indistinct darker variegations; a dark cross band between the eyes; a round yellow spot on the tibia; lower surfaces white or yellowish; throat brown-dotted; a small purple spot on each side of the gular appendage.

Total length	131	millim.
Head	12	„
Width of head	7	„
Body	29	„
Fore limb	15	„
Hind limb	27	„
Tibia	9	„
Tail	90	„

Honduras and Yucatan.

a–b. ♂.	Belize.	(Types.)
c. ♀.	Yucatan.	

72. Anolis cymbops.

Anolis cymbops, *Cope, Proc. Ac. Philad.* 1864, p. 173.

Head once and two thirds as long as broad, slightly longer than the tibia; snout rather pointed; forehead concave, no frontal ridges; upper head-scales keeled; supraorbital semicircles separated by one row of scales of nearly equal size; eight enlarged, keeled supraocular scales, in three longitudinal rows, those of the inner row largest and in contact with the supraoculars; occipital smaller than the ear-opening, separated from the supraorbitals by three series of scales; canthus rostralis angular, canthal scales three;

loreal rows seven; eight upper labials to below the centre of the eye; ear-opening large, vertically oval. Gular appendage scarcely indicated in the female. Body not compressed. Dorso-lateral scales granular, minute, gradually slightly larger and feebly keeled dorsally, two vertebral rows abruptly enlarged, smaller than ventrals; latter smooth or indistinctly keeled, and subimbricate. The adpressed hind limb reaches between the eye and the nostril; digital dilatations rather narrow; eighteen lamellæ under phalanges II. and III. of the fourth toe. Tail slightly compressed, once and a half as long as head and body. Brown above, with a darker vertebral band; a dark cross band on tibia; lower surfaces brownish.

Total length	100	millim.
Head	12	„
Width of head	7	„
Body	28	„
Fore limb	18	„
Hind limb	34	„
Tibia	10·5	„
Tail	60	„

Vera Cruz.

a. ♀. Vera Cruz. (Type.)

73. Anolis concolor.

Anolis refulgens, (*non Schleg.*) *Hallow. Proc. Ac. Philad.* 1860, p. 480.

Anolis (Gastrotropis) concolor, *Cope, Proc. Ac. Philad.* 1862, p. 180.

General form that of *A. lineatopus*. Head elongate, snout rather narrow; forehead concave, frontal ridges not acute, convergent anteriorly; upper head-scales more or less keeled; scales of the supraorbital semicircles in contact, or separated by one series of granules medially; supraoculars numerous, keeled; occipital large, not in contact with superciliaries, in a depression; loreal rows five or six; upper labials eight; ear-opening moderate, vertical. A strong nuchal fold in larger specimens. Lateral scales minute, dorsals larger, keeled, two median rows abruptly largest; ventrals keeled. The adpressed hind limb reaches the anterior part of the orbit; digital expansions moderate. Tail cylindrical or slightly compressed. General colour darker or lighter brownish green, the extremities faintly banded; a few dorsal dots in the young.

Nicaragua.

74. Anolis bourgæi.

Anolis bourgeæi, *Bocourt, Miss. Sc. Mex., Rept.* p. 76, pl. xv. fig. 9.

Head rather large, longer than the tibia; scales on the anterior part of the snout keeled, on the rest of the head smooth; supraorbital semicircles separated by one series of scales; seven or eight enlarged, smooth supraocular scales arranged irregularly; occipital as large as the eye-opening, separated from the supraorbitals by two series of scales; canthus rostralis angular; loreal rows six; ear-opening much smaller than the occipital. Gular appendage moderately developed. Body short. Dorsal scales small, keeled, two vertebral series slightly enlarged; lateral scales smaller, granular; ventral scales larger than dorsals, keeled. Tail cylindrical. No enlarged postanal scales. Brownish yellow above, with a single or double vertebral series of small brown spots; a transverse brown streak between the eyes; lower surfaces yellowish white, lips and chin dotted with reddish brown.

Total length	113	millim.
Head, to ear	11	,,
Width of head	7	,,
From ear to vent	32	,,
Tibia	9	,,
Tail	70	,,

Mexico.

75. Anolis salvini.

Head much longer than the tibia, its width contained once and two thirds in its length; forehead slightly concave, no frontal ridges; upper head-scales smooth; supraorbital semicircles separated by one series of scales; about ten enlarged, smooth supraocular scales, irregularly arranged; occipital not enlarged; canthus rostralis obtuse, canthal scales four; loreal rows four; seven labials to below the centre of the eye; ear-opening small, oval, oblique. Gular appendage small; gular scales keeled. Body not compressed; no dorso-nuchal fold. Dorsal scales small, subrhomboidal, juxtaposed, feebly keeled, a little larger than the granular and smooth lateral ones; ventrals a little larger, rhomboidal, imbricate, keeled. The adpressed hind limb reaches the shoulder; digital expansions well developed; sixteen lamellæ under phalanges II. and III. of the fourth toe. Tail a little longer than head and body, feebly compressed, with a vertebral series of larger scales forming a very slightly serrated edge. Male with enlarged postanal scales. Pinkish grey above, with a few darker spots; a large dark brown spot on each side of the snout, connected by a line from nostril to nostril; dark streaks radiating from the eye; an oblique dark band from the eye to the back of the head; a V-shaped dark band between the fore limbs and another on the base of the tail; limbs and tail with rather indistinct dark cross bands; lower surfaces whitish.

Total length	126	millim.
Head	16	,,
Width of head	9·5	,,
Body	41	,,
Fore limb	23	,,

Hind limb	34 millim.
Tibia	10 „
Tail	69 „

Guatemala.

a. ♂. Guatemala. F. D. Godman, Esq. [P.].

76. Anolis læviventris.

Dactyloa læviventris, *Wiegm. Herp. Mex.* p. 47.
Anolis læviventris, *Peters, Mon. Berl. Ac.* 1863, p. 141; *Bocourt, Miss. Sc. Mex., Rept.* p. 87, pl. xvi. fig. 18.

Head longer than the tibia; forehead concave; upper head-scales smooth; scales of the supraorbital semicircles large, separated by a series of small scales; seven to nine smooth enlarged supraocular scales, arranged in two or three longitudinal rows and separated from the supraorbitals by a series of granules; occipital larger than the ear-opening, separated from the supraorbitals by three series of scales; eight upper labials; ear-opening rather small, suboval. Gular appendage well developed. Dorsal scales rhomboidal, imbricate, distinctly keeled, the four median series a little larger; lateral scales granular; ventrals rhomboidal, feebly keeled, larger than dorsals. Limbs moderate. Tail slightly compressed, feebly keeled. Male with enlarged postanal scales. Greyish yellow above, with small brown spots on the median line of the back and on the limbs; lower surfaces yellowish white.

Total length	110 millim.
Head, to ear	11 „
Width of head	6·5 „
Length to vent	41 „
Tibia	9 „

Mexico.

77. Anolis nebulosus.

Dactyloa nebulosa, *Wiegm. Herp. Mex.* p. 47.
Anolis nebulosus, *Bocourt, Miss. Sc. Mex., Rept.* p. 68, pl. xv. fig. 3.

Head about once and two thirds as long as broad, longer than the tibia; forehead concave, frontal ridges distinct, short, divergent; upper head-scales smooth; scales of the supraorbital semicircles large, in contact medially, produced forwards as frontal series; three or four large, smooth, transverse supraocular scales forming a single longitudinal series, separated from the supraorbitals by a row of granules; occipital much larger than the ear-opening, separated from the supraorbitals by one or two series of scales; canthus rostralis angular, canthal scales three or four; loreal rows four or five; six or seven labials to below the centre of the eye; ear-opening small, oval. Gular appendage large, extending posteriorly beyond the thorax, small in the female; gular scales keeled. Body not or but slightly compressed; no dorso-nuchal fold. Dorsal scales oval, subrhomboidal, subimbricate, keeled, a little smaller than the ventrals, passing gradually into the laterals, which are minutely granular; ventrals rhomboidal, imbricate, strongly keeled. The adpressed hind limb reaches the posterior border of the orbit, or nearly that point; digital expansions moderate; thirteen to sixteen lamellæ under phalanges II. and III. of the fourth toe. Tail cylindrical, not twice as long as head and body, covered with equal strongly keeled scales. Male with enlarged postanal scales. Brownish above, with dark brown spots, sometimes arranged in a double longitudinal series on the back, or forming angular cross bars with the angle pointing backwards; a more or less distinct dark streak from the eye to the nape; sometimes black lines radiating from the eye; limbs with dark cross bands; lower surfaces whitish; gular appendage grey or brown.

Total length	153 millim.
Head	14 ,,
Width of head	8·5 ,,
Body	34 ,,
Fore limb	20 ,,
Hind limb	34 ,,
Tibia	10 ,,
Tail	105 ,,

Mexico.

a. ♂.	Cuernavaca.	Paris Museum [P.].
b–c. ♂.	Jalapa.	Mr. Hoege [C.]; F. D. Godman, Esq. [P.].
d–h. ♂, ♀, & hgr.	Ventanas.	Hr. A. Forrer [C.].
i–k, *l–n*. ♂, ♀, & hgr.	Presidio.	Hr. A. Forrer [C.].
o–q. ♂.	Tres Marias.	Hr. A. Forrer [C.].

78. Anolis nebuloides.

Anolis nebuloides, *Bocourt, Miss. Sc. Mex., Rept.* p. 74, pl. xiii. fig. 10.

Very closely allied to *A. nebulosus*, from which it differs only in the following points:—Scales on the snout feebly tricarinate; supraorbital semicircles usually separated by one series of scales; ear-opening larger, its vertical diameter nearly equalling the length of the occipital; dorsal scales larger, nearly as large as the ventrals.

Mexico.

a–b. ♂.	Huamuchla.

79. Anolis palpebrosus.

Anolis palpebrosus, *Peters, Mon. Berl. Ac.* 1873, p. 740.

Closely allied to *A. nebulosus*. Head as long as the tibia; scales on the anterior half of the snout keeled; supraorbital semicircles in contact; about twenty enlarged supraocular scales, the larger ones keeled, arranged in four longitudinal series, separated from the supraorbitals by a series of small scales; occipital larger than the ear-opening, separated from the supraorbitals by three or four series of scales; six or seven loreal rows in the middle; eight upper labials. A distinct nuchal fold. Dorsal scales distinctly keeled, hardly half as large as the ventrals, which are also keeled, and passing gradually into the small granular scales of the sides. Limbs slender; digital expansions well developed. Tail with a vertebral series of enlarged scales. Upper surfaces with metallic gloss; a brown cross band between the eyes; loreal region black; lips spotted black and yellow; back and tail with broad dark cross bands; tibia with an oblique light line.

Total length	145	millim.
Head	15	„
Length to vent	57	„
Fore limb	28	„
Hind limb	48	„

Chiriqui.

80. Anolis intermedius.

Anolis intermedius, *Peters, Mon. Berl. Ac.* 1863, p. 143; *Bocourt, Miss. Sc. Mex., Rept.* p. 69, pl. xv. fig. 4.
—— nannodes, *Cope, Proc. Ac. Philad.* 1864, p. 173; *Bocourt, l. c.* p. 71, pl. xv. fig. 5.
—— tessellatus, *O'Shaughn. Ann. & Mag. N. H.* (4) xv. 1875, p. 279.

Head about once and two thirds as long as broad, much longer than the tibia; forehead concave, no frontal ridges; upper head-scales smooth, or more or less distinctly tricarinate; scales of the supraorbital semicircles large, in contact medially; supraocular scales very variable in shape and in size, smooth or feebly keeled, arranged in two longitudinal series, those of the inner series the largest and sometimes much dilated transversely; occipital much larger than the ear-opening, separated from the supraorbitals by one or two series of scales; canthus rostralis angular, canthal scales three or four; loreal rows three or four; six or seven labials to below the centre of the eye; ear-opening rather small, suboval. Gular appendage moderately large, absent in the female; gular scales obtusely keeled. Body not compressed, no dorso-nuchal fold. Dorsal scales small, rhomboidal, imbricate, keeled, passing gradually into the laterals, which are smaller and granular; ventrals larger than dorsals, rhomboidal, imbricate, sharply keeled. The adpressed hind limb reaches the neck or the ear-opening; digital expansions moderate; fifteen or sixteen lamellæ under phalanges II. and III. of the fourth toe; some enlarged granules on the hinder side of the thighs. Tail cylindrical, not twice as long as head and body, covered with equal, strongly keeled scales. Male with enlarged postanal scales. Yellowish or reddish brown above, with more or less distinct darker spots or markings; frequently a reddish angular cross band between the eyes; female sometimes with a broad light vertebral band; lower surfaces whitish or brownish.

Total length	118	millim.
Head	13	„
Width of head	8	„
Body	33	„
Fore limb	17	„
Hind limb	28	„
Tibia	8	„
Tail	72	„

Central America.

a–b. ♂ ♀.	Coban, Guatemala.	O. Salvin, Esq. [C.]. (Types of *A. nannodes.*)
c. ♀.	—— ?	
d. ♀.	Costa Rica.	O. Salvin, Esq. [C.]. (Type of *A. tessellatus.*)
e. ♀.	Irazu, Costa Rica.	F. D. Godman, Esq. [P.].
f–l. ♂, ♀, & yg.	Jalapa, Mexico.	Mr. Hoege [C.]; F. D. Godman, Esq. [P.].

81. Anolis sallæi.

Anolis sallæi, *Günth. Proc. Zool. Soc.* 1859, p. 421; *Bocourt, Miss. Sc. Mex., Rept.* p. 90, pl. xiii. fig. 3, & pl. xvi. fig. 21.
—— williamsii, *Bocourt, Nouv. Arch. Mus.* vi. 1870, *Bull.* p. 16, *and l. c.* pl. xiii. fig. 9.
—— jacobi, *Bocourt, l. c.* p. 73, pl. xiii. fig. 8.

Head about once and two thirds as long as broad, longer than the tibia; forehead concave, frontal ridges short and divergent; upper head-scales more or less distinctly keeled; scales of the supraorbital semicircles large, in contact medially or separated by one series of scales; enlarged supraocular scales variable in number and in size, usually keeled and arranged in two or three longitudinal series, sometimes strongly dilated transversely, and in a single series, as in *A. nebulosus*; occipital as large as the eye-opening, separated from the supraorbitals by one to three series of scales; canthus rostralis angular, canthal scales three; loreal rows four to six; six or seven upper labials to below the centre of the eye; ear-opening small, oval or subtriangular. Gular appendage large, extending posteriorly beyond the thorax, slightly indicated in the female; gular scales keeled. Body slightly compressed in the male; no dorso-nuchal fold. Dorsal scales rather large, rhomboidal or hexagonal, subimbricate and sharply keeled, passing gradually into the very small granules of the sides; ventrals larger than dorsals, rhomboidal, imbricate, sharply keeled. The adpressed hind limb reaches the ear or a little beyond; digits moderately dilated; fifteen or sixteen lamellæ under phalanges II. and III. of the fourth toe. Tail cylindrical, once and three fourths to twice and one fourth the length of head and body, covered with large, equal, strongly keeled scales. No enlarged postanal scales. Greyish or brownish above, with golden reflections; females frequently with a broad light vertebral band; hind limbs usually with a few large yellowish spots; lower surfaces yellowish white; gular appendage with a large round purplish black (dark blue) spot.

Total length	140	millim.
Head	14	„
Width of head	8·5	„
Body	36	„
Fore limb	20	„
Hind limb	34	„
Tibia	11	„
Tail	90	„

Central America.

a. ♀.	Mexico.	M. Sallé [C.]. (Type.)
b–d. ♂ & hgr.	Guatemala.	F. D. Godman, Esq. [P.].
e. ♀.	Lower Vera Paz.	O. Salvin, Esq. [C.].
f. ♂.	South slope of the Atitlan Volcano.	Paris Museum [P.].
g. ♀.	Hacienda del Hobo, S. Mexico.	P. Geddes, Esq. [P.].
h–i. ♀.	Oaxaca.	
k–m. ♂ ♀.	Mexico.	

82. Anolis binotatus.

Anolis binotatus, *Peters, Mon. Berl. Ac.* 1863, p. 140; *Bocourt, Miss. Sc. Mex., Rept.* p. 92, pl. xvi. figs. 22, 23.

Very closely allied to, and perhaps not specifically distinct from, *A. sallœi*, from which it is distinguished by a rather larger ear-opening (its diameter measuring somewhat more than half the length of the occipital) and smaller dorsal scales.

Guayaquil and Central America (Salvador).

83. Anolis cumingii.

Anolis cumingii, *Peters, Mon. Berl. Ac.* 1863, p. 140; *Bocourt, Miss. Sc. Mex., Rept.* p. 89, pl. xvi. fig. 20.

Also very near *A. sallœi*. Ear-opening nearly as long as the occipital. Male with enlarged postanal scales, without spot on the gular appendage.

Mexico.

84. Anolis cupreus.

Anolis cupreus, *Hallow. Proc. Ac. Philad.* 1860, p. 481.
—— hoffmanni, *Peters, Mon. Berl. Ac.* 1863, p. 142; *Bocourt, Misc. Sc. Mex., Rept.* p. 86, pl. xv. figs. 15, 16.
? Anolis dollfusianus, *Bocourt, l. c.* p. 84, pl. xvi. fig. 19.

Head about once and two thirds as long as broad, as long as or slightly longer than, the tibia; forehead concave, frontal ridges very short, or indistinct; upper head-scales strongly keeled; scales of the supraorbital semicircles enlarged, separated by one or two series of scales; a few enlarged, strongly keeled supraocular scales, irregularly arranged; occipital as large as, or a little larger than, the ear-opening, separated from the supraorbitals by two or three series of scales; canthus rostralis angular, canthal scales three; loreal rows six; six or seven upper labials to below the centre of the eye; ear-opening rather large, vertically oval. Gular appendage large, extending posteriorly beyond the thorax, scarcely indicated in the female; gular scales keeled. Body slightly compressed in the male; no dorso-nuchal crest. Dorsal scales rhomboidal or hexagonal, subimbricate and keeled, graduating into the small keeled granules of the sides; ventrals larger than dorsals, rhomboidal, imbricate, sharply keeled. The adpressed hind limb reaches the eye or the anterior border of the orbit; digits rather feebly dilated; thirteen or fourteen lamellæ under phalanges II. and III. of the fourth toe. Tail cylindrical, not quite twice as long as head and body, covered with large, equal, strongly keeled scales. No enlarged postanal scales. Brownish above, with metallic gloss; usually a paler vertebral band; one specimen with a round black spot above the shoulder; no spot on the gular appendage.

	millim.		millim.
Total length	127	Fore limb	18
Head	13	Hind limb........	33
Width of head	7·5	Tibia............	11·5
Body	30	Tail	84

Central America.

a–f. ♂, ♀, and hgr.	Guatemala.	F. D. Godman, Esq. [P.].
g–h. ♂.	Neighbourhood of Irazu, Costa Rica.	F. D. Godman and O. Salvin, Esqrs. [P.].

85. Anolis crassulus.

Anolis crassulus, *Cope, Proc. Ac. Philad.* 1864, p. 173; *Bocourt, Miss. Sc. Mex., Rept.* p. 82, pl. xvi. fig. 17.

Head once and two thirds as long as broad, considerably longer than the tibia; forehead concave, no frontal ridges; upper head-scales smooth or feebly keeled; scales of the supraorbital semicircles large, in contact medially, or separated by one or two series of scales; a few enlarged, smooth or feebly keeled supraocular scales arranged in two or three longitudinal series; occipital rather small, not, or but slightly, larger than the ear-opening, separated from the supraorbitals by two or three series of scales; canthus rostralis angular, canthal scales three; loreal rows four; six or seven upper labials to below the centre of the eye; ear-opening moderate, oval, oblique. Gular appendage large, indicated in the female; gular scales keeled. Body short, slightly compressed in the male; no dorso-nuchal fold. Dorsal scales large, rhomboidal or hexagonal, imbricate, strongly keeled, rather irregular, arranged in about fourteen longitudinal series; the dorsal scales abruptly pass into the laterals, which are small, subgranular, keeled, and intermixed with a few scattered enlarged rhomboidal ones; ventral scales very large, larger than dorsals, rhomboidal, imbricate, strongly keeled. The adpressed hind limb reaches the eye or the posterior border of the orbit; digits moderately dilated; fifteen or sixteen lamellæ under phalanges II. and III. of the fourth toe. Tail slightly compressed, nearly twice as long as head and body, covered with large, strongly keeled scales. Male with enlarged postanal scales. Bronzy or coppery above, uniform or with small ill-defined darker spots; a brown indistinct band on each side of the head and neck, passing through the eye; female with a broad yellowish vertebral band; lower surfaces yellowish.

	millim.		millim.
Total length	137	Fore limb	22
Head	14	Hind limb	37
Width of head	8·5	Tibia	11
Body	33	Tail	90

Central America.

a. ♀.	Central America.	O. Salvin, Esq. [C.].	(Types.)
b–c. ♂.	Coban, Guatemala.		
d. ♂.	Plateau of Guatemala.	Paris Museum [P.].	

86. **Anolis humilis.**

Anolis humilis, *Peters, Mon. Berl. Ac.* 1863, p. 138; *Bocourt, Miss. Sc. Mex., Rept.* p. 105, pls. xiii. fig. 7, and xvi. fig. 31.

Head once and two thirds to once and three fourths as long as broad, longer than the tibia; forehead slightly concave; frontal ridges feebly marked, short and divergent; all the upper head-scales strongly keeled; scales of the supraorbital semicircles scarcely enlarged, separated by two or three series of scales; a few enlarged supraoculars; occipital hardly distinguishable; canthus rostralis angular; loreal rows seven or eight; seven to nine upper labials to below the centre of the eye; ear-opening large, vertically oval. Gular appendage moderately developed; gular scales keeled. Body short, slightly compressed; no dorso-nuchal fold. Dorsal scales very large, larger than ventrals, hexagonal, subimbricate, strongly keeled, forming about ten longitudinal series; lateral scales small, subgranular, keeled; ventrals roundish-rhomboidal, imbricate, strongly keeled, the keels not forming continuous lines. The adpressed hind limb reaches the anterior border of the orbit, or between the latter and the nostril; digits very slightly dilated; thirteen or fourteen lamellæ under phalanges II. and III. of the fourth toe. Tail slightly compressed, a little longer than head and body. No enlarged postanal scales. Brownish above, females sometimes with a broad, yellow, dark-edged vertebral band; lower surfaces whitish; a small purplish spot on each side of the throat in the male.

	millim.		millim.
Total length	90	Fore limb	17
Head	12	Hind limb........	32
Width of head	7	Tibia	10
Body	24	Tail	54

Central America.

a. Several spec., ♂ ♀.	Vera Paz.	O. Salvin, Esq. [C.].
b–c. ♂.	—— ?	

87. Anolis quaggulus.

Anolis quaggulus, *Cope, Proc. Am. Phil. Soc.* xxii. 1885, p. 391.

Closely allied to *A. humilis*. Seven rows of truncate and keeled dorsal scales, much larger than the ventrals. General colour golden green; a series of blackish chevrons on the back, with the angle directed posteriorly, and with the lateral branches thickened anteriorly; from the apex of each branch a delicate blackish line descends vertically, stopping on the side below its middle, thus dividing the side into vertical areas.

From snout to vent 32 millim.

Nicaragua.

88. Anolis tropidonotus.

Anolis tropidonotus, *Peters, Mon. Berl. Ac.* 1863, p. 135, *and Ann. & Mag. N. H.* (4) iv. 1869, p. 273; *Bocourt, Miss. Sc. Mex., Rept.,* p. 103, pls. xiii. fig. 6, and xvi. fig. 30.

Norops auratus, (*non Daud.*) *O'Shaughn. Ann. & Mag. N. H.* (4) iii. 1869, p. 188, and iv. p. 274.

—— tropidonotus, *O'Shaughn. eod. loc.* (4) xv. 1875, p. 277.

Head about once and two thirds as long as broad, not, or scarcely, longer than the tibia; forehead concave; frontal ridges feebly marked, short and divergent: all the upper head-scales strongly keeled; scales of the supraorbital semicircles scarcely enlarged, separated by two series of scales; a few enlarged supraoculars; occipital scarcely enlarged; canthus rostralis angular; loreal rows six or seven; six or seven labials to below the centre of the eye; ear-opening large, vertically oval. Gular appendage large; gular scales keeled. Body short, compressed; no dorso-nuchal fold. Dorsal scales very large, larger than ventrals, rhomboidal, imbricate, sharply keeled, forming about twelve longitudinal series; lateral scales small, imbricate, keeled; ventrals rhomboidal, pointed, sharply keeled. The adpressed hind limb reaches the end of the snout; digits very slightly dilated; fourteen or fifteen lamellæ under phalanges II. and III. of the fourth toe. Tail slightly compressed, not twice as long as head and body. No enlarged postanal scales. Bronzy or golden above, uniform, or with ill-defined dark brown spots; a rather indistinct dark cross band between the eyes; lower surfaces pale golden.

	millim.		millim.
Total length	135	Fore limb	24
Head	15	Hind limb	45
Width of head	9	Tibia	14
Body	34	Tail	86

Mexico.

a. ♂.	Mexico.	Prof. Peters [P.].
b. ♂.	Mexico.	
c. ♂.	Oaxaca.	M. Sallé [C.].
d. ♂.	——?	M. Sallé [C.].

89. **Anolis uniformis.**

Anolis uniformis, *Cope, Proc. Am. Phil. Soc.* xxii. 1885, p. 392.

Very near *A. tropidonotus*. All the scales of the frontal region equal, so that the supraorbitals cannot be distinguished by size. Dorsal scales in ten to twelve rows, imbricate, not truncate. Tibia a little shorter than the short head. Colour reddish brown above and greenish below; limbs and head above brown.

From snout to vent 36 millim.

Yucatan and Guatemala.

90. **Anolis metallicus.**

Anolis metallicus, *Bocourt, Ann. Sc. Nat.* (5) xvii. 1872, art. 2, and *Miss. Sc. Mex., Rept.* pl. xvii. bis. fig. 1.

Differs from *A. tropidonotus* in having the dorsal scales not larger than the ventrals.

Mexico.

91. **Anolis pœcilopus.**

Anolis (Dracontura) pœcilopus, *Cope, Proc. Ac. Philad.* 1862, p. 179.

Head broad, muzzle full, rather prominent; frontal ridges slightly developed, bounding a rather deep concavity; upper head-scales minute, subgranular, rugulose; supraorbital semicircles separated by granules; supraoculars small, keeled; occipital minute or wanting; canthus rostralis weakly pronounced, soon obsolete; loreal series nine or ten; superior labials nine; auricular orifice small, vertical. Gular appendage large. No dorsal or nuchal fold. Abdominal scales small, ovate, keeled; lateral minute, gradually merging into the dorsal, which are larger, flat, keeled, and in numerous rows. The adpressed hind limb reaches the end of the snout; digital expansions narrow. Tail compressed, cylindrical, its scales keeled, the median row largest. General colour above brown; the extremities and digits with numerous light cross bands; sides darker, with numerous longitudinal light lines, one commencing above the axilla most distinct; light vertical bands ascend from this to a superior obsolete longitudinal band; in female specimens dark chevron-shaped spots cross the back; beneath pale yellowish.

Head, to ear	17 millim.
From ear to vent	51 ,,
Hind limb	53 ,,
Tail	119 ,,

Colombia.

92. Anolis oxylophus.

Anolis oxylophus, *Cope, Journ. Ac. Philad.* (2) viii. 1876, p. 123, pls. xxiv. fig. 4, and xxviii. fig. 5.

Resembles the preceding, except in the following points:—Frontal ridges much more prominent; canthus rostralis distinct to near the nostril; occipital bounded laterally and in front by a number of scuta of the same size and form; superior labials thirteen. Dorsal scales about equal to the ventrals. The adpressed hind limb reaches the middle of the orbit. Colour dark brown, the limbs and feet with pale cross bands; a distinct whitish band extends from the scapular region to near the groin; abdomen yellowish, with a broad brown border on each side.

Total length	217 millim.
Head, to ear	19 ,,
Width of head	11 ,,
Length, to vent	76 ,,
Hind limb	60 ,,

Costa Rica.

93. Anolis godmani.

Head once and three fourths or twice as long as broad, slightly longer than the tibia; snout rather pointed; forehead and interorbital region concave, frontal ridges short and divergent; upper head-scales very small, uni- or tricarinate; scales of the supraorbital semicircles enlarged, separated by two to four series of scales; a few enlarged, keeled supraocular scales; occipital larger than the ear-opening, separated from the supraorbitals by three to five series of scales; canthus rostralis sharp, canthal scales three or four; loreal rows six to eight; six upper labials to below the centre of the eye; ear-opening moderate, vertically oval. Gular appendage moderate; gular scales keeled. Body not or scarcely compressed; no dorso-nuchal fold. Dorsal scales very small, subrhomboidal, keeled, graduating into the minute granules of the sides; ventrals larger, roundish-hexagonal, imbricate, keeled. Limbs slender; the adpressed hind limb reaches nearly the end of the snout; digits feebly dilated; thirteen or fourteen lamellæ under phalanges II. and III. of the fourth toe. Tail cylindrical, nearly twice as long as head and body, without vertebral series of enlarged scales. Male with slightly enlarged postanal scales. Greenish-brown above; upper lip yellowish, bordered above by a sharply defined brown band passing through the eye and above the ear, vanishing behind the shoulder; a dark cross band between the eyes; sometimes a few small distant brown spots on the vertebral line.

	millim.		millim.
Total length	107	Fore limb	16
Head	12	Hind limb........	34
Width of head	6	Tibia	11
Body	26	Tail	69

Central America.

a–d. ♂ & hgr.	Guatemala.	F. D. Godman, Esq. [P.].
e–g. ♂ ♀.	Irazu, Costa Rica.	F. D. Godman and O. Salvin, Esqrs. [P.].

94. **Anolis radulinus.**

Anolis (Gastrotropis) radulinus, *Cope, Proc. Ac. Philad.* 1862, p. 180.

Size small, head not short. Facial rugæ weak, enclosing a shallow concavity, which is covered with minute keeled scales; canthus rostralis acute, soon obliterated; supraorbital semicircles separated by three or four rows of minute scales; occipital distinct, isolated; loreal rows six or seven; ear-opening rather large, vertical. Gular appendage large. Lateral scales minute; the dorsal scales larger, flat, keeled, becoming larger medially, where two series are abruptly enlarged; ventral scales ovate, keeled. Digital dilatations very narrow. Posterior extremity reaching to beyond the orbit. Tail cylindrical, or slightly compressed, median row of scales a little larger. Above golden brown, with eight or ten narrow chevron-shaped cross lines, the angles directed posteriorly; a lateral series of small white spots, most distinct anteriorly, beneath which is a light longitudinal band, obsolete posteriorly; a narrow brown band between the eyes; extremities banded; beneath brownish white.

From snout to vent	45 millim.
Head, to ear	13 „
Tail	90 „

Truando district, Colombia.

95. **Anolis lemurinus.**

Anolis (Gastrotropis) lemurinus, *Cope, Proc. Ac. Philad.* 1861, p. 213.

Size small. Head short, broad, front concave, eyes very large; upper head-scales keeled; supraorbital semicircles in contact medially or separated by one row of very small scales; supraocular scales fifteen to twenty, more or less keeled, separated from the supraorbitals by small scales; occipital small, elliptic, separated from the supraorbitals by two series of scales; frontal ridges obsolete; canthus rostralis very short; loreal rows six; six or seven upper labials. Gular appendage small. A weak nuchal fold. Inferior lateral scales smaller than dorsal, the latter, with the superior lateral and ventral, keeled. The adpressed hind limb reaches the posterior or the anterior border of the orbit. Tail entirely cylindrical, covered with strongly keeled scales, about twice the length of head and body. Colour leek-green, tinged with brown anteriorly; a brown band between the eyes; a crescentic brown band upon the occiput, angularly bordered with black anteriorly, and having a median longitudinal black spot; a similar small black spot upon the nape at the confluence of two oblique pale brown bands, one from each orbit, often paler-bordered inferiorly; the pale border sometimes extending to near the groin as a narrow longitudinal band; a third small black spot upon the interscapular region; three or four large brown spots upon the posterior part of the dorsal region; sides and labial region varied with brown; extremities and digits obliquely banded with the same; belly immaculate, the reflections golden and green.

Total length 72 millim.

Veragua.

96. Anolis trachyderma.

Anolis trachyderma, *Cope, Journ. Ac. Philad.* (2) viii. 1876, p. 168.

Head oval, as long as the tibia; forehead concave, frontal ridges obsolete; upper head-scales keeled; ten keeled, enlarged supraocular scales; occipital small, nearly equal to ear, separated from the supraorbitals by two rows of scales; six loreal rows in the middle; nine upper labials. Gular appendage large. Three or four median dorsal series, nearly as large as the ventrals, flat, keeled, and quickly graduating into the granular angulate scales of the sides; ventral scales small, obtusely keeled. The adpressed hind limb reaches the middle of the orbit. Tail round, covered with flat, keeled scales. Colour black above and below, excepting thorax, abdomen, and inferior middle line of tail, which are fulvous; an indistinct light band across the chin, half way between eye and nostril.

From snout to vent	55	millim.
Head, to ear	19	,,
Width of head	8	,,
Hind limb	45	,,

Nauta, Upper Amazons.

97. Anolis stigmosus.

Anolis stigmosus, *Bocourt, Nouv. Arch. Mus.* v. 1869, *Bull.* p. 43, *and Miss. Sc. Mex., Rept.* pl. xv. fig. 13.

Head once and three fourths as long as broad, as long as or slightly shorter than the tibia; snout rather pointed; forehead slightly concave, frontal ridges very short; upper head-scales keeled; scales of the supraorbital semicircles a little enlarged, separated by two or three series of scales; ten or twelve enlarged, strongly keeled supraocular scales, separated from the supraorbitals by one series of granules; occipital not quite as large as the ear-opening, separated from the supraorbitals by three or four series of scales; canthus rostralis feebly marked; seven or eight loreal rows; seven to ten labials to below the centre of the eye; ear-opening rather large, vertically oval. Gular appendage moderately large; gular scales keeled. Body slightly compressed; no dorso-nuchal fold. Dorsal scales rather small, hexagonal, strongly keeled, subimbricate, passing gradually into the minute, granular, keeled laterals; ventrals larger than dorsals, rounded, imbricate, strongly keeled. The adpressed hind limb reaches the nostril, or between the latter and the orbit: digits feebly dilated; fifteen or sixteen lamellæ under phalanges II. and III. of the fourth toe. Tail cylindrical or very slightly compressed, covered with equal keeled scales; its length nearly twice that of head and body. Male with enlarged postanal scales. Male pale golden brown above, with the vertebral region lighter, and a series of dark brown spots partly confluent into a band along each side of the back; a yellow spot on the tibia; lower surfaces yellowish white. Female with a broad coppery-brown band on each side of the head, passing through the eye, widening and covering the whole of the body save a dark brown vertebral band.

Total length	180	millim.
Head	17	„
Width of head	9·5	„
Body	43	„
Fore limb	26	„
Hind limb	51	„
Tibia	15·5	„
Tail	120	„

Colombia, Ecuador.

a. ♂.	——?	Mme. Pfeiffer [C.].
b. ♀.	Tanti, Ecuador, 2000 feet.	E. Whymper, Esq. [C.].

98. **Anolis biporcatus.**

Dactyloa biporcata, *Wiegm. Herp. Mex.* p. 47.
Anolis (Dracontura) vittigerus, *Cope, Proc. Ac. Philad.* 1862, p. 179.
—— biporcatus, *Bocourt, Miss. Sc. Mex., Rept.* p. 98, pl. xv. fig. 8; *O'Shaughn. Ann. & Mag. N. H.* (4) xv. 1875, p. 274.

Head about once and two thirds as long as broad, as long as or slightly shorter than the tibia; forehead slightly concave; frontal ridges very short and feeble; upper head-scales keeled; scales of the supraorbital semicircles enlarged, separated by one or two series of scales; strongly keeled enlarged supraocular scales, separated from the supraorbitals by one or two rows of granules; occipital as large as, or a little larger than, the ear-opening, separated from the supraorbitals by two to four series of scales; canthus rostralis angular, canthal scales three; loreal rows six to eight; six or seven labials to below the centre of the eye; ear-opening moderately large, roundish or suboval. Gular appendage moderately large, small but present in the female; gular scales keeled. Body slightly compressed; no dorso-nuchal fold. Dorsal scales very small, rhomboidal, imbricate, keeled, graduating into the granular laterals; ventral scales much larger than dorsals, rhomboidal, imbricate, strongly keeled. The adpressed hind limb reaches the anterior border of the orbit, or between the latter point and the end of the snout; digits rather feebly dilated; seventeen or eighteen (nineteen) lamellæ under phalanges II. and III. of the fourth toe. Tail roundish, covered with strongly keeled scales, the median dorsal series scarcely enlarged; length of tail twice, or a little more, that of head and body. No enlarged postanal scales. Brownish golden above, with large dark brown spots or angular markings; frequently one or two more or less defined white lateral bands; female sometimes with a broad light, dark-edged vertebral band; lower surfaces white, with golden and greenish metallic reflections.

Total length	190	millim.
Head	16	"
Width of head	10	"
Body	44	"
Fore limb	27	"
Hind limb	51	"
Tibia	15	"
Tail	130	"

Central America to Venezuela and Ecuador.

a–f, g. ♂ ♀.	Lower Vera Paz.	O. Salvin, Esq. [C.].
h. ♀.	Dueñas, Guatemala.	O. Salvin, Esq. [C.].
i. ♂.	Rio Polochic, Guatemala.	O. Salvin, Esq. [C.].
k–l. ♂.	Leghoon.	
m. ♀.	Venezuela.	
n. ♀.	Trinidad.	C. Taylor, Esq. [P.].
o. ♀.	W. Ecuador.	Mr. Fraser [C.].

99. Anolis chrysolepis.

Draconura chrysolepis, *Gray, Cat.* p. 207.
Anolis chrysolepis, *Dum. & Bibr.* iv. p. 94; *Bocourt, Miss. Sc. Mex., Rept.* p. 99, pl. xvi. fig. 26; *Peters, Mon. Berl. Ac.* 1863, p. 142.
—— planiceps, part., *Trosch. in Schomb. Reise Brit. Guian.* iii. p. 649.
—— nummifer, *O'Shaughn. Ann. & Mag. N. H.* (4) xv. 1875, p. 278.
—— turmalis, *O'Shaughn. l. c.*
—— (Draconura) chrysolepis, part., *O'Shaughn. Proc. Zool. Soc.* 1881, p. 241.

Head about once and two thirds as long as broad, as long as or a little shorter than the tibia; forehead scarcely concave; frontal ridges absent or short and very feeble; upper head-scales uni- or tricarinate; scales of the supraorbital semicircles slightly enlarged, separated by one or two series of scales; numerous enlarged, strongly keeled supraocular scales; occipital considerably larger than the ear-opening, separated from the supraorbitals by two or three series of scales; canthus rostralis angular, canthal scales three; loreal rows five to seven; seven to nine labials to below the centre of the eye; ear-opening small, roundish or subtriangular. Gular appendage present, small, nearly as much developed in the female as in the male. Gular scales keeled. Body not or but feebly compressed; no dorso-nuchal fold. Dorsal scales small, larger than the minute, granular, keeled laterals, rhomboidal or hexagonal, juxtaposed or subimbricate, keeled; ventrals large, rounded or rhomboidal, imbricate, strongly keeled. The adpressed hind limb reaches the tip of the snout or a little beyond; digits very feebly dilated; fourteen to seventeen lamellæ under phalanges II. and III. of the fourth toe. Tail rounded, covered with equal keeled scales, at least twice as long as head and body. Male without or with slightly enlarged postanal scales. Brownish above, with metallic reflections, with indistinct symmetrical darker markings; flanks frequently with round black dots; a dark cross band between the eyes; female frequently with a light, dark-margined vertebral band ending in a point on the nape, widening towards the sacrum; lower surfaces whitish with metallic reflections, sometimes with scattered black dots, or a rather indistinct median longitudinal blackish streak on the belly.

Total length	195	millim.
Head	18	„
Width of head	12	„
Body	47	„
Fore limb	32	„
Hind limb	60	„
Tibia	19	„
Tail	130	„

Central and South America.

a–c. ♀.	Demerara Falls.	(Types of *A. nummifer.*)
d–f. ♀ & hgr.	Surinam?	
g. ♀.	Interior of Brazil.	
h. ♀.	Esmeraldas.	Mr. Fraser [C.].
i–k. ♂ ♀.	Island of Grenada.	(Types of *A. turmalis.*)
l–n. ♂ ♀.	Venezuela.	
o. ♂.	Honduras.	

100. **Anolis scypheus.**

Anolis chrysolepis, (*non D. & B.*) *Guich. in Casteln. Amér. Sud, Rept.* p. 15, pl. iv. fig. 1; *Bocourt, Miss. Sc. Mex., Rept.* p. 101; *O'Shaughn. Proc. Zool. Soc.* 1880, p. 491.

—— scypheus, *Cope, Proc. Ac. Philad.* 1864, p. 172.

Draconura chrysolepis, *O'Shaughn. Ann. & Mag. N. H.* (4) xv. 1875, p. 278.

Anolis (Draconura) chrysolepis, part., *O'Shaughn. Proc. Zool. Soc.* 1881, p. 241.

Head once and two thirds as long as broad, as long as or a little shorter than the tibia; forehead slightly concave; frontal ridges distinct, short; upper head-scales small, keeled; scales of the supraorbital semicircles slightly enlarged, separated by three or four (two) series of scales; numerous slightly enlarged, keeled supraocular scales; occipital nearly as large as the ear-opening, separated from the supraorbitals by three or four series of scales; canthus rostralis angular, canthal scales two or three; loreal rows eight to ten; eight to ten labials to below the centre of the eye; ear-opening moderately large, oval, horizontal, slightly oblique. Gular appendage small, nearly as much developed in the female as in the male; gular scales keeled. Body not or but feebly compressed; no dorso-nuchal fold. Dorsal scales very small, slightly larger than the minute, granular, keeled laterals, rhomboidal, juxtaposed, keeled; ventrals large, rhomboidal, imbricate, strongly keeled. The adpressed hind limb reaches the tip of the snout or a little beyond; digits very feebly dilated; seventeen to nineteen lamellæ under phalanges II. and III. of the fourth toe. Tail rounded, covered with equal keeled scales, usually at least twice as long as head and body. No enlarged postanal scales. Olive or brownish above, with metallic reflections; back usually with large angular dark spots, sometimes confluent into a zigzag band or series of rhombs; female sometimes with a broad light, dark-edged vertebral band; a dark transverse band between the eyes; lower surfaces whitish, with metallic reflections; a large dark blue spot on each side of the gular appendage; usually a blackish streak along the middle of the belly in the female.

Total length	241	millim.
Head	21	,,
Width of head	14	,,
Body	57	,,
Fore limb	45	,,
Hind limb	78	,,
Tibia	24	,,
Tail	163	,,

Venezuela to Upper Amazons.

a. ♀.	Caracas.	(Type.)
b, c, d. ♀.	Canelos, Ecuador.	Mr. Buckley [C.].
e. ♀.	Pallatanga, Ecuador.	Mr. Buckley [C.].
f. ♂.	Peruvian Amazons.	Mr. Bartlett [C.].

101. Anolis nitens.

Draconura nitens, *Gray, Cat.* p. 207.
Draconura nitens, *Wagl. Syst. Rept.* p. 149.
Anolis refulgens (*Schleg.*), *Dum. & Bibr.* iv. p. 91.
—— planiceps, part., *Trosch. in Schomb. Reise Brit. Guian.* iii. p. 649.
—— (Draconura) nitens, *Peters, Mon. Berl. Ac.* 1863, p. 142; *Bocourt, Miss. Sc. Mex., Rept.* pl. xvi. fig. 25.

Head short, shorter than the tibia; forehead slightly concave; upper head-scales small, keeled; occipital larger than the ear-opening, which is round. A small gular appendage. Body slightly compressed; no dorso-nuchal fold. Dorsal scales subhexagonal, juxtaposed, smooth, the two vertebral series slightly larger; ventral scales larger than dorsal, rhomboidal, strongly keeled. The adpressed hind limb reaches beyond the tip of the snout; digits very feebly dilated. Tail rounded, more than twice as long as head and body. Greyish above, with metallic gloss; posterior part of the body with oblique brown cross bands; sides of the head and neck marbled with brown; limbs and tail with brown cross bands; lower surfaces greenish white.

Total length	108	millim.
Head	18	„
Body	37	„
Fore limb	32	„
Hind limb	57	„
Tail	125	„

Surinam.

102. **Anolis leptoscelis.** (Plate V. fig. 3.)

Draconura nitens, *(non Wagl.) O'Shaughn. Ann. & Mag. N. H.* (4) xv. 1875, p. 277.

Head small, about once and two thirds as long as broad, as long as the tibia; snout short, swollen between the nostrils; forehead and interorbital region concave; no frontal ridges; upper head-scales very small, strongly keeled; scales of the supraorbital semicircles enlarged, separated by two or three series of scales; numerous enlarged, keeled supraocular scales; occipital smaller than the ear-opening, separated from the supraorbitals by two or three series of scales; canthus rostralis short; loreal series six or seven; seven or eight upper labials to below the centre of the eye; ear-opening large, vertically oval. Gular appendage very small; gular scales obtusely keeled. Body not compressed; no dorso-nuchal fold. Dorsal scales very small, subrhomboidal, juxtaposed, feebly keeled, two vertebral series somewhat more enlarged, graduating into the minute granules of the flanks; ventral scales larger, also small, irregular hexagonal, imbricate, feebly keeled. Limbs long and slender; the adpressed hind limb reaches the tip of the snout; digital expansions very narrow; sixteen lamellæ under phalanges II. and III. of the fourth toe. Tail cylindrical, twice as long as head and body, without dorsal series of enlarged scales. Male with enlarged postanal scales. Upper surfaces brownish or greenish, back with small darker spots; a broad brown band from the eye to the side of the body; an angular transverse brown band between the eyes; limbs with regular, narrow, oblique, light, dark-edged cross bands; belly whitish; throat brown; one or two whitish lines across the chin.

Total length	138 millim.
Head	12 „
Width of head	7 „
Body	34 „
Fore limb	22 „
Hind limb	41 „
Tibia	12·5 „
Tail	92 „

Upper Amazons.

a–c. ♂.	Pebas.	H. W. Bates, Esq. [C.].
d. ♂.	Yurimaguas, Huallaga River, N.E. Peru.	Dr. Hahnel [C.].

103. **Anolis lentiginosus.** (Plate VI. fig. 1.)

Anolis lentiginosus, *O'Shaughn. Ann. & Mag. N. H.* (4) xv. 1875, p. 279.

Head once and two thirds as long as broad, shorter than the tibia; snout short, subacuminate, swollen between the nostrils; forehead very slightly concave, no frontal ridges; upper head-scales uni- or tri-carinate; scales of the supraorbital semicircles slightly enlarged, separated in the middle by one scale; numerous enlarged, keeled supraocular scales; occipital scarcely enlarged; canthus rostralis short, angular; loreal rows seven; ten or eleven upper labials to below the centre of the eye; ear-opening rather large, suboval. Gular appendage very small; gular scales keeled. Body slightly compressed; no dorso-nuchal fold. Scales on the vertebral region nearly as large as the ventrals, hexagonal, juxtaposed, feebly keeled, graduating on the sides into the minute granules of the flanks; ventrals moderately large, roundish-subrhomboidal, strongly imbricate, sharply keeled. Limbs long and slender; the adpressed hind limb reaches beyond the tip of the snout; digits very feebly dilated; fourteen lamellæ under phalanges II. and III. of the fourth toe. Tail cylindrical, once and a half as long as head and body, covered with equal, keeled scales. No enlarged postanal scales. Golden brown above, freckled with black; a few black cross lines on the back; a dark cross band between the eyes; lower surfaces whitish; the skin of the gular appendage black.

Total length	142 millim.
Head	15 „
Width of head	9 „
Body	42 „
Fore limb	29 „
Hind limb	55 „
Tibia	17 „
Tail	85 „

Surinam.

a. ♂.	Surinam.	Hr. Kappler [C.]. (Type.)

104. **Anolis bombiceps.**

Anolis bombiceps, *Cope, Journ. Ac. Philad.* (2) viii. 1876, p. 168.

Head, short, wide, and convex above, shorter than the tibia; forehead concave, frontal ridges distinct; upper head-scales smooth; supraorbital semicircles separated by two rows of scales; supra-ocular scales smooth or nearly so; occipital subround, larger than tympanum, surrounded by numerous smaller scales; loreal rows six; ear-opening round. Gular appendage small. Dorsal scales much smaller than ventrals, smooth, and a little larger than the almost granular laterals; ventrals keeled. The adpressed hind limb reaches beyond the front of the orbit. Tail subround, with equal scales. General colour bright olive, shaded with brown above; a dark band from ear to shoulder, and vertical blackish bars on the sides; fan dark indigo?

Total length	130	millim.
Head, to ear	11	,,
Width of head	6	,,
Hind limb	40	,,
Tail	84	,,

Nauta, Upper Amazons.

105. **Anolis capito.**

Anolis (Draconura) capito, *Peters, Mon. Berl. Ac.* 1863, p. 142; *Bocourt, Miss. Sc. Mex., Rept.* p. 101, pl. xvi. fig. 27.
—— carneus, *Cope, Proc. Ac. Philad.* 1864, p. 169.

Head about once and a half as long as broad, a little shorter than the tibia; forehead deeply concave, no frontal ridges; upper head-scales keeled; scales of the supraorbital semicircles slightly enlarged, separated by two or three series of scales; numerous slightly enlarged, keeled supraocular scales; occipital scarcely distinct; canthus róstralis short, angular; loreal rows eight; nine or ten labials to below the centre of the eye; ear-opening large, vertically oval. Gular appendage slightly indicated in the female; gular scales keeled. Dorsal scales small, hexagonal, juxtaposed, smooth, not distinctly larger than the laterals, which are round; ventral scales larger, rounded, imbricate, strongly keeled. The adpressed hind limb reaches the anterior border of the orbit or the nostril; digits very feebly dilated; fifteen or sixteen lamellæ under phalanges II. and III. of the fourth toe. Tail rounded, covered with equal keeled scales; its length not twice that of head and body. Pale brown or yellowish above, with irregular dark brown spots or longitudinal bands on the neck and back; lower surfaces yellowish white.

Total length	263	millim.
Head	25	,,
Width of head	17	,,

Body	68	millim.
Fore limb	43	,,
Hind limb	82	,,
Tibia	27	,,
Tail	170	,,

Central America (Tabasco to Costa Rica).

a. ♀.	Guatemala.	F. D. Godman, Esq. [P.].
b. ♀.	Vera Paz, low forest.	O. Salvin, Esq. [C.].
c. Hgr.	Lower Vera Paz.	O. Salvin, Esq. [C.]. (Type of *A. carneus.*)

106. **Anolis obtusirostris.**

Anolis obtusirostris, *Peters, Mon. Berl. Ac.* 1873, p. 741.

Head short, as in *A. capito*, but longer than the tibia; supraorbital semicircles separated by three or four series of scales; occipital very small; six or seven loreal rows; twelve upper labials; ear-opening vertical. Dorsal scales very small, granular, convex, smooth, a little larger than the laterals; ventrals larger, keeled. Olive-brown above, greenish inferiorly.

Total length	150	millim.
Head	13	,,
Width of head	9	,,
Length to vent	50	,,
Fore limb	18	,,
Hind limb	32	,,
Tibia	10	,,

Chiriqui.

4. **NOROPS.**

Norops, *Wagler, Syst. Amph.* p. 149; *Dum. & Bibr.* iv. p. 81; *Fitzing. Syst. Rept.* p. 62; *Gray, Cat. Liz.* p. 207; *Bocourt, Miss. Sc. Mex., Rept.* p. 108.

Tympanum distinct. Body slightly compressed, covered with imbricate scales. No dorsal crest. Male with a gular appendage. Digits depressed or slightly dilated, inferiorly with smooth transverse lamellæ, the distal joint narrower but not raised. No femoral or præanal pores. Tail long, cylindrical. Lateral teeth tricuspid; palate toothless. No sternal fontanelle. Abdominal ribs.

Tropical America.

1. **Norops auratus.**

Norops auratus, *Gray, Cat.* p. 207.

Anolis auratus, *Daud. Rept.* iv. p. 89; *Peters, Mon. Berl. Ac.* 1863, p. 135, *and Ann. & Mag. N. H.* (4) iv. 1869, p. 273.

Norops auratus, *Wagl. l. c.*; *Dum. & Bibr.* p. 82, pl. xxxvii. fig. 2; *Bocourt, l. c.* pl. xiii. fig. 5, and xvi. fig. 33.

Draconura nitzschii, *Wiegm. Herp. Mex.* p. 16.
—— 12-striata, *Berthold, Abh. Ges. Götting,* 1840, p. 62, pl. ii. fig. 7.
Læmanctus (Norops) auratus, *Fitzing. l. c.*
Norops macrodactylus, *Hallow. Proc. Ac. Philad.* 1856, p. 222.
Anolis (Dracontura) duodecimstriatus, *Cope, Proc. Ac. Philad.* 1861, p. 212.
Norops duodecimstriatus, *O'Shaughn. Ann. & Mag. N. H.* (4) iii. 1869, p. 191, and iv. 1869, p. 274.

Habit slender. Head nearly twice as long as broad, longer than the tibia; upper head-scales subequal, uni- or tricarinate; occipital enlarged, larger than the ear-opening; five labials to below the centre of the eye; ear-opening oval, slightly oblique. Gular appendage moderately large, slightly indicated in the female; gular scales keeled. Dorsal and ventral scales large, equal, rhomboidal, imbricate, sharply keeled, the former in ten or twelve longitudinal series; lateral scales very small, keeled. The adpressed hind limb reaches the eye, or between the latter and the ear; digits not distinctly dilated. Tail more than twice as long as head and body, covered with equal, sharply keeled scales. Golden brown above; a dark band, usually bordered inferiorly by a white one, extends from the eye to the side of the body, passing above the ear; lower surfaces yellowish; the male's gular appendage dark blue or with a bluish spot.

Total length	180	millim.
Head	14	"
Width of head	7	"
Body	36	"
Fore limb	21	"
Hind limb	37	"
Tibia	11	"
Tail	130	"

South and Central America.

a–b. ♂ ♀.	Surinam.	
c. ♂.	Vryheids Lust, Demerara.	Rev. W. T. Turner [C.].
d. ♀.	Santarem, Upper Amazons.	H. W. Bates, Esq. [C.].
e. ♂.	Granada.	F. D. Godman, Esq. [P.].
f. ♂.	——?	Zoological Society.
g. ♀.	W. Indies (?).	

2. **Norops ophiolepis.**

Anolis (Dracontura) ophiolepis, *Cope, Proc. Ac. Philad.* 1861, p. 211.
Norops ophiolepis, *Bocourt, Miss. Sc. Mex., Rept.* pl. xvi. fig. 34.

Distinguished from the preceding chiefly in the following points:—Scales on forehead larger; occipital not larger than the ear-opening. Dorsal scales forming eighteen longitudinal series. Digits slightly dilated. Tail less than twice the length of head and body. "Under surfaces yellowish, unspotted. Above, the ground colour is golden olivaceous in the males, in the female golden brown. There is a light median dorsal line extending to the sacral region; another similar one extends from the superior posterior part of the orbit upon the anterior part of the tail, and a third from the posterior superior labials, through the tympanic orifice, upon the base of the tail. The superior lateral band is bordered above by a deep brown band." Length to vent 27 millim.

Cuba.

a. ♀.	Cuba.	Prof. Peters [P.].

5. TROPIDODACTYLUS.

Tympanum distinct. Body compressed, covered with imbricate scales. No dorsal crest. Male with a large, non-inflatable, folding gular appendage. No transverse gular fold. Digits cylindrical, with pluricarinate lamellæ inferiorly. No femoral or præanal pores. Tail long, cylindrical. Lateral teeth tricuspid; pterygoids toothed. No sternal fontanelle. Abdominal ribs.

Venezuela and West Indies.

1. **Tropidodactylus onca.** (Plate VI. fig. 2.)

Norops onca, *O'Shaughn. Ann. & Mag. N. H.* (4) xv. 1875, p. 280.

Head about once and two thirds as long as broad, longer than the tibia; snout acuminate, canthus rostralis very short; upper head-scales subequal, pluricarinate; occipital slightly enlarged, not larger than the ear-opening; eleven or twelve labials to below the centre of the eye; ear-opening vertically elliptic. Gular appendage very large; gular scales keeled. Dorsal scales subhexagonal, subimbricate, strongly keeled, not quite as large as ventrals, much larger than laterals; ventrals rhomboidal, imbricate, strongly keeled. The adpressed hind limb reaches the posterior border of the orbit. Tail not twice as long as head and body, covered with subequal keeled scales. Yellowish brown above, spotted or marbled with dark brown; a series of more or less defined dark rhomboidal spots with lighter centre along each side of the back; a dark cross band between the eyes; the skin of the gular appendage blackish.

Total length	213	millim.
Head	22	"
Width of head	13	"
Body	53	"
Fore limb	36	"
Hind limb	56	"
Tibia	17	"
Tail	138	"

Venezuela; Dominica.

a. ♂.	Venezuela.	(Types.)
b. ♂.	Dominica.	(Types.)

6. POLYCHRUS.

Polychrus, *Cuv. R. A.* ii. p. 40; *Merrem. Tent.* p. 48; *Fitzing. N. Classif. Rept.* p. 17; *Wagl. Syst. Amph.* p. 149; *Wiegm. Herp. Mex.* p. 16; *Dum. & Bibr.* iv. p. 63; *Fitzing. Syst. Rept.* p. 62; *Gray, Cat. Liz.* p. 183; *Bocourt, Miss. Sc. Mex., Rept.* p. 111.
Ecphymotes, part., *Fitzing. N. Classif. Rept.* p. 17; *Gray, l. c.* p. 184.
Læmanctus, part., *Dum. & Bibr.* iv. p. 70.
Ecphymatotes, *Fitzing. Syst. Rept.* p. 62.
Sphærops, *Gray, l. c.* p. 183.
Ecphymotes, part., *Gray, l. c.* p. 184.
Chaunolæmus, *Peters, Mon. Berl. Ac.* 1869, p. 786.

Tympanum distinct. Body compressed, covered with small scales; no nuchal or dorsal crest. Head with enlarged scales. No transverse gular fold; male with a small gular pouch. Digits compressed, with carinated lamellæ inferiorly; four large scales at the base of the claw; third and fourth toes equal. Femoral pores present. Tail very long, round. Lateral teeth tricuspid; pterygoid teeth. No sternal fontanelle. Abdominal ribs.

Tropical America.

Synopsis of the Species.

Ventral scales unicarinate; a serrated ridge in front of the throat	1. *marmoratus*, p. 98.
Ventral scales unicarinate; lateral scales larger than median dorsals	2. *acutirostris*, p. 99.
Ventral scales multicarinate, smaller than the gulars	3. *gutturosus*, p. 100.

1. Polychrus marmoratus.

Polychrus marmoratus, *Gray, Cat.* p. 183.
Lacerta marmorata, *Linn. S. N.* i. p. 368.
Iguana marmorata, *Latr. Rept.* i. p. 265.
Agama marmorata, *Daud. Rept.* iii. p. 433.
Polychrus marmoratus, *Merr. Tent.* p. 48; *Spix, Spec. Nov. Lacert.* p. 14, pl. xiv.; *Guér. Icon. R. A., Rept.* pl. xi. fig. 3; *Wied, Beitr.* p. 110, *and Abbild.*; *Dum. & Bibr.* p. 65; *Bocourt, Miss. Sc. Mex., Rept.* pl. xvii. fig. 6.
—— virescens, (*Wied*) *Wagler, Icon. Amph.* pl. xii.
—— strigiventris, *Wagl. l. c.*

Snout obtusely pointed; nostril equally distant from the orbit and the tip of the snout; eye-opening small, smaller than the tympanum; upper head-scales smooth or feebly granulate or striated, very large on the snout; regular supraorbital semicircles of large scales, separated by one row of scales; supraocular scales small; labials large, few; mental with a median cleft posteriorly. Gular scales smooth or feebly keeled, smaller than ventrals, separated by granules on the gular pouch; a median denticulation along the front half of the throat. Scales on the body unicarinate, juxtaposed on the flanks, a little larger and imbricate on the belly and the middle of the back. Limbs moderately elongate; the adpressed hind limb reaches the elbow in the female, the axilla or the shoulder in the male. Nine to eleven femoral pores on each side. Tail nearly three times as long as head and body, covered with equal, keeled scales. Colour and markings very variable; upper surfaces usually reddish-brown in spirit, with blackish spots or symmetrical transverse markings; the sutures between the head-shields black; black lines radiating from the eye, the two most constant being directed to above the tympanum and to the angle of the mouth; lower surfaces paler, uniform or brown-spotted.

	♂.	♀.	
Total length	395	497	millim.
Head	26	32	„
Width of head ..	18	20	„
Body	79	110	„
Fore limb	42	50	„
Hind limb	62	69	„
Tail	290	355	„

South America.

a, b. ♀ & hgr.	Venezuela.	
c, d. ♀ & hgr.	Demerara.	
e–g. ♂ ♀.	Berbice.	
h–l. ♂ ♀.	Para.	R. Graham, Esq. [P.].
m. ♀.	Bahia.	Dr. Wucherer [C.].
n. ♀.	Brazil.	Lord Stuart [P.].
o. ♂.	Brazil.	T. Bell, Esq. [P.].
p, q, r. ♂ ♀.	Brazil.	
s. Ad., skeleton.	Brazil.	
t, u. Ad., stuffed.	—— ?	

2. Polychrus acutirostris.

Sphærops anomalus, *Gray, Cat.* p. 184.
Ecphymotes acutirostris, *Gray, l. c.* p. 185.
Polychrus acutirostris, *Spix, Spec. Nov. Lacert.* p. 15, pl. xiv. A.; *Cope, Proc. Ac. Philad.* 1864, p. 176; *Peters, Mon. Berl. Ac.* 1877, p. 410.
Ecphymotes acutirostris, *Fitzing. N. Class. Rept.* p. 49.
Polychrus anomalus, *Wiegm. Herp. Mex.* p. 16; *Dum. & Bibr.* p. 69; *Reinh. & Lütk. Vidensk. Meddel.* 1861, p. 221; *Bocourt, Miss. Sc. Mex., Rept.* pl. xvii. fig. 7; *Boettg. Zeitschr. f. Naturw.* lviii. 1885, p. 214.
Læmanctus acutirostris, *Dum. & Bibr.* p. 75.

Snout more pointed than in the preceding; nostril a little nearer the orbit than the tip of the snout; eye-opening very small; upper head-scales as in the preceding. Gular scales smooth, elongate, smaller than ventrals, forming longitudinal series separated by granules on the gular pouch; no denticulation. Scales on the body unicarinate, very feebly so on the back and sides, distinctly on the belly; the scales largest on the flanks, smallest on the middle of the back. Limbs short; the adpressed hind limb reaches the elbow in the male, not so far in the female. Nine femoral pores on each side. Tail twice or twice and a half as long as head and body, covered with strongly keeled scales. Pale olive above, uniform or with irregular angular black cross bands on the body; black lines radiating from the eye, the two most marked extending to the fore limb, the lower passing through the tympanum.

	♂.	♀.	
Total length	293	322	millim.
Head	22	25	,,
Width of head ..	11	14	,,
Body	66	92	,,
Fore limb	31	39	,,
Hind limb	40	50	,,
Tail	205	205	,,

Brazil; Paraguay.

a–b, c, d. ♂ ♀. Brazil.

3. Polychrus gutturosus.

Polychrus gutturosus, *Berthold, Abh. Ges. Götting.* iii. 1847, p. 5, pl. i. fig. 1.
—— (Chaunolæmus) multicarinatus, *Peters, Mon. Berl. Ac.* 1869, p. 786; *Bocourt, Miss. Sc. Mex., Rept.* p. 112, pl. xvii. fig. 8.

Snout obtusely pointed; nostril equally distant from the orbit and the tip of the snout; eye-opening a little larger than the tympanum; upper head-scales as in the preceding. Gular scales oval, striated, much larger than ventrals, those on the gular pouch widely separated by granules; no denticulation. Scales on the body of nearly equal size, the dorsals and laterals unicarinate, the ventrals tri- or quinquecarinate. Limbs moderately elongate, the scales mostly quinquecarinate. Six to eighteen femoral pores on each side. Tail long, round, covered with unicarinate scales. Olive-brown above, greenish yellow inferiorly; an irregular yellow band along the back of the thighs; tail with indistinct darker cross bands.

Total length	153	millim.
Head	35	,,
Width of head	22	,,
Fore limb	63	,,
Hind limb	78	,,
Tail	338	,,

Colombia; Costa Rica.

7. CORYTHOPHANES.

Corythophanes, (*Boie*) *Gray*, *Griff. A. K.* ix. *Syn.* p. 55; *Wiegm. Herp. Mex.* p. 15; *Dum. & Bibr.* iv. p. 172; *Fitzing. Syst. Rept.* p. 52; *Gray, Cat. Liz.* p. 194; *Bocourt, Miss. Sc. Mex., Rept.* p. 117.
Chamæleopsis, (*Wiegm.*) *Gray*, *Griff. A. K.* ix. *Syn.* p. 45; *Wiegm. Isis*, 1831, p. 296, *and Herp. Mex.* pp. 15, 37; *Fitzing. l. c.*; *Gray, Cat.* p. 194.

Tympanum distinct. Head with two sharp ridges uniting posteriorly into an elevated bony crest directed upwards and backwards; a bony supraorbital arch formed by the præ- and postfrontal. Body compressed, covered with imbricate smooth or keeled scales; a nuchal crest, attached to the cranial crest; a low dorsal crest. A gular pouch and a transverse gular fold. Infradigital lamellæ keeled. No femoral pores. Tail long, feebly compressed. Lateral teeth tricuspid; pterygoid teeth. No sternal fontanelle. No abdominal ribs.
Central America; Ecuador?

Synopsis of the Species.

Nuchal crest continuous with dorsal, descending obliquely; upper head-scales smooth	1. *cristatus*, p. 101.
Nuchal crest continuous with dorsal, descending obliquely; upper head-scales keeled	2. *percarinatus*, p. 102.
Nuchal crest not continuous with dorsal.	3. *hernandesii*, p. 103.

1. Corythophanes cristatus.

Corythophanes cristatus, *Gray, Cat.* p. 194.
Agama cristata, *Merr. Tent.* p. 50.
Corythophanes cristatus, *Gray, Griff. A. K.* ix. *Syn.* p. 55; *Gravenh. Acta Ac. Leop.-Carol.* xvi. ii. 1833, p. 938, pl. lxv. figs. 6–10; *Dum. & Bibr.* p. 174; *A. Dum. Cat. Méth. Rept.* p. 60, *and Arch. Mus.* viii. p. 517, pl. xx. fig. 1; *Bocourt, Miss. Sc. Mex., Rept.* p. 118.

The upper surface of the head between the angular ridges rhomboidal, twice as long as the posterior (single) ridge or crest; the supraciliary ridge continuous with the parietal; nostril a little nearer the orbit than the tip of the snout; tympanum large, vertically oval; upper head-scales small and smooth, irregular; no enlarged occipital; enlarged scales on the ridges; a series of feebly enlarged, not projecting, slightly keeled scales from the orbit to above the tympanum; eight to ten upper and as many lower labials. Gular region with longitudinal series of small scales separated by granules; a strong crest of triangular scales along the throat. Nuchal crest with serrated edge, descending gradually from the cranial crest to the dorsal ridge, which is feebly serrated. A more or less distinct fold along each side of the back. Dorso-lateral scales irregular, unequal, smooth, the enlarged ones sometimes feebly keeled; ventral scales moderately large, strongly keeled, imbricate, arranged in transverse series. Limbs long, covered with very large keeled scales; the adpressed hind limb reaches the tip of the snout or a little beyond. Tail more than twice as long as head and body, slender, compressed, above with feebly keeled scales, inferiorly with larger strongly keeled ones. Olive- or reddish-brown above, with irregular blackish-brown spots or cross bands; a white spot on the elbow and another at the base of the arm; blackish streaks radiate from the eye, extending down to the lower lip, one directed obliquely towards the lower border of the tympanum and another to its upper border; lower surfaces pale brown, with a few darker spots.

Total length	338	millim.
Head (to extremity of helmet)	53	,,
Width of head	18	,,
Body	73	,,
Fore limb	52	,,
Hind limb	98	,,
Tail	235	,,

Guatemala to Costa Rica.

a–b. ♂.	Vera Paz.	O. Salvin, Esq. [C.].
c. ♀.	Costa Rica.	O. Salvin, Esq. [C.].
d. Yg.	Costa Rica.	

2. **Corythophanes percarinatus.**

Corythophanes percarinatus, *A. Dum. Arch. Mus.* viii. p. 518, pl. xx. fig. 3; *Bocourt, Miss. Sc. Mex., Rept.* p. 120, pl. xvii. fig. 2.

The upper surface of the head constricted behind the supraciliary ridge; the parietal ridges uniting only a short distance in front of the extremity of the helmet; nostril equally distant from the orbit and the tip of the snout; upper head-scales keeled or striated; a short series of enlarged keeled scales from the orbit to above the tympanum, the posterior projecting, tubercle-like. Gular scales elongate, strongly keeled, forming longitudinal series, separated by granules; a strong crest of triangular scales along the throat. Nuchal crest with serrated edge, descending gradually from the cranial crest to the serrated dorsal ridge. A more or less distinct fold along each side of the back. Dorso-lateral scales irregular, unequal, all the larger ones distinctly keeled; ventral scales large, strongly keeled. Limbs as in the preceding. All the caudal scales strongly keeled. Olive-brown above, with irregular blackish-brown spots or cross bands; broad black bands radiating from the eye, not extending beyond the orbit, except one which reaches the anterior border of the tympanum; lower surfaces whitish olive.

Total length	300	millim.
Head (to extremity of helmet)	42	,,

Width of head	17 millim.
Body	73 „
Fore limb	50 „
Hind limb	87 „
Tail	200 „

Guatemala; Ecuador?

a. ♀.	Lanquin, Vera Paz.	O. Salvin, Esq. [C.].
b–d. ♂ ♀.	Ecuador (?).	Mr. Fraser [C.].

3. Corythophanes hernandesii.

Chamæleopsis hernandesii, *Gray, Cat.* p. 194.
Chamæleopsis hernandesii, (*Wiegm.*) *Gray, Griff. A. K.* ix. *Syn.* p. 45; *Wiegm. Isis*, 1831, p. 298, *and Herp. Mex.* p. 38, pl. vi.; *Gravenh. Acta Ac. Leop.-Carol.* xvi. ii. 1833, p. 948, pl. lxv. figs. 1–5; *Gray, Zool. Beechey's Voy.* p. 94, pl. xxx. fig. 1.
Corythophanes chamæleopsis, *Dum. & Bibr.* iv. p. 176; *Sumichrast, Arch. Sc. Phys. et Nat.* xix. 1864, p. 58.
—— (Chamæleopsis) hernandesii, *Fitzing. Syst. Rept.* p. 52.
—— mexicanus, *Bocourt, Miss. Sc. Mex., Rept.* p. 122, pl. xvii. fig. 1.

The upper surface of the head constricted behind the supraciliary ridge, which ends in an angle; the parietal ridges uniting only a short distance in front of the extremity of the helmet; nostril equally distant from the orbit and the tip of the snout; upper head-scales keeled or striated; a large bony tubercle or spine above the tympanum. Gular scales elongate, strongly keeled, forming longitudinal series, separated by granules; a feeble denticulation along the throat. Nuchal crest small, descending vertically from the cranial crest, not continuous with the serrated dorsal ridge. A distinct longitudinal fold on the scapular and iliac regions. Dorso-lateral scales irregular, unequal, the larger ones keeled; ventral scales large, strongly keeled. Limbs very long, covered with large keeled scales; the adpressed hind limb reaches considerably beyond the tip of the snout. Tail as in the preceding. Olive-brown above, with irregular blackish-brown spots or cross bands; the eyelids and the space between the orbit and the tympanum blackish brown; lips with or without dark vertical bars; lower surfaces brownish, with darker spots.

Total length	345 millim.
Head (to extremity of helmet)	44 „
Width of head	20 „
Body	65 „
Fore limb	55 „
Hind limb	108 „
Tail	250 „

Mexico.

a. ♀.	Orizaba.	
b. Hgr.	Yucatan.	
c. Hgr.	Mexico.	Dr. J. E. Gray [P.].
d. Hgr.	Mexico.	M. Sallé [C.].

8. LÆMANCTUS.

Læmanctus, part., *Wiegm. Herp. Mex.* pp. 16, 45; *Dum. & Bibr.* iv. p. 70.

Læmanctus, *Fitzing. Syst. Rept.* p. 63; *Gray, Cat. Liz.* p. 185; *Bocourt, Miss. Sc. Mex., Rept.* p. 114; *Bouleng. Bull. Soc. Zool. France,* 1877, p. 460.

Tympanum distinct. Head plane above, shelving forwards, the occipital region more or less raised and produced beyond the occiput. Body compressed, covered with imbricate keeled scales; dorsal crest feebly developed or absent. A strong transverse gular fold; no gular pouch. Limbs very long; infradigital lamellæ with a median tubercle-like keel. No femoral pores. Tail very long, round. Lateral teeth tricuspid; pterygoid teeth. Clavicle loop-shaped proximally. No sternal fontanelle. No abdominal ribs.

Central America.

Synopsis of the Species.

I. Scales on upper surface of snout much larger than those on occipital region.

A. A series of projecting triangular scales on the posterior contour of the head; a feeble dorsal crest.

57 to 61 scales round the middle of the body .... 1. *serratus*, p. 104.

45 to 51 scales round the middle of the body .... 2. *alticoronatus*, p. 105.

B. No distinct serration on the posterior contour of the head; no dorsal crest.. 3. *longipes*, p. 105.

II. Scales on upper surface of snout not distinctly larger than those on occipital region; no dorsal crest .... 4. *deborrii*, p. 106.

1. Læmanctus serratus.

Læmanctus longipes, (*non Wiegm.*) *A. Dum. Arch. Mus.* viii. p. 512, pl. xxi. fig. 4; *Bocourt, l. c.* p. 114, pl. xvii. fig. 4.

—— serratus, *Cope, Proc. Ac. Philad.* 1864, p. 176; *Bocourt, l. c.* p. 116; *Bouleng. l. c.* p. 463, pl. vii. fig. 3.

Upper head-scales rugose, very large on upper surface of snout, where they usually form three regular pairs; the posterior part of the head more elevated in the males than in the females, bordered by projecting triangular ribbed scales; labials large; tympanum larger than the eye-opening. Gular scales small, suboval, feebly keeled, enlarged and hexagonal near the labials. Dorsal ridge more or less strongly serrated; scales on the body unicarinate, smallest on the sides, largest on the abdomen, 57 to 61 round the middle of the body. The adpressed hind limb reaches beyond the tip of the snout; digits very long, very unequal. Tail round, more than three times as long as head and body, covered with keeled scales. Reddish or purplish above (in spirit), with dark-brown bands across the body, most distinct on the vertebral region; a dark brown band from the eye to the tympanum, sometimes continued along the side of the back; a whitish streak from below the eye to the fore limb, and from axilla to groin; a white spot on each side of the base of the tail, close to the thigh; lower surfaces pale purplish or greenish; tail with indistinct dark annuli.

Total length	610	millim.
Head (to extremity of helmet)	47	,,
Width of head	25	,,
Body (from occiput)	105	,,
Fore limb	80	,,
Hind limb	140	,,
Tail	460	,,

Southern Mexico; Yucatan.

a–c. ♂ ♀.	Mexico.
d. Yg.	Vera Cruz.
e. Ad., stuffed.	Mexico.

2. **Læmanctus alticoronatus.**

Læmanctus alticoronatus, *Cope, Proc. Ac. Philad.* 1865, p. 192, *and Proc. Am. Phil. Soc.* 1869, pl. ii.

Very closely allied to the preceding, differing in the larger scales (45 to 51 round the middle of the body), the more elongate gulars, and the less distinct vertebral serration. Pale purplish, with anastomosing dark lines; no white streak along the flanks, but a white spot in front and another behind the thigh.

Yucatan.

a. ♀.	Yucatan.

3. **Læmanctus longipes.**

Læmanctus longipes, *Gray, Cat.* p. 185.

Læmanctus longipes, *Wiegm. Herp. Mex.* p. 46, pl. iv.; *Dum. & Bibr.* iv. p. 172; *Bouleng. l. c.* p. 464, pl. vii. fig. 2.

The scales on the forehead much larger than those on the back of the head, but smaller and less regular than in *L. serratus*; posterior border of head angular, without projecting triangular scales. Gular scales mostly bi- or tricarinate. No dorsal crest. Otherwise as in

L. serratus. Reddish or olive above, with darker bands across the back; a broad dark band from the eye to the fore limb, passing through the tympanum, bordered inferiorly by a yellowish streak beginning at the anterior angle of the eye; a yellowish streak, sometimes broken up, from axilla to groin; a yellowish spot on each side of the base of the tail; lower surfaces usually pink.

Total length	730	millim.
Head	48	„
Width of head	23	„
Body	102	„
Fore limb	76	„
Hind limb	140	„
Tail	540	„

Mexico.

a–b, c–d. ♀ & yg.	Mexico.	
e. Hgr.	Mexico.	Dr. Kaup.
f. Yg.	Mexico.	M. Sallé [C.].

4. Læmanctus deborrii.

Læmanctus de borrei, *Bouleng. l. c.* p. 465, pl. vii. fig. 1.

The scales on the forehead not or scarcely larger than those on the back of the head; posterior contour of the head without projecting scales. Gular scales large (20 from mental plate to gular fold), bi- or tricarinate. No dorsal crest. Part of the scales on the body bi- or tricarinate. Head yellowish above, the posterior contour black; body above dark purple or dark green, with blackish cross bands; a light lateral streak, as in the preceding species; lower surfaces lilac.

Total length	690	millim.
Head	45	„
Width of head	23	„
Body	95	„
Fore limb	70	„
Hind limb	128	„
Tail	510	„

Tabasco; Vera Paz; Guatemala.

a–b. ♂ & yg.	Vera Paz.	O. Salvin, Esq. [C.].
c. ♀.	Coban.	O. Salvin, Esq. [C.].

9. BASILISCUS.

Basiliscus, *Laur. Syn. Rept.* p. 50; *Wagler, Syst. Amph.* p. 148; *Wiegm. Herp. Mex.* p. 15; *Dum. & Bibr.* iv. p. 179; *Fitzing. Syst. Rept.* p. 53; *Gray, Cat. Liz.* p. 192, *and Ann. & Mag. N. H.* (2) x. 1852, p. 438; *Bocourt, Miss. Sc. Mex., Rept.* p. 124.

Corythæolus, *Kaup, Isis*, 1828, p. 1147; *Wiegm. l. c.* pp. 15, 39; *Fitzing. l. c.*; *Gray, Cat.* p. 192, *and Ann. & Mag. N. H.* (2) x. p. 439.
Œdicoryphus, *Wagl. l. c.*
Thysanodactylus, *Gray, Cat.* p. 193, *and Ann. & Mag. N. H.* (2) x. p. 439.
Ptenosaura, *Gray, Ann. & Mag. N. H.* (2) x. p. 438.
Lophosaura, *Gray, l. c.*
Cristasaura, *Gray, l. c.*
Craneosaura, *Gray, in Richards. Zool. Herald*, p. 148.
Paraloma, *Cope, Proc. Ac. Philad.* 1862, p. 181.

Tympanum distinct. Head, in the adult male, produced posteriorly into a large cartilaginous lobe. Body compressed, covered with imbricate small scales; male with a dorsal crest. No gular pouch; a transverse gular fold. Toes bordered on the outer side by a much developed dermal lobe. No femoral pores. Tail long, strongly compressed, crested in the male. Lateral teeth tricuspid; pterygoid teeth. Clavicle loop-shaped proximally. No sternal fontanelle. No abdominal ribs.

Tropical America.

All species agree in the following characters:—Head largest in the male, with swollen cheeks and a large crest on the back of the head; this crest hardly developed in the females and not at all in the young. Snout pointed; eye-opening and tympanum large; upper head-scales small, irregular, keeled; labials large; a row of large chin-shields on each side; gular scales smooth. Dorsal scales smaller than ventrals, imbricate, more or less distinctly keeled. Limbs long, covered above with keeled scales larger than the ventrals; the adpressed hind limb reaches the tip of the snout, or a little beyond. Caudal scales keeled, much enlarged inferiorly.

Synopsis of the Species.

I. Males with very high, rayed, dorsal and caudal crests; ventral scales perfectly smooth.

Head-crest undivided, rising from an imaginary line connecting the upper borders of the ear-openings .... 1. *americanus*, p. 108.
Head-crest bisected, rising on a line with the posterior borders of the orbits ...... 2. *plumifrons*, p. 108.

II. Males with a low, not rayed, caudal crest.

Dorsal crest rayed; head-crest rising from the occiput; ventral scales keeled* .... 3. *vittatus*, p. 109.
Dorsal crest not rayed; head-crest rising on a line with the posterior borders of the orbits; ventral scales smooth .... 4. *galeritus*, p. 110.

* The keel sometimes very indistinct in adult males.

1. Basiliscus americanus.

Basiliscus americanus, *Gray, Cat.* p. 192.
Thysanodactylus bilineatus, *Gray, l. c.* p. 194.
Lacerta basiliscus, *Linn. S. N.* i. p. 366.
Basiliscus americanus, *Laur. Syn. Rept.* p. 50; *Bocourt, Miss. Sc. Mex., Rept.* p. 127; *Günth. Biol. C.-Amer., Rept.* p. 54.
Iguana basiliscus, *Latr. Rept.* ii. p. 258.
Basiliscus mitratus, *Daud. Rept.* iii. p. 310, pl. xlii.; *Guér. Icon. R. A., Rept.* pl. xi. fig. 2; *Dum. & Bibr.* iv. p. 181; *Duvern. R. A., Rept.* pl. xviii. fig. 1; *Cope, Journ. Ac. Philad.* (2) viii. 1876, p. 125, pl. xxv. fig. 2.
Ophryoessa bilineata, *Gray, Zool. Beechey's Voy.* p. 94.
Lophosaura goodridgii, *Gray, Ann. & Mag. N. H.* (2) x. 1852, p. 438; *Richards. Zool. Herald*, p. 143, pl. xxvi.
Basilicus guttulatus, *Cope, l. c.* pp. 125, 156.
—— goodridgii, *Cope, l. c.* p. 125.

Head-crest more or less regularly hood-shaped, rising from an imaginary line connecting the superior borders of the ear-openings, preceded by a more or less distinct keel to between the posterior borders of the orbits. Six or seven labials to below the centre of the eye. The three or four anterior chin-shields in contact with the infralabials. Dorsal crest much developed, sometimes quite as high as the body, supported by bony rays, covered with large thin scales. The lateral fringe of the toes very strongly developed. Caudal crest like the dorsal, though not quite so high. Ventral scales perfectly smooth. Olive-brown above, with more or less distinct angular black cross bars on the back; a more or less distinct lighter band on the temple to the neck, and another, broader and more distinct, from the loreal region to the fore limb.

Total length	800	millim.
Head	60	,,
Width of head	35	,,
Body	180	,,
Fore limb	95	,,
Hind limb	205	,,
Tail	560	,,

Panama and Costa Rica; Martinique (?); Guianas (?).

a, b. ♂.	Tropical America.	
c. ♂, stuffed.	Tropical America.	
d. ♂, stuffed.	Costa Rica.	F. D. Godman, Esq. [P.].
e. Yg.	Fernando de Noronha (?).	H.M.S. 'Chanticleer.' (Type of *Ophryoessa bilineata.*)
f. Yg.	—— ?	

2. Basiliscus plumifrons.

Basiliscus plumifrons, *Cope, Journ. Ac. Philad.* (2) viii. 1876, p. 125, pl. xxv. fig. 1.

Head-crest consisting of a principal posterior portion and an accessory anterior portion; the former commences above a point a little in front of the anterior margin of the ear-opening, the latter rises abruptly from a point on a line with the posterior border of the orbit. Six or seven labials to below the centre of the eye. The three anterior chin-shields in contact with the infralabials. Dorsal crest much developed, a little higher than the body, supported by bony rays, covered with large thin scales. Caudal crest like the dorsal. Ventral scales perfectly smooth. Colour everywhere green, becoming blue on the different crests; a row of distant round yellow spots from the axilla to the groin, and, usually, a similar row along the upperside of the back; usually three vertical pale-bordered black spots on the dorsal crest; crested part of the tail green, the remainder yellowish, with brown rings.

Total length	655	millim.
Head (to ear)	45	,,
Fore limb	87	,,
Hind limb	175	,,
Tail	478	,,

Costa Rica.

3. Basiliscus vittatus.

Corythæolus vittatus, *Gray, Cat.* p. 193.
Basiliscus vittatus, *Wiegm. Isis,* 1828, p. 373; *Dum. & Bibr.* iv. p. 187; *Sumichr. Arch. Sc. Phys. Nat.* xix. 1864, p. 56; *Bocourt, Miss. Sc. Mex., Rept.* p. 129, pl. xvii. fig. 3; *Cope, Journ. Ac. Philad.* (2) viii. 1876, p. 125; *Günth. Biol. C.-Amer., Rept.* p. 55, tab. 28.
Corythæolus vittatus, *Kaup, Isis,* 1829, p. 1147; *Wiegm. Herp. Mex.* p. 40, pl. v.
Œdicoryphus vittatus, *Wagl. Syst. Amph.* p. 148.
Cristasaura mitrella, *Gray, Ann. & Mag. N. H.* (2) x. 1852, p. 439.
Basiliscus americanus, *Salvin, Proc. Zool. Soc.* 1860, p. 453.
Draconura bivittata, *Hallow. Proc. Ac. Philad.* 1860, p. 482.
Paraloma bivittata, *Cope, Proc. Ac. Philad.* 1862, p. 181.
Basiliscus (Cristasaura) nuchalis, *Cope, l. c.*

Head-crest subtriangular, rising from a line connecting the superior borders of the ear-openings, and extending as a strong keel or ridge to between the posterior borders of the orbits. Six or seven labials to below the centre of the eye. Usually only the first or the first two chin-shields in contact with the infralabials. Dorsal crest low, supported by bony rays. The lateral fringe of the toes less developed than in *B. americanus*. Caudal crest reduced to a serrated ridge. Ventral scales keeled, less distinctly in the adult males. Olive or brownish above, usually with more or less distinct black cross bars on the back, and a reddish longitudinal band on each side, commencing from the orbit; a reddish or whitish band from the loreal region to the angle of the jaws, or to the base of the fore limb.

Total length	720	millim.
Head	52	,,
Width of head	32	,,
Body	118	,,
Fore limb	73	,,
Hind limb	165	,,
Tail	550	,,

Mexico to Ecuador.

a, b, c, d. ♂, hgr., & yg.	Mexico.	
e–f. Yg.	Mexico.	M. Sallé [C.].
g. ♂.	Putla, Mexico.	M. Boucard [C.].
h. ♂.	Orizaba.	
i. ♀.	Hacienda del Hobo, S. Mexico.	P. Geddes, Esq. [P.].
k–m, n. ♂, hgr., & yg.	Vera Paz.	O. Salvin, Esq. [C.].
o, p, q, r. ♂, ♀, & yg.	Lanquin, Guatemala.	O. Salvin, Esq. [C.].
s. ♀.	Dueñas, Guatemala.	O. Salvin, Esq. [C.].
t, u, v. ♀ and hgr.	Honduras.	
w. ♂.	Honduras.	(Type of *Cristasaura mitrella.*)
x–y, z. ♂ ♀.	Ecuador.	Mr. Fraser [C.].
a. Skeleton.	Guatemala.	

4. Basiliscus galeritus.

Basiliscus galeritus, *A. Dum. Cat. Méth. Rept.* p. 61, *and Arch. Mus.* viii. p. 522, pl. xxi. fig. 1; *Cope, Proc. Ac. Philad.* 1866, p. 123.
Ptenosaura seemanni, *Gray, Ann. & Mag. N. H.* (2) x. 1852, p. 438.
Craneosaura seemanni (*Gray*), *Richards. Zool. Herald,* p. 148, pl. xxv.
Basiliscus seemanni, *Günth. Proc. Zool. Soc.* 1860, p. 234; *Bocourt, Miss. Sc. Mex., Rept.* p. 131; *Cope, Journ. Ac. Philad.* (2) viii. 1876, p. 125.

Head-crest rounded in profile, rising from between the posterior borders of the orbits, its basal occipital part thick, fleshy. Five labials to below the centre of the eye. The two or three anterior chin-shields in contact with the infralabials. Dorsal crest reduced to a series of compressed trihedral tubercles, separated from one another by two small scales. The lateral fringe of the toes much developed. Caudal crest a serrated ridge. Ventral scales perfectly smooth. Bright green or olive above, uniform or with brown spots; sides sometimes with whitish spots; tail with more or less distinct brown annuli; a white, black-edged band from below the eye to the neck; usually another black band, parallel to the former, on each side of the throat.

Total length	670	millim.
Head	53	,,
Width of head	31	,,
Body	117	,,

Fore limb	78	millim.
Hind limb	170	„
Tail	500	„

Colombia; Ecuador.

a–b, c–d, e–g. ♂, ♀, and yg.	Esmeraldas.	Mr. Fraser [C.].
h. ♂.	—— ?	Mdme. Pfeiffer [C.].

10. OPHRYOESSA.

Uraniscodon, part., *Kaup, Isis,* 1826, p. 89.
Ophryoessa, part., *Fitzing. N. Classif. Rept.* p. 17; *Wiegm. Herp. Mex.* p. 15.
Ophryoessa, *Wagl. Syst. Amph.* p. 149; *Dum. & Bibr.* iv. p. 237; *Fitzing. Syst. Rept.* p. 57; *Gray, Cat. Liz.* p. 196.

Tympanum distinct. Body slightly compressed; a dorsal crest; dorsal scales small, equal. A transverse gular fold; gular pouch absent. Digits denticulated laterally, with sharply keeled lamellæ inferiorly. No femoral pores. Tail long, compressed. Lateral teeth tricuspid; pterygoid teeth. No sternal fontanelle. Abdominal ribs none.

South America.

1. Ophryoessa superciliosa.

Ophryoessa superciliosa, *Gray, Cat.* p. 196.
Lacerta superciliosa, *Linn. S. N.* i. p. 360.
Agama superciliosa, *Daud. Rept.* iii. p. 336; *Kuhl, Beitr. Zool. Vergl. Anat.* p. 105.
Lophyrus xiphosurus, *Spix, Spec. Nov. Lacert. Bras.* p. 9, pl. x.
—— auronitens, *Spix, l. c.* p. 12, pl. xiii *a*.
Uraniscodon superciliosa, *Kaup, Isis,* 1826, p. 90.
Ophryoessa superciliosa, (*Boie*) *Fitzing. N. Classif. Rept.* p. 48; *Guér. Icon. R. A., Rept.* pl. viii. fig. 1; *Dum. & Bibr.* p. 238; *Guichen. in Casteln. Amér. Mér., Rept.* p. 22; *Peters, Mon. Berl. Ac.* 1877, p. 407.
Hypsibatus (Ophryoessa) superciliosus, *Fitzing. Syst. Rept.* p. 57.

Head moderately large, short, rounded, its length equalling nearly once and one third its width; canthus rostralis and supraciliary edge angular, projecting; nostril pierced above the canthus rostralis, nearer the end of the snout than the orbit; tympanum vertically oval, a little larger than the eye-opening. Upper head-scales small, irregular, strongly keeled or tubercular; enlarged subconical tubercles surrounding the large, rugose occipital; five or six upper and as many lower labials. Throat with strong transverse folds; gular scales strongly keeled. Nuchal and dorsal crests a low serrated ridge, the former a little more developed. Dorsal and lateral scales small, equal, rhomboidal, imbricate, strongly keeled; ventrals larger, also strongly keeled. Limbs long, covered with equal keeled scales; the adpressed hind limb reaches the tympanum or the angle of the mouth; digits long and slender, the toes strongly denticulated laterally; in the young the digital fringe is almost as much developed as in *Basiliscus*. Tail strongly compressed, crested like the back; its length about twice that of head and body; caudal scales subequal, keeled, the lower larger. Brownish above, uniform or spotted with darker; frequently a light festooned band from axilla to groin; lower surfaces brownish white.

Total length	450	millim.
Head	30	,,
Width of head	24	,,
Body	110	,,
Fore limb	63	,,
Hind limb	110	,,
Tail	310	,,

Guianas; Brazil.

a–b. ♂.	Demerara Falls.	
c. ♀.	Vryheids Lust, Demerara.	Rev. W. T. Turner [C.].
d. ♂.	Surinam.	Mr. E. Bartlett [C.].
e, f. ♂.	Surinam.	
g, h. ♂ and yg.	Brazil.	
i. Ad., stuffed.	——?	

11. ENYALIOIDES.

Tympanum distinct. Body compressed; a dorsal crest; dorsal scales small. A transverse gular fold; a more or less developed gular sac in the male. Digits compressed, with sharply-keeled lamellæ inferiorly. Males with a few femoral pores *. Tail long, compressed. Lateral teeth tricuspid; pterygoid teeth. No sternal fontanelle. Abdominal ribs.

Veragua to Peru.

Synopsis of the Species.

I. Caudal scales subequal, not forming regular segments; dorsal scales equal.

Ventral scales smooth or indistinctly keeled; tail feebly compressed, not crested .. 1. *laticeps*, p. 113.

II. Tail with rings of larger scales; dorsal scales unequal.

A. Dorsal and nuchal crests continuous; supraciliary edge simple.

Ventral scales smooth or very feebly keeled; scales on the sides scarcely intermixed with larger ones .... 2. *præstabilis*, p. 113.

Ventral scales keeled; two longitudinal series of larger scales on each side of the back .... 3. *heterolepis*, p. 114.

Ventral scales sharply keeled; a single series of larger scales on each side;

* Except in *E. palpebralis*.

about 12 longitudinal series of scales between the dorsal crest and the lateral series	4. *oshaughnessyi*, p. 115.
Ventral scales sharply keeled; a single series of larger scales on each side; dorsal scales extremely small	5. *microlepis*, p. 115.
B. Nuchal crest not continuous with dorsal; supraciliary edge much produced; ventral scales sharply keeled	6. *palpebralis*, p. 116.

1. Enyalioides laticeps.

Enyalius laticeps, *Guichen. in Casteln. Amér. Mér., Rept.* p. 20, pl. v.; *A. Dum. Arch. Mus.* viii. p. 529.
—— planiceps, *Guichen. l. c.* p. 21, pl. vi.

Head large, short, rounded, its length equalling nearly once and a third its width; canthus rostralis and supraciliary edge angular, projecting; nostril pierced below the canthus rostralis, nearer the orbit than the tip of the snout; tympanum vertically oval, as large as the eye-opening. Upper head-scales small, equal, tubercular, strongly keeled or subconical; occipital scarcely enlarged; eleven or twelve upper, and ten or eleven lower labials. Gular scales, uniform large oval smooth granules. Nuchal crest formed of large conical scales, the largest being nearly as long as the vertical diameter of the tympanum; on the back, the crest becomes very low, and is reduced to a feebly serrated ridge. Dorsal and lateral scales very small, equal, subrhomboidal, juxtaposed or feebly imbricate, feebly keeled; ventrals larger, squarish, smooth or indistinctly keeled. Limbs long, covered with equal, keeled scales; the adpressed hind limb reaches the anterior border of the orbit; digits long and slender, feebly denticulated. Three femoral pores on each side. Tail feebly compressed, oval in section, not crested, covered with subequal keeled scales, largest inferiorly; the length of the tail equals nearly twice that of head and body. Green above, with irregular brown spots; tail with regular brown annuli; throat green, belly white.

Total length	352	millim.
Head	32	„
Width of head	25	„
Body	90	„
Fore limb	66	„
Hind limb	113	„
Tail	230	„

Upper Amazons.

a. ♀.	Pallatanga, Ecuador.	Mr. Buckley [C.].

2. Enyalioides præstabilis.

Enyalius præstabilis, *O'Shaughn. Proc. Zool. Soc.* 1881, p. 240, pl. xxv. fig. 1.

Head large, nearly once and a half as long as broad; canthus rostralis and supraciliary edge angular, projecting; nostril pierced below the canthus rostralis, nearer the orbit than the tip of the snout; tympanum vertically oval, nearly as large as the eye-opening. Upper head-scales small, very convex or pointed and erect, largest and conical above the temporal region; no distinct occipital; ten upper and nine lower labials. Male with a gular pouch; gular scales as large as ventrals, convex, elongate, the central erect. Dorsal and nuchal crests continuous, formed of conical scales, gradually decreasing in height to the base of the tail; the longest spines, on the nape, measure less than the vertical diameter of the tympanum. Back and sides covered with small almost uniform pointed and keeled scales, from thirty-four to forty in a vertical series between the ventral scales and the dorsal crest; only a very indistinct lateral series of larger scales on each side; ventral scales smooth or indistinctly keeled. The adpressed hind limb reaches the eye in the male, the temple in the female; digits shorter than in the preceding species. A single femoral pore on each side. Tail compressed, ringed with keeled scales of unequal size, forming segments; a double keel along its upper surface; the length of the tail about once and a half that of head and body. Reddish or purplish-brown above, dotted over with green; a round yellowish spot behind the ear; tail with more or less distinct broad brown and narrow lighter annuli; lower surfaces yellowish; male with the gular pouch black.

Total length	326	millim.
Head	36	„
Width of head	26	„
Body	90	„
Fore limb	63	„
Hind limb	100	„
Tail	200	„

Ecuador.

a. ♂.	Pallatanga.	Mr. Buckley [C.].	(Types.)
b. ♀.	Canelos.	Mr. Buckley [C.].	

3. Enyalioides heterolepis.

Enyalius heterolepis, *Bocourt, Ann. Sc. Nat.* (5) xix. 1874, art. 4, *and Miss. Sc. Mex., Rept.* pl. xx. c. fig. 14.

Very similar to the preceding, from which it differs in the following points:—Head shorter; labials twelve; gular scales keeled. Nuchal and dorsal crests lower, the spines of the former measuring only half the diameter of the tympanum. Dorsal and lateral scales strongly heterogeneous, the larger scales forming two longitudinal series on each side of the back, and irregular vertical series on the flanks; about thirty scales in a vertical series between the ventrals and the dorsal crest; ventrals keeled. Limbs with transverse rows of enlarged scales. Three or four femoral pores on each side. Light brown above, varied with greenish; lower surfaces yellowish white;

male with a black spot on the gular pouch. From snout to vent 138 millim.; tail 172 millim.

Veragua.

4. Enyalioides oshaughnessyi.

Enyalius heterolepis (*non Bocourt*), *Bouleng. Bull. Soc. Zool. France*, 1880, p. 43.
—— oshaughnessyi, *Bouleng. Proc. Zool. Soc.* 1881, p. 246, pl. xxvi.

Head large, once and two thirds as long as broad; canthus rostralis and supraciliary edge angular, projecting; nostril pierced below the canthus rostralis, nearer the orbit than the tip of the snout; tympanum vertically oval, nearly as large as the eye-opening. Upper head-scales small, strongly keeled; a series of larger conical scales above the temple; a small but distinct occipital; fourteen labials. Male with a gular pouch; gular scales keeled. Dorsal and nuchal crests continuous, formed of very large conical scales, gradually diminishing in height to the base of the tail, where it becomes double and soon vanishes; the highest of these scales on the neck measure more than the vertical diameter of the tympanum. Scales of the back and sides small, smooth or slightly keeled, irregularly intermixed with larger but not conical ones; a single lateral series of conical scales, beginning at a small distance behind the ear, and extending to the sacral region; there are about twelve scales in a vertical series between the dorsal crest and the lateral series, and thirty between the latter and the belly; ventral scales strongly keeled. Scales on the limbs subequal, strongly keeled. A single femoral pore on each side. Tail compressed, with rings of strongly keeled scales of unequal size, forming regular segments. Dark brown, with scattered greenish dots; a light rounded spot behind the ear; throat greenish, gular fold black.

Total length (tail injured) ..	310	millim.
Head (to ear-opening)	43	„
From snout to vent	135	„
Fore limb	72	„
Hind limb	95	„

Ecuador.

5. Enyalioides microlepis.

Enyalius laticeps (*non Guich.*); *Günth. Proc. Zool. Soc.* 1859, p. 407.
—— microlepis, *O'Shaughn. Proc. Zool. Soc.* 1881, p. 238, pl. xxiv. fig. 2.

Head large, about once and one third as long as broad; canthus rostralis and supraciliary edge angular, projecting; nostril pierced below the canthus rostralis, nearer the orbit than the tip of the snout; tympanum vertically oval, nearly as large as the eye-opening. Upper head-scales small, very convex, keeled or pointed; large erect scales above the temple; occipital scarcely enlarged; ten to twelve upper and as many lower labials. Male with a gular pouch; gular scales smaller than ventrals, strongly keeled, trihedral, the central erect. Dorsal and nuchal crests continuous, formed of short conical scales; those of the former largest, measuring about half the vertical diameter of the tympanum, and frequently arranged in three longitudinal closely-set rows. Back and sides with very small scales of somewhat unequal size; a lateral series of larger scales, from the shoulder to the base of the tail; the scales on the flanks minute and granular, those on the back a little larger, pointed, and keeled; ventral scales strongly keeled. Limbs and tail as in *E. præstabilis*, the former, however, a little shorter. One or two femoral pores on each side. Pale brown, variegated with darker; a yellowish spot behind the ear, sometimes replaced by a longitudinal band of the same colour; tail with rather irregular dark brown annuli; lower surfaces yellowish, the male's throat bluish-black.

Total length	263	millim.
Head	33	„
Width of head	25	„
Body	80	„
Fore limb	59	„
Hind limb	88	„
Tail	150	„

Ecuador.

a, b. ♂ ♀.	W. Ecuador.	Mr. Fraser [C.].	(Types.)
c. Hgr. ♀.	Guayaquil.	Mr. Fraser [C.].	(Types.)
d–e. ♂ and hgr.	Sarayacu.	Mr. Buckley [C.].	(Types.)

6. Enyalioides palpebralis.

Enyalius palpebralis, *Bouleng. Proc. Zool. Soc.* 1883, p. 46, pl. x.

Head large, once and one third as long as broad; canthus rostralis angular; supraciliary edge much produced posteriorly, with a strong denticulation formed by several large pointed scales; nostril pierced below the canthus rostralis, nearer the orbit than the tip of the snout; tympanum vertically oval, nearly as large as the eye-opening. Upper head-scales small, irregular, strongly keeled or conical; three large conical tubercles behind the supraciliary border, followed by an oblique series of smaller conical tubercles; no distinct occipital; twelve to fifteen upper and as many lower labials. Male with a very slightly developed gular pouch; gular scales smaller than ventrals, keeled, the median subconical. A nuchal crest formed of four to six large conical scales, the highest measuring the vertical diameter of the tympanum, separated from the dorsal crest by an interspace nearly equal to its length; a strong dorsal crest, gradually diminishing in height to the base of the tail, where it becomes double and indistinct; the highest scale of the dorsal crest measuring three fourths the vertical diameter of the tympanum. Scales of the back and sides very small, irregular, keeled, intermixed with larger conical ones; the four or five series contiguous to the dorsal crest larger, rhomboidal, strongly keeled, the keels obliquely directed upwards. Limbs moderately long, covered with unequal strongly keeled scales; the adpressed hind limb reaches the tympanum or the temple; digits rather short, sharply denticulated. No femoral pores. Tail compressed, as long as or a little longer than head and body, covered with strongly keeled scales of unequal size forming regular segments. Yellowish or greyish above, with more or less distinct darker markings, viz. irregular oblique bands down the sides, or a network; tail with brown annuli; lower surfaces brownish white; the male's throat bright yellow.

Total length	235	millim.
Head	32	,,
Width of head	25	,,
Body	88	,,
Fore limb	59	,,
Hind limb	78	,,
Tail	115	,,

Eastern Peru.

a. ♂.	Cashiboya.	Messrs. Veitch [P.].	(Type.)
b–c. ♂ ♀.	Yurimaguas, Huallaga River.	Dr. Hahnel [C.].	

12. ENYALIUS.

Uraniscodon, part., *Kaup, Isis*, 1826, p. 89.
Ophryoessa, part., *Fitzing. N. Classif. Rept.* p. 17; *Wiegm. Herp. Mex.* p. 15.
Enyalius, *Wagler, Syst. Amph.* p. 150; *Dum. & Bibr.* iv. p. 230; *Fitzing. Syst. Rept.* p. 57; *Gray, Cat. Liz.* p. 195.
Læmanctus, part., *Wiegm. l. c.* pp. 16, 46; *Dum. & Bibr.* p. 70.
Dryophilus, *Fitzing. l. c.*
Urostrophus, part., *Fitzing. l. c.* p. 62; *Cope, Proc. Ac. Philad.* 1864, p. 176.
Ecphymotes, part., *Gray, Cat.* p. 184.

Tympanum distinct. Body feebly compressed; a slight dorso-nuchal crest; dorsal scales small. Head-scales small. A transverse gular fold; no gular pouch. Digits scarcely compressed, not denticulated laterally, with smooth or obtusely keeled lamellæ inferiorly. No femoral pores. Tail long, round or slightly compresssd. Lateral teeth tricuspid; pterygoid teeth*. No sternal fontanelle. Abdominal ribs.

South America.

Synopsis of the Species.

I. Lateral scales not distinctly keeled.

Nostril considerably nearer the tip of the snout than the orbit; tibia longer than the head; ventral scales strongly keeled . . 1. *catenatus*, p. 118.

* Absent in *E. fitzingeri*?

Nostril considerably nearer the tip of the snout than the orbit; tibia nearly as long as the head; ventral scales nearly smooth 2. *bibronii*, p. 119.
Nostril nearly equally distant from the orbit and the tip of the snout; tibia shorter than the head; ventral scales strongly keeled . . . . 3. *iheringii*, p. 120.
Nostril nearer the orbit than the tip of the snout; ventral scales keeled . . . . 4. *cærulescens*, p. 120.

II. Lateral scales sharply keeled, very much smaller than those of the middle of the back.

Sides with large black vertical spots, separated from the vertebral zone by a light longitudinal band; belly with white and brown longitudinal bands . . . . 5. *fitzingeri*, p. 121.
Brownish or olive, spotted with black, or with two wavy blackish lateral bands . . 6. *undulatus*, p. 121.

1. **Enyalius catenatus.**

Enyalius rhombifer, *Gray, Cat.* p. 195.
Agama catenata, *Wied, Reise n. Bras.* ii. p. 247, *and Abbild.*
Lophyrus rhombifer, *Spix, Spec. Nov. Lacert. Bras.* p. 9, pl. xi.
—— margaritaceus, *Spix, l. c.* p. 10, pl. xii. fig. 1.
—— albomaxillaris, *Spix, l. c.* p. 11, pl. xiii. fig. 2.
—— brasiliensis, *Less. & Garn. Voy. Coquille, Rept.* pl. i. fig. 3.
Uraniscodon rhombifer, *Kaup, Isis,* 1826, p. 90.
—— margaritaceus, *Kaup, l. c.*
—— catenata, *Kaup, l. c.* p. 91.
Ophryoessa catenata, *Fitzing. N. Class. Rept.* p. 48; *Peters, Mon. Berl. Ac.* 1877, p. 407.
—— margaritacea, *Fitzing. l. c.*
Enyalius rhombifer, *Wagl. Syst. Amph.* p. 150; *Guichen. in Casteln. Amér. Mér., Rept.* p. 19.
Ophyessa rhombifer, *Gray, Griff. A. K.* ix. *Syn.* p. 40.
—— brasiliensis, *Gray, l. c.*
Hypsibatus (Enyalius) catenatus, *Fitzing. Syst. Rept.* p. 57.
—— (Enyalius) margaritaceus, *Fitzing. l. c.* p. 58.
Enyalius bilineatus, part., *A. Dum. Cat. Méth. Rept.* p. 64.

Head moderately large, longer than broad; snout short, as long as or a little shorter than the diameter of the orbit; canthus rostralis angular; nostril nearer the tip of the snout than the orbit; eye large; tympanum large, vertically oval. Upper head-scales small, smooth or feebly keeled; occipital scarcely enlarged; supraocular scales either all very minute or with three or four longitudinal rows of larger ones; ten or eleven upper and as many lower labials. Gular scales small, granular, enlarged and keeled on the sides near the jaws, and in front of the gular fold. Nape and back with a series of oval keeled scales, forming a more or less marked denticulation. Dorsal scales very small, granular, smooth, smallest on the sides; ventral scales rather large, squarish, strongly keeled. Limbs long, covered with imbricate keeled scales; digits long and slender; tibia longer than the head; the adpressed hind limb reaches beyond the tip of the snout. Tail round or very feebly compressed, about twice as long as head and body, covered with keeled scales which are largest inferiorly. Coloration very variable. Females brown, with symmetrical darker, light-edged markings consisting generally of a series of rhombs or alternating triangular spots forming a zigzag band along the middle of the back; usually a dark brown band, widening gradually, from the eye to the shoulder; lower surfaces brownish-white, with scattered brown dots. Males sometimes brown with angular blackish markings, but more usually uniform purplish brown, with the head bluish.

	♂.	♀.	
Total length	297	293	millim.
Head	25	28	„
Width of head ..	20	21	„
Body	72	75	„
Fore limb	52	53	„
Hind limb	105	98	„
Tail	200	190	„

Brazil.

a. ♂.	Bahia.	
b. ♀.	Bahia.	Dr. O. Wucherer [C.].
c. ♂.	Rio Janeiro.	Mrs. Fry [P.].
d. ♂.	Brazil.	Lord Stuart [P.].
e. ♂.	Brazil.	Mrs. Mauger [P.].
f, g–h. ♂ ♀.	Brazil.	
i. Ad., stuffed.	—— ?	

2. Enyalius bibronii.

Enyalius rhombifer (*non Spix*), *Dum. & Bibr.* iv. p. 231.

Distinguished from the preceding in the following points:—All the scales a little larger, ventrals smooth or very obtusely keeled. Limbs shorter; tibia nearly as long as the head; the adpressed hind limb reaches the posterior corner of the eye. The unique specimen in the collection is pale brownish above, with an alternating series of squarish transverse brown spots along the middle of the back, each with a small round white spot in front and behind; a brown band from eye to shoulder; lower surfaces uniform whitish.

Total length (tail broken)	220	millim.
Head	28	„
Width of head	21	„
Body	72	„
Fore limb	46	„
Hind limb	82	„

Guianas, Brazil.

a. ♀.	Bahia.

3. Enyalius iheringii. (Plate VII.)

Enyalius iheringii, *Bouleng. Ann. & Mag. N. H.* (5) 1885, xv. p. 192.

Distinguished from *E. catenatus* in the following points:—Nostril nearly equally distant from the orbit and the tip of the snout. The scales on the vertebral region relatively larger, subrhomboidal, flat, distinctly keeled. Limbs shorter; tibia shorter than the head; the adpressed hind limb reaches the anterior border of the orbit in the male, the angle of the mouth in the female. Purplish-brown above, the male uniform, the female with an alternating series of black spots on the vertebral region, and on each side from nape to base of tail, a broad yellow band; lower surfaces brownish or purplish.

	♂.	♀.	
Total length	292	297	millim.
Head	25	26	,,
Width of head ..	18	20	,,
Body	67	71	,,
Fore limb	50	50	,,
Hind limb	80	76	,,
Tail	200	200	,,

Southern Brazil.

a–b. ♂ ♀.	Rio Grande do Sul.	Dr. H. v. Ihering [C.]. (Types.)

4. Enyalius cærulescens.

Enyalius cærulescens, *Cope, Journ. Ac. Philad.* (2) viii. 1876, p. 169.

"A dentellated nuchal crest, a vertebral band of linear scales; no caudal crest; tail subround in section. Head wide, the muzzle parabolic, its entire surface, including the supraorbital region, covered with equal, acutely tubercular scales. Three rows intervene between the superciliary rows; occipital scarcely distinct. Nostril pierced in a small round shield, nearer orbit than end of muzzle. Loreal region very short, with ten scales in a vertical line. Supraocular scales eighteen, counting to nostril. Labials 11–11; a short series of distinct, smooth infralabials. Auricular opening large. Dorsal scales very small, keeled, in transverse rows; abdominal scales larger, in cross series, interrupted on the middle line, keeled. The hind leg and wrist extend to front of orbit. Colour black, with numerous indistinct blue cross bands. Limbs, especially the fore-arm, and the sides of the neck, dark blue. Lower surfaces white, except throat and chin, which are blackish-blue."

Total length	185	millim.
Length to vent	72	,,
,, to ear..........	18	,,

Amazons, Peru?

5. Enyalius fitzingeri.

Ecphymotes fitzingeri, *Gray, Cat.* p. 184.
Enyalius bilineatus, *Gray, l. c.* p. 196.
Læmanctus fitzingeri, *Wiegm. Herp. Mex.* p. 46; *Dum. & Bibr.* iv. p. 74.
Enyalius bilineatus, *Dum. & Bibr.* p. 234; *Reinh. & Lütk. Vidensk. Meddel.* 1861, p. 229.
Læmanctus (Urostrophus) fitzingeri, *Fitzing. Syst. Rept.* p. 62.

Head moderately large; snout rounded, as long as the diameter of the orbit, with angular canthus rostralis; nostril nearer the end of the snout than the orbit; tympanum nearly as large as the eye-opening; upper head-scales small, smooth, irregular; occipital slightly enlarged, smaller than the tympanum; eight upper and seven or eight lower labials. Gular scales rather small, suboval, juxtaposed, strongly keeled. Scales on the back small, irregular, imbricate, strongly keeled, on the sides minute, juxtaposed, keeled; a series of enlarged keeled scales on the vertebral line, forming a very indistinct serration on the nape and the anterior part of the back; ventral scales large, imbricate, strongly keeled, forming longitudinal and transverse series. Limbs and digits long; tibia as long as the head; the adpressed hind limb reaches the eye, or between the latter and the nostril. Tail more than twice as long as head and body, covered with uniform keeled scales, largest inferiorly. Pale reddish-brown above, head and vertebral region dark brown; a vertebral series of large, roundish, black spots; a series of large, elongate, vertical black spots on each side, the anterior of which extends, as a band, to the eye; lower surfaces buff; two white bands along the belly, and a blackish line on each side; a series of rhomboidal black spots on the anterior part of the tail.

Total length	306 millim.
Head	23 „
Width of head	16 „
Body	63 „
Fore limb	40 „
Hind limb	75 „
Tail	220 „

Brazil.

a. ♀.	Brazil.	
b. ♀.	—— ?	St. G. Mivart, Esq. [P.].

6. Enyalius undulatus.

Ecphymotes undulatus, *Gray, Cat.* p. 185.
—— obtusirostris, *Gray, l. c.*
Læmanctus undulatus, *Wiegm. Herp. Mex.* p. 46; *Dum. & Bibr.* iv. p. 75; *Peters, Mon. Berl. Ac.* 1877, p. 410.
—— obtusirostris, *Wiegm. l. c.*; *Dum. & Bibr. l. c.*
—— (Urostrophus) undulatus, *Fitz. Syst. Rept.* p. 62.

Although never properly characterized, this species is introduced here on the authority of Peters.

Upper surface of body brownish or olive, spotted with black, or with two wavy blackish lateral bands, edged beneath with white; belly white or yellow ashy.

Brazil.

13. ANISOLEPIS.

Anisolepis, *Bouleng. Ann. & Mag. N. H.* (5) xvi. 1885, p. 85.

Tympanum distinct. Body cylindrical: no dorso-nuchal crest. Dorsal lepidosis heterogeneous, keeled; ventral scales large and keeled. Head-scales small. A strong transverse gular fold; no gular sac. Digits subcylindrical, with smooth lamellæ inferiorly. No femoral pores. Tail long, round. Lateral teeth tricuspid; no pterygoid teeth. No sternal fontanelle. Abdominal ribs.

Southern Brazil.

1. **Anisolepis iheringii.** (Plate IX. fig. 3.)

Anisolepis iheringii, *Bouleng. l. c.* p. 86.

Head small, body elongate. Nostril lateral, near the end of the snout; ear-opening moderately large, oval. Upper head-scales small and smooth, smallest on the supraocular region; occipital enlarged, suboval, about as large as the tympanum; upper labials eight, very low. Gular scales granular and keeled medially, larger and smooth anteriorly, gradually larger, rhomboidal, imbricate, and strongly keeled towards the gular fold, which is strong and straight. Median dorsal scales larger, irregular, imbricate, strongly keeled, the largest forming one or two indistinct longitudinal series on each side of the vertebral line; dorso-lateral scales very small, granular, keeled, unequal, intermixed with irregularly scattered enlarged scales. Ventral scales much larger than dorsals, equal, rhomboidal, imbricate, strongly keeled, the keels forming straight longitudinal series. The adpressed hind limb reaches the axilla or the shoulder. Tail more than twice as long as head and body, covered with equal keeled scales. Olive-brown above, with a series of triangular dark brown spots on each side of the vertebral line, forming a zigzag band; this is bordered externally with yellowish or reddish; the triangular spots may send forth narrow dark brown lines obliquely directed posteriorly down the sides; lower surfaces yellowish or coppery, the throat with a few blackish dots or longitudinal lines; tail above with a series of rhomboidal, dark, light-edged spots.

Total length	245	millim.
Head	15	„
Width of head	10	„
Body	54	„
Fore limb	27	„
Hind limb	41	„
Tail	176	„

Southern Brazil.

a–b. ♀.	S. Lorenzo, Rio Grande do Sul.	Dr. H. v. Ihering [C.]. (Types.)

14. UROSTROPHUS*.

Urostrophus, *Dum. & Bibr.* iv. p. 77.; *Gray, Cat. Liz.* p. 184.
Urostrophus, part., *Fitzing. Syst. Rept.* p. 62.

Tympanum distinct. Body subcylindrical or slightly compressed, covered with small smooth scales, granular on the back; no nuchal or dorsal crest. Head with feebly enlarged scales. A transverse gular fold; no gular pouch. Digits slightly compressed, with smooth lamellæ inferiorly. No femoral pores. Tail long, round. Lateral teeth tricuspid; pterygoid teeth. No sternal fontanelle. Abdominal ribs.

South America.

1. Urostrophus vautieri.

Urostrophus vautieri, *Gray, Cat.* p. 184.
Urostrophus vautieri, *Dum. & Bibr.* p. 78, pl. xxxvii. fig. 1.
Læmanctus (Urostrophus) vautieri, part., *Fitz. l. c.*
Leiosaurus vautieri, *Reinh. & Lütk. Vidensk. Meddel.* 1861, p. 223.

Head moderately large; snout rounded, with obtuse canthus rostralis; nostril nearer the end of the snout than the orbit; tympanum oval, as large as the eye-opening; upper head-scales smooth, largest on the snout, smaller on supraorbital region; scales forming regular supraorbital semicircles, separated by one or two rows of scales; occipital enlarged, a little smaller than the tympanum; a series of enlarged infraorbitals, second largest; eight or nine upper and seven or eight lower labials. Gular scales small and granular, strongly enlarged and polygonal near the labials. Scales on upper surface of body and limbs very small, granular, juxtaposed, of lower surfaces larger, flat, slightly imbricate, all smooth. Limbs and digits moderately elongate; the adpressed hind limb reaches the shoulder or the neck in males, not to the axilla in females. Tail about once and a half the length of head and body, curling at the end, covered with small, keeled scales. Brown above, marbled with darker, dirty white inferiorly; tail above with equidistant black spots.

	♂.	♀.	
Total length	158	227	millim.
Head	17	21	„
Width of head ..	12	14	„
Body	50	71	„
Fore limb	25	32	„
Hind limb	36	47	„
Tail	91	135	„

* ? *Leiosaurus valdivianus*, Philippi, Arch. f. Nat. 1861, p. 298.—Valdivia, Chili.

Brazil.

a, b. ♂ ♀.	Brazil.	
c–d. ♂ ♀.	Rio Grande do Sul.	Dr. v. Ihering [C.].

2. Urostrophus torquatus.

Leiosaurus torquatus, *Philippi, Arch. f. Nat.* 1861, p. 295.

Head larger and more convex than in the preceding species; tympanum much larger than the eye-opening; occipital slightly enlarged, not measuring half the vertical diameter of the tympanum. Body slightly compressed; dorsal scales minutely granular; ventrals much larger, flat, hexagonal, subimbricate. Limbs long, the adpressed hind limb reaching nearly the posterior border of the orbit. Tail not curly, measuring about once and a half the length of head and body, covered with rings of small smooth scales. Greyish olive, lighter inferiorly; gular fold black.

Total length	235	millim.
Head	25	,,
Width of head	19	,,
Body	65	,,
Fore limb	42	,,
Hind limb	66	,,
Tail	145	,,

Chili.

a. ♂.	Chili.

15. LIOSAURUS *.

Leiosaurus, part., *Dum. & Bibr.* iv. p. 241; *Fitzing. Syst. Rept.* p. 58; *Gray, Cat. Liz.* p. 224; *Bocourt, Miss. Sc. Mex., Rept.* p. 152.

Tympanum distinct. Body depressed, covered with small smooth scales, granular on the back; no dorsal crest. Upper head-scales small. A transverse gular fold; no gular pouch. Digits slightly depressed, with smooth lamellæ inferiorly. No femoral pores. Tail moderately long, round. Lateral teeth tricuspid; pterygoid teeth. No sternal fontanelle. Abdominal ribs.

South America.

1. Liosaurus bellii.

Leiosaurus bellii, *Gray, Cat.* p. 224.

Leiosaurus bellii, *Dum. & Bibr.* p. 242, pl. xxxi. fig. 1; *A. Dum. Arch. Mus.* viii. pl. xxii. fig. 2; *Bocourt, l. c.* p. 152, pl. xvii. bis. fig. 3.

---

* ? *Leiosaurus scapulatus*, Burmeister, Reise La Plata, ii. p. 522.—Sierra de Uspallata, 5000 ft.

Habit stout, *Phrynocephalus*-like. Head rather large, slightly longer than broad; snout short, rounded, with very short and obtuse canthus rostralis; nostril considerably nearer the end of the snout than the orbit; ear-opening vertically oval, a little smaller than the eye-opening. Upper head-scales small, convex, smooth, largest on the forehead and between the orbits, where they form four longitudinal series in the middle; occipital scarcely enlarged; supraocular scales very small, with a group of slightly enlarged ones; a series of infraorbitals, that immediately under the eye the largest; labials small, rostral and mental much larger. Dorsal scales very small, equal, granular, flat, smooth; ventrals larger, though also very small, roundish-subquadrangular, subimbricate, smooth. Limbs short; the adpressed hind limb reaches hardly the ear; digits slightly depressed, with long falciform claws. Tail tapering gradually, nearly as long as head and body, covered with small, juxtaposed, flat, smooth scales. Pale greyish above, ornamented with symmetrical blackish, white-edged spots; a black band on the middle of the snout, bifurcating between the orbits, the branches terminating on the occiput; a black band on each side of the head, passing through the eye and above the ear, and extending to above the axilla; another black band from the angle of the mouth; a series of posteriorly-directed tricuspid black spots on the middle of the back; three or four series of oval spots along the flanks; tail with blackish annuli; lower surfaces whitish, with interrupted longitudinal blackish lines.

Total length	133	millim.
Head	19	"
Width of head	17	"
Body	46	"
Fore limb	28	"
Hind limb	42	"
Tail	68	"

South America.

a. ♀	S. America.	T. Bell, Esq. [P.].

16. DIPLOLÆMUS.

Diplolæmus, *Bell, Zool. Beagle, Rept.* p. 19; *Gray, Cat. Liz.* p. 224.

Tympanum distinct. Body depressed, covered with small smooth scales; no dorsal crest. Upper head-scales small. A transverse gular fold; no gular pouch. Digits cylindrical, with smooth lamellæ inferiorly. No femoral pores. Tail rather short, round. Lateral teeth with obtuse or subspherical crowns; no teeth on the palate. No sternal fontanelle. Abdominal ribs.

Patagonia.

1. Diplolæmus darwinii.

Diplolæmus darwinii, *Gray, Cat.* p. 225.
—— bibronii, *Gray, l. c.*
Diplolæmus darwinii, *Bell, l. c.* p. 20, pl. x.
—— bibronii, *Bell, l. c.* p. 21, pl. xi.; *A. Dum. Cat. Méth. Rept.* p. 68, *and Arch. Mus.* viii. p. 534.

Habit stout, *Phrynocephalus*-like. Head rather large, convex, slightly longer than broad; snout short, rounded, with very short and obtuse canthus rostralis; nostril directed backwards and upwards, a little nearer the end of the snout than the orbit; ear-opening vertical, subcrescentic, a little smaller than the eye-opening. Upper head-scales small, more or less convex, smooth, largest on the forehead and on the narrow interorbital space, where they form only two series; occipital scarcely enlarged; supraocular scales very small, enlarged towards the supraorbital border; a series of infraorbitals, that immediately under the eye the largest; labials small; rostral and mental much larger, pentagonal. Dorsal scales very small, equal, granular, flat, smooth; ventral scales larger, rhomboidal, imbricate, smooth; gular scales smaller than ventrals, juxtaposed or subimbricate, except those in front of the gular fold, which resemble the ventrals. Limbs short; the adpressed hind limb reaches the axilla or the neck; digits short, round, not at all denticulated laterally, with long falciform claws. Tail round, tapering gradually, a little shorter or a little longer than the head and body, covered with small, juxtaposed, flat, smooth scales. Pale greyish olive above, tinged with rusty; back usually with regular festooned broad dark cross bands, composed principally of numerous close, small, dark brown spots; they are bordered posteriorly with yellowish; these bands sometimes less regular, broken up by a vertebral or a pair of lateral bands of the light ground-colour; limbs with transverse dark spots or bands; tail with regular dark cross bands; lower surfaces yellowish white, uniform or brown-dotted.

	♂.	♀.	
Total length......	187	201	millim.
Head	26	30	„
Width of head....	24	28	„
Body	64	76	„
Fore limb	36	38	„
Hind limb	56	55	„
Tail	97	95	„

Patagonia.

a–b. ♂ & hgr.	Port Desire.	C. Darwin, Esq. [C.].	(Types.)
c–g, h. ♂, ♀, hgr., & yg.	Port Desire.	C. Darwin, Esq. [C.].	(Types of *D. bibronii.*)
i, k. ♂ ♀.	Port Desire.	T. Bell, Esq. [P.].	
l–m. ♀.	——?		

17. PRISTIDACTYLUS.

Leiosaurus, part., *Dum. & Bibr.* iv. p. 241.
Pristidactylus, *Fitzing. Syst. Rept.* p. 58.
Ptenodactylus, *Gray, Cat. Liz.* p. 224.

Tympanum distinct. Body depressed, covered with small smooth scales, granular on the back; no dorsal crest. Two transverse gular folds; no gular pouch. Digits depressed, with bi- or tricarinate lamellæ inferiorly, the toes serrated on their inner border. No femoral pores. Tail long, rounded. Lateral teeth tricuspid; pterygoid teeth.

Patagonia.

1. Pristidactylus fasciatus.

Leiosaurus fasciatus, *Gray, Cat.* p. 224.
Leiosaurus fasciatus, *D'Orb. Amér. Mér., Rept.* p. 8, pl. iii. figs. 5–7; *Dum. & Bibr.* p. 244.
—— (Pristidactylus) fasciatus, *Fitzing. l. c.*

In habit similar to *Liosaurus bellii*, but the tail nearly twice as long as the body. Only two series of scales between the orbits. Greyish above, with broad, transverse, dark brown bands; a large deep black spot on each shoulder; lower surfaces white, chest and belly with traces of brown cross streaks.

From snout to vent 44 millim.

Rio Negro, N. Patagonia.

18. SCARTISCUS.

Scartiscus, *Cope, Proc. Ac. Philad.* 1862, p. 182.

Tympanum distinct. Body depressed; a median dorsal keel. Dorsal scales large and imbricate. Upper head-scales small. No antepectoral fold. Digits strongly pectinate inferiorly. No femoral or præanal pores. Tail slender, compressed, covered with appressed equal verticils of weakly keeled scales. No pterygoid teeth. Abdominal ribs.

Paraguay.

1. Scartiscus caducus.

Scartiscus caducus, *Cope, l. c.*

Head-scales numerous, small, keeled; nostrils lateral, below the canthus rostralis. A slight preaxillary fold. Median dorsal keel forming a low crest on the nape. Scales of the body large, flat, caducous, keeled everywhere; the dorsal in posteriorly convergent series. The adpressed hind limb reaches the anterior border of the orbit. Third and fourth toes nearly equal. Colour above brown, with numerous dark brown chevrons, the angles directed posteriorly; a narrow, vertical, yellow line from the axilla; extremities and tail cross-banded; beneath yellowish brown, the throat darker, light-varied.

From snout to vent	83	millim.
Tail	160	„
Fore limb	40	„
Hind limb	70	„

Paraguay.

19. CHALARODON.

Chalarodon, *Peters, Mon. Berl. Ac.* 1854, p. 616, *and Reise n. Mossamb.* iii. p. 33.

Tympanum distinct. Body depressed; a dorsal crest; dorsal scales small. Upper head-scales slightly enlarged; occipital large. A transverse gular fold. Digits denticulated laterally, with sharply-keeled lamellæ inferiorly. No femoral pores. Tail long, round. Lateral teeth tricuspid; pterygoid teeth. No sternal fontanelle. Abdominal ribs.

Madagascar.

1. Chalarodon madagascariensis.

Chalarodon madagascariensis, *Peters, ll. cc., Mossamb.* p. 34, pl. vi. fig. 3.

Head moderately large, short, rounded, once and one third as long as broad; canthus rostralis and supraciliary edge angular; nostril pierced above the canthus rostralis, nearer the end of the snout than the orbit; tympanum vertically oval, as large as the eye-opening. Upper head-scales uni- or tricarinate, smaller on the supraocular region; occipital large and round; a series of large infraorbitals; eight or nine upper labials. Gular fold strong. Scales on the body small, subimbricate, keeled, largest on the belly, smallest on the flanks; a series of enlarged scales, indication of a dorsal crest, along the nape and back. The adpressed hind limb reaches the nostril; digits long and slender, claws long, falciform. Tail roundish, not quite twice as long as head and body, covered with small, equal, keeled scales. Pinkish grey above, with darker angular transverse bands on the back, separated by round whitish spots; a large black spot on the occipital; limbs with dark cross bands; a white, black-edged band on the posterior side of the thigh; lower surfaces whitish.

The unique specimen in the collection having lost a portion of the tail, the following dimensions are taken from Peters's description:—

Total length	142	millim.
Head	14	„
Width of head	9·5	„
Body	36	„
Fore limb..	25	„
Hind limb	43	„
Tail	92	„

South-western coast of Madagascar.

a. Yg.	St. Augustin's Bay.	Prof. Peters [P.]. (One of the types.)

20. HOPLURUS*.

Oplurus, *Cuv. R. A.* 2nd ed. ii. p. 47; *Dum. & Bibr.* iv. p. 359; *Gray, Cat. Liz.* p. 221.
Tropidurus, part., *Wagler, Syst. Amph.* p. 146; *Wiegm. Herp. Mex.* p. 18.
Hoplurus, *Fitzing. Syst. Rept.* p. 76.
Doryphorus (*non Cuv.*), *Fitzing. l. c.* p. 77.
Tropidurus (*non Wied*), *Gray, l. c.* p. 225.
Centrura, part., *A. Dum. Arch. Mus.* viii. p. 556.

Tympanum distinct. Body depressed; no dorsal crest. Dorsal scales uniform, imbricate or juxtaposed. Upper head-scales slightly enlarged; occipital more or less enlarged. A strong transverse gular fold; no gular sac. Digits compressed, with keeled lamellæ inferiorly. No femoral or præanal pores. Tail moderately long, subcylindrical, with whorls of spinose scales. Lateral teeth tricuspid; pterygoid teeth. No sternal fontanelle. Abdominal ribs.

Madagascar.

Synopsis of the Species.

Dorsal scales larger than ventrals, imbricate, strongly keeled; the whorls of large spinose caudal scales separated from each other by a whorl of smaller scales . . . . 1. *sebæ*, p. 129.
Dorsal scales scarcely larger than ventrals, slightly imbricate, smooth or indistinctly keeled; caudal scales equal 2. *cyclurus*, p. 130.
Dorsal scales granular, smaller than ventrals; caudal scales equal . . . . 3. *quadrimaculatus*, [p. 131.

1. Hoplurus sebæ.

Oplurus braziliensis, *Gray, Cat.* p. 222.
Oplurus torquatus, (*non Cuv.*) *Guérin, Icon. R. A., Rept.* pl. xii. fig. 3.
—— sebæ, *Dum. & Bibr.* p. 361.
Hoplurus sebæ, *Fitzing. Syst. Rept.* p. 76; *Peters, Reise n. Mossamb.* iii. p. 31, pl. vi. fig. 1.
—— barnardi, part., *Peters, Mon. Berl. Ac.* 1854, p. 616.
—— torquatus, *Peters, in Decken's Reisen O. Afr.* iii. i. p. 14; *Boettger, Abh. Senck. Ges.* xii. 1881, p. 476.

Upper head-scales keeled; occipital large; anterior border of ear strongly denticulated. A small nuchal crest. Dorsal and nuchal scales rhomboidal, imbricate, strongly keeled, larger than laterals and ventrals; latter as well as gulars feebly keeled. Scales on the limbs strongly keeled, as large as or a little larger than the dorsals. Tail about once and a half as long as head and body, with whorls of large spinose keeled scales, separated from each other by

* *Oplurus saxicola*, Grandid. Rev. et Mag. Zool. (2) xxi. 1869, p. 340.
Oplurus fierinensis, id. ibid. p. 341.—Fierin, Madagascar.

a whorl of much smaller scales, which are smooth and shorter on the upper surface of the tail, longer and keeled on the lower surface; the large caudal scales at least four times as large as the largest tibials. Olive above, with one to seven uninterrupted transverse black bands on the nape and back, that between the fore limbs being constant; lower surfaces and tail yellowish, throat bluish grey or with bluish-grey reticulation.

Total length	260	millim.
Head	29	,,
Width of head	23	,,
Body	91	,,
Fore limb	56	,,
Hind limb	82	,,
Tail (reproduced)	140	,,

Madagascar.

a, b, c. Ad. & yg. —— ? Sir E. Belcher [P.].

2. Hoplurus cyclurus.

Tropidurus cyclurus, *Gray, Cat.* p. 225.
Cordylus brasiliensis, var. β, *Laur. Syn. Rept.* p. 52.
Uromastyx cyclurus, *Merr. Tent.* p. 56; *Wied, Nova Acta Ac. Leop.-Carol.* xiv. 1828, i. p. 129, pl. xv.
Oplurus torquatus, *Cuv. R. A.* 2nd ed. ii. p. 48.
Tropidurus cuvieri, *Gray, Griff. A. K.* ix. *Syn.* p. 41.
Oplurus maximiliani, *Dum. & Bibr.* iv. p. 365.
Doryphorus maximiliani, *Fitzing. Syst. Rept.* p. 77.
Hoplurus barnardi, part., *Peters, Mon. Berl. Ac.* 1854, p. 616.
—— cyclurus, *Peters, Reise n. Mossamb.* iii. p. 32, pl. vi. fig. 2.

Upper head-scales very feebly keeled; occipital moderate; anterior border of ear not denticulated. Nuchal crest merely indicated by a few tubercular scales. Dorsal scales very slightly larger than ventrals, roundish, subimbricate, smooth or indistinctly keeled; gular and ventral scales smooth. Scales on the limbs keeled, larger than dorsals. Tail little longer than head and body, with whorls of large spinose keeled scales which are larger than in *H. sebæ*, and have the posterior edge finely serrated. Olive above, with more or less distinct dark cross bands on the nape and back, that between the fore limbs black, and interrupted on the dorsal line; lower surfaces whitish, throat variegated with grey.

Total length	139	millim.
Head	18	,,
Width of head	13	,,
Body	46	,,
Fore limb	27	,,
Hind limb	41	,,
Tail	75	,,

Madagascar.

a. Hgr. —— ?

3. Hoplurus quadrimaculatus.

Oplurus quadrimaculatus, *A. Dum. Cat. Méth. Rept.* p. 83.
Centrura quadrimaculata, *A. Dum. Arch. Mus.* viii. p. 558, pl. xxii. fig. 4.
? Oplurus montanus, *Grandid. Rev. & Mag. Zool.* (2) xxi. 1869, p. 340.
Hoplurus quadrimaculatus, *Peters, Reise n. Mossamb.* iii. p. 31.

Upper head-scales small, smooth; occipital slightly enlarged; anterior border of ear denticulated. No nuchal crest. Dorsal scales very small, granular, flat, smooth, smaller than ventrals; latter small, subimbricate, smooth; gular scales granular. Body and base of tail much depressed. Scales on the limbs keeled and much larger than those on the body. Tail about once and two thirds as long as head and body, with whorls of rather small, spinose, keeled scales, the size of which equals hardly twice that of the tibials. Dark olive-brown, with small round lighter spots and transverse series of irregular roundish black spots, which are however sometimes reduced to one pair on each side behind the shoulder; a series of more or less confluent light spots on each side of the back; throat and breast grey, or marbled with grey; belly and lower surface of thighs whitish.

Total length	350	millim.
Head	30	,,
Width of head	24	,,
Body	100	,,
Fore limb	52	,,
Hind limb	98	,,
Tail	220	,,

Madagascar.

a. Ad.	Madagascar.	Col. Lyoll [C]. Zool. Soc.
b–e, f–g. Ad., hgr., & yg.	E. Betsileo.	Rev. D. Cowan [C.].

21. STENOCERCUS.

Stenocercus, *Dum. & Bibr.* iv. p. 350; *Fitzing. Syst. Rept.* p. 71; *Gray, Cat. Liz.* p. 219.
Trachycyclus, *Dum. & Bibr. l. c.* p. 355; *Fitzing. l. c.*; *Gray, l. c.*
Microphractus, *Günther, Proc. Zool. Soc.* 1859, p. 90.

Tympanum distinct. Body subcylindrical or depressed; dorsal crest absent, or represented by a slightly serrated ridge; dorsal scales more or less imbricate. Upper head-scales small, smooth. A curved fold in front of each shoulder, not extending across the throat. Digits compressed, with keeled lamellæ inferiorly. No femoral or anal pores. Tail subcylindrical or compressed, with whorls of strongly keeled or spinose scales. Lateral teeth tricuspid; pterygoid teeth present or absent. A large sternal fontanelle. No abdominal ribs.

Western South America.

Synopsis of the Species.

I. Posterior dorsal scales larger than ventrals, and smaller than the caudals, which are strongly spinose.

Fifth toe extending as far as second; posterior dorsal scales only a little larger than the anterior	1. *marmoratus*, p. 132.
Fifth toe extending as far as second; posterior dorsal scales very much larger than the anterior	2. *torquatus*, p. 133.
Fifth toe not extending as far as second; a dorsal denticulation	3. *roseiventris*, p. 133.

II. Posterior dorsal scales not larger than ventrals.

Dorsal scales considerably smaller than ventrals..........................	4. *humeralis*, p. 134.
Dorsal scales nearly as large as ventrals..	5. *varius*, p. 134.

III. Dorsal scales large, as large as or scarcely smaller than the caudals.

The adpressed hind limb reaches slightly beyond the tympanum; scales on the side of the neck smaller than those on the temple......................	6. *cupreus*, p. 135.
The hind limb reaches the eye; scales on the side of the neck larger than those on the temple	7. *mœstus*, p. 136.

1. Stenocercus marmoratus.

Trachycyclus marmoratus, *Gray, Cat.* p. 219.
Trachycyclus marmoratus, *Dum. & Bibr.* iv. p. 356, pl. xxxix. bis. fig. 1; *D'Orb. Voy. Am. Mér.* v. *Rept.* p. 8, pl. iv. figs. 4–8.
Heterotropis (Trachycyclus) marmoratus, *Fitzing. Syst. Rept.* p. 71.

No pterygoid teeth. Ear-opening with two tubercular scales in front. Sides of neck folded, studded with small tubercles. Body subcylindrical. No dorsal denticulation. Dorsal scales transversely rhomboidal, keeled, not mucronate, the keels obliquely converging towards the dorsal line; gular and ventral scales imbricate, smooth, the latter a little broader than long. Hind limb with large, strongly keeled, spinose scales; when stretched forwards, it reaches hardly the ear; fifth toe reaching as far as second. Tail a little longer than head and body, thick, conical, slightly depressed at the base, covered with whorls of large spinose scales. Upper surfaces fulvous; back marbled with brown; limbs and tail with transverse brown bands; lower surfaces white.

Total length	116 millim.
Head	6 „
Body	31 „

Fore limb	22 millim.
Hind limb	33 „
Tail	65 „

Pampa-Ruiz, Province Laguna, Bolivia.

2. Stenocercus torquatus. (Plate VIII. fig. 1.)

No pterygoid teeth. Anterior border of ear tubercular. Median supraocular scales slightly broader than long; occipital indistinct. Sides and lower surface of neck irregularly folded. Body subtriangular. No dorsal denticulation. Nuchal, lateral, and anterior dorsal scales very small, subgranular, obtusely keeled; on the posterior half of the back the scales become gradually much larger, rhomboidal, broader than long, keeled, not mucronate. Gular scales granular; ventrals rhomboidal, imbricate, smooth, much larger than the anterior dorsals, but smaller than the posterior. The adpressed hind limb reaches the antehumeral fold; fifth toe reaching as far as second. Tail scarcely longer than head and body, feebly compressed, with whorls of large spinose scales. Upper surfaces and throat dark olive; a black transverse band on the dorsal surface from one antehumeral fold to the other; pectoral region blue; belly pinkish white.

Total length	166 millim.
Head	21 „
Width of head	14 „
Body	59 „
Fore limb	33 „
Hind limb	47 „
Tail	86 „

Peru.

a. ♂. Peru.

3. Stenocercus roseiventris.

Stenocercus roseiventris, *Gray, Cat.* p. 219.
Stenocercus roseiventris, *Dum. & Bibr.* iv. p. 350; *D'Orb. Voy. Am. Mér.* v. *Rept.* p. 8, pl. iv. figs. 1-3.
Steironotus (Stenocercus) roseiventris, *Fitzing. Syst. Rept.* p. 71.

Pterygoid teeth. No denticulation on the anterior border of the ear. Interorbital space slightly grooved. Supraocular region with five or six longitudinal series of scales, the two median of which are slightly broader than long; occipital scale scarcely larger than the surrounding ones. Sides of neck longitudinally folded. A vertebral denticulation. Body subtriangular. Dorsal scales transversely rhomboidal, keeled, the keels obliquely converging towards the median line: lateral and ventral scales smooth. The adpressed hind limb reaches the ear; fifth toe not extending as far as second. Tail a little longer than head and body, thick, compressed, with sharp upper edge; it is covered with whorls of alternately larger and smaller keeled scales, the larger of which are spinose. Olive above; throat marbled grey and white; belly pink.

Total length	182	millim.
Head	24	,,
Body	62	,,
Fore limb	34	,,
Hind limb	57	,,
Tail	96	,,

Irupana Mountains, Province Yungas, Bolivia.

4. Stenocercus humeralis. (Plate VIII. fig. 2.)

Microphractus humeralis, *Günth. Proc. Zool. Soc.* 1859, p. 90.

No pterygoid teeth. Anterior border of ear denticulated. Median supraocular scales slightly broader than long; occipital indistinct. Sides and lower surface of neck irregularly folded. Body depressed. A vertebral denticulation. Dorsal scales very small, rhomboidal, imbricate, keeled, not mucronate, the keels obliquely converging towards the vertebral line; nuchal and lateral scales minute, granular, obtusely keeled; gular scales granular; ventral scales considerably larger than the largest dorsals, rhomboidal, imbricate, smooth. The adpressed hind limb reaches the ear or not quite so far; fifth toe reaching nearly as far as second. Tail a little longer than head and body, rounded, tapering, covered with rings of strongly keeled scales about the size of the ventrals. Upper parts bluish-green, marbled with dark brown; a black transverse band, interrupted on the vertebral line, extends from one antehumeral fold to the other; throat bluish-green, with round lighter spots; lower surfaces greenish yellow.

Total length	182	millim.
Head	19	,,
Width of head	14	,,
Body	61	,,
Fore limb	32	,,
Hind limb	49	,,
Tail	102	,,

Ecuador.

a–b. ♂ ♀.	Guayaquil.	Mr. Fraser [C.].	(Types.)
c. ♂.	Cuenca.	Mr. Fraser [C.].	

5. Stenocercus varius. (Plate VIII. fig. 3.)

No pterygoid teeth. Anterior border of ear slightly denticulated. Supraocular scales, median series much broader than long; occipital not enlarged. Sides and lower surface of neck strongly folded. Body depressed. No dorsal denticulation. Dorsal scales small, rhomboidal, imbricate, keeled, not mucronate, nearly as large as the ventrals; the keels forming slightly oblique lines. Gular scales smaller than ventrals, obtusely keeled, imbricate; ventral scales rhomboidal, imbricate, smooth. The adpressed hind limb reaches halfway between the ear and the eye; fifth toe extending nearly as far as second. Tail nearly twice as long as head and body, rounded, tapering; caudal scales strongly keeled, not mucronate, of moderate size, larger than those on the body, arranged in rings. Upper surfaces dark bluish-grey, marbled with black; tail with black cross bars; lower surfaces bluish-grey; belly a little lighter.

Total length	184	millim.
Head	17	,,
Width of head	12	,,
Body	47	,,
Fore limb	31	,,
Hind limb	46	,,
Tail	120	,,

—— ?

a. ♂.	—— ?	J. Brenchley, Esq. [P.].

6. Stenocercus cupreus. (Plate IX. fig. 1.)

Pterygoid teeth. Anterior border of ear denticulated. Median supraocular scales slightly broader than long; occipital indistinct; temporal scales obtusely keeled. Side of neck with two oblique folds, the antehumeral strongest and with a serrated edge on its lower half. Body round. A scarcely distinct dorsal denticulation. Dorsal scales large, strongly imbricate, sharply keeled, ending in a short spine; the keels forming continuous lines, which are slightly oblique on the posterior part of the back; lateral scales smaller, feebly keeled. Scales under the neck larger than gulars and ventrals, which are rhomboidal, imbricate, smooth, and much smaller than dorsals. The adpressed hind limb reaches a little beyond the ear; fifth toe not extending as far as second. Tail nearly twice as long as head and body, tapering, slightly compressed; caudal scales a little larger than dorsals, keeled, spinose, forming rings. Greyish-brown above with coppery gloss, and with small irregular dark brown spots; lower surfaces yellowish white.

Total length	110	millim.
Head	19	,,
Width of head	13	,,
Body	56	,,
Fore limb	30	,,
Hind limb	53	,,
Tail	135	,,

Peru.

a. ♂.	Huanuco.	H. J. Veitch, Esq. [P.].

7. Stenocercus mœstus. (Plate IX. fig. 2.)

No pterygoid teeth. Anterior border of ear denticulated. A series of broad supraoculars; occipital very small; temporal scales obtusely keeled. Scales on the sides of the neck strongly keeled and a little larger than those on the temples; a slight longitudinal fold behind the ear, and a small granular pit in front of the antehumoral fold; latter short, and with a denticulation formed by three or four large scales. Body subquadrangular. No dorsal denticulation. Dorsal scales large, rhomboidal, strongly imbricate, sharply keeled, not mucronate; the keels forming continuous straight lines; lateral scales a little smaller and less strongly keeled. Gular and ventral scales smaller than dorsals, rhomboidal, imbricate, smooth. The adpressed hind limb reaches the eye; toes very long, fifth not extending as far as second. Tail rounded, long, with rings of scales similar to, and not longer than, the dorsals. Dark olive above, spotted all over with whitish; supraciliary edge, and a streak from below the eye to above the ear, whitish; lower surfaces greenish white.

Total length	154	millim.
Head	15	„
Width of head	10	„
Body	42	„
Fore limb	26	„
Hind limb	45	„
Tail (reproduced, bifid)	97	„

Peru.

a. ♂. Lima. J. M. Cowper, Esq. [P.].

22. CTENOBLEPHARIS.

Ctenoblepharys, *Tschudi, Faun. Per., Rept.* p. 36.

Tympanum distinct. Body depressed; no dorsal crest. Dorsal scales smooth, flat, juxtaposed. Upper head-scales small. A curved fold in front of each shoulder, not extending across the throat. Digits subcylindrical, with smooth lamellæ inferiorly. No femoral pores. Male with anal pores. Tail short, conical. Lateral teeth tricuspid; pterygoid teeth present. No sternal fontanelle. No abdominal ribs.

Peru.

1. Ctenoblepharis adspersus.

Ctenoblepharys adspersa, *Tschudi, l. c.* pl. i. fig. 2.

Snout short, subtriangular; canthus rostralis indistinct, nostril superior; edge of eyelids fringed; ear-opening moderately large, without denticulation. Upper head-scales small, subequal, smooth. Dorsal scales transversely suboval, with small granules in the intervals between them; scales on the lower surfaces slightly larger, rhomboidal, imbricate, smooth, largest under the neck. The adpressed hind limb reaches the neck. Digits rather short. Tail a little shorter than head and body, with rings of smooth scales. Four anal pores. Pale greyish olive above, with darker marblings enclosing roundish cream-coloured spots; lower surfaces uniform whitish.

Total length..........	105	millim.
Head	15	„
Width of head........	13	„
Body	45	„
Fore limb............	27	„
Hind limb	38	„
Tail	45	„

Peru.

a. ♂. Arequiba, 7500 feet. W. Tegetmeier, Esq. [P.].

23. HELOCEPHALUS *.

Helocephalus, *Philippi, Reise Atacama*, p. 167.
Leiosaurus, part., *Burmeister, Reise La Plata*, ii. p. 523.

This genus is apparently closely related to the preceding, from which it differs in having the infradigital lamellæ bicarinate, and lacking the pterygoid teeth.

Atacama and North-western La Plata.

1. Helocephalus nigriceps.

Helocephalus nigriceps, *Philippi, l. c.* pl. vi. fig. 1.

Head short, covered with small smooth scales, the supraoculars being the smallest of all; ear-opening rather small, without denticulation. Dorsal scales hexagonal, juxtaposed, smooth, forming oblique lines. The adpressed hind limb does not quite reach the axil. Tail nearly as long as head and body, conical, with rings of small squarish smooth scales. Pale grey above, sides orange with vertical black bars; head and neck deep black; lower surfaces white, black-spotted.

Total length..........	147	millim.
Head	21	„
Body	56	„
Fore limb............	34	„
Hind limb	50	„
Tail	70	„

Pajonal, Atacama, 10,500 feet.

* *Leiosaurus multipunctatus*, Burmeister, *l. c.* p. 524.—Sierra de Uspallata, 8000 feet.
L. marmoratus, id. ibid.—Desert west of Catamarca.

24. LIOLÆMUS *.

Liolæmus, *Wiegm. Nova Acta Ac. Leop.-Carol.* xvii. i. 1835, p. 227, *and Herp. Mex.* p. 18; *Gravenh. Nova Acta Ac. Leop.-Carol.* xviii. 1838, p. 717; *Fitzing. Syst. Rept.* p. 73; *Gray, Cat. Liz.* p. 212; *Girard, U.S. Explor. Exped., Herp.* p. 334.
Proctotretus, part., *Dum. & Bibr.* iv. p. 266.
Ptychodeira, *Fitzing. l. c.* p. 74; *Girard, l. c.* p. 340.
Liodeira, *Fitzing. l. c.* p. 74; *Gray, l. c.* p. 210; *Girard, l. c.* p. 339.
Sauridis, *Tschudi, Faun. Per., Rept.* p. 33.
Proctotretus, *Gray, l. c.* p. 216.
Chrysosaurus, *Gay, Hist. Chile, Rept.* p. 24.
Rhytidodeira, *Girard, Proc. Ac. Philad.* 1857, p. 198, *and U.S. Explor. Exped., Herp.* p. 349.
Eulæmus, *Girard, ll. cc.* pp. 198, 357.
Ortholæmus, *Girard, ll. cc.* pp. 198, 368.

Tympanum distinct. Body depressed; no dorsal crest. Dorsal scales more or less imbricate, keeled. Upper head-scales small or rather large. No fold across the throat; no gular appendage. Digits subcylindrical, with keeled lamellæ inferiorly. No femoral pores. Male with anal pores. Tail conical or slightly depressed. Lateral teeth tricuspid; pterygoid teeth. Sternal fontanelle present or absent †. No abdominal ribs.

America south of the Equator.

Synopsis of the Species.

I. Sides of the neck without any folds, covered with scales similar in every respect to those of the upper surface.

28 or 30 scales round the middle of the body; nostril sublateral, well visible from above	1. *nitidus*, p. 140.
30 or 32 (or less) scales round the midddle of the body; nostril perfectly lateral	2. *chilensis*, p. 141.
40 scales round the middle of the body	3. *gravenhorstii*, p. 142.

II. Sides of neck more or less distinctly folded, covered with scales smaller than those of the upper surface.

A. Scales on the middle of the back larger than those on the sides of the belly; scales of the sides of the neck rhomboidal, imbricate, keeled.

---

* *Liolæmus elegans*, Tschudi, Faun. Per., Rept. p. 33.—Chancay, Peru.
L. (*Sauridis*) *modestus*, id. l. c. p. 34.—Miraflores, Peru.
Chrysosaurus morio, Gay, Hist. Chile, Rept. p. 47.—Valdivia.

† Absent in *L. wieqmanni* and *L. multimaculatus*.

1. Ventral scales smooth.

42 or 44 scales round the middle of the body; interparietal smaller than parietals; a long band-like scale on the supero-anterior border of the ear	4. *lemniscatus*, p. 143.
50 scales round the middle of the body; interparietal a little larger than parietals	6. *fuscus*, p. 144.
40 scales round the middle of the body; keels of the dorsal scales forming continuous straight lines..	7. *gracilis*, p. 145.
2. Ventral scales slightly keeled.	5. *stantoni*, p. 144.

B. Scales on the middle of the back not larger than those on the sides of the belly.

1. Ventral scales not larger than dorsals.

a. Lateral scales as large as or but little smaller than dorsals.

α. Dorsal scales elongate rhomboidal, ending in a small spine or mucro.

40 scales round the middle of the body; the distance from the base of the fifth toe to the extremity of the fourth toe exceeds the length of the head	8. *cyanogaster*, p. 145.
50 to 58 scales round the middle of the body	9. *bibronii*, p. 146.
40 to 44 scales round the middle of the body; the distance from the base of the fifth toe to the extremity of the fourth does not exceed the length of the head	11. *magellanicus*, p. 148.

β. Dorsal scales simply pointed.

48 to 54 scales round the middle of the body	10. *nigromaculatus*, p. 147.
More than 60 scales round the middle of the body; hinder side of thighs uniformly granulate	13. *kingii*, p. 149.
More than 60 scales round the middle of the body; hinder side of thighs with a patch of enlarged rhomboidal scales	14. *fitzingeri*, p. 150.

b. Lateral scales considerably smaller than dorsals.

α. A small azygos frontal scale.

Dorsal scales pointed posteriorly, strongly keeled; two or three pointed scales project in front of the arm . . 12. *lineomaculatus*, p. 149.
Dorsal scales pointed posteriorly, strongly keeled; no projecting scales in front of the arm . . . . 15. *pictus*, p. 151.
Dorsal scales truncate or rounded posteriorly, feebly keeled . . . . 16. *tenuis*, p. 152.

β. No azygos scale between the orbits.

A single series of scales between the labials and the infraorbital . . . . . . 19. *darwinii*, p. 155.
Two series of scales between the labials and the infraorbitals . . . . 20. *wiegmanni*, p. 156.

2. Dorsal scales smaller than ventrals.

a. Nostril latero-superior.

Dorsal and lateral scales small, roundish-hexagonal, imbricate . . . . 17. *multiformis*, p. 153.
Dorsal scales small, hexagonal, scarcely imbricate; laterals smaller still, juxtaposed . . . . 18. *signifer*, p. 154.

b. Nostril superior.

Interparietal at least as large as the tympanum . . . . 21. *occipitalis*, p. 156.
Interparietal much smaller than the tympanum . . . . 22. *multimaculatus*, p. 157.

1. **Liolæmus nitidus.**

Leiodera chilensis, *Gray, Cat.* p. 210.
Tropidurus chilensis, (*non Less.*) *Wiegm. Nov. Acta Ac. Leop.-Carol.* xvii. 1835, i. p. 233.
—— nitidus, *Wiegm. l. c.* p. 234, pl. xvii. fig. 2.
—— olivaceus, *Wiegm. l. c.* p. 268.
Proctotretus chilensis, var. A, *Dum. & Bibr.* iv. p. 269; *Guichen. in Gay, Hist. Chile, Rept.* p. 24; *Dum. Voy. Venus, Zool. Rept.* pl. i. fig. 1.
Liolæmus lineatus, *Gravenh. Nov. Acta Ac. Leop.-Carol.* xviii. 1838, ii. p. 723, pl. liv. figs. 1-7.
—— nitidus, *Gravenh. l. c.* p. 727, figs. 8–10; *Girard, U.S. Explor. Exped., Herp.* p. 334.
—— unicolor, *Gravenh. l. c.* p. 728.
—— marmoratus, *Gravenh. l. c.* p. 729, pl. liv. fig. 11.
Proctotretus chilensis, part., *Bell, Zool. Beagle, Rept.* p. 2, pl. ii. fig. 1.
Liolæmus olivaceus, *Fitzing. Syst. Rept.* p. 74.
Liodeira chilensis, part., *Girard, l. c.* p. 340.
? Proctotretus marmoratus, *Philippi, Reise Atacama*, p. 165.

In habit similar to a *Tropidosaura*. Nostril sublateral. Upper head-scales rather large: an azygos frontal; interparietal smaller than parietals, separated from the frontals by a pair of fronto-parietals; a series of three or four enlarged supraoculars; a single series of scales between the labials and the infraorbital; temporal scales keeled; two or three projecting scales on the anterior border of the ear, the uppermost large and band-like, as if formed of the fusion of two scales. Sides of neck not plicate, covered with large scales similar to those on the dorsal surface. Nuchal and dorsal scales large, rhomboidal, very strongly keeled, the keel ending in a spine; ventral scales a little smaller, rounded or obtusely pointed, smooth; twenty-eight or thirty scales round the middle of the body. The adpressed hind limb reaches the axil or the neck; hinder side of thighs uniformly granulate. Two or four anal pores in the male. Tail once and a half to twice as long as head and body; caudal scales smaller than dorsals. Female and young coppery brown above, with waved dark-brown, yellowish-edged cross bands on the back; adult male dark olive-brown above, the head dotted with yellowish, and the dorsal scales edged with yellowish, the two colours forming irregular variegations; lower surfaces whitish in the female, the throat marbled or obliquely striated with blackish; in the males they are greenish white, olive, or blackish, with irregular darker spots or lines; the throat, if not entirely black, marbled or striated with black.

Total length..........	243	millim.
Head	21	"
Width of head........	17	"
Body	72	"
Fore limb	32	"
Hind limb	54	"
Tail...	150	"

Chili.

a, b. Many specs.: ♂, ♀, hgr., & yg.	Valparaiso.	'Challenger' Exped.
c, d–e. ♂ & yg.	Valparaiso.	Dr. Cunningham [P.].
f–h. ♂.	Huasco.	
i–k. ♂.	N. Chili.	C. Darwin, Esq. [C.].
l. Hgr. ♂.	Chili.	

2. Liolæmus chilensis.

Leiodera gravenhorstii, part., *Gray, Cat.* p. 211.
Calotes chiliensis, *Lesson, Voy. Coquille, Rept.* p. 36, pl. i. fig. 2.
Tropidurus chilensis, *Wiegm. Nov. Acta Ac. Leop.-Carol.* xvii. 1835, i. p. 268.
Proctotretus chilensis, var. B, *Dum. & Bibr.* iv. p. 269; *Guichen. in Gay, Hist. Chile, Rept.* p. 24, pl. ii. fig. 1; *Dum. Voy. Venus, Zool. Rept.* pl. i. fig. 2.
—— chilensis, part., *Bell, Voy. Beagle, Rept.* p. 2.
Liolæmus chilensis, *Fitzing. Syst. Rept.* p. 75.
Liodeira chilensis, part., *Girard, U.S. Explor. Exped., Herp.* p. 340.
? Proctotretus modestus, *Philippi, Reise Atacama*, p. 166.

Closely allied to the preceding. Nostril quite lateral. Upper head-scales somewhat smaller, usually feebly striated; frontal frequently divided longitudinally; if undivided, hardly larger than the frontoparietals; interparietal and parietals subequal in size, or latter a little larger; temporal scales very feebly keeled; projecting scales in front of the ear-opening very small, scarcely distinct. The scales usually form 30 or 32 series round the middle of the body (24 and 26 in species *f–g*). Yellowish or greenish above, with metallic gloss and irregular dark brown marblings; a more or less distinct light band on each side, from the eye to the base of the tail; tail usually with three dark brown longitudinal lines, the median most defined; lower surfaces yellowish or greenish, with strong metallic gloss, constantly immaculate.

Total length..........	268 millim.
Head	20 "
Width of head........	16 "
Body	73 "
Fore limb	33 "
Hind limb	52 "
Tail	175 "

Chili.

a–b. ♂.	Chili.		(Types of *Leiodera gravenhorstii.*)
c–d. Hgr. ♂.	Chili.	C. Darwin, Esq. [C.].	(Types of *Leiodera gravenhorstii.*)
e. Hgr. ♂.	Chili.	Mr. Bridges [C.].	
f–g. ♂ ♀.	Chili.		
h, i. ♀ & yg.	Talcahuano.	Dr. Coppinger [C.].	

3. **Liolæmus gravenhorstii.** (PLATE X. fig. 1.)

Leiodera gravenhorstii, part., *Gray, Cat.* p. 211.

Nostril quite lateral. Upper head-scales moderate, three pairs between the orbits; interparietal smaller than parietals; a series of three or four enlarged supraoculars; a single series of scales between the labials and the infraorbital; temporal scales keeled; one or two small projecting scales on the anterior border of the ear. Sides of neck not plicate, covered with large scales similar to those on the dorsal surface. Nuchal and dorsal scales large, rhomboidal, very strongly keeled, the keel ending in a short spine; ventral scales smaller than dorsals, rounded, smooth; forty scales round the middle of the body. The adpressed hind limb reaches the axil in the female, between the shoulder and the ear in the male; hinder side of thighs uniformly granulate. Two anal pores in the male. Tail more than twice as long as head and body; caudal scales a little smaller than dorsals. Coppery brown above, with four light golden bands, the upper extending from above the temple to the base of the tail, the lower from the ear to the groin; in the female a few paired blackish spots between the light dorsal bands; lower surfaces pale golden.

Total length	173	millim.
Head	13	„
Width of head	9	„
Body	40	„
Fore limb	20	„
Hind limb	33	„
Tail	120	„

Chili.

a-b. ♂ ♀. Chili. (Types.)

4. Liolæmus lemniscatus.

Leiodera gracilis, part., *Gray, Cat.* p. 211.
Liolæmus lemniscatus, *Gravenh. Nova Acta Ac. Leop.-Carol.* xviii. 1838, ii. p. 731, pl. liv. fig. 12.
—— hieroglyphicus, *Gravenh. l. c.* p. 732.
Proctotretus mosaicus, *Hombr. & Jacq. in Dumont d'Urv. Voy. Pôle Sud, Rept.* pl. ii. fig. 1; *Guichen. in Gay, Hist. Chile, Rept.* p. 26; *A. Dum. Cat. Méth. Rept.* p. 72, *and Arch. Mus.* viii. p. 541.
—— femoratus, *Girard, Proc. Ac. Philad.* 1854, p. 227, *and in Gillis, U.S. Nav. Astron. Exped.* ii. p. 219, pl. xl. figs. 5–12.
Ptychodeira femorata, *Girard, U.S. Explor. Exped., Herp.* p. 344.
—— mosaica, *Girard, l. c.* p. 349.
Liolæmus (Ptychodeira) femoratus, *Steindachn. Novara, Rept.* p. 37.

Nostril lateral. Upper head-scales rather large, usually more or less distinctly striated; an azygos frontal, seldom divided longitudinally, separated from the interparietal by a pair of fronto-parietals; interparietal smaller than parietals; a series of three or four enlarged supraoculars; a single series of scales between the labials and the infraorbital; temporal scales keeled; one or two projecting scales on the anterior border of the ear, and a long narrow band-like scale on the supero-anterior border. Sides of neck with a feeble longitudinal fold, the scales imbricate, keeled, pointed, smaller than those on the upper surface. Dorsal scales large, rhomboidal, very strongly keeled, ending in a sharp point; ventral scales smaller, rounded, smooth; forty-two or forty-four scales round the middle of the body. The adpressed hind limb reaches hardly the axil in the female, the shoulder or a little beyond in the male; hinder side of thighs uniformly granulate. Two to four anal pores in the male. Tail about twice as long as head and body; caudal scales (except the basal ones) smaller than dorsals. Brownish above, with golden or coppery gloss; two light yellowish streaks on each side from the eye to the hind limb, the lower passing on the upper border of the tympanum; a series of transverse or angular blackish-brown spots on each side of the back, separated by the lighter vertebral zone, which is immaculate; seldom a black vertebral line; usually a series of black spots between the light lateral bands; a whitish streak or a series of whitish spots on the back of the thighs; lower surfaces yellowish white, immaculate or spotted with brown on the throat and breast.

Total length	136	millim.
Head	11	„
Width of head	8·5	„
Body	35	„
Fore limb	16	„
Hind limb	27	„
Tail	90	„

Chili; Mendoza.

a–b. ♂ & hgr.	Valparaiso.	C. Darwin, Esq. [P.].
c. Many spec., ♂ ♀.	Valparaiso.	'Challenger' Exped.
d. ♂.	Talcahuano.	Dr. Coppinger [C.].
e. ♂.	Chili.	C. Darwin, Esq. [P.].
f–g. ♀.	Chili.	Mr. Bridges [C.].
h–m, n. ♂ ♀.	Chili.	
o. ♀.	Mendoza.	

5. Liolæmus stantoni.

Proctotretus stantoni, *Girard, Proc. Ac. Philad.* 1854, p. 227, *and in Gillis, U.S. N. Astron. Exped.* ii. p. 221, pl. xl. figs. 13–20.
Ptychodeira stantoni, *Girard, U.S. Expl. Exped., Herp.* p. 347.

Appears to be very closely allied to, if distinct from, the preceding species. However, it is said to have the ventral scales slightly keeled. Ground-colour deep brown, with a reddish tint posteriorly; two parallel light streaks on each side; abdomen unicolor, with metallic reflections.

Santiago, Chili.

6. Liolæmus fuscus. (Plate X. fig. 2.)

Upper head-scales perfectly smooth; interparietal a little larger than parietals, in contact with the frontals; frontoparietals very small and widely separated; the scale on the upper anterior border of the ear only a little larger than the preceding temporals. Fifty scales round the middle of the body. Otherwise as in *L. lemniscatus*. Brown above, with two light lateral bands on each side, the interspace between them with closely-set blackish-brown spots; a black vertebral line and small black spots on the sides of the back; lower surfaces brownish, chin and sides of throat mottled with darker.

Total length	136	millim.
Head	13	„
Width of head	9	„
Body	35	„
Fore limb	17	„
Hind limb	29	„
Tail	88	„

Chili.

a. ♂.	Valparaiso.

7. Liolæmus gracilis.

Leiodera gracilis, part., *Gray, Cat.* p. 211.
Proctotretus gracilis, *Bell, Zool. Beagle, Rept.* p. 4, pl. i. fig. 2; *A. Dum. Arch. Mus.* viii. p. 542.
Ptychodeira gracilis, *Girard, U.S. Explor. Exped., Herp.* p. 341, pl. xvii. figs. 22–28.

Nostril lateral. Upper head-scales rather large, smooth; an azygos frontal, separated from the interparietal by a pair of fronto-parietals; interparietal smaller than parietals; a series of three or four enlarged supraoculars; a single series of scales between the labials and the infraorbital; temporal scales feebly keeled; two or three small projecting scales on the anterior border of the ear. Sides of neck with a very indistinct longitudinal fold, the scales imbricate, keeled, pointed, smaller than those on the upper surface. Dorsal scales large, rhomboidal, strongly keeled, pointed, the keels forming continuous longitudinal lines; ventral scales smaller, rounded, smooth; forty scales round the middle of the body. The adpressed hind limb reaches halfway between the shoulder and the ear; hinder side of thighs uniformly granular. Four anal pores. Tail about twice as long as head and body; caudal scales nearly as large as dorsals. Pale greyish olive above, uniform, or with small blackish dots on the sides; two light yellowish streaks on each side, the upper extending from the supraciliary edge to the base of the tail, the lower from the eye to the hind limb, passing on the upper border of the ear; a light line along the hind side of the thighs; lower surfaces whitish, throat mottled with brown.

Total length	158	millim.
Head	13	„
Width of head	9	„
Body	40	„
Fore limb	19	„
Hind limb............	31	„
Tail	105	„

Patagonia; Argentine Republic.

a. ♂.	Port Desire.	C. Darwin, Esq. [P.].	(Type.)
b. ♂.	Bahia Blanca.		
c. ♂.	Tucuman.		

8. Liolæmus cyanogaster.

Leiolæmus cyanogaster, *Gray, Cat.* p. 212.
Proctotretus cyanogaster, *Dum. & Bibr.* iv. p. 273; *Bell, Zool. Beagle, Rept.* p. 12, pl. v. fig. 2; *Guichen. in Gay, Hist. Chile, Rept.* p. 28, pl. ii. fig. 2; *Dum. Voy. Venus, Zool.* p. 292, pl. ii. fig. 2.
Liolæmus (Liodeira) cyanogaster, *Fitzing. Syst. Rept.* p. 74.
Proctotretus intermedius, *Dum. l. c.* p. 290, pl. ii. fig. 1.
Ptychodeira cyanogaster, *Girard, U.S. Explor. Exped., Herp.* p. 347.
—— intermedia, *Girard, l. c.* p. 348.

Nostril lateral. Upper head-scales rather small, striated; interorbital scales in one or two rows; interparietal small; parietals usually indistinct, if distinct larger than interparietal; a series of four or five enlarged supraoculars; a single series of scales between the labials and the infraorbitals; temporal scales feebly keeled; one or two small scales project very slightly on the anterior border of the ear. Sides of neck covered with very small scales, with a strong longitudinal fold bifurcating towards the ear; a curved antehumeral fold. Dorsal scales large, rhomboidal, strongly keeled, ending in a sharp point or mucro; ventral scales a little smaller, rounded, smooth; forty scales round the middle of the body. The adpressed hind limb reaches the axil in the female, the ear or slightly beyond in the male; hinder side of thighs uniformly granulate. Male with two anal pores. Tail once and a half to twice as long as head and body; caudal scales nearly as large as dorsals. Olive or coppery brown above, with a few small blackish spots, and a light laterodorsal band; lower surfaces with strong metallic gloss, the throat and breast greenish or blue-green.

Total length	142	millim.
Head	15	,,
Width of head	11	,,
Body	42	,,
Fore limb	25	,,
Hind limb	40	,,
Tail	85	,,

Chili.

a. ♀.	Valparaiso.	C. Darwin, Esq. [P.].
b–c. ♂ & hgr.	Valdivia.	C. Darwin, Esq. [P.].

9. Liolæmus bibronii.

Leiolæmus bellii, *Gray, Cat.* p. 212.
—— bibronii, *id. l. c.*
—— lineatus, *id. l. c.* p. 213.
Proctotretus bibronii, *Bell, Zool. Beagle, Rept.* p. 6, pl. iii. fig. 1.
Rhytidodeira bibroni, *Girard, Proc. Ac. Philad.* 1857, p. 198, *and U.S. Expl. Exped., Herp.* p. 352.

Nostril lateral. Upper head-scales rather small, smooth; a small azygos frontal, separated from the interparietal by a pair of frontoparietals; interparietal as large as, or a little smaller than, the parietals; a series of three enlarged supraoculars; a single series of scales between the labials and the infraorbital; temporal scales indistinctly keeled; ear-opening with a very indistinct denticulation in front. Sides of neck covered with very small scales, with an irregular longitudinal fold; a curved antehumeral fold. Dorsal scales moderate, rhomboidal, strongly keeled, ending in a sharp point; ventral scales nearly as large, but rounded and smooth; fifty to fifty-eight scales round the middle of the body. The adpressed hind limb reaches the axil in the female, in front of the shoulder in the male; hinder side of thighs uniformly granulate. Male with two or three anal pores. Tail about once and a half as long as head and body; caudal scales as large as dorsals. Greyish olive above, spotted with dark brown, and with or without a black vertebral line, or dark olive with small pale greenish spots or dots; lower surfaces greyish, mottled with blackish, or blackish.

Total length	175	millim.
Head	16	,,
Width of head	12	,,
Body	54	,,
Fore limb	25	,,
Hind limb	40	,,
Tail	105	,,

Chili; Patagonia.

a. Hgr.	Port Desire.	C. Darwin, Esq. [P.].	(Type.)
b–c. ♂.	Chili.	(Types of *L. lineatus.*)	
d. ♀.	Chili.	(Type of *L. bellii.*)	
e. ♀.	Chili.		

10. **Liolæmus nigromaculatus.**

Leiolæmus nigromaculatus, *Gray, Cat.* p. 213.
—— inconspicuus, *Gray, l. c.*
Tropidurus nigromaculatus, *Wiegm. Nova Acta Ac. Leop.-Carol.* xvii. 1835, i. p. 229.
? Tropidurus oxycephalus, *Wiegm. l. c.* p. 232.
Proctotretus nigromaculatus, *Dum. & Bibr.* iv. p. 281; *Bell, Zool. Beagle, Rept.* p. 10, pl. iv. fig. 2; *Guichen. in Gay, Hist. Chile, Rept.* p. 34; *Dum. Voy. Venus, Zool.* p. 293, pl. iii. fig. 1.
Liolæmus oxycephalus, *Gravenh. Nova Acta Ac. Leop.-Carol.* xviii. 1838, ii. p. 735, pl. liv. fig. 13.
—— conspersus, *Gravenh. l. c.* p. 737, fig. 14.
Ptychodeira nigromaculata, *Fitzing. Syst. Rept.* p. 73.
Rhytidodeira nigromaculata, *Girard, Proc. Ac. Philad.* 1857, p. 198, *and U.S. Explor. Exped., Herp.* p. 353.
—— oxycephala, *Girard, ll. cc.* pl. xxviii. figs. 26–33.
Proctotretus pallidus, *Philippi, Reise Atacama,* p. 166, pl. vi. fig. 3.
—— bisignatus, *Philippi, l. c.* fig. 2.

Nostril lateral. Upper head-scales rather large, smooth; a small azygos frontal, separated from the interparietal by a pair of fronto-parietals; interparietal smaller than, or nearly as large as, parietals; a series of three or four broad supraoculars; a single series of scales between the labials and the infraorbital; temporal scales smooth or very feebly keeled; three or four large, pointed scales projecting in front of the ear, largest and nearly covering the latter in the full-grown male. Sides of neck covered with very small scales, which are lanceolate and somewhat raised on the longitudinal fold; an indistinct curved antehumeral fold. Dorsal scales moderate, rhomboidal, strongly keeled, pointed; ventral scales smaller, rounded or obtusely pointed, smooth; forty-eight to fifty-four scales round the middle of the body. The adpressed hind limb reaches the shoulder or between the latter and the ear; hinder side of thighs uniformly granulate. Male with two to four anal pores. Tail once and a fourth to once and a half as long as head and body; caudal scales (except the basal ones) smaller than dorsals. Females and young pale brownish above, with darker and lighter spots sometimes forming undulated cross bands on the body, limbs, and tail; male variegated pale olive and blackish; a black spot or vertical bar in front of the shoulder. Lower surfaces whitish in the female, greenish in the male; throat usually spotted or marbled with blackish.

Total length	223	millim.
Head	23	,,
Width of head	19	,,
Body	70	,,
Fore limb	34	,,
Hind limb	53	,,
Tail	130	,,

Chili; Atacama.

a–c. ♂.	Coquimbo.	Dr. Cunningham [C.].
d–e. ♂ ♀.	Coquimbo.	Dr. Coppinger [C.].
f. Yg.	Coquimbo.	C. Darwin, Esq. [P.].
g. Hgr.	Chili.	J. E. Gray, Esq. [P.]. (Type of *L. inconspicuus.*)
h. Hgr. ♂.	Chili.	
i. Hgr. ♀.	——?	

11. Liolæmus magellanicus.

Proctotretus magellanicus, *Hombr. & Jacq. in Dumont d'Urv. Voy. Pôle Sud, Rept.* pl. ii. fig. 2; *A. Dum. Cat. Méth. Rept.* p. 75, *and Arch. Mus.* viii. p. 543.

Nostril lateral. Upper head-scales moderate, convex, smooth; a small azygos frontal, separated from the interparietal by a pair of fronto-parietals; interparietal as large as or a little smaller than the parietals; a series of three or four enlarged supraoculars; a single series of scales between the labials and the infraorbitals; temporal scales keeled; ear-opening small, with one or two small projecting scales in front. Sides of neck irregularly folded; a small curved antehumeral fold; the scales pointed, imbricate, smaller than those on the upper surface; two or three larger pointed scales project in front of the arm. Dorsal scales moderate, elongate rhomboidal, very strongly keeled, ending in a short spine; scales on the flanks as large as the dorsals; ventrals nearly as large, but rounded and smooth; forty to forty-four scales round the middle of the body. The adpressed hind limb seldom reaches the axil; digits very short; hinder side of thighs uniformly granulate. Male with three to five anal pores. Tail not or but slightly longer than head and body; caudal scales nearly as large as dorsals. Brownish or olive above, with four or five light yellowish or greenish streaks, that extending on each side of the back, from the eye to the base of the tail, widest and best defined; between the light bands, series of large squarish black spots; a black longitudinal streak on the upper surface of the snout, and another, curved one, bordering on each side the interorbital region, which is usually yellowish; lower surfaces greyish white, with metallic gloss, mottled with bluish black, sometimes nearly entirely black.

Total length	128 millim.
Head	13 ,,
Width of head	11 ,,
Body	49 ,,
Fore limb	21 ,,
Hind limb	32 ,,
Tail	66 ,,

Straits of Magellan.

a. ♂.	Cape Gregory.	C. Darwin, Esq. [P.].
b–e. ♂ ♀.	Philip Bay.	Dr. Cunningham [C.].
f–k. ♂, ♀, & hgr.	Sts. of Magellan.	Dr. Cunningham [C.].

12. Liolæmus lineomaculatus.

Leiolæmus kingii, part., *Gray, Cat.* p. 215.
Proctotretus kingii, part., *Bell, Zool. Beagle, Rept.* p. 13, pl. vi. fig. 2.

Very closely related to the preceding, which it resembles in proportions, size, and coloration, but to be distinguished by the following characters:—Temporal scales not distinctly keeled. Scales on the sides of the neck minute, granular. Scales of the body smaller, fifty-four to sixty round the middle; lateral scales smaller than dorsals and ventrals; dorsal scales shorter, simply pointed. The adpressed hind limb reaches nearly the axil in the female, the shoulder in the male. Pale olive or greyish above, with pale longitudinal streaks and blackish squarish spots arranged as in the preceding species; lower parts more or less abundantly spotted or marbled with blackish.

Total length	121 millim.
Head	13 ,,
Width of head	11 ,,
Body	43 ,,
Fore limb	20 ,,
Hind limb	31 ,,
Tail	65 ,,

Patagonia.

a–d. ♂ ♀.	Patagonia.	C. Darwin, Esq. [P.].

13. Liolæmus kingii.

Leiolæmus kingii, part., *Gray, Cat.* p. 215.
Proctotretus fitzingeri, part., *Dum. & Bibr.* iv. p. 286; *Guichen. in Gay, Hist. Chile, Rept.* p. 38.

Proctotretus kingii, part., *Bell, Zool. Beagle, Rept.* p. 13, pl. vi. fig. 1.
Rhytidodeira kingii, *Girard, Proc. Ac. Philad.* 1857, p. 198, *and U.S. Explor. Exped., Herp.* p. 350.

Nostril lateral. Upper head-scales moderate, smooth or rugulose; one or two small azygos frontals, separated from the interparietal by a pair of fronto-parietals; interparietal as large as or a little smaller than parietals; a series of four or five enlarged supraoculars; a single series of scales between the labials and the infraorbitals; temporal scales smooth; one or two very small lobules in front of the ear. Sides of neck with a longitudinal fold bifurcating towards the ear; a small curved antehumeral fold; scales small, granular. Dorsal scales small, rhomboidal, pointed, keeled; lateral scales as large or a little smaller; ventral scales nearly as large as dorsals, smooth, rounded; sixty-six to seventy-six scales round the middle of the body. The adpressed hind limb reaches the shoulder in the male, the axil in the female; hinder side of thighs uniformly granular. Male with six to eight anal pores. Tail about as long as head and body; caudal scales as large as dorsals. Brown above, with dark and light transverse spots sometimes forming cross bands on the back; these may be interrupted by a light vertebral line. Lower surfaces whitish, mottled or marbled with blackish, sometimes entirely bluish-black.

Total length	155	millim.
Head	20	,,
Width of head	17	,,
Body	60	,,
Fore limb	28	,,
Hind limb	43	,,
Tail	75	,,

Chili; Patagonia.

a, b. ♀ & yg.	Port Desire.	C. Darwin, Esq. [P.].	(Types.)
c, d. ♂.	Port Desire.	T. Bell, Esq. [P.].	
e. ♂.	—— ?	Haslar Collection.	

14. **Liolæmus fitzingeri.**

Leiolæmus fitzingerii, *Gray, Cat.* p. 215.
Proctotretus fitzingeri, part., *Dum. & Bibr.* iv. p. 286; *Guichen. in Gay, Hist. Chile, Rept.* p. 38.
—— fitzingeri, *Bell, Zool. Beagle, Rept.* p. 11, pl. v. fig. 1.
Ptychodeira fitzingeri, *Fitzing. Syst. Rept.* p. 73.
Eulæmus fitzingeri, *Girard, Proc. Ac. Philad.* 1857, p. 198, *and U.S. Explor. Exped., Herp.* p. 365.
—— affinis, *Girard, ll. cc.* pp. 198, 365.

Very closely allied to the preceding species, from which it differs in a few points only:—Dorsal scales less strongly keeled and very obtusely pointed (76 scales round the middle of the body). A patch of enlarged rhomboidal scales on the hinder side of the thigh. Dark olive above, with blackish and whitish spots; lower surface of neck, breast, and middle of belly black.

Total length	144	millim.
Head	18	,,
Width of head	15	,,
Body	66	,,
Fore limb	29	,,
Hind limb	47	,,
Tail (reproduced)	60	,,

Chili; Patagonia.

a. ♂. Patagonia. C. Darwin, Esq. [P.].

15. Liolæmus pictus.

Leiolæmus pictus, *Gray, Cat.* p. 213.
Proctotretus pictus, *Dum. & Bibr.* iv. p. 276; *Bell, Zool. Beagle, Rept.* p. 5, pl. ii.; *Hombr. & Jacq. in Dumont D'Urv. Voy. Pôle Sud, Rept.* pl. ii. bis. fig. 1; *Guichen. in Gay, Hist. Chile, Rept.* p. 30, pl. i. fig. 2.
Liolæmus (Liodeira) pictus, *Fitzing. Syst. Rept.* p. 74.
Eulæmus pictus, *Girard, Proc. Ac. Philad.* 1857, p. 198, *and U.S. Explor. Exped., Herp.* p. 364.
Proctotretus prasinus, *Cope, Proc. Ac. Philad.* 1868, p. 120.

Nostril lateral. Upper head-scales rather small, smooth; a small azygos frontal, separated from the interparietal by a pair of fronto-parietals; interparietal smaller than parietals; a series of more or less distinctly enlarged supraoculars; a single series of scales between the labials and the infraorbital; temporal scales smooth or obtusely keeled; ear-opening without, or with one or two very small projecting scales in front. Sides of neck granular, irregularly plicate; an antehumeral fold. Dorsal scales small, somewhat variable in size, rhomboidal, pointed, strongly keeled; lateral scales smaller, keeled; ventrals as large as dorsals, squarish, imbricate, smooth; fifty-two to seventy scales round the middle of the body. The adpressed hind limb reaches the ear in the male, the shoulder or a little beyond in the female; hinder side of thighs uniformly granulate. Male with two to four anal pores. Tail about once and two thirds as long as head and body; caudal scales a little larger than dorsals. Greenish or olive above, variously spotted or marbled with black, a latero-dorsal band of the lighter ground colour remaining immaculate: the black spots may be large and angular, and form three or four zigzag bands on the back; the upper parts sometimes dotted all over with pale green; lower surfaces whitish or greenish, chin and sides of throat reticulated or marbled with black.

Total length	184	millim.
Head	16	,,
Width of head	13	,,
Body	51	,,

Fore limb	25 millim.
Hind limb	42 „
Tail	117 „

Chili.

a–e. ♂, ♀, & hgr.	Chiloe.	Dr. Cunningham [C.].
f. ♀.	Valparaiso.	C. Darwin, Esq. [P.].
g, h. ♀ & yg.	Chili.	C. Darwin, Esq. [P.].
i–l. ♂ ♀.	Chili.	

Var. **major**.

Leiolæmus tenuis, part., *Gray, Cat.* p. 214.

Much larger than the typical form. Sixty-six to seventy scales round the middle of the body. Two or three projecting pointed scales in front of the ear in the adult. Black or bluish-black, with a few small irregular lighter spots on the upper surfaces.

Total length	244 millim.
Head	24 „
Width of head	19 „
Body	80 „
Fore limb	40 „
Hind limb	65 „
Tail (reproduced)	140 „

a–b, c. ♂ & hgr. Chili.

Proctotretus niger, Hallow. Proc. Ac. Philad. 1856, p. 233, is probably also a variety of *L. pictus*:—Very dark brown, or black above; a very narrow and somewhat indistinct yellow band, commencing at the axilla and extending along the flanks; a broader brownish one and less distinct above; underparts bluish mingled with copper colour, with metallic reflections; chin and throat conspicuously black-spotted, the larger spots in front; minute black dots distributed over abdomen and under surface of extremities. Total length 82 millim.

Quinquina Island.

16. Liolæmus tenuis.

Leiolæmus tenuis, *Gray, Cat.* p. 213.
—— maculatus, *Gray, l. c.*
Proctotretus tenuis, *Dum. & Bibr.* iv. p. 279; *Bell, Zool. Beagle, Rept.* p. 7, pl. iii. fig. 2; *Hombr. & Jacq. in Dum. d'Urv. Voy. Pôle Sud, Rept.* pl. ii. bis. fig. 2; *Guichen. in Gay, Hist. Chile, Rept.* p. 32, pl. i. fig. 1; *Girard, in Gillis, U.S. Nav. Astron. Exped.* ii. p. 217, pl. xl. figs. 1–4.
Liolæmus (Liodeira) tenuis, *Fitzing. Syst. Rept.* p. 74.
Eulæmus tenuis, *Girard, Proc. Ac. Philad.* 1857, p. 198, *and U.S. Explor. Exped., Herp.* p. 358, pl. xxviii. figs. 9–16.
Liolæmus (Ptychodeira) tenuis, *Steindachn. Novara, Rept.* p. 37.

Nostril sublateral. Upper head-scales rather small, smooth; a small azygos frontal, separated from the interparietal by a pair of frontoparietals; interparietal usually smaller than parietals; a series of three or four broad supraoculars; a single series of scales between the labials and the infraorbital; temporal scales feebly keeled; anterior border of ear not distinctly denticulated. Sides of neck granular, irregularly plicate; an antehumeral fold. Dorsal scales small, feebly imbricate, hexagonal, as broad as or a little broader than long, and shortly and feebly keeled; lateral scales smaller; ventrals as large as dorsals, squarish, smooth; sixty-six to seventy-two scales round the middle of the body. The adpressed hind limb reaches the ear in the male, the shoulder in the female; hinder side of thighs uniformly granular. Male with two or three anal pores. Tail once and two thirds to twice as long as head and body; caudal scales larger than dorsals. Olive or coppery brown above, with blackish spots sometimes forming angular cross bands on the back and tail, more or less profusely dotted with bluish-green or pale olive; lower surfaces pale golden, throat frequently bluish-green.

Total length	179	millim.
Head	15	,,
Width of head	13	,,
Body	44	,,
Fore limb	24	,,
Hind limb	38	,,
Tail	120	,,

Chili.

a–b. ♂.	Lota, Chili.	Dr. Cunningham [C.].
c–d. ♀ & hgr.	Valparaiso.	C. Darwin, Esq. [P.].
e. ♂.	Chili.	C. Darwin, Esq. [P.].
f. Several spec.: ♂, ♀, and hgr.	Chili.	Dr. J. E. Gray [P.].
g, h–k. ♂ ♀.	Chili.	
l. Yg.	Peru (?).	Dr. J. E. Gray [P.]. (Type of *L. maculatus*.)

17. Liolæmus multiformis.

Proctotretus multiformis, *Cope, Journ. Ac. Philad.* (2) viii. 1876, p. 173.

Nostril latero-superior. Upper head-scales small; two small azygos frontals; interparietal and parietals small, subequal; four or five transverse supraoculars; a single series of scales between the labials and the infraocular; anterior border of ear with two small projecting scales. Sides of neck granular, strongly folded. Dorsal scales very small, roundish-hexagonal, imbricate, weakly keeled, with smooth ones intermixed; latero-dorsals as small, smooth; ventrals much larger, rhomboidal, obtusely pointed; sixty to seventy scales round the middle of the body. The adpressed hind limb reaches the axil or the shoulder; hinder side of thighs uniformly granular. Male with five or six anal pores. Tail about once and a half as long as head and body; caudal scales larger than dorsals. Adult dark olive above, the scales edged with lighter; lower surfaces olive or blackish. Young olive-brown, with transverse blackish markings; lower surfaces whitish, throat marbled with grey.

According to Cope, this species occurs in three varieties:—1. Bright green, paler posteriorly, with two rows of large transverse black bars with irregular edges; head brownish above, pale-speckled; lower surfaces olive. 2. Olive-brown above, with faint darker cross bands, pale-bordered behind; sides and below blue to whitish. 3. Light brown with a row of black spots on each side, divided by a longitudinal pale band; below yellowish, marbled with bluish.

Total length	175	millim.
Head	22	,,
Width of head	20	,,
Body	73	,,
Fore limb	30	,,
Hind limb	51	,,
Tail (reproduced)	80	,,

Andes of Peru.

a–e. ♂ & yg., not good state. Guascona. J. B. Pentland, Esq. [P.].

18. Liolæmus signifer.

Leiolæmus signifer, *Gray, Cat.* p. 214.
Proctotretus signifer, *Dum. & Bibr.* iv. p. 288, pl. xxxix. fig. 2; *Bell, Zool. Beagle, Rept.* p. 8, pl. iv. fig. 1; *Guichen. in Gay, Hist. Chile, Rept.* p. 40.
Ptychodeira signifera, *Fitzing. Syst. Rept.* p. 73.
Eulæmus signifer, *Girard, Proc. Ac. Philad.* 1857, p. 198, *and U.S. Explor. Exped., Herp.* p. 366.

Nostril latero-superior. Upper head-scales small, smooth; no frontal; interparietal smaller than parietals; six transverse supra-oculars; a single (or double) series of scales between the labials and the infraocular; temporal scales smooth; anterior border of ear with two slightly projecting small scales. Sides of neck granular, strongly folded. Dorsal scales very small, hexagonal, scarcely imbricate, feebly and tubercularly keeled; lateral scales smaller still; ventrals much larger than dorsals, rounded, smooth; seventy-eight scales round the middle of the body. The adpressed hind limb reaches between the shoulder and the ear; hinder side of thighs uniformly granular. Male with four anal pores. Tail a little longer than head and body; caudal scales much larger than dorsals. Grey above, with small black undulate transverse markings arranged in four longitudinal series; lower surfaces whitish, speckled with grey.

Total length	107	millim.
Head	12	,,
Width of head	10	,,
Body	33	,,

Fore limb	19 millim.
Hind limb	29 „
Tail	62 „

Chili.

a. ♂. Chili.

19. **Liolæmus darwinii.**

Leiolæmus darwinii, *Gray, Cat.* p. 215.
Proctotretus darwinii, *Bell, Zool. Beagle, Rept.* p. 14, pl. vii.
Eulæmus darwini, *Girard, Proc. Ac. Philad.* 1857, p. 198, *and U.S. Explor. Exped., Herp.* p. 361, pl. xvii. figs. 8–14.
? Proctotretus melanopleurus, *Philippi, Reise Atacama,* p. 166.

Nostril latero-superior. Upper head-scales small, convex, smooth; interparietal smaller than parietals; a series of five to seven transverse supraoculars; a single series of scales between the labials and the infraorbital; temporal scales smooth; anterior border of ear not distinctly denticulated. Sides of neck granular, strongly folded; a curved antehumeral fold. Dorsal scales moderate, rhomboidal, obtusely pointed, strongly keeled, the keels forming continuous longitudinal lines; lateral scales smaller than dorsals; ventrals nearly as large as dorsals, obtusely pointed, smooth; fifty-two to sixty scales round the middle of the body. The adpressed hind limb reaches the axil in the female, between the shoulder and the ear in the male; hinder side of thighs with a patch of rhomboidal keeled scales. Male with seven anal pores. Tail nearly once and a half as long as head and body; caudal scales as large as dorsals. Pale greyish brown above, with a series of transverse blackish spots on each side of the vertebral line, each of which is followed by a more or less distinct whitish spot or border; a more or less distinct light streak on each side of the back, and, in the female, another, edged with black, extending from axil to groin; a black spot above the shoulder and another in front of the arm; a white, black-edged streak on the hinder side of the thighs. Lower surfaces white, uniform or spotted with blackish.

Total length	125 millim.
Head	12 „
Width of head	10 „
Body	38 „
Fore limb	19 „
Hind limb	31 „
Tail	75 „

Northern Patagonia; Mendoza.

a–b. ♂ ♀.	Bahia Blanca.	C. Darwin, Esq. [P.]. (Types.)
c–d. ♂ ♀.	High Pampas of S. Luis, Mendoza.	E. W. White, Esq. [C].
e. Hgr.	Mendoza.	

20. Liolæmus wiegmanni.

Leiolæmus wiegmannii, *Gray, Cat.* p. 216.
Proctotretus wiegmannii, *Dum. & Bibr.* iv. p. 284; *Bell, Zool. Beagle, Rept.* p. 15, pl. viii. figs. 1, 2; *Guichen. in Gay, Hist. Chile, Rept.* p. 36; *Dum. Voy. Venus, Zool.* p. 295, pl. iii. fig. 2.
Ptychodeira wiegmanni, *Fitzing. Syst. Rept.* p. 73.
Rhytidodeira wiegmanni, *Girard, Proc. Ac. Philad.* 1857, p. 198, *and U.S. Explor. Exped., Herp.* p. 352.
Ortholæmus fitzroii, *Girard, l. c., and U.S. Explor. Exp.* p. 373.

Nostril sublateral. Upper head-scales very small, striated or feebly keeled in the adult; no frontal; interparietal and parietals subequal; a series of four to six enlarged supraoculars; two series of scales between the labials and the infraorbital; temporal scales keeled; anterior border of ear not denticulated. Sides of neck granular, strongly plicate. Dorsal scales moderate, rhomboidal, strongly keeled, the keels forming continuous lines; lateral scales much smaller, imbricate, keeled; ventrals smaller than dorsals, roundish, smooth; forty-six to fifty-four scales round the middle of the body. The adpressed hind limb reaches the axil in the female, a little beyond the shoulder in the male; hinder side of thighs with a patch of enlarged rhomboidal keeled scales. Anal pores present in both sexes, five to eight in the male, three to eight in the female. Tail a little longer than head and body; caudal scales not larger than dorsals. Pale grey or greyish olive above, with two or four series of transverse, white-edged dark spots, separated by longitudinal light lines; lower surfaces white.

Total length	135	millim.
Head	15	,,
Width of head	12	,,
Body	44	,,
Fore limb	22	,,
Hind limb	34	,,
Tail	76	,,

Chili and North Patagonia to Uruguay.

a–d. ♂, hgr., & yg.	Maldonado.	C. Darwin, Esq. [P.].
e. ♂.	Chili.	
f–i. ♂ ♀.	Bahia Blanca.	C. Darwin, Esq. [P.].
k–l. ♂ & hgr.	Uruguay.	
m. Hgr.	Peru (?).	J. E. Gray, Esq. [P.].
n–o. ♂.	—— ?	

21. Liolæmus occipitalis. (Plate X. fig. 3.)

Liolæmus occipitalis, *Bouleng. Ann. & Mag. N. H.* (5) xv. 1885, p. 192.

Nostril superior. Upper head-scales very small, convex, smooth; no frontal; interparietal larger than the surrounding scales, as large as or larger than the tympanum; a series of three or four enlarged supraoculars; two series of scales between the labials and the infraorbital; temporal scales smooth; anterior border of ear not denticulated. Sides of neck strongly plicate, covered with minute granular scales. Dorsal scales small, subhexagonal, feebly imbricate, keeled; lateral scales smaller, roundish, juxtaposed, smooth; ventrals larger than dorsals, rhomboidal, smooth; sixty-six to seventy-two scales round the middle of the body. The adpressed hind limb reaches the tympanum, or not quite so far; hinder side of thighs uniformly granular, or with a patch of enlarged granules. Eight anal pores in the male. Tail a little depressed, slightly longer than head and body; caudal scales larger than dorsals. Pale greyish above; a series of darker spots on each side of the vertebral line; two darker bands on each side from axilla to groin, separated by a white streak; lower surfaces uniform white, or with small greyish spots on the throat.

Total length	112 millim.
Head	13 „
Width of head	9·5 „
Body	39 „
Fore limb	20 „
Hind limb	31 „
Tail	60 „

Southern Brazil.

a. Hgr.	Rio Grande do Sul.	Dr. H. v. Ihering [C.].	(Type.)
b–f. ♂, ♀, & hgr.	Rio Grande do Sul.	Dr. H. v. Ihering [C.].	

22. Liolæmus multimaculatus.

Proctotretus multimaculatus, *Gray, Cat.* p. 217.
Proctotretus multimaculatus, *Dum. & Bibr.* iv. p. 290; *Bell, Zool. Beagle, Rept.* p. 17, pl. ix. fig. 1; *Guichen. in Gay, Hist. Chile, Rept.* p. 42.
Liolæmus (Liodeira) multimaculatus, *Fitzing. Syst. Rept.* p. 74.
Ortholæmus beaglii, *Girard, Proc. Ac. Philad.* 1857, p. 198, *and U.S. Explor. Exped., Herp.* p. 369, pl. xvii. figs. 15–21.
—— multimaculatus, *Girard, ll. cc.*

Nostril superior. Upper head-scales very small, convex, smooth; no frontal; interparietal much smaller than the tympanum; a series of six or seven transversely enlarged supraoculars; two or three series of scales between the labials and the infraorbital; temporal scales smooth; anterior border of ear not denticulated. Sides of neck granular, strongly plicate. Dorsal scales very small, subhexagonal, scarcely imbricate, feebly keeled; lateral scales a little smaller, separated from one another by minute granules; ventrals much larger than dorsals, squarish, smooth; seventy-two or seventy-four scales round the middle of the body, of which twenty-six or twenty-eight are ventrals. The adpressed hind limb reaches a little beyond the shoulder; hinder side of thighs with a patch of enlarged rhomboidal scales; the digits, which are very slender, and provided with very long falciform claws, are very feebly keeled inferiorly, and the fourth toe is fringed laterally. Eight to ten anal pores. Tail depressed, as long as or slightly longer than head and body; caudal scales much larger than dorsals. Grey above, with numerous small black spots, some of which are bordered with white; lower surfaces white, uniform or spotted with blackish.

Total length	136	millim.
Head	16	,,
Width of head	13	,,
Body	52	,,
Fore limb	25	,,
Hind limb	39	,,
Tail	68	,,

Chili; Northern Patagonia.

a-b. ♂ & yg. Bahia Blanca. C. Darwin, Esq. [P.].

These specimens differ from Bibron's description in having two or three instead of four series of scales between the labials and the infraorbital; also in having a patch of large scales on the hind side of the thighs. Should they be regarded as a distinct species, they must bear the name *L. beaglii*, Girard.

25. SACCODEIRA.

Proctotretus, part., *Dum. & Bibr.* iv. p. 266; *Fitzing. Syst. Rept.* p. 74.
Ptygoderus, (*non Fitz.*) *Gray, Cat. Liz.* p. 216.
Saccodeira, *Girard, Proc. Ac. Philad.* 1857, p. 197, *and U.S. Explor. Exped., Herp.* p. 322.
Proctotretus, *Girard, l. c.* p. 329.
Tropidocephalus, *F. Müller, Verh. Nat. Ges. Basel,* vii. 1882, p. 160.

Tympanum distinct. Body slightly depressed. Dorsal scales imbricate, keeled, the vertebral series simulating a low crest. Upper head-scales small, keeled. No transverse gular fold. Digits subcylindrical or compressed, with keeled lamellæ inferiorly. No femoral or anal pores. Tail subcylindrical; caudal scales not verticillate. Lateral teeth tricuspid; pterygoid teeth. A small sternal fontanelle. No abdominal ribs.

South America.

Synopsis of the Species.

Upper head-scales multicarinate; ventral scales smooth . . . . 1. *ornatissima*, p. 159.
Upper head-scales unicarinate; ventral scales smooth . . . . 2. *pectinata*, p. 159.
Upper head-scales unicarinate; ventral scales keeled . . . . 3. *azurea*, p. 160.

1. Saccodeira ornatissima.

Saccodeira ornatissima, *Girard, ll. cc.* pp. 198, 323, pl. xxviii. figs. 1–8.

Upper head-scales multicarinate; front edge of ear with two projecting scales. Sides of neck pouched, covered with small scales and granules within the pouches; a vertico-oblique fold in front of the shoulder. Dorsal scales moderate, rhombic, conspicuously keeled, forming longitudinal series, the median exhibiting an inconspicuous crest or ridge, formed by somewhat more developed keels than those of the adjoining series; lateral scales smaller, with the keels directed obliquely upwards; ventral scales somewhat smaller than dorsals, subrhomboidal, smooth. The adpressed hind limb reaches the ear. Tail once and a half as long as head and body, slender and tapering, thick and depressed at the base, then slightly compressed and conical posteriorly. Greyish brown above, with a double series of dark brown subtriangular spots along the dorsal region; limbs maculated; upper surface of head blackish; sides and under surface of latter whitish, with small jet-black spots; abdomen uniform dull yellow.

Total length 135 millim.

Peru.

2. Saccodeira pectinata.

Ptygoderus pectinatus, *Gray, Cat.* p. 216.
Proctotretus pectinatus, *Dum. & Bibr.* iv. p. 292; *Bell, Zool. Beagle, Rept.* p. 18, pl. ix. fig. 2; *Guichen. in Gay, Hist. Chile, Rept.* p. 44; *Girard, U.S. Explor. Exped., Herp.* p. 333.
Liolæmus (Proctotretus) pectinatus, *Fitzing. Syst. Rept.* p. 74.
Proctotretus splendidus, *Girard, Proc. Ac. Philad.* 1857, p. 198, *and l. c.* p. 329, pl. xvii. figs. 1–7.

Habit rather stout. Snout short. Upper head-scales imbricate, strongly keeled; front edge of ear with two or three projecting scales. Sides of neck not distinctly plicate, covered with large, imbricate, pointed keeled scales; a very small oblique fold or pit in front of the shoulder. Dorsal scales large, rhomboidal, pointed, sharply keeled, the keels ending in short spines, and forming continuous longitudinal lines; those of the vertebral series and of the fifth or sixth series on each side of the latter stronger, and forming three slight pectinate crests. Gular and ventral scales smaller, rhomboidal, imbricate, smooth, emarginate posteriorly. The adpressed hind limb reaches the axil in the female, halfway between the shoulder and the ear in the male; digits rather short, fourth finger considerably shorter than third. Tail a little longer than head and body, conical, tapering, scaled like the body. Brownish above; back with three rows of regular light-edged, dark brown spots; some of the dorsal scales emerald-green; a transverse brown band between the eye; lateral crest yellowish white; flanks with one or two series of large dark-brown spots, and a more or less distinct whitish streak from axilla to groin; a white dark-edged streak along the back of the thigh; lower surfaces whitish.

Total length	170	millim.
Head	18	"
Width of head	15	"
Body	62	"
Fore limb	29	"
Hind limb	46	"
Tail	90	"

Chili; Patagonia; Argentine Republic.

a–c. ♂ ♀.	Patagonia.	C. Darwin, Esq. [P.].
d. ♂.	High Pampas of S. Luis, Mendoza.	E. W. White, Esq. [C.].
e–g. ♂ & yg.	13 miles off Cape S. Antonio, Buenos Ayres.	E. Gibson, Esq. [P.].

3. Saccodeira azurea.

Tropidocephalus azureus, *Müller, Verh. Nat. Ges. Basel,* vii. 1882, p. 160, pl. —, and 1885, p. 710.
Liolæmus azureus, *Bouleng. Ann. & Mag. N. H.* (5) xv. 1885, p. 192.

Very closely allied to the preceding species, with which it agrees entirely in coloration; differing in the following characters:—Gular and ventral scales keeled, ending in a short spine. The adpressed hind limb reaches the ear. Digits longer; third and fourth fingers nearly equal. Tail longer, about once and two thirds as long as head and body.

Total length	205	millim.
Head	17	"
Width of head	14	"
Body	58	"
Fore limb	31	"
Hind limb	53	"
Tail	130	"

Uruguay; Southern Brazil.

a. ♂.	Dept. of Soriano, Uruguay.	— Havers, Esq. [P.].
b. ♀.	Rio Grande do Sul.	Dr. H. v. Ihering [C.].

26. LIOCEPHALUS *.

Tropidurus, part., *Fitzing. N. Classif. Rept.* p. 17.
Leiocephalus, *Gray, Phil. Mag.* (2) ii. 1827, p. 207; *Fitzing. Syst. Rept.* p. 70.
Holotropis, *Dum. & Bibr.* iv. p. 259.
Steironotus, *Fitzing. l. c.*
Leiocephalus, part., *Gray, Cat. Liz.* p. 217.

---

* 1. *Leiocephalus microlepis*, Gray, Cat. p. 274.—Specimen missing.
2. *Scelotrema formosum*, Tschudi, Faun. Per., Herp. p. 27, pl. i. fig. 1.—Peru.
3. *Scelotrema crassicaudatum*, id. l. c.—Peru.

Pristinotus, *Gravenh. N. Acta Ac. Leop.-Carol.* xviii. 1838, ii. p. 738.
Scelotrema, *Tschudi, Faun. Per., Herp.* p. 27*.
Ophryoessoides, *A. Dum. Cat. Méth. Rept.* p. 66, *and Arch. Mus.* viii. p. 530.
Brachysaurus, *Hallow. Proc. Ac. Philad.* 1856, p. 232.

Tympanum distinct. Body more or less compressed; a more or less developed dorsal crest; dorsal scales uniform, imbricate, keeled. Upper head-scales small or enlarged, smooth or keeled; no large occipital. No transverse gular fold; no gular sac. Digits compressed, with keeled lamellæ inferiorly. No femoral or præanal pores. Tail long, more or less compressed and crested. Lateral teeth tricuspid; pterygoids usually toothed. A sternal fontanelle. Abdominal ribs.

West Indies; South America.

Synopsis of the Species.

I. Parietal plates distinct; a series of band-like transverse supraoculars.

A. Parietal plates longitudinally divided; nasal in contact with the rostral.

1. Ventral scales smooth.

a. Outer parietals not twice as large as inner.

α. Caudal crest not higher than dorsal.

Side of neck granular; adpressed hind limb not reaching the eye .... 1. *schreibersii*, p. 162.
Side of neck, for a short distance posterior to the ear-opening, granular; adpressed hind-limb quite or nearly reaching end of snout .... 2. *macropus*, p. 163.
Side of neck entirely covered with imbricate, pointed, keeled scales; adpressed hind limb not reaching beyond the eye 3. *vittatus*, p. 163.

β. Caudal crest higher than dorsal.

Keels of the dorsal scales unusually strong 4. *melanochlorus*, p. 164.
Upper head-scales striated; a black band on the side of the head, passing through the eye .... 5. *personatus*, p. 164.
Upper head-scales smooth, except the supraoculars; dorsal scales larger than ventrals .... 6. *eremitus*, p. 165.

* The small pouch between the tail and the base of the hind limb, noticed by Tschudi, is more or less distinct in specimens of *L. aculeatus*, *trachycephalus*, and *guentheri*; it is most developed in *L. ornatus*.

b. Outer parietals at least twice as large as inner .... 7. *carinatus*, p. 165.

2. Ventral scales keeled .... 8. *herminieri*, p. 166.

B. Parietal plates transversely divided.

Upper head-scales smooth or feebly striated; a dorso-lateral crest; ventral scales strongly keeled .... 9. *aculeatus*, p. 167.
Upper head-scales smooth; ventrals smooth or feebly keeled .... 10. *iridescens*, p. 167.
Upper head-scales unicarinate; ventrals keeled .... 11. *erythrogaster*, p. 168.

II. Head-scales small, parietals indistinct.

A. No supraciliary crest.

1. Upper head-scales not keeled; a series of transversely enlarged supraoculars .... 12. *formosus*, p. 168.

2. Upper head-scales keeled.

43 scales round the middle of the body; gular scales as large as dorsals ...... 13. *ornatus*, p. 168.
47 scales round the middle of the body; gular scales smaller than dorsals .... 14. *trachycephalus*, [p. 169.
61 to 75 scales round the middle of the body .... 15. *guentheri*, p. 169.

B. A supraciliary crest.

Posterior supraciliary scales compressed; temporal scales imbricate .... 16. *tricristatus*, p. 170.
Supraciliary crest ending in a large pyramid-shaped tubercle; temporal scales tubercular....................... 17. *dumerilii*, p. 170.

1. Liocephalus schreibersii.

Pristinotus schreibersii, *Gravenh. N. Acta Ac. Leop.-Carol.* xviii. 1838, ii. p. 739, pl. liv. figs. 15, 16.
Steironotus schreibersii, *Fitzing. Syst. Rept.* p. 70.
Liocephalus schreibersii, *Cope, Proc. Ac. Philad.* 1868, p. 123.

Upper head-scales large, striated; nasal in contact with the rostral; supraorbitals in contact; supraoculars seven or eight, band-like, separated from the supraorbitals by a series of small scales; parietals two pairs on one row, subequal, or outer pair a little larger. Sides of neck strongly plicate, covered with very small granular scales. Dorsal crest low. Dorsal scales small, not mucronate, the keels forming nearly straight longitudinal lines;

lateral scales very small; ventrals a little larger than dorsals, smooth, rounded, broader than long; seventy-one to seventy-nine scales round the middle of the body. The adpressed hind limb reaches the ear or not quite so far. Tail strongly compressed, crested like the back. Olive-brown above, with small lighter spots, with or without transverse dark brown spots on the back and tail; hinder side of thighs frequently with a white, dark-edged longitudinal band; lower surfaces pale olive, the male's throat darker; transverse curved series of small whitish dark-edged round spots from side to side across the belly.

	millim.		millim.
Total length	215	Fore limb	34
Head	20	Hind limb	68
Width of head	17	Tail	110
Body	85		

San Domingo.

a–c. ♂ & hgr. S. Domingo. — Parker, Esq. [P.]

2. Liocephalus macropus.

Liocephalus macropus, *Cope, Proc. Ac. Philad.* 1862, p. 184, and 1868, p. 122.

Much like *L. vittatus*, but differs in the greater breadth and smoothness of the head-shields, and in the granular nature of the scales for some distance posterior to the ear-opening. Hind limb quite or nearly reaching end of snout. Above olive, often with a metallic lustre; sometimes brownish; a transverse interscapular and crural pale-bordered spot, which is often indistinct; a blackish band extends from the orbit to the groin, mostly varied with white near the axilla, obsolete posteriorly; beneath yellowish or greenish, the gular region sometimes streaked with black.

From snout to vent 70 millim.; tail 110 millim.

Eastern Cuba.

3. Liocephalus vittatus.

Leiocephalus schreibersii, (*non Gravh.*) *Gray, Cat.* p. 218.
Holotropis vittatus, *Hallow. Proc. Ac. Philad.* 1856, p. 151.
Liocephalus vittatus, *Cope, Proc. Ac. Philad.* 1862, p. 184, and 1868, p. 122.
—— raviceps, *Cope, ll. cc.* pp. 183, 122.
Leiocephalus (Holotropis) vittatus, *Reinh. & Lütk. Vidensk. Meddel.* 1862, p. 271.

Upper head-scales large, striated; nasal in contact with the rostral; supraorbitals in contact; supraoculars six, band-like, separated from the supraorbitals by a series of small scales; parietals two pairs on one row, subequal, or outer pair a little larger. Sides of neck strongly plicate, covered with pointed, imbricate, keeled scales. Dorsal crest very low. Dorsal scales shortly mucronate, the keels obliquely turned towards the vertebral line; laterals a little smaller; ventrals nearly as large as dorsals, rounded, broader than long, smooth; forty-three to fifty-five scales round the middle of the body. The adpressed hind limb reaches the eye or between the latter and the ear. Tail feebly compressed, crested like the back. Olive-brown or greenish above, with a more or less distinct lighter latero-dorsal band, and a series of whitish spots, frequently confluent into a band, from the ear to the groin; back with or without angular dark brown spots; lower surfaces greenish or whitish, uniform or black-spotted.

	millim.		millim.
Total length	155	Body	45
Head	15	Fore limb	23
Width of head	12	Tail	95

Cuba, San Domingo.

a. ♂.	Cuba.	W. S. MacLeay, Esq. [P.].
b. ♀.	S. Domingo.	M. Sallé [C.].
c. ♂.	S. Domingo.	

4. Liocephalus melanochlorus.

Liocephalus melanochlorus, *Cope, Proc. Ac. Philad.* 1862, p. 184, and 1868, p. 122.

Apparently very closely related to *L. vittatus*. Scales of the back large, the keels unusually strong. Crest highest on the tail. General colour light green, principally appearing above in two dorsal bands and in the interspaces of a series of about ten transverse black bands extending between them; these become more or less confluent in old males, as do also sometimes the vertical bars on the sides, which are often present; posteriorly beneath spotted with greenish white; gular region coarsely reticulated with black.

From snout to vent 90 millim.; tail 180 millim.

San Domingo.

5. Liocephalus personatus.

Liocephalus personatus, *Cope, Proc. Ac. Philad.* 1862, p. 182, and 1868, p. 122.

? Liocephalus trigeminatus, *Cope, ll. cc.* pp. 183, 122.

Upper head-scales large, striated; nasal in contact with the rostral; supraorbitals in contact; supraoculars six or seven, band-like, separated from the supraorbitals by a series of small scales; parietals two pairs on one row, outer pair slightly larger than inner. Sides of neck with strong angular folds; scales imbricate, pointed, keeled. Dorsal crest strong. Dorsal scales shortly mucronate, the keels obliquely turned towards the vertebral line; laterals a little smaller; ventrals nearly as large as dorsals, rounded, broader than long, smooth; forty-one to fifty-three scales round the middle of the body. The adpressed hind limb reaches the eye, or between the latter and the ear. Tail strongly compressed, the crest more developed than the dorsal. Olive above; head yellowish brown, with a broad black band from the nostril through the eye to above the ear, sometimes continued to the fore limb; lips whitish; lower surfaces greenish.

	millim.		millim.
Total length	276	Fore limb	42
Head	21	Hind limb........	80
Width of head	16	Tail	168
Body	87		

Perhaps not specifically distinct from *L. vittatus.*
San Domingo.

a–c. ♂.	S. Domingo.	M. Sallé [C.].
d, e. ♂.	S. Domingo.	

6. Liocephalus eremitus.

Liocephalus eremitus, *Cope, Proc. Ac. Philad.* 1868, p. 122.

Head-scales large, smooth, except the supraoculars; latter weakly keeled, six, separated from the supraorbitals by one row of scales; parietals two pairs on one row, outer pair larger than inner. Folds of side of neck strong; two oblique, one nearly horizontal. Dorsal crest high on tail, elsewhere moderate. Dorsal scales large, larger than laterals and ventrals; latter smooth, præanals smaller, keeled. The adpressed hind limb reaches the front of the orbit. Tail moderately compressed. Dark olive-brown above, with a deep brown dorso-lateral longitudinal shade, connected by numerous indistinct cross bars, which are light-margined behind; a dark brown spot between eye and ear; lower surfaces brown, with numerous scattered whitish scales, which are most thickly gathered on the pectoral region.

	millim.
From snout to vent	63
Head, to ear	17
Width of head	15

Navassa Island, near San Domingo.

7. Liocephalus carinatus.

Leiocephalus carinatus, *Gray, Cat.* p. 217.
—— macleayii, *Gray, l. c.* p. 218.
Leiocephalus carinatus, *Gray, Phil. Mag.* (2) ii. 1827, p. 208; *Cope, Proc. Ac. Philad.* 1862, p. 184, and 1868, p. 122.
Holotropis microlophus, *Dum. & Bibr.* iv. p. 264; *Cocteau, in R. de la Sagra, Hist. Cuba, Rept.* p. 82, pl. vii.
Steironotus (Leiocephalus) carinatus, *Fitzing. Syst. Rept.* p. 70.

Upper head-scales large, smooth, or feebly striated; nasal in contact with the rostral; supraorbitals in contact; supraoculars five or six, band-like, separated from the supraorbitals by a series of small scales; parietals two pairs on one row, outer pair at least twice as large as inner. A strong <-shaped fold on the side of the neck, the scales of which are pointed, imbricate, and keeled. Dorsal crest, a feebly serrated ridge. Dorsal scales large, broader than long, not or but very slightly mucronate, feebly serrated posteriorly, the keels forming continuous slightly oblique lines; lateral scales smaller than dorsals; ventrals nearly as large as dorsals, smooth, rounded, broader than long; forty-three scales round the middle of the body. The adpressed hind limb reaches the eye. Tail strongly compressed, the crest more developed than on the body. Olive above, with small lighter spots, with or without blackish marblings on the flanks; throat marbled with dark olive.

	millim.		millim.
From snout to vent	105	Fore limb	47
Head	25	Hind limb	82
Width of head	20		

Cuba.

a. ♂.	W. Indies.	(Type.)
b, c. ♂ & yg.	Cuba.	W. S. MacLeay, Esq. [P.]. (Types of *L. macleayii.*)

8. Liocephalus herminieri.

Leiocephalus herminieri, *Gray, Cat.* p. 217.
Holotropis herminieri, *Dum. & Bibr.* iv. p. 261, pl. xliv.
Liocephalus herminieri, *Cope, Proc. Ac. Philad.* 1868, p. 122.

Upper head-scales large, more or less distinctly striated; nasal in contact with the rostral; supraorbitals separated by a series of smaller scales; supraoculars six or seven, band-like; seven or eight elongate shields across the occipital region. A strong oblique fold in front of the arm. Dorsal crest well developed, its height equalling the diameter of the eye-opening. Dorsal scales large, the keels forming continuous oblique series; lateral and ventral scales smaller, keeled. The adpressed hind limb reaches the angle of the mouth. Tail strongly compressed, crested like the back. Greenish brown above, with or without irregular yellowish cross-bands on the back; head yellowish, with four or five vertical black bars on each side; lower surfaces yellowish; throat with oblique transverse black bands.

	millim.		millim.
Total length	319	Fore limb	56
Head	37	Hind limb	104
Body	102	Tail (reproduced)	180

Trinidad; Martinique.

a. Adult, skeleton. Martinique.

9. **Liocephalus aculeatus.** (Plate XI.)

Leiocephalus aculeatus, *O'Shaughn. Ann. & Mag. N. H.* (5) iv. 1879, p. 303, *and Proc. Zool. Soc.* 1881, p. 243.

Upper head-scales large, smooth or feebly striated; nostril perfectly lateral; nasal usually separated from the rostral; supraorbitals in contact; supraoculars three or four large transverse, separated from the supraorbitals by a series of small scales; a smaller anterior and a larger posterior parietal on each side, separated by three shields, viz. a pair of interparietals and an occipital; supratemporal edge with two projecting scales. Sides of neck not folded, covered with sharply pointed, imbricate, keeled scales. Dorsal crest a low serrated ridge; a weaker lateral crest on each side of the back. Dorsal scales keeled only on their posterior half; lateral scales similar; ventrals a little larger, strongly keeled, ending in a long sharp point; thirty-three to forty-one scales round the middle of the body. The adpressed hind limb reaches the anterior border of the orbit. Tail very long, compressed, with serrated upper edge. Bronzed green or brown above, with more or less distinct angular darker cross lines on the back and vertical bars on the sides: lower surfaces green or bronzy, belly sometimes pinkish; the male's throat black.

	millim.		millim.
Total length	363	Fore limb	49
Head	24	Hind limb	85
Width of head	20	Tail	255
Body	84		

Ecuador; Northern Peru.

a–d. ♂, ♀, & hgr.	Moyobamba.	Mr. A. H. Roff [C.]. (Types.)
e–g. ♂ & hgr.	Pallatanga, Ecuador.	Mr. Buckley [C].
h. Hgr.	Canelos, Ecuador.	Mr. Buckley [C.].

10. **Liocephalus iridescens.**

Liocephalus iridescens, *Günth. Proc. Zool. Soc.* 1859, p. 409, pl. xx. fig. B; *Cope, Proc. Ac. Philad.* 1868, p. 121.

Upper head-scales large, smooth; nostril superior; nasal separated from the rostral; supraorbitals in contact; four or five large transverse supraoculars, usually separated from the supraorbitals by a series of small scales; a smaller anterior and a larger posterior parietal on each side; an interparietal and an occipital, separated by a pair of larger shields. Sides of neck not folded, covered with sharply pointed, imbricate, keeled scales. Dorsal crest low. Dorsal scales not very strongly keeled; lateral scales scarcely smaller; ventrals a little larger, pointed, smooth, or feebly keeled; thirty-seven to forty-three scales round the middle of the body. The adpressed hind limb reaches the eye or the anterior border of the orbit. Tail strongly compressed, crested like the back. Bronzy green above, darker on the sides, with a more or less distinct lighter dorso-lateral streak; back usually with angular black markings, pointing backwards; a black collar or two large black spots under the neck; throat and belly rose-coloured; chest yellow.

	millim.		millim.
Total length	210	Fore limb	35
Head	19	Hind limb	62
Width of head	14	Tail	145
Body	46		

Ecuador.

a–f. ♂, hgr., & yg. W. Ecuador. Mr. Fraser [C.]. (Types.)

11. Liocephalus erythrogaster.

Brachysaurus erythrogaster, *Hallow. Proc. Ac. Philad.* 1856, p. 232.

Upper head-scales large, unicarinate; nostril superior; four or five large transverse supraoculars. Neck slightly folded upon its sides. Dorsal crest low. Scales upon back, sides, and abdomen large, broader than long, strongly keeled. Tail cyclo-tetragonal at the base. Head, back, and extremities brownish; beneath of the same colour, except upon chin and throat and sides of abdomen, which are of a beautiful pink-red colour.

From snout to vent 60 millim.

Colombia.

12. Liocephalus formosus.

Liocephalus formosus, *Bouleng. Bull. Soc. Zool. France*, 1880, p. 43.

Upper head-scales small, convex, smooth; four transversely dilated supraoculars. Sides of neck strongly plicate. Dorsal crest a slight ridge. Dorsal scales rather large, the keels directed obliquely towards the vertebral line; ventrals as large as dorsals, smooth. Tail roundish. Dark olive above, with a few large black spots on the sides of the vertebral line and across the tail; throat black; chest, flanks, and limbs inferiorly metallic blue-green; belly greyish.

	millim.		millim.
Total length	166	Fore limb	32
Head	16	Hind limb	44
Width of head	12·5	Tail	101
Body	49		

Andes of Ecuador.

13. Liocephalus ornatus. (Plate XII.)

Liocephalus ornatus, *Gray, Cat.* p. 219.
Liocephalus ornatus, *Cope, Proc. Ac. Philad.* 1868, p. 121.

Upper head-scales small, keeled or striated; nasal separated from the rostral; no large supraoculars. Sides of neck not plicate, covered with pointed, imbricate, keeled scales. A well-developed dorsal crest. Dorsal scales large, strongly mucronate, strongly keeled, the keels forming continuous straight longitudinal lines; laterals and ventrals a little smaller; latter smooth; forty-three scales round the middle of the body. Gular scales as large as dorsals. The adpressed hind limb reaches the eye. Tail strongly compressed, the crest not quite as much developed as the dorsal. Olive above, with ill-defined darker spots; a large black spot on the shoulder; throat and breast greenish; chin and belly black.

	millim.		millim.
Total length	235	Fore limb	37
Head	19	Hind limb	59
Width of head	14	Tail	155
Body	61		

Ecuador.

a. ♂. Guayaquil. (Type.)

14. **Liocephalus trachycephalus.**

Holotropis trachycephalus, *A. Dum. Cat. Méth. Rept.* p. 70, *and Arch. Mus.* viii. p. 539, pl. xxiii. fig. 1.

Upper head-scales small, unicarinate; nasal separated from the rostral; no large supraoculars; no distinct denticulation on the anterior border of the ear. Sides of neck not distinctly plicate, covered with pointed, imbricate, keeled scales. Dorsal crest a low serrated ridge. Dorsal scales moderately large, scarcely mucronate, strongly keeled, the keels forming continuous straight longitudinal lines; laterals a little smaller; ventrals nearly as large as dorsals, rhomboidal, smooth; forty-seven scales round the middle of the body. Gular scales smaller than dorsals. The adpressed hind limb reaches the ear. Tail moderately compressed, with slightly serrated ridge. Bronze-olive above, metallic green inferiorly; belly and a band across the throat black.

	millim.		millim.
Total length	241	Fore limb	34
Head	17	Hind limb	54
Width of head	12	Tail	165
Body	59		

Colombia.

a. ♂. Bogotá.

15. **Liocephalus guentheri.** (PLATE XIII.)

Liocephalus ornatus, (*non Gray*) *Günth. Proc. Zool. Soc.* 1859, p. 408.
—— trachycephalus, (*non A. Dum.*), *O'Shaughn. Proc. Zool. Soc.* 1881, p. 244; *Bouleng. Ann. & Mag. N. H.* (5) ix. 1882, p. 458.

Upper head-scales small, unicarinate; nasal separated from the rostral; no large supraoculars; ear-opening with a more or less distinct denticulation in front. Sides of neck not distinctly plicate, covered with pointed, imbricate, keeled scales. Dorsal crest a low serrated ridge. Dorsal scales small, not or but scarcely mucronate, strongly keeled, the keels forming slightly oblique continuous lines; laterals and ventrals as large as dorsals; latter rhomboidal, smooth; sixty-one to seventy-five scales round the middle of the body. The adpressed hind limb reaches the shoulder or the ear. Tail moderately compressed, with feebly serrated ridge. Olive- or bronze-green, uniform or with lighter or darker spots; lower surfaces metallic, sometimes marbled with blackish: throat usually green; males usually with a black band across the throat, sometimes extending to the chest; sometimes a small black spot on the shoulder.

	millim.		millim.
Total length	250	Fore limb	34
Head	20	Hind limb	57
Width of head	16	Tail	165
Body	65		

Ecuador.

a–b, c. ♂ ♀.	Guayaquil.	Mr. Fraser [C.].
d, e, f. ♂ ♀.	W. Ecuador.	Mr. Fraser [C.].
g. ♀.	Sarayacu, Ecuador.	Mr. Buckley [C.].
h. ♂.	Colombia?	
i. Several specs.: ♂, ♀, & hgr.	——?	J. Brenchley, Esq. [P.].
k–l, m, n. ♂ ♀.	——?	

16. Liocephalus tricristatus.

Ophryoessoides tricristatus, *A. Dum. Cat. Méth. Rept.* p. 60, *and Arch. Mus.* viii. p. 531, pl. xxii. fig. 1.

A strong supraciliary crest; upper head-scales small, keeled; nostril lateral. Sides of neck not folded. Dorsal and ventral scales large, keeled. Dorsal crest a low serrated ridge; another similar ridge on each side of the back. Tail feebly compressed. Light brown above; back with transverse, brown, yellow-edged bands; similar vertical bands on the sides.

Head and body 60 millim.; tail 100 millim.

Brazil.

17. Liocephalus dumerilii.

Ophryoessoides dumerilii, *Steindachn. Novara, Rept.* p. 33, pl. ii. fig. 5.

Apparently allied to the preceding. The supraciliary crest ends in a large pyramid-shaped tubercle; temple with large tubercular shields. Light yellowish-brown above; a series of triangular dark reddish-brown spots on each side of the dorsal line.

Pará.

27. TROPIDURUS.

Tropidurus, *Wied, Abbild. Nat. Bras.*; *Fitzing. Syst. Rept.* p. 71; *Peters, Mon. Berl. Ac.* 1871, p. 644.
Tropidurus, part., *Fitzing. N. Classif. Rept.* p. 17; *Wagler, Syst. Amph.* p. 146; *Wiegm. Herp. Mex.* p. 18.
Ecphymotes (*non Fitz.*), *Cuv. R. A.* 2nd ed. ii. p. 47; *Dum. & Bibr.* iv. p. 342.
Platynotus, *Wagler, l. c.*
Microlophus, *Dum. & Bibr.* p. 334; *Gray, Cat. Liz.* p. 221; *Peters, l. c.*
? Tropidogaster, *Dum. & Bibr.* p. 329; *Gray, l. c.* p. 227.
Steirolepis, *Fitzing. l. c.* p. 72.
? Tritropis, *Fitzing. l. c.* p. 59.
Taraguira, *Gray, l. c.* p. 219.
Eulophus, *Tschudi, Faun. Per., Herp.* p. 25.
Craniopeltis, *Peters, l. c.* p. 645.
Læmopristus, *Peters, l. c.*
Aneuporus, *Bocourt, Miss. Sc. Mex., Rept.* p. 215.

Tympanum distinct. Body more or less compressed or depressed, without or with a more or less developed dorsal crest. Dorsal scales uniform. Upper head-scales enlarged, occipital very large. A strong curved or oblique fold on each side in front of the shoulder, not extending across the throat; frequently a transverse fold in front of the latter; no gular sac. Digits compressed, with keeled lamellæ inferiorly. No femoral or præanal pores. Tail long, subcylindrical or compressed. Lateral teeth tricuspid; pterygoid teeth. A sternal fontanelle. No abdominal ribs.

South America.

Synopsis of the Species.

I. Nostril above the canthus rostralis; ventral scales smooth.

A. A dorsal denticulation or crest, distinct at least on the nape.

1. Dorsal scales larger than ventrals.

a. A strong oblique antehumeral fold, not meeting its fellow on the breast; no transverse gular fold.

Dorsal scales more than twice as large as ventrals . . . . 1. *grayi*, p. 172.
Dorsal scales little larger than ventrals . . 2. *pacificus*, p. 173.

b. A transverse gular fold in front of the oblique antehumeral fold.

Scales on the snout keeled . . . . 3. *occipitalis*, p. 173.
Scales on the snout smooth . . . . 4. *bocourtii*, p. 173.

2. Dorsal scales not larger than ventrals.

Dorsal scales smaller than ventrals; sides of neck uniformly granular . . . . 5. *peruvianus*, p. 174.

Dorsal scales nearly as large as ventrals; sides of neck with groups of erect spinose scales .... 6. *spinulosus*, p. 175.

B. No dorso-nuchal crest.

1. Dorsal scales keeled.

Dorsal scales scarcely larger than ventrals; enlarged supraocular scales not occupying more than half the width of the supraocular region .... 7. *torquatus*, p. 176.

Dorsal scales scarcely larger than ventrals; supraocular region nearly entirely covered by a series of transverse band-like scales .... 8. *hygomi*, p. 177.

Dorsal scales at least twice as large as ventrals .... 9. *hispidus*, p. 177.

2. Dorsal scales smooth.......... 10. *semitæniatus*, p. 178.

II. Nostril lateral; ventral scales tricarinate .... 11. *blainvillii*, p. 178.

1. Tropidurus grayi.

Leiocephalus grayii, *Gray, Cat.* p. 218.
Leiocephalus grayii, *Bell, Zool. Beagle, Rept.* p. 24, pl. xiii. fig. 1.
Holotropis grayii, *A. Dum. Cat. Méth. Rept.* p. 70, *and Arch. Mus.* viii. p. 538.
Craniopeltis grayii, *Peters, Mon. Berl. Ac.* 1871, p. 645.
—— bivittata, *Peters, l. c.*
Tropidurus (Craniopeltis) grayii, *Steindachn. Festschr. zool.-bot. Ges. Wien,* 1876, p. 310, pl. ii. fig. 1.

Upper head-scales smooth; a series of five to seven transverse band-like supraoculars, covering the greater part of the supraocular region; occipital as broad as or broader than the latter, usually broader than long; ear-opening with a more or less developed fringe of pointed scales anteriorly. A strong curved antehumeral fold; sides of neck more or less strongly plicate, the folds forming a pouch between the ear and the antehumeral fold. A dorso-nuchal crest, well developed in the male, reduced to a mere serrated ridge in the female. Dorsal scales large, much larger than ventrals, very strongly keeled, ending in a mucro or spine; the keels forming continuous slightly oblique longitudinal series; lateral scales smaller, also strongly keeled; ventrals smooth; fifty-five to sixty-five scales round the middle of the body. The adpressed hind limb reaches the ear, or between the latter and the eye. Tail once and a half to once and two thirds as long as head and body, compressed, especially in the male, crested or serrated like the back. Olive above, spotted with black; sometimes a light yellowish band on each side of the back and another from axilla to groin; antehumeral fold black in the male; lower surfaces yellowish or greenish white, throat and chest frequently spotted with black; the male's throat usually black.

	millim.		millim.
Total length	250	Fore limb	44
Head	20	Hind limb	72
Width of head	16	Tail	160
Body	70		

Galapagos Islands.

a, b, c–g. ♂, ♀, & yg.	Chatham and Charles Islands.	C. Darwin, Esq. [P.]. (Types.)
h. ♂.	Albemarle Island.	Commander Cookson [P.].
i. Hgr. ♂.	Abingdon Island.	Commander Cookson [P.].

2. Tropidurus pacificus.

Tropidurus (Craniopeltis) pacificus, *Steindachn. Festschr. zool.-bot. Ges. Wien,* 1876, p. 313, pl. ii. figs. 2, 3.

Distinguished from *T. grayi* by the much smaller and less strongly keeled dorsal scales; they little exceed the ventrals in size and are never mucronate; ninety-one to ninety-five scales round the middle of the body. Dark olive above, with small black and greenish spots; antehumeral fold black; lower surfaces greenish or bluish-grey, throat and chest spotted with black.

Size and proportions as in the preceding.

Galapagos Islands.

a, b. ♂.	Abingdon Island.	Commander Cookson [P.].

3. Tropidurus occipitalis.

Læmopristus occipitalis, *Peters, Mon. Berl. Ac.* 1871, p. 645.

Scales on the snout keeled; seven or eight broad supraoculars; occipital very large, twice as broad as long. A strong transverse gular fold, with denticulated edge, in front of the antehumeral fold. A dorsal crest of very pointed scales, extending on the tail. Dorsal scales keeled, larger than the smooth ventrals; lateral scales smallest, keeled. Grey-brown above, with four irregular rhomboidal transverse spots, the first on the middle of the nape, the second between the shoulders, the two others on the anterior part of the back; on each side of the back a yellowish-green metallic longitudinal band; antehumeral fold black; lower surfaces yellowish.

Peru.

4. Tropidurus bocourtii.

Aneuporus occipitalis, *Bocourt, Miss. Sc. Mex., Rept.* p. 215, pl. xviii. fig. 1.

Craniopeltis occipitalis, *Cope, Journ. Ac. Philad.* (2) viii. 1876, p. 173.

Upper head-scales smooth; a series of four or five transverse supraoculars not measuring more than half the width of the supraocular region; occipital much broader than the latter, nearly twice as broad as long. A short curved fold above each shoulder, not descending on the chest; an oblique fold or pouch on each side of the neck connected with its fellow by a transverse gular fold. A vertebral denticulation. Dorsal scales moderate, rhomboidal, strongly keeled, the keels forming slightly oblique lines; lateral and ventral scales a little smaller, latter smooth. Limbs rather short. Tail a little longer than head and body, roundish. Olive above, variegated with dark yellow; lower surfaces yellowish white.

From snout to vent 60 millim.; tail 81 millim.

Peru.

5. Tropidurus peruvianus.

Microlophus peruvianus, *Gray, Cat.* p. 221.
Lophyrus araucanus, *Lesson, Voy. Coquille, Rept.* pl. ii. fig. 1.
Stellio peruvianus, *Less. l. c.* fig. 2.
Tropidurus microlophus, *Wiegm. Nova Acta Ac. Leop.-Carol.* xvii. 1835, i. p. 223, pl. xvi.
—— heterolepis, *Wiegm. l. c.* p. 225, pl. xvii. fig. 1.
Microlophus lessonii, *Dum. & Bibr.* iv. p. 336; *Guichen. in Gay, Hist. Chile, Rept.* p. 48.
Steirolepis microlophus, *Fitzing. Syst. Rept.* p. 72.
—— heterolepis, *Fitzing. l. c.*
—— peruviana, *Fitzing. l. c.* p. 73.
—— xanthostigma, *Tschudi, Faun. Per., Herp.* p. 29.
—— tigris, *Tschudi, l. c.* p. 30.
—— thoracica, *Tschudi, l. c.* p. 31.
—— quadrivittata, *Tschudi, l. c.* p. 32.
Microlophus peruvianus, *Girard, U.S. Explor. Exped., Herp.* p. 314; *Cope, Journ. Ac. Philad.* (2) viii. 1876, p. 173.
—— heterolepis, *Cope, l. c.* p. 171.
—— inguinalis, *Cope, l. c.* p. 172.

Upper head-scales smooth; a series of four to six more or less transversely enlarged supraoculars; occipital usually as long as broad or a little broader than long, nearly as broad as the supraocular region; ear-opening with a fringe of short pointed scales; temples granulate. A strong curved antehumeral fold, nearly meeting its fellow on the throat; latter with more or less distinct cross folds; sides of neck minutely granulate, strongly plicate. Body more or less depressed; one or two slight folds along the flanks. A dorso-nuchal crest of slightly enlarged keeled or compressed scales, most developed on the nape, sometimes absent on the back. Dorsal scales very small, juxtaposed, smooth or feebly keeled, smaller still on the flanks; ventrals larger, imbricate, smooth. The adpressed hind limb reaches the eye or a little beyond. Tail once and a half to twice as long as head and body, round or feebly compressed, with a small serrated dorsal ridge; caudal scales much larger than dorsals, strongly keeled, spinose. Colour extremely variable.

	millim.		millim.
Total length	370	Fore limb	65
Head	29	Hind limb........	112
Width of head	26	Tail	230
Body	111		

Peru ; Chili.

A. Upper surfaces blackish-olive, more or less abundantly dotted with pale olive, with or without a black lateral band; lower surfaces bluish or greenish grey; throat and chest black or blackish.

a–b. ♂.	On rocks at Pescadores Bay.	Mr. H. Whitely [C.].
c–e. ♂ & hgr.	On rocks near Yslay.	Mr. H. Whitely [C.].
f–g. Yg.	Peru.	C. Darwin, Esq. [P.].

B. Dark olive above, hind limbs with whitish spots; a white band, as long as the femur, on each side in front of the hind limb; lower surfaces white, gular region with black angular bands pointing backwards.

h. Yg.	Peru.	Smithsonian Institution [P.].

C. Yellowish grey or olive above, with yellowish or pale olive dots and black cross lines or reticulations; lower surfaces dirty white, belly sometimes black; throat with black angular bands pointing backwards; chest black or with black cross bands.

i, k, l. ♂ & hgr.	Peru.	
m. ♂.	Chili.	
n–p. ♂ & hgr.	——?	Sir A. Smith [P.].

D. Pale olive-grey, vertebral line lighter; back with transverse blackish spots or cross bands, sometimes reticulated; limbs with blackish cross bands; lower surfaces whitish, throat and sides of neck with round black spots.

q, r. ♂ & yg.	Near Lima.	Prof. W. Nation [P.].
s. ♂.	Lima.	J. M. Cowper, Esq. [P.].
t. Hgr. ♂	Peru.	

6. Tropidurus spinulosus.

Microlophus spinulosus, *Cope, Proc. Ac. Philad.* 1862, p. 351.
Leiocephalus (Craniopeltis) variegatus, *O'Shaughn. Ann. & Mag. N. H.* (5) iv. 1879, p. 301.
Tropidurus (Microlophus) spinulosus, *Boettg. Zeitschr. f. Naturw.* lviii. 1885, p. 216.

Upper head-scales smooth, convex; a series of four or five feebly dilated transverse supraoculars; occipital a little longer than broad, narrower than the supraocular region; temporal scales conical; a group of small erect spine-like scales in front of the ear. A strong, slightly curved antehumeral fold, nearly meeting its fellow on the throat; a strong transverse fold in front of the latter, forming a small pouch on each side of the throat. Sides of neck irregularly folded, with groups of small spinose scales. A small dorsal crest, most developed on the nape. Dorsal scales small, rhomboidal, imbricate, keeled, ending in a spine, obliquely directed towards the vertebral line; lateral and ventral scales as large as dorsals, latter smooth. The adpressed hind limb reaches the shoulder, or between the latter and the ear. Tail roundish, with serrated upper ridge; caudal scales larger than dorsals. Upper surfaces olive, lighter-dotted, with ill-defined dark brown crescentic cross bands; sides bluish; antehumeral fold black; lower surfaces yellowish, throat clouded with blackish.

	millim.		millim.
Total length	284	Fore limb	59
Head	28	Hind limb........	80
Width of head	25	Tail (reproduced) ..	150
Body	106		

Paraguay; Argentine Republic.

a–b. ♂.	Paraguay.		
c, d. ♂.	Cosquin, Cordova.	E. W. White, Esq. [C.].	(Types of *Leiocephalus variegatus.*)

7. Tropidurus torquatus.

Taraguira torquata, *Gray, Cat.* p. 220.
—— darwinii, *Gray, l. c.*
Stellio torquatus, *Wied, Reise n. Bras.* i. p. 139.
Tropidurus torquatus, *Wied, Beitr. Nat. Bras.* i. p. 137, *and Abbild.*; *Reinh. & Lütk. Vidensk. Meddel.* 1861, p. 227; *Peters, Mon. Berl. Ac.* 1877, p. 409.
Agama tuberculata, *Spix, Spec. Nov. Lacert. Bras.* p. 12, pl. xv. fig. 1.
Ecphymotes tuberculata, *Guér. Icon. R. A., Rept.* pl. xii. fig. 2.
—— torquatus, *Dum. & Bibr.* iv. p. 344.
Tropidurus microlepidotus, *Fitzing. Syst. Rept.* p. 72.
Taraguira torquata, *Girard, U.S. Explor. Exped., Herp.* p. 318.

Upper head-scales smooth; a series of transversely enlarged supraoculars, the width of which never exceeds half the width of the supraocular region; occipital usually longer than broad, and narrower than the supraocular region; temporal scales flat and feebly keeled; a fringe of long scales in front of the ear. A strong, slightly curved antehumeral fold, widely separated from its fellow; one or two more or less marked transverse gular folds; sides of neck granular, with one or two oblique folds or pouches. No dorsonuchal ridge. Dorsal scales small, slightly larger than ventrals, keeled, rhomboidal, imbricate, and ending in a short spine in the males, subhexagonal in the females and young; keels of the dorsal scales forming oblique lines; ventrals smooth. The adpressed hind limb reaches the ear or the eye. Tail once and one fourth to once and a half as long as head and body, round or slightly compressed; caudal scales much larger than dorsals, ending in a spine. Olive or brownish above, usually with darker spots and lighter dots; usually a more or less distinct light and a dark lateral dorsal band; a black band on the antehumeral fold, usually extending on the scapular region; lower surfaces yellowish or greyish, throat and breast often marbled with blackish; the male's throat usually black.

	millim.		millim.
Total length	250	Fore limb	48
Head	25	Hind limb........	76
Width of head	21	Tail	150
Body	75		

Guianas; Brazil.

a–c. Hgr. & yg.	Rio Janeiro.	C. Darwin, Esq. [P.].
d. ♀.	Rio Janeiro.	Dr. Cunningham [C.].
e. ♀.	Rio Janeiro.	The Lords of the Admiralty [P.].
f, g. Yg.	Rio Janeiro.	G. Busk, Esq. [P.].
h–i. ♂.	Abrolhos Island.	C. Darwin, Esq. [P.]. (Types of *Taraguira darwinii.*)
k–l. Hgr. & yg.	Brazil.	W. Wright, Esq. [P.].
m–o, p–q, r. ♂ & hgr.	Brazil.	
s. ♂.	S. America.	

8. Tropidurus hygomi.

Tropidurus hygomi, *Reinh. & Lütk. Vidensk. Meddel.* 1861, p. 228, pl. v. fig. 9.

The only important difference distinguishing this species from the preceding is the greater width of the band-like supraoculars, which cover nearly entirely the supraocular region, being separated from the supraorbitals by one and from the supraciliaries by one or two series of very small scales. The scales on the body a trifle larger than in *T. torquatus.*

Brazil.

a. ♀.	Brazil.	Copenhagen Museum [P.]. (One of the types.)

9. Tropidurus hispidus.

Taraguira smithii, *Gray, Cat.* p. 221.
Agama hispida, *Spix, Spec. Nov. Lacert. Bras.* p. 12, pl. xv. fig. 2.
—— nigrocollaris, *Spix, l. c.* p. 13, pl. xvi. fig. 2.
—— cyclurus, *Spix, l. c.* p. 14, pl. xvii. fig. 1.
Proctotretus tœlsneri, *Berthold, Götting. Nachr.* 1859, p. 179.
Trachycyclus superciliaris, *Günth. Proc. Zool. Soc.* 1861, p. 16.
Tropidurus macrolepis, *Reinh. & Lütk. Vidensk. Meddel.* 1861, p. 227, pl. v. fig. 8.
—— hispidus, *Peters, Mon. Berl. Ac.* 1877, pp. 409, 413.

Distinguished from *T. torquatus* by the size of the dorsal scales, which are at least twice as large as the ventrals. No light bands along the body.

Brazil; Venezuela.

a, b. ♂.	Bahia.	Dr. Wucherer [C.].

c. ♂.	Bahia.	Dr. Wucherer [C.]. (Type of *Trachycyclus superciliaris*.)
d–f, *g*. ♂ ♀.	Pernambuco.	J. P. Smith, Esq. [P.]. (Types of *Taraguira smithii*.)
h. ♂.	Pará.	
i. ♂.	Brazil.	C. Darwin, Esq. [P.]. } (Types of *T. smithii*.)
k–l. ♂.	Brazil.	
m. ♂.	Brazil.	Copenhagen Museum [P.]. (As typical of *T. macrolepis*.)
n. ♂.	Venezuela.	

10. Tropidurus semitæniatus.

Agama semitæniata, *Spix*, *Spec. Nov. Lacert. Bras.* p. 13, pl. xvi. fig. 1.
Platynotus semitæniatus, *Wagl. Syst. Amph.* p. 146.
Tropidurus (Platynotus) semitæniatus, *Peters*, *Mon. Berl. Ac.* 1877, pp. 409, 414.

Distinguished from *T. torquatus* in having perfectly smooth dorsal scales.

Bahia.

11. Tropidurus? blainvillii.

Tropidogaster blainvillii, *Gray*, *Cat.* p. 227.
Tropidogaster blainvillii, *Dum. & Bibr.* iv. p. 330, pl. xxxix. bis. fig. 2.
Ptychosaurus (Tritropis) blainvillii, *Fitzing. Syst. Rept.* p. 59.

Nostril lateral. Upper head-scales keeled; no enlarged supraoculars; occipital moderately large; anterior border of ear denticulated. Sides of neck irregularly folded; two transverse gular folds, the anterior straight, the posterior arched. A vertebral denticulation, extending on the tail. A fold along the flanks. Dorsal scales small, scarcely imbricate, keeled; ventral scales squarish-hexagonal, tricarinate. The adpressed hind limb reaches the end of the snout. Tail round or slightly compressed. Buff above, with small confluent brown lines; a greyish dorso-lateral band; hinder side of thighs with a longitudinal series of yellowish-white spots encircled with black, and above them a yellowish black-edged line; throat brownish, chest greyish, belly yellowish.

	millim.		millim.
Head	15	Fore limb	26
Body	38	Hind limb	45

—— ?

28. URANISCODON.

Uraniscodon, part., *Kaup*, *Isis*, 1826, p. 89.
Ecphymotes, part., *Fitzing. N. Classif. Rept.* p. 17.
Hypsibatus, *Wagler*, *Syst. Amph.* p. 150; *Wiegm. Herp. Mex.* p. 15; *Dum. & Bibr.* iv. p. 253; *Fitzing. Syst. Rept.* p. 58.

Plica, *Gray, Griff. A. K.* ix. *Syn.* p. 40, *and Cat. Liz.* p. 223.
Uperanodon, *Dum. & Bibr.* p. 247; *Fitzing. l. c.*
Ptychosaurus, *Fitzing. l. c.* p. 59.
Ptychopleura, *Fitzing. l. c.*
Uraniscodon, *Gray, Cat.* p. 222.

Tympanum distinct. Body subcylindrical or depressed; a slightly serrated dorsal ridge; dorsal scales small, imbricate, uniform. Upper head-scales enlarged, occipital very large. Neck strongly plicate inferiorly, the folds forming a pair of pouches on each side; no gular sac. Digits compressed, bent at the articulations, with keeled lamellæ inferiorly. No femoral or præanal pores. Tail long, cylindrical or compressed. Lateral teeth tricuspid, the anterior ones slightly enlarged and canine-like; pterygoid teeth present or absent. A large sternal fontanelle. No abdominal ribs.
South America.

1. Uraniscodon umbra.

Uraniscodon umbra, *Gray, Cat.* p. 222.
—— pictum, part., *Gray, l. c.* p. 223.
Lacerta umbra, *Linn. S. N.* i. p. 367.
Agama picta, *Wied, Abbild.*
Lophyrus ochrocollaris, *Spix, Spec. Nov. Lacert. Bras.* p. 10, pl. xii. fig. 2.
Uraniscodon umbra, *Kaup, Isis,* 1826, p. 90.
—— pictum, *Kaup, l. c.*
Hypsibatus umbra, *Wagl. Syst. Amph.* p. 150.
Uperanodon ochrocollare, *Dum. & Bibr.* iv. p. 248.
—— pictum, part., *Dum. & Bibr.* p. 251.
Hyperanodon peltigerus, *Cope, Journ. Ac. Philad.* (2) viii. 1876, p. 170.
—— umbra, *Peters, Mon. Berl. Ac.* 1877, pp. 408, 413.

Snout short and very broadly rounded; nostril small, turned upwards; upper head-scales usually with small granular asperities; anterior supraorbital scales transversely dilated, forming two diverging series; a series of large transverse supraoculars; occipital broader than long, as broad as or broader than the supraocular region; labials large, four or five upper and five or six lower; ear-opening large, its anterior border not or but very slightly denticulated. A strong oblique antehumeral fold; a feeble longitudinal and a strong transverse gular fold, the latter forming a small pouch on each side; sides of neck feebly plicate, without spines. Dorsal crest most developed on the nape, reduced to a very slight ridge on the back. Body not depressed; dorsal scales as large as or a little larger than ventrals, rhomboidal, keeled, the keels forming oblique lines; ventrals rhomboidal, more or less distinctly keeled or shortly mucronate. The size of the scales varies considerably. The adpressed hind limb reaches the ear or the eye. Tail round or slightly compressed, slender, at least twice as long as head and body. Reddish or purplish brown above, with more or less distinct blackish cross bands; a broad black band on the antehumeral fold, sometimes extending across the nape; frequently a large yellowish spot on the side of the neck, in front of the black band; lower surfaces brownish or yellowish, uniform or clouded with brown.

	millim.		millim.
Total length........	297	Fore limb	50
Head	22	Hind limb	75
Width of head	18	Tail	205
Body..............	70		

South America.

a–b. ♂.	Demerara Falls.	
c. Yg.	Pernambuco.	J. P. Smith, Esq. [P.].
d. ♂.	Interior of Brazil.	
e. ♂.	Brazil.	
f. ♀.	Sarayacu, Ecuador.	Mr. Buckley [C.].
g–k. ♂, ♀, & hgr.	Moyobamba, Peru.	Mr. A. H. Roff [C.].
l, m. ♂.	South America.	

2. Uraniscodon plica.

Plica umbra, *Gray, Cat.* p. 223.
Lacerta plica, *Linn. S. N.* i. p. 367.
Iguana chalcidica, *Laur. Syn. Rept.* p. 48.
Agama umbra, *Daud. Rept.* iii. p. 375.
—— plica, *Merr. Tent.* p. 55.
Lophyrus panthera, *Spix, Spec. Nov. Lacert. Bras.* p. 11, pl. xiii. fig. 1.
Uraniscodon plica, *Kaup, Isis,* 1826, p. 91.
Lophyrus agamoides, *Gray, Phil. Mag.* (2) ii. 1827, p. 208.
Hypsibatus plica, *Wagl. Syst. Amph.* p. 150.
Ophyessa plica, *Gray, Griff. A. K.* ix. *Syn.* p. 41, pl. —.
Hypsibatus agamoides, *Wiegm. Herp. Mex.* p. 15; *Dum. & Bibr.* iv. p. 254.
? Hypsibatus punctatus, *Dum. & Bibr.* p. 258.
? Ptychosaurus punctatus, *Fitzing. Syst. Rept.* p. 59.
Ptychosaurus (Ptychopleura) plica, *Fitzing. l. c.*
Hypsibatus panthera, *Peters, Mon. Berl. Ac.* 1877, pp. 408, 413.

Head large, short snout; nostril latero-superior; supraciliary edge projecting, angular; upper head-scales imbricate, more or less distinctly keeled; a series of large transverse supraoculars; occipital a little broader than long, narrower than the supraocular region; labials large, four or five upper and five or six lower. Neck constricted and strongly plicate and pouched inferiorly; posterior gular fold continuous with a lateral fold curving above the shoulder and ending above the hind limb; sides of neck with tufts of small, erect spine-like scales. A small dorsal crest, highest on the nape. Body depressed. Dorsal and lateral scales very small, strongly keeled; latero-dorsal fold covered with small erect spinose scales; a second more or less distinct fold on each side from axilla to groin; ventral scales a little larger than dorsals, smooth. The adpressed hind limb reaches the eye or the tip of the snout. Tail slender, a little compressed, with serrated upper ridge; its length twice or a little less than twice the length of head and body. Greyish-olive or greenish above, spotted or marbled with dark brown, which colour forms more or less distinct cross bands on the back and limbs; a collar of black marblings; lower surfaces whitish, throat black.

	millim.		millim.
Total length	394	Fore limb	82
Head	33	Hind limb	123
Width of head	33	Tail	250
Body	111		

South America.

a–b, c. ♂, ♀, & yg.	Island of Granada.	
d–e. ♂ ♀.	Trinidad.	
f, g. ♂ ♀.	Venezuela.	
h. ♀.	Demerara Falls.	
i. ♀.	Pebas.	H. W. Bates, Esq. [C.].
k. Yg.	Pallatanga, Ecuador.	Mr. Buckley [C.].
l–m. ♂ & hgr.	Cayaria, Peru.	Messrs. Veitch [P.].
n. ♂.	S. America.	

29. STROBILURUS.

Strobilurus, *Wiegm. Herp. Mex.* p. 18; *Dum. & Bibr.* iv. p. 353; *Fitzing. Syst. Rept.* p. 71; *Gray, Cat. Liz.* p. 222.

Tympanum distinct. Body depressed; a slightly serrated dorsal ridge; dorsal scales imbricate, uniform. Upper head-scales slightly enlarged, occipital very large. A curved fold in front of each shoulder, not extending across the throat. Digits compressed, bent at the articulations, with keeled lamellæ inferiorly. No femoral or præanal pores. Tail rather short, subcylindrical, with whorls of large spinose scales. Lateral teeth tricuspid; no pterygoid teeth. A large sternal fontanelle. No abdominal ribs.

Brazil.

1. Strobilurus torquatus.

Strobilurus torquatus, *Gray, Cat.* p. 222.
Strobilurus torquatus, *Wiegm. l. c.*; *Dum. & Bibr.* p. 354.
Steironotus (Strobilurus) torquatus, *Fitzing. l. c.*
Doryphorus spinosus, *Guichen. in Casteln. Am. Mér., Rept.* p. 27, pl. vii. fig. 1.

Snout short; nostril small, turned upwards; upper head-scales with small granular asperities; a series of transversely enlarged supraocular scales; occipital as long as broad or a little broader than long, as broad as the supraocular region; seven or eight upper and six or seven lower labials; temporal scales keeled; ear-opening rather large, with two or three small projecting lobes anteriorly. Sides of neck strongly plicate; a strong antehumeral fold; one or two irregular feeble folds across the throat. Body short, depressed, back tectiform. Dorsal scales larger than ventrals, rhomboidal, imbricate, keeled, the keels obliquely converging towards the vertebral line; ventrals smooth. Digits slender. The adpressed hind limb reaches the ear. Tail as long as head and body, conical, strongly depressed at the base, covered with very large rhomboidal spinose scales. Greenish above, tail yellowish olive; head with symmetrical black markings; a crescentic black band across scapular region, and sometimes a second across the nape; lower surfaces greenish white.

	millim.		millim.
Total length........	114	Fore limb	29
Head	16	Hind limb	38
Width of head......	12	Tail............	57
Body..............	41		

Brazil.

a. ♂.	Bahia.	Dr. Wucherer [C.].
b. Ad., stuffed.	Bahia.	M. de Castelnau [P.].
c. Ad., stuffed.	Brazil.	Mrs. Parker [P.].

30. UROCENTRON.

Urocentron, *Kaup, Isis*, 1827, p. 612; *Wagler, Syst. Amph.* p. 145; *Wiegm. Herp. Mex.* p. 18; *Fitzing. Syst. Rept.* p. 77.
Doryphorus, *Cuv. R. A.* 2nd ed. ii. p. 34; *Dum. & Bibr.* iv. p. 369.
Uranocentron, *Gray, Cat. Liz.* p. 225.

Tympanum distinct. Body depressed; no dorsal crest; dorsal scales small, imbricate, uniform. Upper head-scales little enlarged. Two transverse gular folds; no gular pouch. Digits compressed, bent at the articulations, with keeled lamellæ inferiorly. No femoral or præanal pores. Tail short, flat, with whorls of spinose scales. Lateral teeth tricuspid; no pterygoid teeth. A large sternal fontanelle. No abdominal ribs.

South America.

1. Urocentron azureum.

Uranocentron azureum, *Gray, Cat.* p. 226.
Lacerta azurea, *Linn. S. N.* i. p. 362.
Stellio azureus, *Daud. Rept.* iv. p. 36, pl. xlvi.
—— brevicaudatus, *Daud. l. c.* p. 40, pl. xlvii.
Uromastyx cæruleus, *Merr. Tent.* p. 56.
—— azureus, *Merr. l. c.* p. 57.
Urocentron azureum, *Kaup, Isis*, 1827, p. 612.
Ophyessa (Uranocentron) azureus, *Gray, Griff. A. K.* ix. *Syn.* p. 42.
—— (——) brevicaudatus, *Gray, l. c.*
Doryphorus brevicaudatus, *Guér. Icon. R. A., Rept.* pl. vi. fig. 3.
Urocentron brevicaudatum, *Wiegm. Herp. Mex.* p. 18.
Doryphorus azureus, *Dum. & Bibr.* iv. p. 371, pl. xlii. fig. 2.

Snout short, broadly rounded, lips somewhat swollen; nostril turned upwards; upper head-scales small, with small granular asperities; a series of slightly dilated transverse supraoculars; occipital a little longer than broad, narrower than the supraocular region; five or six upper and six or seven lower labials; ear-opening moderate, its anterior border without denticulation. Sides of neck irregularly plicate; a strong transverse gular fold, and a second, less regular, in front. Lateral and anterior dorsal scales very small, granular, smooth; posterior dorsal scales larger, subrhomboidal, obtusely keeled; ventrals larger, squarish, imbricate, smooth. Digits slender. The adpressed hind limb reaches the shoulder or slightly beyond. Tail shorter and narrower than the body, flat inferiorly, about twice as broad as deep, with whorls of very large, obtusely keeled spinose scales, the spines strongest on the sides. Azure blue or olive above, with crescentic black cross bands; lower surfaces greenish or yellowish.

	millim.		millim.
Total length	124	Fore limb	39
Head	20	Hind limb	49
Width of head	17	Tail	46
Body	58		

Guianas; Brazil.

a. ♀.	Brazil.
b. ♀.	British Guiana.
c, d, e–f. ♂ ♀.	S. America.

2. Urocentron flaviceps.

Doryphorus azureus, var., *A. Dum. Cat. Méth. Rept.* p. 85.
—— flaviceps, *Guichen. in Casteln. Am. Mér., Rept.* p. 26, pl. iii. fig. 2; *A. Dum. Arch. Mus.* viii. p. 560.
Uranocentrum flaviceps, *O'Shaughn. Proc. Zool. Soc.* 1881, p. 245.

Upper head-scales very convex. Dorsal scales small, rhomboidal, imbricate, strongly keeled, increasing in size posteriorly; lateral scales smaller; ventrals as large as dorsals, feebly keeled. Tail quite flat, nearly as broad as the body, lanceolate; caudal scales large, rhomboidal, forming longitudinal and transverse series, strongly keeled, spinose, the spines strongest on the sides. Otherwise as in the preceding species. Black, uniform (males) or with numerous round greenish spots and the lower surfaces greenish-white or pale olive; upper surface of head sometimes yellowish.

	millim.		millim.
Total length	168	Fore limb	53
Head	28	Hind limb	68
Width of head	21	Tail	62
Body	78		

Upper Amazons.

a–c. ♂, ♀, & yg.	Canelos, Ecuador.	Mr. Buckley [C.].
d. ♂.	Sarayacu, Ecuador.	Mr. Buckley [C.].
e. ♂.	Peruvian Amazons.	Mr. E. Bartlett [C.].

3. Urocentron castor.

Doryphorus castor, *Cope, Proc. Am. Phil. Soc.* xi. 1870, p. 556.

Apparently very closely related to the preceding. Nostril between two scales. Scales of the sides scarcely smaller than dorsals and ventrals. Tail wider than the body. Colour blackish olive, gradually lighter to the head, which is thickly black-spotted on a green ground; a broad crescentic black collar, which is broadly light-margined behind, much more indistinctly before; the posterior border is again broadly margined with black behind; limbs closely green, spotted on a blackish ground; tail uniform brown; belly black, on the sides dark green, femora black.

Length to vent 60 millim.; tail 33 millim.

Pebas, Ecuador.

31. PHYMATURUS.

Phymaturus, *Gravenh. Nova Acta Ac. Leop.-Carol.* xviii. 1838, ii. p. 749; *Gray, Cat. Liz.* p. 226.
Centrura, *Bell, Zool. Beagle, Rept.* p. 25; *A. Dum. Arch. Mus.* viii. p. 556.

Tympanum distinct. Body depressed; no dorsal crest; dorsal scales small, juxtaposed or subimbricate, uniform. Upper head-scales small. A transverse gular fold; no gular pouch. Digits slightly compressed, with keeled lamellæ inferiorly. No femoral pores; males with a series of pores on the front edge of the vent. Tail rather short, subcylindrical, with whorls of spinose scales. All the teeth trilobate; pterygoid teeth. A sternal fontanelle. No abdominal ribs.

Chili.

1. Phymaturus palluma.

Phymaturus palluma, *Gray, Cat.* p. 226.
Lacerta palluma, *Molina, Saggio sulla Storia nat. del Chile,* 1810, p. 189.
Stellio palluma, *Daud. Rept.* iv. p. 46.
Phymaturus palluma, *Gravenh. l. c.* p. 750, pl. lv. fig. 2.
Centrura flagellifer, *Bell, l. c.* pl. xiv. fig. 2; *A. Dum. l. c.* p. 557, pl. xxii. fig. 5.
Oplurus bibronii, *Guichen. in Gay, Hist. Chile, Rept.* p. 53, pl. iii. fig. 2; *A. Dum. Cat. Méth. Rept.* p. 84.

Head small, snout short; canthus rostralis indistinct, nostril superior; ear-opening moderately large. Head-scales small, equal, subconical on the temples. Throat and sides of neck strongly plicate. Body much depressed. Dorsal scales very small, granular, smooth, a little larger and flat on the middle of the back; ventral scales a little larger, squarish, smooth, in regular transverse series. Limbs short; the adpressed hind limb reaches the axilla; digits short and thick. Male with a series of nine or ten pore-like callosities in front of the vent. Tail about as long as head and body, conical, with whorls of rather large spinose scales. Dark olive, variegated with blackish.

	millim.		millim.
Total length	202	Fore limb	42
Head	21	Hind limb	56
Width of head	20	Tail	100
Body	81		

Chili.

a. ♀.	Chili.	C. Darwin, Esq. [P.]. (Type of *Centrura flagellifer*.)
b, c. ♂.	Chili.	
d. Skeleton.	Chili.	

32. AMBLYRHYNCHUS.

Amblyrhynchus *Bell, Zool. Journ.* ii. 1825, p. 206; *Wiegm. Herp. Mex.* p. 16; *Dum. & Bibr.* iv. p. 193; *Fitzing. Syst. Rept.* p. 55.
Oreocephalus, *Gray, Cat. Liz.* p. 189.

Tympanum distinct. Body slightly compressed; a dorsal crest; dorsal scales small. Head with large tubercular shields. No gular pouch; a feeble transverse gular fold. Digits compressed, with keeled lamellæ inferiorly, united at the base by a rudiment of web. A long series of femoral pores. Tail long, strongly compressed. All the teeth trilobate; pterygoid teeth. No sternal fontanelle. No abdominal ribs.

Galapagos Islands.

1. Amblyrhynchus cristatus.

Oreocephalus cristatus, *Gray, Cat.* p. 189.
Amblyrhynchus cristatus, *Bell, Zool. Journ.* ii. 1825, p. 206, *Suppl.* pl. xii.; *Dum. & Bibr.* iv. p. 195; *Darwin, Journ. Beagle*, p. 466; *Bell, Zool. Beagle, Rept.* p. 23; *A. Dum. Cat. Méth. Rept.* p. 62; *Steind. Festschr. zool.-bot. Ges. Wien*, 1876, p. 316, pls. iii., v., vi., vii.
Iguana (Amblyrhynchus) cristatus, *Gray, Griff. A. K.* ix. *Syn.* p. 37.
—— (——) ater, *Gray, l. c.*
Amblyrhynchus ater, *Dum. & Bibr.* p. 196.
Hypsilophus (Amblyrhynchus) cristatus, *Fitzing. Syst. Rept.* p. 55.
—— (——) ater, *Fitz. l. c.*

Habit stout. Head small, very convex, nearly as broad as long, with swollen lips; nostril oval, pierced in a small swollen shield nearer the end of the snout than the orbit; eye and tympanum small; upper surface of head covered with large convex or conical and spinose, rough, polygonal shields; rostral and mental very small; ten to twelve upper and as many lower labials. Throat strongly plicate, covered with minute granular scales. A dorso-nuchal crest of lanceolate spines, highest on the nape, sometimes subinterrupted between the shoulders. Scales on the body very small, keeled on the back, smooth on the belly. Limbs strong; digits rather short, the third toe strongly denticulated along the inner side of the basal joint. A series of twenty to thirty pores on each side; frequently a second parallel series of a few pores. Tail about once and a half the length of head and body, strongly compressed, crested above, covered with cross series of equal keeled scales. Colour black or blackish brown; abdomen and lower surface of thighs frequently dirty white; young brown above, with pale spots or dots, and more or less distinct darker cross bands on the back.

	centim.		centim.
Total length	133	Fore limb	22
Head	9	Hind limb	27
Width of head	9	Tail	80
Body	44		

Galapagos Islands.

a, b, c. Ad. & hgr., stuffed.	Galapagos.	Capt. Fitzroy [P.].
d. Yg.	Galapagos.	J. Goodridge, Esq. [P.].
e. Hgr.	Galapagos.	Haslar Collection.
f. Hgr., stuffed.	Galapagos.	
g. Skull of *f.*		
h. Yg.	Galapagos.	
i, k, l. Ad.	Charles Island.	Commander W. E. Cookson [P.].
m. Ad.	Iguana Cove.	Commander W. E. Cookson [P.].
n. Ad., stuffed.	Iguana Cove.	Commander W. E. Cookson [P.].
o. Skeleton of *n.*		
p. Ad.	Tagus Cove, Albemarle.	Commander W. E. Cookson [P.].
q–s. Hgr.	Abingdon Island.	Commander W. E. Cookson [P.].

33. CONOLOPHUS.

Amblyrhynchus, part., *Dum. & Bibr.* iv. p. 193.
Conolophus, *Fitzing. Syst. Rept.* p. 55.
Trachycephalus (*non Tschudi*), *Gray, Cat. Liz.* p. 188.

Tympanum distinct. Body a little depressed; a slight dorsal crest; dorsal scales small. Head-scales slightly enlarged. No gular pouch; a feeble transverse gular fold. Digits compressed, with keeled lamellæ inferiorly. A long series of femoral pores. Tail long, roundish, slightly compressed. All the teeth trilobate; pterygoid teeth*. No sternal fontanelle. No abdominal ribs.

Galapagos Islands.

1. Conolophus subcristatus.

Trachycephalus subcristatus, *Gray, Cat.* p. 188.
Amblyrhynchus subcristatus, *Gray, Zool. Misc.* 1831, p. 6, *and Zool. Beechey's Voy., Rept.* p. 93; *Darwin, Journ. Beagle,* p. 469.
—— demarlii, *Dum. & Bibr.* iv. p. 197; *Bell, Zool. Beagle, Rept.* p. 22, pl. xii.
Hypsilophus (Conolophus) demarlii, *Fitzing. Syst. Rept.* p. 55.
Conolophus subcristatus, *Steind. Festschr. zool.-bot. Ges. Wien,* 1876, p. 322, pls. iv.-vii.

Habit stout. Head rather small, little longer than broad; nostril turned upwards and slightly outwards, slightly nearer the end of the snout than the orbit; eye and tympanum of moderate size; upper surface of head covered with rather large, convex or subconical, keeled polygonal scales; occipital very slightly enlarged; rostral and mental moderately large; labials twelve. Throat longitudinally plicate, covered with minute granules. A low nuchal crest of recurved spines, continued as a feebly serrated ridge along the back. Scales on the body very small, keeled on the back, smooth and slightly larger on the belly. Limbs strong; digits very short; fifth toe not longer than second; third toe strongly denticulated along the inner side of the basal joint. Seventeen to twenty-one femoral pores on each side. Tail scarcely longer than head and body, feebly compressed, with very slightly serrated upper edge, covered with small equal keeled scales. Dark brown, lower surfaces and head lighter.

	centim.		centim.
Total length	107	Fore limb	21
Head	9	Hind limb......	26
Width of head	8	Tail	54
Body	44		

Galapagos, Central Islands.

a. Hgr.	Galapagos.	C. Darwin, Esq. [P.].
b. Ad., stuffed.	Galapagos.	Capt. Fitzroy [P.].
c. Ad., stuffed.	Galapagos.	Haslar Collection.
d. Left mandible.	Iguana Cove.	

34. METOPOCEROS.

Iguana, part., *Daud. Rept.* iii. p. 358; *Fitzing. N. Classif. Rept.* p. 17; *Wiegm. Herp. Mex.* p. 15.
Metopoceros, *Wagler, Syst. Amph.* p. 147; *Dum. & Bibr.* iv. p. 210; *Fitzing. Syst. Rept.* p. 54; *Gray, Cat. Liz.* p. 188.
Aloponotus, *Dum. & Bibr.* iv. p. 189; *Fitzing. l. c.*; *Gray, Cat.* p. 187.

* The presence of pterygoid teeth has been denied by Steindachner. They exist in young specimens.

Tympanum distinct. Body compressed: a dorsal crest; dorsal scales small. A small gular pouch and a feeble transverse gular fold. Digits compressed, with keeled lamellæ inferiorly. A long series of femoral pores. Tail long, compressed. Lateral teeth denticulated; pterygoid teeth.

West Indies.

1. Metopoceros cornutus.

? Aloponotus ricordi, *Gray, Cat.* p. 187.
Metopoceros cornutus, *Gray, l. c.* p. 188.
Lacép. Quadr. Ov. ii. p. 493.
Iguana cornuta, *Daud. Rept.* iii. p. 282.
Metopoceros cornutus, *Wagl. Syst. Amph.* p. 148; *Dum. & Bibr.* iv. p. 211; *Cope, Proc. Ac. Philad.* 1866, p. 124; *Günth. Trans. Zool. Soc.* xi. 1882, p. 218, pls. xliii., xliv.
? Aloponotus ricordii, *Dum. & Bibr.* p. 190, pl. xxxvii.
? Hypsilophus (Aloponotus) ricordii, *Fitzing. Syst. Rept.* p. 54.
Hypsilophus (Metopoceros) cornutus, *Fitz. l. c.*
Cyclura nigerrima, *Cope, Am. Nat.* 1885, p. 1006.
—— onchiopsis, *Cope, l. c.*

Head large, swollen below the ears, covered with unequal-sized scales; snout obtuse, with obtuse canthus rostralis, with three enlarged scales, which in the adult male are conical, horn-like; occipital slightly enlarged; nostril large, oval, near the end of the snout, pierced in a convex nasal which is either in contact with the rostral or separated from it by small scales; ten to twelve upper and as many lower labials; a row of enlarged scutes, conical in the male, above the upper labials, from below the front of the orbit to below the tympanum; a row of similar keeled scutes bordering the orbit inferiorly, and extending to above the tympanum; latter large, vertically oval. Gular region covered with very small keeled scales, intermixed on the sides with enlarged tubercles; a row of large chin-shields on each side, the anterior in contact with the labials, the others strongly keeled and separated from the latter by smaller scutes. Dorsal crest formed of lanceolate lobes, the longest of which (in the male) measure a little less than the diameter of the tympanum; a nuchal crest of very low spines, widely separated from the dorsal. Dorsal scales minute, flat, granular, surrounded by finer granulations; these scales sometimes very indistinct, the skin appearing almost naked; ventral scales a little larger, flat, smooth, juxtaposed. Limbs strong; digits rather short, the third toe strongly denticulated along the inner side of the basal joint. A series of thirteen to eighteen femoral pores on each side; frequently a second, somewhat shorter, parallel series of pores. Tail long, compressed, crested above, with verticils of keeled scales forming very indistinct segments. Blackish, uniform or with lighter variegations.

	centim.		centim.
Total length (tail broken)	98	Body	36
Head	12	Fore limb	21
Width of head	9	Hind limb	30

San Domingo, Navasa.

a. ♂.	——?	Zoological Society.
b. ♀, stuffed.	——?	Zoological Society.
c. Skeleton of *b*.		

35. IGUANA.

Iguana, part., *Laur. Syn. Rept.* p. 47; *Daud. Rept.* iii. p. 358; *Fitzing. N. Classif. Rept.* p. 17; *Wiegm. Herp. Mex.* p. 15.
Iguana, *Gray, Ann. Phil.* (2) x. 1825, p. 197; *Dum. & Bibr.* iv. p. 199; *Gray, Cat. Liz.* p. 186; *Bocourt, Miss. Sc. Mex., Rept.* p. 133.
Hypsilophus, *Wagler, Syst. Amph.* p. 147; *Fitzing. Syst. Rept.* p. 54.
Amblyrhynchus (*non Bell*), *Wagl. l. c.* p. 148.

Tympanum distinct. Body compressed; a dorsal crest; dorsal scales small. Upper head-scales large. A transverse gular fold, and a large non-dilatable gular appendage. Digits compressed, with keeled lamellæ inferiorly. A long series of femoral pores. Tail long, compressed. Lateral teeth denticulated; pterygoid teeth. A small sternal fontanelle. One or two poststernal ribs.

Tropical America.

1. Iguana tuberculata.

Iguana tuberculata, *Gray, Cat.* p. 186.
Iguana tuberculata, *Laur. Syn. Rept.* p. 49; *Dum. & Bibr.* iv. p. 203; *A. Dum. Arch. Mus.* viii. p. 526.
Lacerta iguana, *Shaw, Zool.* iii. p. 199; *Duvern. R. A., Rept.* pl. xvii.
Iguana delicatissima (*non Laur.*), *Daud. Rept.* iii. p. 263, pl. xl.
—— cærulea, *Daud. l. c.* p. 286; *Spix, Spec. Nov. Lacert. Bras.* p. 7, pl. vii.
—— sapidissima, *Merr. Tent.* p. 47.
—— squamosa, *Spix, l. c.* p. 5, pl. v.
—— viridis, *Spix, l. c.* p. 6, pl. vi.
—— emarginata, *Spix, l. c.* p. 7, pl. viii.
—— lophyroides, *Spix, l. c.* p. 8, pl. ix.
Hypsilophus iguana, *Wagl. Syst. Amph.* p. 147.

Head moderately large; snout rounded, with obtuse canthus rostralis; nostril rather large, pierced in a swollen nasal, near the tip of the snout; tympanum large, oval; head-scales smooth, variable in shape, large on the forehead and between the orbits; the interorbital series either in contact or separated; supraocular scales small; occipital enlarged; rostral large, pentagonal, mental smaller; ten to twelve labials. Anterior border of the gular appendage with a crest of large triangular compressed scales; gular scales very small on the appendage, larger and in several rows on the sides near the labials, increasing in size towards a large circular shield situated below the tympanum. Dorso-nuchal crest continuous, composed of falciform lobes, the length of which varies considerably; most developed in the full-grown males. Dorsal scales very small, equal, keeled; neck with scattered large conical or obtusely keeled tubercles; ventral scales small, larger than dorsals, smooth or indistinctly keeled. Limbs strong, digits long and slender. Twelve to eighteen femoral pores on each side. Tail strongly compressed, crested above, covered with uniform keeled scales which are larger inferiorly. Green or greenish, lighter beneath; upper surfaces uniform, or speckled or variegated with darker and lighter; usually dark, light-edged, undulated vertical bars down the flanks, tail with more or less distinct dark annuli; usually a whitish band in front of the arm.

	centim.		centim.
Total length	139	Fore limb	18
Head............	7·5	Hind limb	28
Width of head	5	Tail	100
Body	31·5		

South America to Nicaragua; West Indies.

a, b. Ad., stuffed.	W. Indies.	
c. Ad., skeleton.	W. Indies.	
d. ♂.	Island of St. Thomas.	
e. Hgr., ♀.	Dominica.	
f. Yg.	Tobago.	A. Ludlam, Esq. [P.].
g. Hgr.	Trinidad.	
h. Yg.	W. coast of America.	Mr. Bridges [C.].
i. ♂.	Nicaragua.	
k. Yg.	Panama.	Capt. T. C. Dow [P.].
l. Ad., stuffed.	Ecuador.	Mr. Fraser [C.].
m. ♂.	W. Ecuador.	Mr. Fraser [C.].
n. Hgr.	Sarayacu, Peru.	W. Davis, Esq. [C.].
o, p. Yg.	Venezuela.	
q. Hgr.	British Guiana.	
r. ♂.	Vryheids Lust, Demerara.	Rev. W. T. Turner [C.].
s–t. Yg.	Surinam.	Mr. Bartlett [C.].
u, v, w, x.	Surinam.	
y, z. Yg.	Brazil.	
α. Ad., stuffed.	Pernambuco.	J. P. G. Smith, Esq. [P.].
β. Yg.	Pernambuco.	J. P. G. Smith, Esq. [P.].
γ. ♀.	Recife, Pernambuco.	W. A. Forbes, Esq. [P.].
δ–η. ♀, hgr., & yg.	Bahia.	
θ, ι. ♂ & yg.	Upper Amazons.	
κ. Ad., stuffed.	——?	Capt. Kellet and Lieut. Wood [P.].
λ, μ. Ad., stuffed.	——?	
ν, ξ, ο, π, ρ, ς, ♂ ♀.	——?	

Var. **rhinolopha.**

Iguana rhinolophus, *Gray, Cat.* p. 186.

Iguana rhinolopha, *Wiegm. Herp. Mex.* p. 44; *Dum. & Bibr.* iv. p. 207; *Sumichrast, Arch. Sc. Phys. Nat.* xix. 1864, p. 50; *Cope, Proc. Am. Phil. Soc.* 1869, p. 161; *Bocourt, Miss. Sc. Mex., Rept.* p. 134.

Metopoceros cornutus (*non Daud.*), *Tyler, Proc. Zool. Soc.* 1850, p. 106, pl. iii.

I regard this form as merely a variety of the preceding, the only difference being the development of two or three of the median scales of the snout, behind the line of the nostrils, into conical, soft tubercles. In adult males these tubercles are more developed and form a small crest, but in some females the crest is quite indistinct, and the young are not always to be surely distinguished from *I. tuberculata.* I must also observe that some of the specimens (*i*, *s*, *t*) which I have referred to the latter show clearly in the scaling of the snout a tendency towards the *rhinolopha* form, so that there is a gradual passage from the one to the other. The other character which has been given as diagnostic, viz. the number of scales in the dorsal crest (from its origin on the nape to the base of the tail), is hardly of importance. Although there is usually a lesser number of these scales in *I. rhinolopha* than in *I. tuberculata*, this distinction is not a reliable one: our series of specimens show variation between 47 and 66 scales in the latter form and between 44 and 58 in the former; and it is remarkable that the specimens of *I. tuberculata* with subconical scales on the snout give the highest numbers (62 to 66), whereas the specimen of *I. rhinolopha* in which the rostral tubercles are the least developed has, on the contrary, the smallest number (44).

Central America and West Indies.

a. Ad., stuffed.	St. Lucia, W. Indies.	
b. Ad., stuffed.	Central America.	O. Salvin, Esq. [C.].
c. Yg.	Central America.	Haslar Collection.
d. Ad., stuffed.	Mexico.	
e. Yg.	Mexico.	
f. ♂.	Cordova, Mexico.	
g, h. Ad., skins.	Cozumel Island, Yucatan.	F. D. Godman and O. Salvin, Esqs. [P.].
i. Yg.	Honduras.	
k–l. Hgr.	Mazatlan.	Hr. A. Forrer [C.].
m, n, o. Ad., skins.	Presidio.	Hr. A. Forrer [C.].
p. ♀.	Chiapas.	O. Salvin, Esq. [C.].
q. ♂.	Pacific coast of C. America.	A. Vidler, Esq. [P.].
r. Yg.	Pacific coast of Guatemala.	O. Salvin, Esq. [C.].
s. ♀.	Panama.	
t. ♀.	—— ?	

2. Iguana delicatissima.

Iguana delicatissima, *Gray, Cat.* p. 187.

Iguana delicatissima, *Laur. Syn. Rept.* p. 48.

—— nudicollis, *Cuv. R. A.* 2nd ed. ii. p. 45; *Guér. Icon. R. A., Rept.* pl. xi. fig. 1; *Dum. & Bibr.* iv. p. 208; *Cope, Proc. Am. Phil. Soc.* 1869, p. 159.

Amblyrhynchus delicatissimus, *Wagl. Syst. Amph.* p. 148.

Differs from *I. tuberculata* in the following points:—Scales on the occiput more or less conical; no large shields below the tympanum, but a regular series of large chin-shields, parallel with and much larger than the labials, on each side; only a few triangular compressed scales in front of the gular appendage, below the chin. Dorso-nuchal crest constantly rather lower. Ventral scales more distinctly keeled. Nineteen or twenty femoral pores on each side. Uniform green or blackish-olive, head yellowish.

	centim.		centim.
Total length	121	Fore limb	16
Head	7	Hind limb........	26
Width of head	5	Tail	85
Body	29		

West Indies; Swan Islands; Brazil.

a. Ad., stuffed.	——?	
b. ♀.	——?	Sir A. Smith [P.].
c, d, e, f, g, h, i. ♂, ♀, & hgr.	——?	

36. BRACHYLOPHUS.

Brachylophes, *Cuv. R. A.* 2nd ed. ii. p. 40.
Brachylophus, *Wagler, Syst. Amph.* p. 151; *Dum. & Bibr.* iv. p. 225; *Fitzing. Syst. Rept.* p. 55; *Gray, Cat. Liz.* p. 187.
Iguana, part., *Wiegm. Herp. Mex.* p. 15.
Chloroscartes, *Günth. Proc. Zool. Soc.* 1862, p. 188.

Tympanum distinct. Body compressed; a dorsal crest; dorsal scales small. Upper head-scales small. A transverse gular fold; male with a gular pouch. Digits compressed, with keeled lamellæ inferiorly. A short series of femoral pores. Tail long, compressed. Lateral teeth tricuspid; pterygoid teeth. No sternal fontanelle. Abdominal ribs.

Fiji and Friendly Islands.

1. Brachylophus fasciatus.

Brachylophus fasciatus, *Gray, Cat.* p. 187.
Iguana fasciata, *Brongn. Bull. Soc. Philom.* ii. 1880, p. 90, pl. vi. fig. 1; *Daud. Rept.* iii. p. 352.
Brachylophus fasciatus, *Wagl. l. c.*; *Guér. Icon. R. A., Rept.* pl. ix. fig. 1; *Dum. & Bibr.* iv. p. 226; *A. Dum. Cat. Méth. Rept.* p. 64; *Girard, U.S. Explor. Exped., Herp.* p. 374, pl. xviii. figs. 8, 9.
Iguana (Brachylophus) fasciatus, *Gray, Griff. A. K.* ix. *Syn.* p. 37.
Chloroscartes fasciatus, *Günth. l. c.* p. 189, pl. xxv.

Head rather small; snout rounded, with obtuse canthus rostralis; nostril near the tip of the snout; tympanum vertically oval, larger than the eye-opening; upper head-scales small, polygonal, smooth, smallest on supraocular region; occipital not enlarged; six or seven upper and as many lower labials; mental with median cleft posteriorly; gular region covered with round or oval smooth granules intermixed with smaller ones. Dorso-nuchal crest a feebly serrated ridge, a little more developed on the nape. Dorsal scales minute, equal, keeled; ventrals larger, though very small, strongly keeled. Limbs covered with uniform small keeled scales; the adpressed hind limb reaches the tympanum, or between the latter and the shoulder; digits long and slender, the third toe strongly denticulated along the inner side of the basal joint. Femoral pores ten to fifteen on each side, forming a short series. Tail strongly compressed at the base, with serrated ridge above, covered with uniform keeled scales which are much larger inferiorly; the length of the tail at least three times that of head and body. Females uniform dark emerald green above, lighter beneath; males with three broad dark green cross bands on the body, separated by narrow light green bands, and dark and light annuli on the tail.

	millim.		millim.
Total length	695	Fore limb	79
Head	32	Hind limb........	115
Width of head	26	Tail	530
Body	133		

Fiji and Friendly Islands.

a–b. ♂ & yg.	Fiji.	J. McGillivray, Esq. [C.]. (Types of *Chloroscartes fasciatus.*)
c, d, e. ♀, hgr. ♂, & yg.	Fiji.	J. McGillivray, Esq. [C.].
f–g. ♂ ♀.	Kandavu, Fiji.	H. M. S. 'Challenger.'
h. Ad., stuffed.	Ovalau, Fiji.	
i. ♂.	——?	College of Surgeons.
k–l. ♂ ♀.	——?	

37. CYCLURA.

Cyclura, part., *Harlan, Journ. Ac. Philad.* iv. 1824, p. 250; *Wiegm. Herp. Mex.* pp. 15, 41; *Dum. & Bibr.* iv. p. 214; *Fitzing. Syst. Rept.* p. 56.
Cyclura, *Gray, Cat. Liz.* p. 190.

Tympanum distinct. Body compressed; a dorsal crest; dorsal scales small. Enlarged scales on the snout. A small gular pouch and a feeble transverse gular fold. Digits compressed, with keeled lamellæ inferiorly. A long series of femoral pores. Tail long, compressed. Lateral teeth tricuspid; pterygoid teeth. A small sternal fontanelle. A single poststernal rib.

West Indies.

1. Cyclura carinata.

Cyclura macleayi, *Gray, Cat.* p. 190.
—— nubila, *Gray, l. c.*
—— collei, *Gray, l. c.*
Catesby, Nat. Hist. Carol. ii. p. 64, pl. lxiv.
Cyclura carinata, *Harlan, Journ. Ac. Philad.* iv. 1824, pp. 242, 250, pl. xv.; *Wiegm. Herp. Mex.* p. 42; *Cope, Am. Nat.* 1885, p. 1006.
Iguana cyclura, *Cuv. R. A.* 2nd ed. ii. p. 45.
Lacerta nebulosa, *Griff. A. K.* ix. pl. —.

Iguana (Ctenosaura) bellii, *Gray, Griff. A. K.* ix. *Syn.* p. 38.
—— (Cyclura) carinata, *Gray, l. c.* p. 39.
—— (——) nubila, *Gray, l. c.*
Cyclura harlani, *Dum. & Bibr.* iv. p. 218; *Coct. in Sagra, Hist. Cuba, Rept.* p. 96, pl. vi.
—— lophoma, *Gosse, Proc. Zool. Soc.* 1848, p. 99, pl. i.; *Cope, l. c.*
—— baeolopha, *Cope, Proc. Ac. Philad.* 1861, p. 123, *and l. c.*

Head large, swollen below the ears, covered with unequal sized scales; snout obtuse, with obtuse canthus rostralis, with several large, more or less convex shields, the arrangement of which varies considerably; they usually consist of three pairs which are either in contact or separated by small scales; occipital slightly enlarged; nostril large, oval, near the end of the snout, pierced in a convex nasal in contact with the large pentagonal rostral; nine or ten upper and as many lower labials; a row of enlarged keeled infraorbitals, extending to the upper border of the tympanum; latter large, vertically oval. Gular region covered with minute scales, on the sides intermixed with enlarged subconical tubercles; a row of large chin-shields, on each side, the anterior usually in contact with the labials, the others separated from them by smaller scutes. Dorso-nuchal crest rather low, but varying considerably, formed of triangular or lanceolate compressed spines, more or less distinctly lowered or interrupted between the shoulders and hind limbs; in a full-grown male, the longest spines (on the nape) measure two thirds the diameter of the tympanum. Scales of the body very small, more or less indistinctly keeled, scarcely larger on the belly than on the back. Limbs strong; digits moderately long, the third toe strongly denticulated along the inner side of the basal joint. Twenty to twenty-six femoral pores on each side, in a series extending along the whole length of the thigh. Tail long, compressed, crested above, with verticils of keeled scales forming regular segments on the sides, each segment composed of three to five series of smaller scales and a series of larger, slightly spinose ones. Green or dark olive above, speckled with darker and lighter, and with or without more or less distinct blackish transverse bands.

	centim.		centim.
Total length	121	Fore limb	21
Head	10·5	Hind limb	26
Width of head	8	Tail	70
Body	40·5		

Bahamas; Cuba; Jamaica.

a. Hgr.	Cuba.	W. S. MacLeay, Esq. [P.]. (Type of *C. macleayi.*)
b. Ad., stuffed.	Jamaica.	R. Hill, Esq. [P.]. (Type of *C. lophoma.*)
c. Ad., stuffed.	Jamaica.	Haslar Collection. (Type of *C. collei.*)
d. Ad., stuffed.	Jamaica.	J. H. Gurney, Esq. [P.].
e. Ad., stuffed.	Jamaica.	Dr. McNab & T. Bland, Esq. [P.].

f. Yg.	——?	(Type of *C. nubila.*)
g, h. Ad., stuffed.	——?	Haslar Collection.
i. Ad., stuffed.	——?	
k. Skull.	——?	Sir E. Belcher [P.].

38. CTENOSAURA.

Cyclura, part., *Harlan, Journ. Ac. Philad.* iv. 1824, p. 250; *Wiegm. Herp. Mex.* pp. 15, 41; *Dum. & Bibr.* iv. p. 214; *Fitzing. Syst. Rept.* p. 56.
Ctenosaura, *Wiegm. Isis*, 1828, p. 371; *Fitzing. l. c.*; *Gray, Cat. Liz.* p. 191; *Bocourt, Miss. Sc. Mex., Rept.* p. 136.
Enyaliosaurus, *Gray, l. c.* p. 192.

Tympanum distinct. Body scarcely compressed; a dorsal crest; dorsal scales small. Upper head-scales small. A very strong transverse gular fold; no gular pouch. Digits compressed, with keeled lamellæ inferiorly. A short series of femoral pores. Tail more or less spinose. Lateral teeth tricuspid; pterygoid teeth. A small sternal fontanelle. A single poststernal rib.

Central America.

Synopsis of the Species.

Lateral caudal spines not stronger than upper	1. *acanthura*, p. 195.
Lateral caudal spines the strongest; tail feebly depressed at the base ..	2. *hemilopha*, p. 197.
Lateral caudal spines the strongest; tail strongly depressed at the base.	3. *quinquecarinata*, p. 198.

1. Ctenosaura acanthura.

Ctenosaura acanthura, *Gray, Cat.* p. 191.
—— pectinata, *Gray, l. c.*
Lacerta acanthura, *Shaw, Zool.* iii. p. 216.
Uromastix acanthurus, *Merr. Tent.* p. 56.
Cyclura teres, *Harlan, Journ. Ac. Philad.* iv. 1824, pp. 246, 250, pl. xvi.; *Wiegm. Herp. Mex.* p. 43.
—— acanthura, *Gray, Phil. Mag.* (2) iii. 1827, p. 55; *Dum. & Bibr.* iv. p. 222; *Sumichrast, Arch. Sc. Phys. Nat.* xix. 1864, p. 49; *Günth. Biol. C.-Am., Rept.* p. 56.
Ctenosaura cycluroides, *Wiegm. Isis*, 1828, p. 371; *Bocourt, Miss. Sc. Mex., Rept.* p. 143.
Iguana (Ctenosaura) cycluroides, *Gray, Griff. A. K.* ix. *Syn.* p. 37.
—— (——) acanthura, *Gray, l. c.* p. 38.
—— (——) armata, *Gray, l. c.*
—— (——) similis, *Gray, l. c.*
—— (——) lanceolata, *Gray, l. c.*
—— (Cyclura) teres, *Gray, l. c.* p. 39.
Cyclura pectinata, *Wiegm. Herp. Mex.* p. 42, pl. ii.; *Dum. & Bibr.* iv. p. 221.

Cyclura articulata, *Wiegm. l. c.* p. 43.
—— denticulata, *Wiegm. l. c.* pl. iii.
Ctenosaura pectinata, *Cope, Proc. Ac. Philad.* 1866, p. 124; *Bocourt, l. c.* p. 140.
—— teres, *Bocourt, l. c.* p. 142.
—— completa, *Bocourt, l. c.* p. 145.

Scales on the body small, smaller, or scarcely smaller on the back than on the belly; the dorsal scales may be perfectly smooth or obtusely keeled. Dorsal crest more or less developed, continuous with the caudal or interrupted on the sacrum. Tail cylindrical or feebly compressed, with annuli of more or less strongly spinose large scales separated by rings of smaller scales; the latter may be almost smooth or strongly keeled. Four to eight femoral pores on each side. Greyish, brownish, olive, or green, darker above, with blackish variegations usually forming more or less regular cross bands on the body and rings on the tail; some specimens (*e–f*) entirely blackish.

	millim.		millim.
Total length	680	Fore limb	105
Head	60	Hind limb........	170
Width of head	44	Tail	410
Body	210		

Lower California and Central America.

The specimens enumerated hereafter under *C. acanthura* perhaps belong to two or more distinct species; and the fact is, that the difference between an adult male of the true *C. acanthura* and one of *C. completa* is great, too great perhaps to justify the course I follow. However, as I fail to find any constant peculiarities enabling me to draw satisfactory diagnoses, I have provisionally united the various forms, leaving to future workers to decide whether I am right or wrong.

A. Specimens representing *L. acanthura* of Shaw, *C. pectinata* and *cycluroides* of Wiegmann, and *C. pectinata* and *teres* of Bocourt. Dorsal and subtibial scales very small, the former smooth or very obtusely keeled; caudal spines moderate; two or three series of smaller, smooth or feebly keeled scales between the sixth and seventh verticils of spines.

a. Hgr.	California.	(Type).
b, c, d. Ad., stuffed.	Mexico.	
e–f. ♂.	Mexico.	Mr. Geale [C.].
g, h. ♀, stuffed.	Putla, Mexico.	
i–k. ♂.	Mazatlan.	Hr. Forrer [C.].
l, m, n. Ad., skins.	Presidio.	Hr. Forrer [C.].
o. Yg.	Presidio.	Hr. Forrer [C.].
p–r, s–u. ♂, ♀, hgr., & yg.	Tres Marias Islands.	Hr. Forrer [C.].
v. Yg.	Ciudad.	Hr. Forrer [C.].
w–x. Hgr. & yg.	Ventanas.	Hr. Forrer [C.].

y-z, a-β. Yg.	Vera Cruz.
γ. Hgr., stuffed.	Tehuantepec.
δ-ε. Yg.	Tehuantepec.

B. Scales on the body and limbs as in the preceding; caudal spines stronger, the verticils, from the sixth and seventh, separated from each other by a single series of nearly smooth scales.

ζ. Ad., stuffed.	Mexico.

C. Apparently the *C. cycluroides* and *completa* of Bocourt. Dorsal scales larger, more distinctly keeled; scales on lower surface of leg rather larger; caudal spines usually stronger, the spinose verticils separated from each other by two or three series of smaller, strongly keeled scales.

η. ♀.	Chiapas.	O. Salvin, Esq. [P.].
θ. ♀, stuffed.	Belize.	
ι. Hgr.	Belize.	
κ. ♂, stuffed.	Honduras.	
λ. Yg.	Honduras.	
μ, ν, ξ. Hgr. & yg.	——?	

D. Dorsal scales intermediate as to size between A and C, distinctly keeled; scales on lower surface of leg as small as in A; caudal scales as in C.

ο-π. ♂ ♀.	Colon.

2. Ctenosaura hemilopha.

Iguana (Cyclura) acanthura (*non Shaw*), *Blainv. Nouv. Ann. Mus.* iv. 1835, p. 288, pl. xxiv. fig. 1.
Cyclura acanthura, part., *Dum. & Bibr.* iv. p. 222.
—— (Ctenosaura) hemilopha, *Cope, Proc. Ac. Philad.* 1863, p. 105.
Ctenosaura acanthura, *Bocourt, Miss. Sc. Mex., Rept.* p. 138.
—— interrupta (*Dugès*), *Bocourt, Le Natur.* 1882, p. 47.

Scales on the body very small, smaller on the back than on the belly, smooth. Dorso-nuchal crest extending to the sacrum in the male, not beyond the anterior two thirds of the back in the female, not continuous with caudal. Tail depressed at the base, above with alternate rings of large spinose and small smooth scales; only the first four whorls of spines separated from one another by two or three series of smaller scales; the spines on the vertebral line strong and forming a crest, but less developed than those on the sides; scales on lower surface of tail narrow and keeled, each segment, beginning from the seventh, composed of only two rings of scales. Five femoral pores. Olive-grey or green above, sides, sacral region, and limbs, with fine dark brown network; back with darker cross bands, of which a narrow one between the shoulders and two broader ones behind it are black; lower surfaces whitish; lower lip with vertical dark brown bars; throat marbled with brown; belly and lower surface of limbs with round dark brown spots.

Lower California.

a. Hgr. ♀.	California.	M. Botta [C.]. Paris Museum [E.]. (One of the types of *C. interrupta.*)
b. Hgr. ♂.	—— ?	Zoological Society.

3. Ctenosaura quinquecarinata.

Enyaliosaurus quinquecarinatus, *Gray, Cat.* p. 192.
Cyclura quinquecarinata, *Gray, Zool. Misc.* p. 59.
—— (Ctenosaura) quinquecarinata, *Cope, Proc. Am. Phil. Soc.* 1869, p. 161.
Ctenosaura (Enyalosaurus) quinquecarinata, *Bocourt, Miss. Sc. Mex., Rept.* p. 138, *and Le Natur.* 1882, p. 47.

Scales on posterior part of back a little larger than on the belly, rhomboidal, obtusely keeled. Dorso-nuchal crest very low, extending to the sacrum. Outer side of tibia with large spinose scales. Tail scarcely longer than head and body, much depressed in its anterior third, cylindrical posteriorly; its upper surface with alternate rings of larger and smaller scales, the central, and especially the two lateral series of the former, very large and spinose; its lower surface with transverse series of smaller, pointed, keeled scales. Six femoral pores on each side. Yellowish-olive above, with a dark brown network and more or less indistinct transverse cross bands; lower surfaces yellowish, uniform or spotted with brown.

	millim.		millim.
Total length	387	Fore limb	55
Head	35	Hind limb	100
Width of head	26	Tail	205
Body	147		

Isthmus of Tehuantepec.

a. Ad., stuffed.	—— ?	(Type.)
b. Ad., stuffed.	Zanatepec, Tehuantepec.	
c. ♀.	Oaxaca.	

39. CACHRYX.

Cachryx, *Cope, Proc. Ac. Philad.* 1866, p. 124; *Bocourt, Miss. Sc. Mex., Rept.* p. 148.

Tympanum distinct. Body a little compressed; no dorsal crest; dorsal scales small. Upper head-scales small. A very strong transverse gular fold; no gular pouch. Digits compressed, with keeled lamellæ inferiorly. A short series of femoral pores. Tail very short, flattened, with whorls of large spines on its upper surface. Lateral teeth tricuspid.

Yucatan.

1. Cachryx defensor.

Cachryx defensor, *Cope, l. c., and Proc. Am. Phil. Soc.* 1869, pl. x.; *Bocourt, l. c.*

Head longer than broad; loreal region concave; nostril on canthus rostralis, lateral; tympanum as large as the eye, without marginal serrations; upper head-scales small, uniform, a little enlarged on the snout; no enlarged occipital. Sides of neck plicate. Scales of body small, slightly imbricate, homogeneous, smooth, in transverse and oblique longitudinal series, larger on the rump, smaller on the sides; a slightly larger vertebral series; ventral scales smooth, as large as dorsals. Scales of fore limb moderate, some of those of femur and tibia much larger, spiniferous. Digits short. Femoral pores six to nine. Tail shorter than the body; caudal whorls fifteen, the scales below narrowed, keeled, the carina prolonged into a flat spine; spiniferous superior whorls seven, spines nearly erect, those of the median row smaller. Bright olivaceous; shoulder and two bands on humerus, and the anterior part of dorsum, with interscapular region, black, the latter with two cross series of green spots, more or less distinct on the whole body in younger specimens; in older specimens median dorsal region bright rufous.

	millim.		millim.
Total length	215	From snout to vent ..	127
From snout to gular fold	41	Fore limb	54

Yucatan.

40. HOPLOCERCUS.

Hoplocercus, *Fitzing. Syst. Rept.* p. 78; *A. Dum. Arch. Mus.* viii. p. 561.

Pachycercus (*Dugès & Brac.*), *A. Dum. Rev. & Mag. Zool.* vi. 1854, p. 558.

Tympanum distinct. Body depressed; no dorsal crest; dorsal lepidosis heterogeneous. Upper head-scales small. A strong transverse gular fold; no gular sac. Digits compressed, with keeled lamellæ inferiorly. A few femoral pores. Tail short or moderate, flat, or cylindrical, with whorls of spinose scales. Lateral teeth tricuspid; pterygoid teeth. No sternal fontanelle. Abdominal ribs.

South America.

1. Hoplocercus spinosus.

Hoplocercus spinosus, *Fitzing. l. c.*; *A. Dum. l. c.* p. 562.

Pachycercus aculeatus (*Dugès & Brac.*), *A. Dum. l. c.* pp. 239, 561, pl. xii.

Head longer than broad; loreal region perpendicular; nostril below the canthus rostralis, a little nearer the orbit than the end of the snout; tympanum a little larger than the eye-opening, without marginal denticulation. Upper head-scales very small, convex, obtusely keeled; occipital scarcely enlarged. Sides and lower surface of neck strongly plicate. Body slightly depressed, strongly plicate on the sides. Dorsal scales small, irregular, obtusely keeled, juxtaposed, intermixed with larger, obtusely keeled or spinose tubercles forming irregular transverse series; lateral scales minute, granular, intermixed with small tubercles; ventral scales squarish, smooth, larger than dorsals. Gular scales granular. Limbs short; the adpressed hind limb reaches the axilla; some large spinose tubercles on the femur and tibia; digits short. Three to five femoral pores on each side. Tail about as long as the distance between the end of the snout and the fore limb, flat, as broad as the body, ending in a point; the scales on its upper surface of unequal size, rugose, keeled and spinose, with a series of larger ones along the middle; largest of all, forming long curved spines, are the scales of the outer series on each side; infracaudal scales small, smooth, like the ventrals. Brown above, spotted with blackish and yellowish; an oblique yellowish band on the shoulder; lower surfaces whitish; middle of belly and a ⊥-shaped figure on the throat black.

	millim.		millim.
Total length	137	Fore limb	34
Head	26	Hind limb	47
Width of head	18	Tail	44
Body	67		

Brazil.

a, b. Ad. stuffed.	St. Paul.	
c. Skin.	S. America.	Sir A. Smith [P.].
d. ♂.	S. America.	

2. Hoplocercus annularis.

Hoplocercus annularis, *O'Shaughn. Proc. Zool. Soc.* 1881, p. 244, pl. xxv. fig. 2.

Upper head-scales small, convex, smooth, a little enlarged and flat on supraocular regions; no distinct occipital. Dorsal and lateral scales finely granular, intermixed with obtusely keeled suboval tubercles, forming longitudinal series on the back and irregular vertical series down the flanks. Limbs with large conical tubercles; the adpressed hind limb reaches the ear; digits moderately elongate. Two femoral pores on each side. Tail a little longer than head and body, cylindrical, regularly encircled by whorls of large spinose tubercles, alternating with several series of minute scales; subcaudal scales less unequal and keeled. Brown, with transverse blackish bands on the back; an oblique yellowish band on the shoulder; lower surfaces brownish white, neck dark brown.

	millim.		millim.
Total length	141	Fore limb	26
Head	17	Hind limb	42
Width of head	12	Tail	80
Body	44		

Ecuador.

a. ♂. Canelos. Mr. Buckley [C.]. (Type.)

41. DIPSOSAURUS.

Dipsosaurus, *Hallow. Proc. Ac. Philad.* 1854, p. 92; *Bocourt, Miss. Sc. Mex., Rept.* p. 146.

Tympanum distinct. Body subcylindrical; a small dorsal crest; dorsal scales small. Upper head-scales small. A strong transverse gular fold; no gular pouch. Digits compressed, with keeled lamellæ inferiorly. A long series of femoral pores. Tail long, cylindrical. Lateral teeth tricuspid; pterygoid teeth. No sternal fontanelle. No abdominal ribs.

South-western North America.

1. Dipsosaurus dorsalis.

Crotaphytus dorsalis, *Baird & Gir. Proc. Ac. Philad.* 1852, p. 126.
Dipsosaurus dorsalis, *Hallow. l. c., and Rep. U.S. Expl. Surv.* xiii. pt. iv. p. 7, pl. vi. fig. 1; *Baird, U.S. Mex. Bound. Surv., Herp.* pl. xxxii. figs. 7–13; *Bocourt, l. c.*

Snout short, with obtuse canthus rostralis; nostril lateral, vertically elliptic, nearer the end of the snout than the orbit; upper head-scales small, smooth, convex on the forehead and snout; supraocular scales smaller still; an elongate infraorbital scute; labials small; anterior border of ear denticulated. A dorsal crest of slightly enlarged keeled scales. Dorsal scales small, keeled, rhomboidal, juxtaposed, the keels oblique, directed towards the dorsal line; ventral scales imbricate, smooth. Digits moderately long. Twenty-two or twenty-four femoral pores on each side. Tail [twice as long as head and body] covered with equal small keeled scales, forming rings, and a median series of larger scales, forming a serrated ridge. Brownish above, with black reticulation enclosing whitish subcircular spots on the sides; tail semiannulated with darker and lighter; lower surfaces white, the throat with narrow, oblique bluish streaks.

	millim.		millim.
Total length (tail reproduced)	300	Body	109
Head	26	Fore limb	52
Width of head	21	Hind limb	98

Lower California to Arizona.

a. ♂. La Paz, California. Smithsonian Institution [P.].

42. SAUROMALUS.

Sauromalus, *A. Dum. Arch. Mus.* viii. p. 535; *Cope, Proc. Ac. Philad.* 1864, p. 177; *Bocourt, Miss. Sc. Mex., Rept.* p. 149.
Euphryne, *Baird, Proc. Ac. Philad.* 1858, p. 253.

Tympanum distinct. Body depressed; no dorsal crest; dorsal scales small. Upper head-scales small. A strong transverse gular fold; no gular pouch. Digits with tubercularly keeled lamellæ inferiorly. A long series of femoral pores. Tail rather short, cylindrical. Lateral teeth tricuspid; no pterygoid teeth (?).

South-western North America.

1. Sauromalus ater.

Sauromalus ater, *A. Dum. l. c.* p. 536, pl. xxiii. fig. 3; *Bocourt, l. c.* pl. xvii. bis. fig. 11.
Euphryne obesus, *Baird, l. c., and U.S. Mex. Bound. Surv., Rept.* pl. xxvii.

Habit stout, head and body much depressed. Snout narrowed, obtuse; nostril a little nearer the end of the snout than the orbit, directed upwards and slightly outwards; upper head-scales small, smooth, a little enlarged on the snout; occipital not enlarged; labials small; four conical scales forming a strong denticulation on the anterior edge of the ear. Sides of neck strongly plicate, with some large conical scales. Flanks longitudinally plicate; a slight median groove along the back. Dorsal scales small, squarish, smooth, forming regular transverse series; flanks granulate; ventral scales smooth, a little smaller than dorsals. Limbs strong, covered with scales larger than those on the body; they are smooth on the arms and thighs, tubercular on the forearms and legs; digits short. Fourteen or fifteen femoral pores on each side. Tail scarcely longer than head and body, cylindrical, depressed at the base, covered with moderate-sized, equal, keeled, slightly spinose scales arranged in verticils. Yellowish brown or olive-green, head and limbs somewhat lighter; back dotted with black, with five broad transverse bars in the young.

	millim.		millim.
From snout to vent	128	Tail	132
Head	24		

Lower California and Sonora.

43. CROTAPHYTUS*.

Crotaphytus, *Holbr. N. Amer. Herp.* ii. p. 79; *Baird & Gir. in Stansb. Exped. Gr. Salt Lake*, p. 339; *Bocourt, Miss. Sc. Mex., Rept.* p. 153.
Leiosaurus, part., *A. Dum. Arch. Mus.* viii. p. 532.

Tympanum distinct. Body depressed; no dorsal crest; dorsal scales small. Upper head-scales small. A transverse gular fold; no gular pouch. Digits with sharply keeled lamellæ inferiorly. A long series of femoral pores. Tail long, cylindrical. Lateral teeth tricuspid; pterygoid teeth. No sternal fontanelle. No abdominal ribs.

Southern North America.

1. Crotaphytus collaris.

Agama collaris, *Say, in Long's Exped. Rocky Mount.* ii. p. 252; *Harlan, Med. Phys. Res.* p. 142, pl. —. fig. 4.
Crotaphytus collaris, *Holbr. l. c.* pl. x.; *Wied, Nov. Act. Ac. Leop.-Car.* xxxii. 1865, p. 58; *Baird, Rep. U.S. Expl. Surv.* xiii. pt. iii. p. 17, pl. xxiv. fig. 1; *Cope, Proc. Ac. Philad.* 1866, p. 302; *Bocourt, l. c.* p. 154, pl. xvii. bis. figs. 5, 6.
Leiosaurus collaris, *A. Dum. l. c.*

Head large, much depressed, very distinct from neck, especially in the males, in which it bears resemblance to that of *Gecko verticillatus*; nostril large, turned upwards and outwards, a little nearer the end of the snout than the orbit; ear-opening large, vertically reniform; scales on the snout a little enlarged, irregular, convex; a series of enlarged supraorbital scales, in contact or fusing with each other between the orbits; supraocular scales small; back of the head with small granules; occipital not enlarged; labials very small; a series of enlarged infraorbital scales, very variable in number and in size, the median one sometimes much elongate, owing to the fusion of two or three scales. Throat covered with small granules, which are slightly enlarged and flat in front of the gular fold. Sides of the neck strongly plicate. Dorsal scales uniform small juxtaposed granules; ventral scales larger, flat, hexagonal. Limbs long; the adpressed hind limb reaches the eye or the tip of the snout; digits rather long. Seventeen to twenty femoral pores on each side. Tail slender, cylindrical, nearly twice as long as head and body, covered with uniform, small, smooth or feebly keeled scales. Brownish or olive above, back guttate with whitish; two parallel oblique black bands on each side across scapular region, usually separated on the median line of the nape; lower surfaces white; throat frequently spotted with grey.

---

* *Crotaphytus reticulatus* Baird, Proc. Ac. Philad. 1858, p. 253.—Texas.
Crotaphytus copeii, Yarrow, Proc. U.S. Nat. Mus. v. 1882, p. 441.—La Paz, California.

	♂.	♀.
	millim.	millim.
Total length	302	306
Head	31	27
Width of head	27	22
Body	76	73
Fore limb	44	41
Hind limb	90	82
Tail	195	206

Southern and Central United States; North Mexico.

a–e. ♂ ♀.	N. Texas.	E. D. Cope, Esq. [E.].
f. ♂.	Creek Boundary.	
g. ♀.	Colorado.	W. Treharne, Esq. [P.].

2. Crotaphytus wislezenii.

Crotaphytus wislezenii, *Baird & Gir. Proc. Ac. Philad.* 1852, p. 69, *and in Stansbury's Exped. Gr. Salt Lake* p. 340, pl. iii., *and U.S. Mex. Bound. Surv., Rept.* p. 7, pl. xxxi.; *Bocourt, Miss. Sc. Mex., Rept.* p. 155, pl. xvii. bis. fig. 4.
—— gambelii, *Baird & Gir. Proc. Ac. Philad.* 1852, p. 126.
—— fasciatus, *Hallow. Proc. Ac. Philad.* 1852, p. 206, *and in Sitgreave's Exped. Zuni & Color. Riv.* p. 115, pl. v.
Leiosaurus hallowellii, *A. Dum. Arch. Mus.* viii. p. 533.

This species differs from *C. collaris* in the following points:— Head smaller and narrower; nostril more lateral; upper head-scales smaller, those bordering the orbit above not differentiated, in three or four longitudinal series on the interorbital region; a much elongate scute below the orbit. Male with enlarged postanal scales. Coloration very variable; no black collar. The specimens in the collection are greyish brown above, with large round brown spots and narrow transverse light (vermilion) bands on the back, limbs, and tail; lower surfaces white, throat spotted or streaked with grey.

	♂.	♀.
	millim.	millim.
Total length	314	255
Head	26	26
Width of head	20	20
Body	78	79
Fore limb	43	44
Hind limb	85	87
Tail	210	150

Western North America, from Oregon to North Mexico.

a. Yg.	Oregon.	
b–c. ♂ ♀.	Fresno, California.	Smithsonian Institution [P.].
d. ♀.	California.	

44. PETROSAURUS.

Tympanum distinct. Body depressed; no dorsal crest; dorsal scales small, granular, smooth, uniform. Upper head-scales a little enlarged, occipital very large. Two strong transverse gular folds without denticulation; no gular sac. Digits with keeled lamellæ inferiorly. A long series of femoral pores. Tail long, feebly depressed. Lateral teeth tricuspid; no pterygoid teeth. A sternal fontanelle. No abdominal ribs.

Lower Calfornia.

1. Petrosaurus thalassinus.

Uta thalassina, *Cope, Proc. Ac. Philad.* 1863, p. 104.

Head rather small, snout narrowed; nostril large, superior, equally distant from the orbit and the end of the snout; ear-opening large, oval, with a denticulated lobe of skin anteriorly; head-scales smooth; a single scale in the middle between the orbits; three longitudinal series of supraocular scales, surrounded by granules; occipital nearly as large as the tympanum; a large infraorbital; upper labials seven or eight, very low; lower labials larger. Throat granular, with two strong transverse folds, the anterior forming a small pouch on each side; a third, less marked, anterior fold; sides of neck strongly plicate. Dorsal and lateral scales uniform small, convex, smooth granules; ventrals larger, flat, squarish or hexagonal. Limbs moderate, with larger keeled scales; the adpressed hind limb reaches the tympanum; digits strong, compressed. Sixteen or seventeen femoral pores on each side. Tail slightly depressed, slender, covered with rings of small, squarish, obliquely keeled scales; its length once and a half that of head and body. Olive above, fore limbs and anterior part of back with a fine black network; back with transverse series of pale greenish spots alternating with dark cross bands, of which the three anterior are of an intense black, bordered with pale greenish posteriorly; tail with darker and lighter cross bars; lower surfaces greenish white, throat blackish.

	millim.		millim.
Total length	311	Fore limb	52
Head	30	Hind limb	81
Width of head	23	Tail	185
Body	96		

Lower California.

a. ♂. Cape St. Lucas. Smithsonian Institution [P.].

45. CALLISAURUS.

Callisaurus, *Blainv. Nouv. Ann. Mus.* iv. 1835, p. 286; *Dum. & Bibr.* iv. p. 324; *Gray, Cat. Liz.* p. 226; *Bocourt, Miss. Sc. Mex., Rept.* p. 158.

Megadactylus, *Fitzing. Syst. Rept.* p. 59.
Homalosaurus, *Hallow. Proc. Ac. Philad.* 1852, p. 179.

Tympanum distinct. Body depressed: no dorsal crest; dorsal scales small, uniform. Upper head-scales small, occipital very large. Two transverse gular folds, the posterior with denticulated border; no gular pouch. Digits with sharply keeled lamellæ inferiorly. A long series of femoral pores. Tail rather short, depressed. Lateral teeth indistinctly tricuspid; no pterygoid teeth. A large sternal fontanelle. No abdominal ribs.

South-western North America.

1. Callisaurus draconoides.

Callisaurus draconoides, *Gray, Cat.* p. 227.
Callisaurus draconoides, *Blainv. l. c.* p. 286, pl. xxiv. fig. 2; *Dum. & Bibr.* p. 326; *Bocourt, l. c.* p. 158, pl. xvii. bis. fig. 10.
Megadactylus draconoides, *Fitzing. l. c.* p. 60.
Homalosaurus ventralis, *Hallow. l. c.* p. 179, *and in Sitgreave's Exped. Zuni & Color. Riv.* p. 117, pl. vi.
Callisaurus ventralis, *Baird, U.S. Mex. Bound. Surv., Rept.* p. 8.

Head small, much depressed, upper lip projecting; nostril large, superior; ear-opening large, vertically oval. Head-scales small, irregular and smooth, largest on the snout; occipital large, a little broader than long; a large infraorbital scale; upper labials rather large and oblique. Dorsal scales uniform small juxtaposed granules; ventral scales larger, flat, squarish or subhexagonal, smooth. A slight fold along each side. Limbs very long; the adpressed hind limb reaches the tip of the snout, or beyond; digits long. Sixteen or seventeen femoral pores on each side. Male with enlarged post-anal scales. Tail long and slender, depressed, covered with small smooth scales. Greyish above, dotted with white; tail and limbs with angular dark cross bands; a white and blackish streak along the hind side of the thighs; lower surfaces white; a bluish band (males) on each side of the belly, crossed by two [or three] oblique black spots; lower surface of tail with transverse black spots.

	millim.		millim.
Total length (tail reproduced)	155	Body	76
Head	18	Fore limb ..	45
Width of head	13	Hind limb ..	77

California to Texas.

a–b. ♂. —— ? Smithsonian Institution [P.].

46. UMA.

Uma, *Baird, Proc. Ac. Philad.* 1858, p. 253; *Cope, eod. loc.* 1866, p. 310.

"Ears distinct. A very long infraorbital plate. Palate without teeth. Outer face of upper labials plane and broadly vertical; the labials themselves much imbricated, and very oblique. Scales of body above equal, much smaller than ventral ones. Interorbital space with two series of plates. Claws very long, slender and straight. Sides with a round black spot. Tail?"

According to Cope, "this genus is distinguished from *Callisaurus* by the presence of a series of spines, movable on their bases, on the outer margin of the foot."

Arizona.

1. Uma notata.

Uma notata, *Baird, l. c.*

"Head about two fifths the head and body. Above light pea-green, spotted with darker green. Beneath white. Head and body about two inches long."

Mohave Desert.

47. HOLBROOKIA.

Holbrookia, *Girard, Proc. Am. Assoc. Adv. Sc.* iv. 1851, p. 201, *and in Stansbury's Exped. Gr. Salt Lake*, p. 341; *A. Dum. Arch. Mus.* viii. p. 545; *Bocourt, Miss. Sc. Mex., Rept.* p. 160.

Cophosaurus, *Troschel, Arch. f. Nat.* 1850*, p. 389.

Tympanum concealed. Body depressed; no dorsal crest; dorsal scales small, uniform. Upper head-scales small, occipital large. Two strong transverse gular folds, the posterior with denticulated edge; no gular pouch. Digits with sharply keeled lamellæ inferiorly. A long series of femoral pores. Tail rather short, subcylindrical. Lateral teeth subconical; no pterygoid teeth. A large sternal fontanelle. No abdominal ribs.

Southern North America.

Synopsis of the Species.

I. The adpressed hind limb reaches the tip of the snout; foot nearly as long as the distance between the axil and the groin.

Tail much flattened, with black spots inferiorly .... 1. *texana*, p. 208.

Tail roundish, immaculate inferiorly; tibia not longer than the distance between the end of the snout and the antepectoral fold; a pair of black spots on each side of the belly .... 2. *propinqua*, p. 208.

Tail roundish, immaculate inferiorly; tibia longer than the distance between the end of the snout and the antepectoral fold; no spots on the sides of the belly .... 3. *elegans*, p. 209.

II. The adpressed hind limb does not reach the eye; foot measuring one half to two thirds the distance between the axil and the groin.

---

* Issued 1852.

One or two black latero-ventral spots, none under the tail .... 4. *maculata*, p. 209.
No black spots on sides; under surface of tail black-spotted .... 5. *lacerata*, p. 209.

1. Holbrookia texana.

Cophosaurus texanus, *Trosch. Arch. f. Nat.* 1850, p. 389, pl. vi.
Holbrookia texana, *Baird & Gir. Proc. Ac. Philad.* 1852, p. 125, *and U.S. Mex. Bound. Surv., Rept.* p. 8, pl. xxx.; *Bocourt, Miss. Sc. Mex., Rept.* p. 163; *Cope, Bull. U.S. Nat. Mus.* no. 17, 1880, p. 14.
—— affinis, *Baird & Gir. ll. cc.*; *Boc. l. c.*

Head small, much depressed; nostril large, superior; head-scales flat, smooth; occipital broader than long; a large infraorbital; upper labials six or seven, oblique. Dorsal scales minute, hexagonal, flat, smooth; ventrals much larger. Edge of the gular fold very feebly denticulated. Limbs long; the adpressed hind limb reaches the end of the snout; foot as long as the distance between axilla and groin. Twelve to fourteen femoral pores on each side. Male with enlarged postanal scales. Tail strongly depressed, slightly longer than head and body. Pale grey above, with darker grey and white markings; lower surfaces white; throat and a large patch on each side of the belly, blue in the male, the latter crossed by two crescentic black bands; tail with a series of black transverse spots inferiorly; a dark streak along the hind side of the thighs.

	millim.		millim.
Total length	115	Fore limb	29
Head	12	Hind limb........	48
Width of head	10	Tail	60
Body	43		

Texas to California; Northern Mexico.

a. ♂. N. Texas. E. D. Cope, Esq. [E.].

2. Holbrookia propinqua.

Holbrookia propinqua, *Baird & Gir. Proc. Ac. Philad.* 1852, p. 126; *Bocourt, Miss. Sc. Mex., Rept.* p. 162; *Cope, Bull. U.S. Nat. Mus.* no. 17, 1880, p. 15.

Intermediate between *H. texana* and *H. maculata*. Head and general scaling as in the latter; limbs as in the former. Tail subcylindrical [longer than head and body]. Fifteen or sixteen femoral pores on each side. Grey above, sides dotted with white; a series of large angular dark spots on each side of the vertebral line; a pair of bluish-black spots on each side of the belly; tail immaculate inferiorly; a dark streak along the hind side of the thighs.

From snout to vent 51 millim., hind limb 46 millim.

Texas and Mexico.

a. ♂. Presidio, W. Mexico. Hr. A. Forrer [C.].

3. Holbrookia elegans.

Holbrookia elegans, *Bocourt, Miss. Sc. Mex., Rept.* p. 164, pl. xvii. bis. fig. 8.

Supraocular scales subequal; occipital as long as broad. Dorsal scales keeled. Tibia longer than the distance between the end of the snout and the gular fold. Greyish above, with four longitudinal series of angular brown spots on each side of the vertebral line; those of the upper series largest and white-edged; a blackish streak along the hind side of the thighs; lower surfaces yellowish, no lateral or caudal spots.

Mazatlan.

4. Holbrookia maculata.

Holbrookia maculata, *Girard, Proc. Am. Assoc. Adv. Sc.* iv. 1851, p. 201, *and in Stansbury's Exped. Gr. Salt Lake*, p. 342, pl. iv. figs. 1–3; *A. Dum. Arch. Mus.* viii. p. 546; *Bocourt, Miss. Sc. Mex., Rept.* p. 161, pl. xvii. bis. fig. 7; *Cope, Bull. U.S. Nat. Mus.* no. 17, 1880, p. 15.

—— approximans, *Baird, Proc. Ac. Philad.* 1858, p. 253; *Bocourt, l. c.* p. 163.

Head relatively a little larger than in *H. texana*; upper head-scales more convex; occipital as long as broad, or a little broader than long; upper labials very narrow and very oblique, five or six. Dorsal scales not quite so small, granular or subhexagonal, smooth or feebly keeled. Edge of the gular fold strongly denticulated. Limbs shorter; the adpressed hind limb does not reach the eye; foot measuring one half to two thirds the distance between axilla and groin. Eleven to fifteen femoral pores on each side. Male with enlarged postanal scales. Tail subcylindrical, shorter than head and body. Grey above, vertebral region frequently lighter, with a series of large darker spots on each side; sometimes with scattered whitish dots; one or two black spots on each side of the belly.

	millim.		millim.
Total length	104	Fore limb	24
Head	12	Hind limb........	39
Width of head	11	Tail	49
Body	43		

Southern United States; North Mexico.

a–b. ♀.	Cherokees, Tennessee.	
c–d. ♀.	Kansas.	
e–f. ♂ ♀.	N. Texas.	E. D. Cope, Esq. [E.].
g–h, i. ♂ ♀.	Mexico.	

5. Holbrookia lacerata.

Holbrookia lacerata, *Cope, Bull. U.S. Nat. Mus.* no. 17, 1880, p. 15.

Allied to *H. maculata*. Tail cylindrical, a little longer than the body. Hind foot short, less than one third of head and body. Six or eight supraocular scales surrounded by granules; scales of snout tubercular; labials less elongate, five oblique, one flat. Femoral pores twelve or thirteen. Colour light brown, with six pairs of transverse dark brown bars between the scapular region and groin, which extend downwards and backwards on the abdomen: their posterior border is serrate or digitate, and edged with yellowish, producing a variegated pattern; a pale band on inferior part of side, which is crossed by the ends of the lateral spots; no bluish-black spots on the sides of the belly; transverse bluish-black spots on the inferior side of the tail.

Texas.

48. UTA.

Uta, *Baird & Gir. in Stansb. Exped. Gr. Salt Lake*, p. 344.
Urosaurus, *Hallow. Proc. Ac. Philad.* 1854, p. 92.
Phymatolepis, *A. Dum. Arch. Mus.* viii. p. 548; *Bocourt, Miss. Sc. Mex., Rept.* p. 165.

Tympanum distinct. Body depressed; no dorsal crest; dorsal scales small, keeled, uniform or heterogeneous. Upper head-scales enlarged, occipital very large. A strong transverse gular fold, with denticulated edge; a second anterior fold sometimes present. Digits with keeled lamellæ inferiorly. Femoral pores. Tail moderate, subcylindrical, covered with large keeled scales. Lateral teeth tricuspid; no pterygoid teeth. A large sternal fontanelle. No abdominal ribs.

North America and Mexico.

In all the species the head is relatively small, the snout short, with angular canthus; nostril superior, near the end of the snout; upper head-scales smooth. The digits are slender and pectinate inferiorly. Males possess enlarged postanal scales.

Synopsis of the Species.

I. Dorsal scales uniform, larger than those on the flanks.

Dorsal scales minute, a little larger than those on the sides . . . . 1. *stansburiana*, p. 211.
Dorsal scales much larger than those on the sides . . . . 2. *elegans*, p. 211.

II. A band of six to eight longitudinal series of equal-sized enlarged scales along the middle of the back.

No lateral series of enlarged scales . . . . 3. *nigricauda*, p. 212.
A lateral series of scales imitating the lateral line of fishes . . . . 4. *ornata*, p. 213.

III. Dorsal lepidosis heterogeneous.

A band of six or eight longitudinal series of enlarged dorsal scales, the two

median smaller; an irregular lateral series of widely separated enlarged scales	5. *gratiosa*, p. 213.
Larger dorsal scales in four regular series, two on either side of the median smaller ones	6. *symmetrica*, p. 213.
A series of very large scales on each side of the vertebral line, where they are separated from each other by one or two series of smaller scales; a lateral series of enlarged scales; no supratympanic crest	7. *lateralis*, p. 214.
Six enlarged dorsal rows, of which the median of each three is larger than the others; a lateral series; a supratympanic crest of short acute scales......	8. *auriculata*, p. 214.
A series of enlarged scales on each side of the vertebral line, the scales on which are minute; sides with irregularly scattered enlarged scales	9. *bicarinata*, p. 215.
A narrow vertebral band of irregular enlarged scales; sides with scattered enlarged scales; anterior border of ear without projecting scales	10. *irregularis*, p. 216.

1. Uta stansburiana.

Uta stansburiana, *Baird & Gir. in Stansb. Exped. Gr. Salt Lake*, p. 345, pl. v. figs. 4, 5.

Upper surface of body covered with minute equal polygonal scales, somewhat larger than those on the sides; ventrals larger. Two transverse gular folds. Three or four conical and acute scales project from the anterior border of the ear. Tail nearly once and a half as long as head and body. Blackish brown above, marbled with greenish yellow, or with whitish-yellow irregular dots; sometimes a double row of dorsal patches of a deeper black may be seen along the back; below, uniform greenish yellow, except under the head, where the green predominates, intermingled with brownish narrow bands. From snout to vent 50 millim.

Texas to California.

a. ? Yg.	Cape St. Lucas, California.	Smithsonian Institution [P.].

2. Uta elegans.

? Uta schottii, *Baird, Proc. Ac. Philad.* 1858, p. 253.
Uta elegans, *Yarrow, Proc. U.S. Nat. Mus.* 1882, p. 442.

Frontal transversely divided, in contact with or narrowly separated from the occipital; four large supraoculars; three large projecting pointed scales on the anterior border of the ear. Anterior gular fold not continuous across the throat, forming a pouch on each side.

Dorsal scales equal, rhomboidal, sharply keeled, a little smaller than the ventrals, but much larger than the laterals. The adpressed hind limb reaches the eye; scales on upper surface of limbs larger than dorsals. Twelve to fourteen femoral pores on each side. Tail once and two thirds to once and three fourths as long as head and body. Upper caudal scales much larger than dorsals; subcaudals smooth or very obtusely keeled. Bluish or brownish above, dotted with lighter, with or without a double series of dark transverse spots along the back; female with a more or less distinct yellowish, black-edged dorso-lateral band; a large deep-blue spot on each side behind the axilla; belly whitish or bluish; throat blue in the male, blackish-blue spotted with white in the female.

	millim.		millim.
Total length......	150	Fore limb	25
Head	15	Hind limb	45
Width of head....	10	Tail...............	95
Body	40		

California; New Mexico.

a. ♂.	La Paz.	Smithsonian Institution [P.]. (One of the types.)
b. ♂.	La Paz.	Smithsonian Institution [P.].
c–d. ♀.	Deming, N. Mexico.	Mr. Forrer [C.].

3. Uta nigricauda.

Uta nigricauda, *Cope, Proc. Ac. Philad.* 1864, p. 176.

A single frontal, separated from the occipital by the pair of frontoparietals; four large supraoculars; three or four projecting pointed scales on the anterior border of the ear. A strong pouch on each side of the throat, not connected by a transverse fold. A band of six or seven longitudinal rows of equal-sized, hexagonal, strongly keeled dorsal scales, smaller than ventrals; lateral and dorso-lateral scales minute. The adpressed hind limb reaches between the shoulder and the ear; scales on upper surface of hind limb larger than dorsals. Eleven femoral pores on each side. Upper caudal scales much larger than ventrals, strongly keeled, mucronate; subcaudals sharply keeled. Dark grey above, with transverse blackish bands on the body, interrupted on the vertebral line; the anterior of these bands, on the scapular region, intense black; sides and limbs with whitish dots; tail blackish; belly blue on each side (male), deepest near the narrow whitish median line. [Throat in males orange.]

	millim.		millim.
Total length......	95	Fore limb..........	19
Head	10	Hind limb	28
Width of head	8	Tail (reproduced)....	50
Body	35		

Lower California.

a. ♂.	La Paz.	Smithsonian Institution [P.].

4. Uta ornata.

Uta ornata, *Baird & Gir. Proc. Ac. Philad.* 1852, p. 126, *and U.S. Mex. Bound. Surv., Rept.* p. 7.

"This species may be distinguished from *U. stansburiana*, of which it has the general appearance, by a dorsal space covered with five or six rows of scales larger than those on the sides of the body. Along the middle of the sides there exists one row of small scutellæ imitating the lateral line in fishes. The ground-colour is reddish brown, with transversely elongated black patches all along the upper part of the body and tail. The belly is unicolor in the female, whilst it is blue in the male."

A variety (*linearis*, Baird) is distinguished in having "four linear interrupted black stripes instead of transverse bands."

Texas to California.

5. Uta gratiosa.

Urosaurus graciosus, *Hallow. Proc. Ac. Philad.* 1854, p. 92, *and Rep. U.S. Explor. Surv. R. R.* x. pt. iv. 1859, p. 4, pl. vii. fig. 1.
Uta graciosa, *Baird, U.S. Mex. Bound. Surv., Rept.* p. 7.

Frontal divided transversely, in contact with the occipital; six transversely enlarged supraoculars; two projecting pointed scales on the anterior border of the ear. No anterior gular fold. A band of six or eight longitudinal series of enlarged, roundish-hexagonal, strongly keeled dorsal scales, the two median series smaller; the largest of these scales considerably larger than the ventrals; the rest of the dorsal and lateral surfaces covered with minute smooth granules; an irregular lateral series of widely separated enlarged keeled scales. The adpressed hind limb reaches between the shoulder and the ear; scales on upper surface of limbs nearly as large as the largest dorsals. Ten or eleven femoral pores on each side. Tail more than twice as long as head and body; upper caudal scales larger than the largest dorsals; subcaudals sharply keeled. Greyish above, with transverse darker spots, interrupted on the vertebral line; lower surfaces whitish, dotted with grey; belly blue on each side (male).

	millim.		millim.
Total length......	202	Fore limb	25
Head	14	Hind limb	39
Width of head....	10	Tail..............	143
Body	45		

New Mexico to California.

a. ♂. Fort Bliss, New Mexico. Smithsonian Institution [P.].

6. Uta symmetrica.

Uta symmetrica, *Baird, Proc. Ac. Philad.* 1858, p. 253; *Coues, in Wheeler's Explor. W. 100th Mer.* v. p. 597.

"Larger dorsal scales in four regular series, two on either side of the median smaller ones. Head short, depressed, one and a half times as wide as deep. Tail one and a half times the head and body. General colour light brown above, the belly white. Sides with broad transverse bands of blackish."

California.

7. Uta lateralis.

Uta (Phymatolepis) lateralis, *Bouleng. Ann. & Mag. N. H.* (5) xi. 1883, p. 342.

Frontal transversely divided and in contact with the occipital; three or four large supraoculars; three or four pointed scales form a denticulation on the anterior border of the ear. A deep fold or pouch on each side of the throat in front of the collar, not connected by a distinct transverse fold. Body much depressed. A series of very large strongly keeled scales on each side of the vertebral line, where they are separated from each other by one or two series of smaller keeled scales; they are bordered externally by another series of small keeled scales; the size of one of these large scales much exceeds that of the ventrals, and its length equals more than half the diameter of the ear-opening. The rest of the back and sides covered with minute granules; a lateral series of irregular enlarged keeled scales from the neck to above the hind limb. The adpressed hind limb reaches the ear or slightly beyond; the enlarged scales on the limbs smaller than the largest dorsals. Femoral pores, ten to thirteen on each side. Tail about once and a half as long as head and body; upper caudal scales large, equal; subcaudals keeled. Upper surfaces grey, tinged with brown; a yellowish-white streak from the tip of the snout, along the upper lip and passing through the tympanum, to the fore limb; a Z-shaped black mark in front of the arm, the posterior branch upon the latter, the anterior branch on the neck, above the white streak; a series of five or six oval or subrhomboidal black spots along each side of the back; flanks with blackish and whitish spots; limbs transversely barred with oblique black lines; the male's throat light blue in the middle, and the belly of the same colour but with a median white zone.

	millim.		millim.
Total length	135	Fore limb	24
Head	13	Hind limb........	37
Width of head	9·5	Tail	82
Body	40		

Western Mexico.

a. ♂.	Presidio.	Hr. Forrer [C.].	(Types.)
b–e. ♂ & hgr.	Tres Marias Islands.	Hr. Forrer [C.].	

8. Uta auriculata.

Uta auriculata, *Cope, Proc. Bost. Soc. Nat. Hist.* xiv. 1871, p. 303.

Frontal transversely divided. A crest of short acute scales on the ridge of the os quadratum above the large tympanum. Scales above minute, except six carinate dorsal rows, of which the median of each three is larger than the others; a dorso-lateral dermal line on each side is covered by rounded scales larger than those of the adjacent regions. Colour blue above, with eight pairs of black spots between scapular and crural regions, which are more or less connected across the median line; another series of black spots alternating on the sides; a large black crescent on scapula on each side; gular region cross-banded with blackish.

	millim.		millim.
Total length (tail reproduced)	175	Head (to posterior margin of tympanum)	18
From snout to vent .	74		

Socorro, Revillagigedo Islands.

9. Uta bicarinata.

Phymatolepis bicarinatus, *A. Dum. Arch. Mus.* viii. p. 549, pl. xxiii. fig. 2; *Bocourt, Miss. Sc. Mex., Rept.* p. 165, pl. xvii. bis. fig. 9.
Uta bicarinata, *Cope, Proc. Ac. Philad.* 1864, p. 177.

Frontal usually undivided, separated from the occipital by the pair of frontoparietals; four large supraoculars; anterior border of ear-opening not denticulated, or with a few small projecting tubercles. A deep fold or pouch on each side of the throat in front of the collar, not connected by a distinct transverse fold. Dorsal scales minutely granular, intermixed with irregularly scattered enlarged keeled or subconical scales and a more or less regular series of enlarged keeled scales on each side of the vertebral line, which is minutely scaled; the largest dorsal scales larger than the ventrals. The adpressed hind limb reaches the axilla in the female, between the shoulder and the ear in the male; the largest scales on the limbs a little larger than the largest dorsals. Femoral pores, eight to twelve on each side. Tail a little longer than head and body; upper caudal scales of unequal size; subcaudals keeled. Brownish above, spotted or marbled with darker, or dotted with lighter; lower surfaces whitish, throat brown-dotted; the male's belly entirely blue.

	millim.		millim.
Total length	124	Fore limb	24
Head	12	Hind limb........	34
Width of head	10	Tail	69
Body	43		

Western Mexico.

a. ♀.	Tehuantepec.	
b–c. ♂.	Cuernavaca.	
d. ♂.	Presidio.	Hr. Forrer [C.].
e. ♂.	Ventanas.	Hr. Forrer [C.].
f. Yg.	Ciudad.	Hr. Forrer [C.].

10. Uta irregularis.

Phymatolepis (Uta) irregularis, *Fischer, Abh. Nat. Ver. Brem.* vii. 1882, p. 232, pl. xvii. figs. 1–4.

Frontal undivided, separated from the occipital by the pair of frontoparietals; no denticulation on the anterior border of the ear. Dorsal scales granular, with a narrow band of irregular enlarged keeled scales on the vertebral line; a few scattered enlarged scales on the sides; a series of erect pointed scales on each side of the neck and another on the hinder part of the back to above the hind limbs. The adpressed hind limb reaches the axilla. Ten femoral pores on each side. Greenish-grey above, lighter inferiorly; the nuchal and dorsal ridges, a cross-line between the shoulders, and three narrow cross-bands on the back, black; throat yellowish grey, dotted and marbled with black; belly light grey. From snout to vent 95 millim.

Mexico.

49. SCELOPORUS.*

Sceloporus, *Wiegm. Isis,* 1828, p. 369, *and Herp. Mex.* p. 48; *Fitzing. Syst. Rept.* p. 75; *Bocourt, Miss. Sc. Mex., Rept.* p. 167.
Tropidolepis, *Cuv. R. A.* 2nd ed. ii. p. 38; *Dum. & Bibr.* iv. p. 294; *Fitzing. l. c.* p. 76; *Gray, Cat. Liz.* p. 208.
Tropidurus, part., *Wagler, Syst. Amph.* p. 146.

Tympanum distinct. Body depressed; no dorsal crest; dorsal scales imbricate, keeled, equal. Upper head-scales enlarged, occipital very large. No gular fold; no gular pouch. Digits with keeled lamellæ inferiorly. Femoral pores. Tail moderate or long, cylindrical or compressed. Lateral teeth tricuspid; no pterygoid teeth. A sternal fontanelle. No abdominal ribs.

North and Central America.

Synopsis of the Species.

I. Lateral scales as large as or a little smaller than, and passing gradually into, the dorsals.

A. Lateral scales directed obliquely upwards and backwards; the series of femoral pores not meeting medially.

1. The distance between the base of the fifth toe and the extremity of the fourth less than that between the nostril and the arm.

---

* *Sceloporus marmoratus,* Hallow. Proc. Ac. Philad. 1852, p. 178, and in Sitgr. Exped. Zuni & Color. Riv. p. 110, pl. ii. *S. delicatissimus,* Hallow. *ll. cc.* pp. 178, 109, pl. i.

Texas.

a. Dorsal scales forming parallel longitudinal lines.

α. Occipital broader than the parietals.

* Less than 50 scales on the middle of the back between the occipital shield and the base of the tail.

† The distance between the fifth toe and the extremity of the fourth not exceeding that between the end of the snout and the posterior border of the ear.

Less than 40 scales on a line between occipital and tail; 5 to 7 dorsal scales correspond to the length of the shielded part of the head .... 1. *torquatus*, p. 219.
More than 40 scales between occipital and tail; 9 or 10 correspond to the length of the shielded part of the head .... 12. *consobrinus*, [p. 229.

†† The distance between the fifth toe and the extremity of the fourth exceeds that between the end of the snout and the posterior border of the ear.

‡ More than 30 scales between the occipital and the base of the tail; 7 to 10 scales correspond to the length of the shielded part of the head.

Dorsal scales strongly keeled and mucronate; head-shields constantly smooth........ 4. *formosus*, p. 222.
Dorsal scales strongly keeled, moderately or shortly mucronate; head-shields frequently rugose or striated .... 11. *undulatus*, p. 227.
Dorsal scales weakly keeled, shortly mucronate; head-shields smooth .... 5. *yarrovii*, p. 223.

‡‡ Not more than 30 scales between the occipital and the tail; 5 or 6 scales correspond to the length of the shielded part of the head.

Tibia as long as the shielded part of the head .... 3. *acanthinus*, p. 221.
Tibia longer than the shielded part of the head .... 10. *spinosus*, p. 226.

** 50 or more scales between the occipital and the tail.

Dorsal scales in about 64 transverse series, with but slight carination, mucronation and denticulation .... 6. *ornatus*, p. 224.
Dorsal scales in 50 to 56 transverse series, simply pointed, not denticulated ...... 13. *gratiosus*, p. 230.
Dorsal scales in about 80 transverse series . 14. *couchii*, p. 231.

β. Occipital not broader than the parietals.

Occipital broader than long and as broad as the parietals; anterior border of ear feebly denticulated .... 8. *melanorhinus*, p. 225.
Occipital as long as broad and narrower than the parietals; auricular scales large. 9. *zosteromus*, p. 225.

b. Dorsal scales forming slightly oblique longitudinal rows converging towards the vertebral line.

α. Tail subcylindrical.

Less than 50 scales between the occipital and the tail; 9 scales correspond to the length of the head .... 7. *dugesii*, p. 224.
Less than 60 scales between the occipital and the tail: 12 to 15 correspond to the length of the head .... 15. *grammicus*, p. 231.
More than 60 scales between the occipital and the tail; not larger than the ventrals. 16. *microlepidotus*, [p. 232.

β. Tail much compressed ...... 19. *pyrrhocephalus*, [p. 235.

2. The distance between the base of the fifth toe and the extremity of the fourth equals that between the nostril and the arm.

Head-shields smooth; dorsal scales strongly mucronate .... 2. *serrifer*, p. 221.
Head-shields keeled or striated.......... 20. *chrysostictus*, p. 235.

B. Lateral scales longitudinal or nearly so; the series of femoral pores meeting medially.

Tibia usually a little shorter than the shielded part of the head; the distance between the base of the fifth toe and the extremity of the fourth does not exceed the distance between the end of the snout and the posterior border of the ear .... 17. *æneus*, p. 233.
Tibia usually a little longer than the shielded part of the head .... 18. *scalaris*, p. 234.

II. Lateral scales much smaller than, and abruptly differentiated from, the dorsals.

A. Lateral scales squamous, imbricate; 9 or more femoral pores on each side .... 21. *variabilis*, p. 236.

B. Lateral scales squamous, imbricate; less than 9 femoral pores.

Two canthal scales .... 22. *siniferus*, p. 237.
One canthal scale; ventrals obtusely keeled. 23. *fulvus*, p. 238.
One canthal scale; ventrals perfectly smooth. 24. *squamosus*, p. 238.

C. Lateral scales granular; more than 9 femoral pores .... 25. *utiformis*, p. 238.

1. Sceloporus torquatus.

Tropidolepis torquatus, *Gray, Cat.* p. 208.
Sceloporus torquatus, *Wiegm. Isis*, 1828, p. 369, *and Herp. Mex.* p. 49, pl. vii. fig. 1; *Bocourt, Miss. Sc. Mex., Rept.* p. 171, pl. xviii. fig. 9; *Cope, Proc. Am. Philos. Soc.* xxii. 1885, p. 401.
Agama torquata, *Green & Peale, Journ, Ac. Philad.* vi. 1830, p. 231.
Tropidolepis torquatus, *Gray, Griff. A. K.* ix. *Syn.* p. 43, *and Zool. Beechey's Voy.* p. 95, pl. xxx. fig. 2; *Dum. & Bibr.* iv. p. 301.

Head-shields smooth; a series of four or five transversely dilated supraoculars, bordered internally by a series of moderate-sized scales; usually two canthal scales; occipital as long as broad, much larger than the parietals, which are sometimes divided in two pairs; anterior border of ear with a strong denticulation formed by three or four pointed scales, which are as large as or a little larger than those preceding. Dorsal scales much larger than ventrals, as broad as long, moderately keeled, with more or less distinctly denticulated posterior border, not or but shortly mucronate, forming parallel longitudinal series; twenty-seven to thirty-four scales between the occipital shield and the base of the tail; five to seven scales, taken in the middle of the back, correspond to the length of the shielded part of the head; lateral scales keeled, directed obliquely upwards and backwards, as large as or slightly smaller than the dorsals; ventrals smooth, entire, or bi- or tricuspid: thirty-eight to forty-two scales round the middle of the body. The adpressed hind limb reaches the cervical fold or the ear; tibia as long as the shielded part of the head; the distance between the base of the fifth toe and the extremity of the fourth equals or slightly exceeds the distance between the end of the snout and the ear. Fifteen to twenty femoral pores on each side. Caudal scales about as large as dorsals, strongly mucronate. Male with large postanal scales. Olive above, uniform or black-spotted; a broad, black, yellow-edged band from shoulder to shoulder; lower surfaces whitish or greenish in the females, throat and breast spotted with blackish; males with the throat and the sides of the belly cobalt-blue; the breast and median ventral zone black or whitish.

	millim.		millim.
Total length	212	Fore limb	39
Head	22	Hind limb	60
Width of head	20	Tail	120
Body	70		

Mexico.

a. ♀.	Plateau of Mexico.	P. Geddes, Esq. [P.].
b. ♂.	Hacienda del Hobo, S. Mexico.	P. Geddes, Esq. [P.].
c–d, e. ♂ ♀.	Mexico.	

Var. **poinsettii.**

Sceloporus poinsettii, *Baird & Gir. Proc. Ac. Philad.* 1852, p. 126, *and U. S. Mex. Bound. Surv., Rept.* pl. xxix. figs. 1–3.
—— torquatus, var. C, *Bocourt, l. c.* p. 173.
—— torquatus poinsettii, *Cope, l. c.* p. 402.
—— torquatus cyanogenys, *Cope, l. c.*

The largest supraoculars less dilated, forming two longitudinal series. Femoral pores ten to fifteen. Black collar continuous across scapular region, and light-edged; the body and tail sometimes (specs. *a–e*) with broad blackish and narrower pale greenish cross bands.

From snout to vent 120 millim.; tail 145 millim.

Sonora, Texas, Mexico.

a–e. ♂, hgr. & yg.	Ciudad, Mexico.	Hr. Forrer [C.].
f–h. Hgr. & yg.	Jalapa, Mexico.	Mr. Hoege [C.].
i. ♀.	Mexico.	
k. ♂.	California (?).	

Var. **mucronatus.**

Sceloporus torquatus mucronatus, *Cope, l. c.* p. 402.

"Two rows of large supraocular scales. Dorsal scales strongly mucronate. Anterior collar border divided; chin, throat and sides blue."

Monterey, Nuevo Leon.

Var. **melanogaster.**

Sceloporus melanogaster, *Cope, l. c.* p. 400.

Supraoculars as in the typical form. Dorsal scales strongly mucronate and denticulated; twenty-seven scales from the occipital to the base of the tail, and thirty-four round the middle of the body. The adpressed hind limb reaches the shoulder. Eighteen femoral pores. Greenish bronzy above; collar (a continuation of the black of the throat) reduced to a short oblique band in front of the arm, separated from its fellow by five rows of scales; thorax, middle of belly, and groin, black; throat and sides of belly blue.

From snout to vent 130 millim.

Mexico.

a. ♂.	—— ?	Zoological Society.

Var. **ferrariperezii.**

Sceloporus ferrariperezii, *Cope, l. c.* p. 400.

Supraoculars as in the typical form. Usually a single canthal scale. Dorsal scales strongly mucronate. The adpressed hind limb reaching a little anterior to the ear-opening. Eighteen femoral pores. A black scapular spot which sends a narrow black line upwards and backwards, which sometimes meets its fellow of the opposite side, but generally fails to do so by one or two scales.

About the size of *S. dugesii*.

Mexico.

2. Sceloporus serrifer.

Sceloporus serrifer, *Cope, Proc. Ac. Philad.* 1866, p. 124, *and Proc. Am. Philos. Soc.* xxii. 1885, p. 395.

Head-shields smooth; a series of five very large transversely dilated supraoculars, bordered internally by an incomplete series of minute scales; two canthal scales; occipital nearly as long as broad, much larger than the parietals; anterior border of ear with a denticulation formed by six pointed scales, which are scarcely as large as those preceding. Dorsal scales larger than ventrals, strongly keeled, and ending in a long spine; on the posterior part of the back the scales slightly converge towards the vertebral line; thirty-five scales between the occipital shield and the base of the tail; ten scales correspond to the length of the shielded part of the head; lateral scales keeled, directed obliquely upwards and backwards, nearly as large as dorsals; ventrals smooth, distinctly bicuspid; forty-two scales round the middle of the body. The adpressed hind limb reaches between the ear and the orbit; tibia as long as the shielded part of the head; the distance between the base of the fifth toe and the extremity of the fourth equals the distance between the nostril and the arm. Ten or eleven (nine) femoral pores. Caudal scales like the dorsals. Olive above, brown-spotted; a dark brown collar, interrupted in the middle. "Throat and sides of male blue, the latter broadly black-bordered behind and medially."

	millim.		millim.
Total length	164	Fore limb	38
Head	19	Hind limb	52
Width of head	18	Tail..............	92
Body	53		

Yucatan.

a. ♀. Mexico. M. Sallé [C.].

3. Sceloporus acanthinus.

Sceloporus acanthinus, *Bocourt, Miss. Sc. Mex., Rept.* p. 180, pl. xviii. fig. 10, and pl. xix. fig. 4; *Cope, Proc. Am. Phil. Soc.* xxii. 1885, p. 396.

Head-shields smooth; a series of four or five very large transverse supraoculars, bordered internally by an incomplete series of very small scales; two canthal scales; occipital as long as broad, much larger than the parietals; anterior border of ear with a strong denticulation formed by five pointed scales, which are not larger than those preceding. Dorsal scales much larger than ventrals, strongly keeled, denticulated and ending in a long spine posteriorly, forming parallel longitudinal series. Thirty-two scales between the occipital shield and the base of the tail; six scales correspond to the length of the shielded part of the head (eight or nine in the young, owing to the proportionally larger head); lateral scales keeled, directed obliquely upwards and backwards, gradually merging into the ventrals and dorsals; ventrals smooth, distinctly bicuspid; forty scales round the middle of the body. The adpressed hind limb reaches nearly the ear; tibia as long as the shielded part of the head; the distance between the base of the fifth toe and the extremity of the fourth equals the distance between the end of the snout and the middle of the oblique cervical fold. Twelve to fourteen femoral pores. Caudal scales a little smaller than dorsals. [Male with enlarged postanal scales.] Olive-brown above, with a rather ill-defined black collar; throat green, belly whitish (female). " Male grass-green above, with a rather narrow black collar complete above and below; chin greenish, throat, breast, and side of belly blue; a narrow yellowish, black-edged median ventral line."

	millim.		millim.
Total length	165	Fore limb	32
Head	17	Hind limb	48
Width of head	15	Tail	97
Body	51		

The adult male attains the size of *S. torquatus*.

Guatemala; South Mexico.

a–b. ♀ & yg. Jalapa. Mr. Hoege [C.].

4. Sceloporus formosus.

Tropidolepis formosus, *Gray, Cat.* p. 209.
Sceloporus formosus, *Wiegm. Herp. Mex.* p. 50, pl. vii. fig. 2; *Bocourt, Miss. Sc. Mex., Rept.* p. 182, pl. xviii. fig. 3.
Tropidolepis formosus, *Dum. & Bibr.* iv. p. 303.
Sceloporus malachiticus, *Cope, Proc. Ac. Philad* 1864, p. 178, *and Proc. Am. Philos. Soc.* xxii. 1885, p. 394.
—— lunæi, *Bocourt, l. c.* p. 184, pl. viii. bis. fig. 5; *Cope, l. c.* p. 395.
—— smaragdinus, *Bocourt, l. c.* p. 186, pls. xviii. fig. 6, and xix. fig. 1.
—— viviparus, *Cope, l. c.* p. 398.
—— tæniocnemis, *Cope, l. c.* p. 399.
—— torquatus formosus, *Cope, l. c.* p. 402.

Head-shields smooth; a series of four or five more or less transversely enlarged supraoculars, bordered internally by a series of usually moderate scales; one or two canthal scales; occipital as long as broad, much larger than the parietals, of which there are frequently two on each side; anterior border of ear with a strong denticulation formed by four or five pointed scales, which are not larger than those preceding. Dorsal scales much larger than ventrals, usually longer than broad, strongly keeled, denticulated or tricuspid, and ending in a long spine posteriorly, forming parallel longitudinal series; thirty-five to forty-seven scales between the occipital shield and the base of the tail; seven to ten scales correspond to the length of the shielded part of the head: lateral scales keeled, directed obliquely upwards and backwards, gradually merging into the ventrals and dorsals; ventrals smooth, distinctly bicuspid; thirty-six to fifty-four scales round the middle of the body. The adpressed hind limb reaches the shoulder or the ear: tibia about as long as the shielded part of the head; the distance between the base of the fifth toe and the extremity of the fourth equals the distance between the end of the snout and the anterior extremity or the middle of the oblique cervical fold. Eleven to sixteen femoral pores. Caudal scales as large as or a little larger than dorsals. Male with enlarged postanal scales. Olive or green above, uniform or with blackish spots; a black scapular band, rarely meeting its fellow on the nape, extending across the throat in the males; the latter have the greater part of the throat, the belly except a narrow median whitish zone, and frequently also the pectoral region, dark blue; belly yellowish white in the females, the throat frequently blue-green.

	millim.		millim.
Total length	196	Fore limb	40
Head	20	Hind limb	58
Width of head	19	Tail	110
Body	66		

Central America.

a–b. ♂ and yg.	Jalapa.	Mr. Hoege [C.].
c. ♂.	Yzabal, Guatemala.	O. Salvin, Esq. [C.].
d–g, h–l, m–n. ♂, ♀, & yg.	Dueñas, Guatemala.	O. Salvin, Esq. [C.].
o–q. ♂ ♀.	Guatemala.	O. Salvin, Esq. [C.].
r–v. ♂ ♀.	Irazu, Costa Rica.	F. D. Godman and O. Salvin, Esqrs. [P.].

This form connects *S. torquatus* and *S. undulatus*. From the former, it is easily distinguished by the smaller scales and the longer digits; from the latter, however, no absolutely constant structural character appears to distinguish it, or, rather, I have failed to detect any. But the coloration and various minor peculiarities permit its separation from its northern congener.

5. Sceloporus yarrovii.

Sceloporus jarrovii, *Cope, in Wheeler's Rep. Explor. W. 100th Mer.* v. p. 569, pl. xxiii. fig. 2, *and Proc. Am. Philos. Soc.* xxii. 1885, p. 396.

Head-shields smooth; a series of large transverse supraoculars; two canthal scales; occipital as long as broad, much larger than the parietals; two large auricular scales. Dorsal scales larger than ventrals, weakly keeled, shortly mucronate; about forty-three scales between the occipital shield and the base of the tail: ten scales correspond to the shielded part of the head; ventrals smooth, bicuspid; fifty scales round the middle of the body. The adpressed hind limb reaches the ear; tibia as long as the shielded part of the head; the distance between the base of the fifth toe and the extremity of the fourth slightly exceeds the distance between the end of the snout and the posterior border of the ear. Seventeen (or fifteen) femoral pores. Caudal scales a little larger than dorsals. Olive above; a black, light-edged collar; two light bands on the side of the neck, the upper from the eye and continuous with the anterior border of the collar, the lower commencing at the tip of the snout and passing through the ear; gular region and sides of belly blue; chin and middle of belly greenish-white. (Male above bluish-black; dorsal scales all with yellow centres.)

	millim.		millim.
Total length	163	Fore limb	32
Head	17	Hind limb........	45
Width of head	15	Tail	95
Body	51		

Arizona; North-western Mexico.

a. ♀. Ciudad, Mexico. Hr. A. Forrer [C.]

6. Sceloporus ornatus.

Sceloporus ornatus, *Baird, Proc. Ac. Philad.* 1858, p. 254; *Cope, Proc. Am. Philos. Soc.* xxii. 1885, p. 396.

Dorsal scales in about sixty-four transverse series, with but slight carination, mucronation and denticulation. Femoral pores twelve. A well-marked black cervical collar, complete above and margined with yellowish. Colour dark green above, nearly black towards the median line; back with small yellowish spots.

Coahuila, N. Mexico.

7. Sceloporus dugesii.

Sceloporus dugesii, *Bocourt, Miss. Sc. Mex., Rept.* p. 188, pl. xviii. fig. 7; *Cope, Proc. Am. Philos. Soc.* xxii. 1885, p. 396.
—— intermedius, *A. Dugès, La Naturaleza,* iv. 1876, p. 29, pl. i. figs. 21–32.

Head-shields smooth; supraoculars scarcely enlarged transversely; two canthal scales; occipital as long as broad, much larger than the parietals; anterior border of ear with a denticulation formed by four pointed scales, which are not larger than those preceding. Dorsal scales a little larger than ventrals, broader than long, keeled, obtuse, not denticulated, converging to the middle line posteriorly; forty-four to forty-eight scales between the occipital shield and the base of the tail; nine scales correspond to the shielded part of the head; lateral scales keeled, pointed, directed obliquely upwards and backwards; ventrals smooth, entire. Limbs short, tibia as long as the shielded part of the head. Thirteen to fifteen femoral pores. Caudal scales a little larger than dorsals, strongly keeled. Male with enlarged postanal scales. Brownish olive above, with a double series of narrow dark brown spots on the vertebral zone; sides with oblique light lines; a broad blackish, light-edged scapular collar; limbs and tail with brown cross-bands; lower surfaces yellowish in the female, the throat striped with bluish; males with the throat bluish, with concentric darker lines, and the sides of the belly blue, black-edged internally.

	millim.		millim.
Total length	120	Width of head	10
Head (to posterior border of occipital)	11	From snout to vent.	54

Mexico.

8. Sceloporus melanorhinus.

Sceloporus melanorhinus, *Bocourt, Journ. Zool.* v. 1876, p. 341, *and Ann. Sc. Nat.* (6) iii. 1876, art. 12; *Cope, Proc. Am. Philos. Soc.* xxii. 1885, p. 395.

Head-shields smooth; a series of four transversely dilated supraoculars; two canthal scales; occipital broader than long, and as broad as the parietals; anterior border of ear rather feebly denticulated; dorsal scales much larger than ventrals, keeled and ending in a strong point, not denticulated; six scales correspond to the length of the shielded part of the head. Eighteen to twenty femoral pores. Caudal scales smaller than dorsals. No enlarged postanal scales in the male. Ochreous above, tinged with greenish on the flanks, with or without a dorsal series of rhomboidal brown spots; end of snout, lips, and sides of neck dark brown; males with the throat and the sides of the belly cobalt blue, the chin black.

Size of *S. spinosus*.

Tehuantepec, Colima.

9. Sceloporus zosteromus.

Sceloporus zosteromus, *Cope, Proc. Ac. Philad.* 1863, p. 105, *and Proc. Am. Philos. Soc.* xxii. 1885, p. 395.
—— clarkii zosteromus, *Cope, Check-List N. A. Rept.* p. 49.
—— rufidorsum, *Yarrow, Proc. U.S. Nat. Mus.* v. 1882, p. 442.

Closely allied to *S. spinosus*. Parietals wider than occipital, which is not wider than long. Auricular scales much longer than those adjacent. The adpressed hind limb reaches the orbit. Femoral pores more than ten. Male above bright olivaceous, with a reddish longitudinal dorsolateral band on each side, crossed medially by numerous indistinct brown bars, often obsolete; under surfaces and sides of tail yellowish; anterior face of femur, groin, a large spot anterior to brachium, a broad band connecting the latter with the former, and the latter of each side, across posterior gular region, black, which shades into blue on the sides and throat; chin light green; brachium black in front.

From snout to vent 97 millim.

Lower California.

10. Sceloporus spinosus.

Tropidolepis spinosus, *Gray, Cat.* p. 209.
Sceloporus spinosus, *Wiegm. Isis*, 1828, p. 370, *and Herp. Mex.* p. 50, pl. vii. fig. 3; *Baird, U.S. Mex. Bound. Surv., Rept.* p. 5, pl. xxix. figs. 4–6; *Bocourt, Miss. Sc. Mex., Rept.* p. 174, pl. xviii. fig. 2; *Cope, Proc. Am. Philos. Soc.* xxii. 1885, p. 395.
Tropidolepis spinosus, *Gray, Griff. A. K.* ix. *Syn.* p. 43; *Dum. & Bibr.* iv. p. 304.

Head-shields smooth; a series of three or four large transversely dilated supraoculars; two canthal scales; occipital as long as broad, broader than the parietals, which are rather large; anterior border of ear with a strong denticulation formed of four or five pointed scales, the two upper of which are usually larger than those preceding. Dorsal scales much larger than ventrals, as broad as long, strongly keeled and mucronate, distinctly denticulated, forming parallel longitudinal series; twenty-eight to thirty scales between the occipital shield and the base of the tail; five or six scales (seven in the young, owing to the proportionally larger head) correspond to the length of the shielded part of the head; lateral scales keeled, directed obliquely upwards and backwards, graduating into the dorsals and ventrals; latter smooth, bicuspid; thirty-six to thirty-eight scales round the middle of the body. The adpressed hind limb reaches the shoulder or the ear; tibia, in the adult, longer than the shielded part of the head; the distance between the base of the fifth toe and the extremity of the fourth exceeds the distance between the end of the snout and the posterior border of the ear. Seven to ten femoral pores. Caudal scales as large as or a little smaller than dorsals. Male with enlarged postanal scales. Pale olive or yellowish-brown above, with dark brown transverse spots, and a more or less distinct, broad, lighter dorso lateral band; head frequently with fine brown transverse lines across its upper surface and others radiating from the eye; lower surfaces whitish, throat longitudinally streaked with grey or blackish; male with a black antehumeral spot usually connected with its fellow on the throat; the whole or part of the gular region, and the sides of the abdomen, blue; pectoral region and median ventral zone whitish.

	millim.		millim.
Total length	240	Fore limb	48
Head	25	Hind limb........	72
Width of head	22	Tail	135
Body	80		

Mexico.

a. ♀.	Mexico.	
b. ♂.	Puebla.	
c. Hgr.	Altisco.	
d. ♀.	Guanajuato.	
e–l. ♂ ♀ & hgr.	Presidio, W. Mexico.	Hr. A. Forrer [C.].
m–n. ♀ & hgr.	California (?).	

Var. **horridus.**

Tropidolepis horridus, *Gray, Cat.* p. 209.
Sceloporus horridus, *Wiegm. Herp. Mex.* p. 50; *Bocourt, Miss. Sc. Mex., Rept.* p. 178, pl. xviii. fig. 8; *Cope, Proc. Am. Philos. Soc.* xxii. 1885, p. 394.
Tropidolepis horridus, *Dum. & Bibr.* iv. p. 306.
Sceloporus oligoporus, *Cope, Proc. Ac. Philad.* 1864, p. 177.

Distinguished by the smaller number of femoral pores, viz. two to five on each side.

Mexico.

a, b. ♀ & hgr.	Cuernavaca.	
c. Yg.	Ventanas.	Hr. A. Forrer [C.].

Var. **clarkii.**

Sceloporus clarkii, *Baird & Gir. Proc. Ac. Philad.* 1852, p. 127, *and U.S. Mex. Bound. Surv., Rept.* p. 5; *Yarrow, in Wheeler's Rep. Explor. W. 100th. Mer.* v. p. 575, pl. xxiii. fig. 1.
—— magister, *Hallow. Proc. Ac. Philad.* 1854, p. 93, *and Rep. U.S. Explor. Surv. R. R.* x. part iv. 1859, p. 5.
—— floridanus, *Baird, Proc. Ac. Philad.* 1858, p. 254.
—— thayerii (*non B. & G.*), *Bocourt, Miss. Sc. Mex., Rept.* p. 176, pl. xviii. fig. 5.

Femoral pores twelve to fifteen on each side.

S. United States; N. Mexico.

a. ♀.	Colorado Bottom.	Smithsonian Institution [P.].
b. ♀.	California.	
c. ♂.	Mexico.	

This form appears to be completely linked with *S. undulatus.*

11. Sceloporus undulatus.

Tropidolepis undulatus, *Gray, Cat.* p. 208.
Agama undulata (*Bosc*), *Daudin, Rept.* iii. p. 384; *Harlan, Med. Phys. Res.* p. 140.
Lacerta hyacinthina, *Green, Journ. Ac. Philad.* i. 1818, p. 349.
—— fasciata, *Green, l. c.*
Uromastyx undulatus, *Merr. Tent.* p. 57.
Tropidolepis undulatus, *Gray, Griff. A. K.* ix. *Syn.* p. 43; *Dum. & Bibr.* iv. p. 298; *Holbr. N. Am. Herp.* ii. p. 73, pl. ix.
Sceloporus undulatus, *Fitzing. Syst. Rept.* p. 75; *Girard, U.S. Expl. Exped., Herp.* p. 379, pl. xix. figs. 15–21; *Bocourt, Miss. Sc. Mex., Rept.* p. 195, pl. xviii. bis. fig. 11; *Cope, Proc. Am. Philos. Soc.* xxii. 1885, p. 395.

Sceloporus occidentalis, *Baird & Gir. Proc. Ac. Philad.* 1852, p. 175; *Girard, U.S. Expl. Exped., Herp.* p. 383, pl. xix. figs. 8–14.
—— frontalis, *Baird & Gir. l. c.*; *Gir. l. c.* p. 384, pl. xix. figs. 1–7.
—— longipes, *Baird, Proc. Ac. Philad.* 1858, p. 254.

Head-shields smooth or rugose; a series of four or five large transversely dilated supraoculars; two canthal scales; occipital as long as broad, or a little longer than broad, larger than the parietals; anterior border of ear with a denticulation of pointed scales not larger than those preceding. Dorsal scales much larger than ventrals, strongly keeled, more or less strongly mucronate, entire, tricuspid, or very indistinctly denticulated, forming parallel longitudinal series; thirty-three to forty scales between the occipital shield and the base of the tail; seven to nine (exceptionally six) scales correspond to the length of the shielded part of the head; lateral scales keeled, directed obliquely upwards and backwards, nearly as large as dorsals; ventrals smooth, bicuspid: thirty-eight to fifty scales round the middle of the body. The adpressed hind limb reaches between the cervical fold and the orbit; tibia as long as, or longer than, the shielded part of the head; the distance between the base of the fifth toe and the extremity of the fourth exceeds the distance between the end of the snout and the posterior border of the ear. Thirteen to seventeen femoral pores. Caudal scales as large as or a little larger than dorsals. Male with enlarged postanal scales. Greyish, brownish, or olive above, usually with undulating dark brown bands across the back and fine lines across the head; sometimes a more or less distinct lighter dorso-lateral band. Lower surfaces yellowish-white in the female, often spotted with brown or with a brown median ventral line, and a bluish spot on each side of the throat. Male with the throat and the sides of the belly blue, edged with black, the pectoral and median ventral regions yellowish; the throat sometimes nearly entirely black, the blue reduced to a round spot on each side.

	millim.		millim.
Total length	155	Fore limb	29
Head	16	Hind limb	44
Width of head	14	Tail	85
Body	54		

North America, Mexico, Guatemala.

a. ♀.	New Orleans.	
b, c–d. ♂.	California.	J. K. Lord, Esq. [P.].
e–f. ♀.	California.	M. Sallé [C.].
g–h. ♂.	California.	Mr. F. Gruber [C.].
i–k, l. ♂ ♀.	N. America.	Dr. J. Green [P.].
m. ♂ ♀.	N. America.	E. Doubleday, Esq. [P.].
n. ♂.	N. America.	Lord Ampthill [P.].
o. ♀.	N. America.	Sir R. Murchison [P.].
p. ♂.	Yzabal.	O. Salvin, Esq. [C.].
q. ♂.	——?	

Var. **thayeri.**

Sceloporus thayerii, *Baird & Gir. Proc. Ac. Philad.* 1852, p. 127.
—— undulatus thayerii, *Cope, Check-List N. Am. Rept.* p. 49.

"Colour yellowish-green above, with two longitudinal bands of brown or black, and two yellow ones, on each side; sometimes, instead of bands, a series of small patches of the same colour is observed; the black patches also form transverse and undulating narrow bands. The belly is uniform yellow. The male has on each side an elongated patch of blue, with a narrow band of black along the abdominal margin; under the throat there is a median black patch limiting on each side another blue spot."

Gulf of Mexico to Sonora.

Var. **bocourtii.**

Sceloporus biseriatus (*non Hallow.*), *Bocourt, l. c.* p. 197, pl. xviii. bis. fig. 10.
—— smaragdinus (*non Bocourt*), *Cope, in Wheeler's Rep. Explor. W.* 100*th Mer.* v. p. 572, pl. xxiv. fig. 2.
—— undulatus smaragdinus, *Cope, Proc. Am. Philos. Soc.* xxii. 1885, p. 399.

Scales of the auricular denticulation usually distinctly larger than those preceding. Thirty-six to forty-three transverse series of dorsal scales. Olive or brownish above, with four longitudinal series of dark brown spots; the male's lower surfaces sometimes blackish, with the exception of the throat and the sides of the belly, which are blue; throat and sides of the belly constantly blue or blue-green in the female and young, a mode of coloration which, in other forms, is restricted to the male sex.

From snout to vent 83 millims.

Western North America.

a–e. ♀ & yg.	Monterey, California.	
f–g. ♂.	Mt. Whitney, California.	Smithsonian Institution [P.].
h. ♀.	California.	Smithsonian Institution [P.].
i. ♂.	Santa Cruz.	Mr. Forrer [C.].
k. ♂.	——?	

Sceloporus biseriatus, Hallow. Rep. U.S. Explor. Surv. R. R. x. part iv. 1859, p. 6, pls. vi. fig. 2, & viii.; Cope, Proc. Am. Philos. Soc. xxii. 1885, p. 395.—Is perhaps likewise a variety of *S. undulatus*, characterized by the much smaller lateral scales.

California.

12. Sceloporus consobrinus.

Sceloporus consobrinus, *Baird & Gir. in Marcy's Explor. Red River*, p. 237, pl. x. figs. 5–12; *Cope, in Hayden's Rep. U.S. Geol. Surv. Montana*, p. 468; *Yarrow, in Wheeler's Rep. Explor. W.* 100*th Mer.* v. p. 574; *Cope, Proc. Am. Philos. Soc.* xxii. 1885, p. 395.
—— tristichus, *Cope, in Wheeler's Rep. Explor. W. of* 100*th Mer.* v. p. 571.
—— garmani, *Bouleng. Proc. Zool. Soc.* 1882, p. 761, pl. lvi.

Head-shields smooth: a series of four or five transversely dilated supraoculars, the widest not measuring more than half the width of the supraocular region; two canthal scales; occipital as long as broad, or a little longer than broad, larger than the parietals; anterior border of ear with a strong denticulation, formed of three or four large pointed scales. Dorsal scales larger than ventrals, strongly keeled, slightly mucronate, tricuspid, forming parallel longitudinal series; forty-two to forty-four scales between the occipital shield and the base of the tail; nine or ten scales correspond to the length of the shielded part of the head; lateral scales keeled, directed obliquely upwards, a little smaller than dorsals; ventrals smooth, bi- or tricuspid; forty-two or forty-four scales round the middle of the body. The adpressed hind limb reaches the axilla or the cervical fold; tibia slightly shorter than the shielded part of the head; the distance between the base of the fifth toe and the extremity of the fourth equals the distance between the end of the snout and the posterior border of the ear. Twelve to seventeen femoral pores. Caudal scales as large as, or a little larger than dorsals. Male with enlarged postanal scales. Reddish brown above, with a greyish vertebral band and one or two yellowish lateral bands on each side; dark-brown spots between the longitudinal bands; a blackish streak at base of humerus; lower surfaces yellowish white, uniform in the female; male with a small blue spot on each side of the throat, and an elongate black-margined blue blotch on each side of the belly, the white interspace between them measuring twice their width.

	millim.		millim.
Total length	136	Fore limb	23
Head	14	Hind limb	34
Width of head ...	11	Tail	79
Body	43		

N. America, Mexico.

a–c. ♂ ♀.	Dacota.	Zoological Society. (Types of *S. garmani.*)
d. ♂, bad state.	Putla, Mexico.	

13. Sceloporus gratiosus.

Sceloporus graciosus, *Baird & Gir. Proc. Ac. Philad.* 1852, p. 69, *and in Stansb. Exped. Gr. Salt Lake*, p. 346, pl. v. figs. 1–3; *Cope, Proc. Am. Philos. Soc.* xxii. 1885, p. 395.

—— gracilis, *Baird & Gir. Proc. Ac. Philad.* 1852, p. 175; *Girard, U.S. Explor. Exped., Herp.* p. 386, pl. xx. figs. 1–9; *Bocourt, Miss. Sc. Mex., Rept.* p. 190, pl. xviii. fig. 4.

Head-shields smooth; a series of feebly dilated transverse supraoculars; two canthal scales; occipital very large, as broad as long; parietals very small, two on each side; anterior border of ear with a strong denticulation, formed of four or five pointed scales, which are a little larger than those preceding. Dorsal scales larger than ventrals, keeled, sharply pointed, forming parallel longitudinal lines;

fifty-two to fifty-six scales between the occipital shield and the base of the tail; twelve or thirteen (or fourteen) scales correspond to the length of the shielded part of the head; lateral scales keeled, directed obliquely upwards, a little smaller than dorsals; ventrals smooth, more or less distinctly bicuspid; fifty or fifty-two scales round the middle of the body. The adpressed hind limb reaches between the shoulder and the ear; tibia as long as the shielded part of the head; the distance between the base of the fifth toe and the extremity of the fourth equals the distance between the end of the snout and the middle of the oblique cervical fold. Sixteen to eighteen femoral pores. Caudal scales larger than dorsals. Male with enlarged postanal scales. Olive above, with two light yellowish bands along each side, and series of blackish spots, sometimes confluent into longitudinal bands; male with a blue, black-edged patch on each side of the body and some irregular blackish spots on the throat.

	millim.		millim.
Total length (tail reproduced)	80	Body	31
Head	11	Fore limb	17
Width of head	8·5	Hind limb	28

Western North America and Mexico.

a. ♂.	California.	J. K. Lord, Esq. [P.].
b, *c*. ♀.	Putla, Mexico.	

14. Sceloporus couchii.

Sceloporus couchii, *Baird, Proc. Ac. Philad.* 1858, p. 254.

"Cephalic plates smooth. Scales very small. About eighty oblique dorsal rows from head to tail. Femoral pores twenty-five. Colour above dark green, with two lateral light stripes, separated by eighteen rows of scales. Back with irregular spots. Sides with a white band from groin. An obsolete dark indigo patch on each side of the belly, widely separated below. Sides of jaws banded transversely with blue and whitish. A subcircular indigo patch in front of shoulder, surrounded by light blotches."

Santa Caterina, N. Leon.

15. Sceloporus grammicus.

Tropidolepis grammicus, *Gray, Cat.* p. 209.

Sceloporus grammicus, *Wiegm. Isis*, 1828, p. 370, *and Herp. Mex.* p. 51; *Bocourt, Miss. Sc. Mex., Rept.* p. 192, pl. xviii. bis. fig. 12; *Cope, Proc. Am. Philos. Soc.* xxii. 1885, p. 395.

—— pleurostictus, *Wiegm. Isis*, 1828, p. 370.

Tropidolepis grammicus, *Gray, Griff. A. K.* ix. *Syn.* p. 43; *Dum. & Bibr.* iv. p. 308.

Sceloporus heterurus, *Cope, Proc. Ac. Philad.* 1866, p. 322.

Head-shields smooth or slightly rugose; a series of feebly dilated transverse supraoculars; two canthal scales; occipital a little broader than long; parietals very small, usually one pair on each side; anterior border of ear feebly denticulated. Dorsal scales larger than ventrals, strongly keeled, not or but slightly mucronate, forming oblique series; forty-five to fifty scales between the occipital shield and the base of the tail; twelve to fifteen scales correspond to the length of the shielded part of the head; lateral scales keeled, directed obliquely upwards, gradually merging into the dorsals and ventrals; latter smooth, mostly entire; forty-six to fifty scales round the middle of the body. The adpressed hind limb reaches the ear; tibia as long as the shielded part of the head; the distance between the base of the fifth toe and the extremity of the fourth slightly exceeds the distance between the end of the snout and the posterior border of the ear. Fourteen or fifteen (thirteen to sixteen) femoral pores. Caudal scales much larger than dorsals. Male with enlarged postanal scales. Olive above, with transverse black spots; a narrow black scapular collar, interrupted medially; male with a blue, black-edged elongate patch on each side of the belly.

	millim.		millim.
From snout to vent.	40	Fore limb	18
Head	11	Hind limb	26
Width of head	8		

Reaches a size of 66 millim. from snout to vent.

Mexico, Guatemala.

a–b. Hgr. ♂ ♀. Guatemala. F. D. Godman, Esq. [P.].

16. Sceloporus microlepidotus.

Tropidolepis microlepidotus, *Gray, Cat.* p. 200.
Sceloporus grammicus, var. *α*, *Wiegm. Isis*, 1828, p. 370.
—— microlepidotus, *Wiegm. Herp. Mex.* p. 51; *Bocourt, Miss. Sc. Mex., Rept.* p. 194, pl. xviii. bis, fig. 13; *Cope, Proc. Am. Philos. Soc.* xxii. 1885, p. 395.
Tropidolepis microlepidotus, *Dum. & Bibr.* iv. p. 308.
Sceloporus dispar, *Baird & Gir. Proc. Ac. Philad.* 1852, p. 127.

Head-shields usually smooth; one or more series of feebly dilated transverse supraoculars; two canthal scales; occipital usually a little broader than long; parietals very small, usually one pair on each side; anterior border of ear with a denticulation of slightly enlarged pointed scales. Dorsal scales as large as or a little smaller than ventrals, strongly keeled, not or but slightly mucronate, forming oblique longitudinal lines; sixty-two to seventy-eight scales between the occipital shield and the base of the tail; sixteen to twenty (fifteen to twenty-two) scales correspond to the length of the shielded part of the head; ventrals smooth, entire or bicuspid; sixty-six to seventy-eight scales round the middle of the body. The adpressed hind limb reaches the shoulder or the ear; tibia about as long as the shielded part of the head; the distance between the base of the fifth toe and the extremity of the fourth exceeds the distance between the end of the snout and the posterior border of the ear. Fourteen to eighteen (twenty-two) femoral pores. Caudal scales much larger than dorsals. Male with enlarged postanal scales. Brownish, olive, or green above, the sides marbled with black; dorsal region immaculate, or irregularly spotted with black and whitish, more often with a double series of angularly curved black markings pointing forwards; head frequently with transverse black lines; usually a large black spot in front of the shoulder, sometimes connected with its fellow across the throat; lower surfaces yellowish or greenish, throat sometimes marbled with grey: male with the throat and the sides of the belly blue, broadly edged with black near the narrow whitish median ventral line.

	millim.		millim.
Total length	155	Fore limb	30
Head	15	Hind limb	44
Width of head	13	Tail	88
Body	52		

Mexico.

a–g. ♂ ♀.	Ciudad.	Hr. A. Forrer [C.].
h. Hgr.	Ventanas.	Hr. A. Forrer [C.].
i–k, *l*. ♀ & hgr.	Puebla.	
m–p. ♂ ♀.	Orizaba.	
q–t. ♂ & hgr.	Jalapa.	Mr. Hoege [C.].
u–v. ♂.	Mexico.	M. Sallé [C.].
w, *x*. ♂ ♀.	Mexico.	

17. Sceloporus æneus.

Tropidolepis æneus, *Gray, Cat.* p. 210.
Sceloporus æneus, *Wiegm. Isis*, 1828, p. 370, *and Herp. Mex.* p. 50; *Bocourt, Miss. Sc. Mex., Rept.* p. 205, pl. xviii. bis, fig. 4; *Cope, Proc. Am. Philos. Soc.* xxii. 1885, p. 394.
Tropidolepis æneus, *Dum. & Bibr.* iv. p. 309.

Head-shields keeled or striated; an irregular series of feebly dilated transverse supraoculars; one or two canthal scales; occipital as long as broad; parietals very small; anterior border of ear very slightly denticulated. Dorsal scales larger than ventrals, strongly keeled, sharply pointed, forming parallel longitudinal series; forty to forty-four scales between the occipital and the base of the tail; six to ten scales correspond to the length of the shielded part of the head; lateral scales nearly as large as the dorsals, more or less distinctly keeled, forming straight longitudinal or slightly oblique series: ventrals smooth, mostly bicuspid; forty to forty-six scales round the middle of the body. The adpressed hind limb reaches hardly the axilla in the female, the shoulder or a little beyond in the male; tibia not quite as long as the shielded part of the head; the distance between the base of the fifth toe and the extremity of the fourth equals or slightly exceeds the distance between the end of the snout and the ear. Fourteen to nineteen femoral pores on each side, the two series meeting medially. Caudal scales as large as, or a little smaller than dorsals. Male with enlarged postanal scales. Metallic green or coppery above, usually with one or two lighter dorso-lateral bands on each side, with or without four longitudinal series of transverse dark-brown spots; a blackish spot in front of the shoulder; lower surfaces greenish white in the female, the throat and breast mottled with grey; in the male, the throat, breast, and belly entirely blackish-blue, or largely and closely marbled with blackish-blue.

	millim.		millim.
Total length	135	Fore limb	21
Head	14	Hind limb	33
Width of head	12	Tail	75
Body	46		

Mexico.

a–b, c–d. ♂ ♀.	Puebla.
e. ♀.	Orizaba.
f–g. ♀.	Mexico.

18. **Sceloporus scalaris.**

Tropidolepis scalaris, *Gray, Cat.* p. 210.
Sceloporus scalaris, *Wiegm. Isis,* 1828, p. 370, *and Herp. Mex.* p. 50, pl. viii. fig. 2; *Bocourt, Miss. Sc. Mex., Rept.* p. 202, pl. xviii. bis. fig. 9; *Cope, Proc. Am. Philos. Soc.* xxii. 1885, p. 394.
Tropidolepis scalaris, *Gray, Zool. Beechey's Voy.* p. 95, pl. xxx. fig. 3; *Dum. & Bibr.* iv. p. 310.

Head-shields keeled or striated; a more or less regular series of transversely dilated supraoculars; two canthal scales; occipital as long as broad, or a little longer than broad; parietals very small; anterior border of ear very slightly denticulated. Dorsal scales larger than ventrals, strongly keeled, sharply pointed, forming parallel longitudinal series; forty-three to forty-eight scales between the occipital and the base of the tail; seven to ten scales correspond to the length of the shielded part of the head; lateral scales nearly as large as dorsals, more or less distinctly keeled, forming straight longitudinal series; ventrals smooth, mostly bicuspid; thirty-eight to forty-six scales round the middle of the body. The adpressed hind limb reaches the axilla or the shoulder in the female, between the shoulder and the ear in the male; tibia usually a little longer than the shielded part of the head; the distance between the base of the fifth toe and the extremity of the fourth exceeds the distance between the end of the snout and the posterior border of the ear. Sixteen to nineteen femoral pores on each side, the two series meeting medially. Caudal scales as large as or larger than dorsals. Male with enlarged postanal scales. Yellowish brown or olive above, with four regular series of large crescentic brown spots, the series separated by more or less well-defined light longitudinal lines; head with symmetrical dark-brown markings; a black or deep-blue spot in front of the arm; lower surfaces yellowish white, the throat frequently with oblique blackish lines converging posteriorly; an elongate deep-blue patch on each side of the belly in the male.

	millim.		millim.
Total length	148	Fore limb	25
Head	15	Hind limb	40
Width of head	12·5	Tail	85
Body	48		

Mexico to Sonora and South-western Texas.

a–b. ♀.	Altisco.	
c–d. Hgr. & yg.	Hacienda del Hobo.	P. Geddes, Esq. [P.].
e, f–g, h–k. ♂ ♀.	Mexico.	

19. Sceloporus pyrrhocephalus.

Sceloporus pyrrhocephalus, *Cope, Proc. Ac. Philad.* 1864, p. 177, *and Proc. Am. Philos. Soc.* xxii. 1885, p. 394.

Head-shields smooth; one canthal scale; a series of five transverse supraoculars; occipital large, much broader than long; parietals exceedingly small; scales of ear-fringe a little larger than those anterior to them. Dorsal scales much larger than ventrals, strongly keeled, unimucronate, converging to the middle line, in about twenty-five transverse series from nape to crural region; lateral scales larger than abdominals; of the latter a portion only slightly emarginate. Femoral pores twelve. Tail much compressed. Greenish brown, with a broad black band from the scapular region to the groin, light bordered above; below yellowish, sides bluish ash to near the median line, on each side of which is a series of from seven to nine transverse blue bars; upper labial and gular region striped with a series of black or bluish lines, which converge posteriorly on a paler or deeper yellow ground; top of head bright chestnut-red or brown.

	millim.		millim.
Head	13	Hind limb	36
Body	45	Foot	15
Fore limb	24		

Colima, Mexico.

20. Sceloporus chrysostictus.

Sceloporus chrysostictus, *Cope, Proc. Ac. Philad.* 1866, p. 125, *and Proc. Am. Philos. Soc.* xxii. 1885, p. 394.
? Sceloporus cupreus, *Bocourt, Miss. Sc. Mex., Rept.* p. 210, pl. xviii. bis. fig. 2.

Head-shields keeled or striated; two (or one) canthal scales; a series of four or five large transverse supraoculars; occipital broader than long; parietals very small, one or two on each side; anterior border of ear very feebly denticulated. Dorsal scales much larger than ventrals, strongly keeled, simply pointed, forming slightly oblique longitudinal lines; forty to foryt-five scales between the occipital shield and the base of the tail; eight to ten dorsal scales correspond to the length of the shielded part of the head; lateral scales smaller than, but graduating into, the dorsals, keeled and directed obliquely upwards; thirty-six scales round the middle of the body. The adpressed hind limb reaches the anterior border of the orbit or the nostril; tibia longer than the shielded part of the head; the distance between the base of the fifth toe and the extremity of the fourth equals the distance between the arm and the nostril or the tip of the snout. Thirteen to fifteen femoral pores. Caudal scales nearly as large as dorsals. Male with enlarged postanal scales. Brownish or greenish above, strongly metallic; a golden dorso-lateral band; a blackish spot at axilla; hinder side of thighs with a transverse and several vertical blackish bars; lower surfaces yellowish white, throat sometimes brownish; male without distinctive ventral markings.

	millim.		millim.
Total length	157	Fore limb	29
Head	16	Hind limb	52
Width of head	11·5	Tail	95
Body	46		

Guatemala and South Mexico.

a–f. ♂, ♀, & hgr. Yucatan.

21. Sceloporus variabilis.

Tropidolepis variabilis, *Gray, Cat.* p. 209.
Sceloporus variabilis, *Wiegm. Herp. Mex.* p. 51; *Bocourt, Miss. Sc. Mex., Rept.* p. 200, pl. xviii. bis, fig. 1, and xix. fig. 2; *Cope, Proc. Am. Philos. Soc.* xxii. 1885, p. 394.
Tropidolepis variabilis, *Dum. & Bibr.* iv. p. 308.

Head-shields keeled or striated; two canthal scales; a series of four or five large transverse supraoculars; occipital as long as broad; parietals very small; anterior border of ear with three or four slightly enlarged pointed scales. Dorsal scales much larger than ventrals, strongly keeled, sharply pointed, forming twelve to fifteen parallel longitudinal series; fifty-six to seventy-eight scales between the occipital shield and the base of the tail; twelve to fifteen scales correspond to the length of the shielded part of the head; lateral scales much smaller than dorsals, keeled, directed obliquely upwards; fifty-six to seventy-eight scales round the middle of the body. The adpressed hind limb reaches the eye, or between the latter and the ear; tibia usually a little longer than the shielded part of the head; the distance between the base of the fifth toe and the extremity of the fourth usually equals the distance between the arm and the nostril. Eleven to thirteen femoral pores. Caudal scales as large as or a little larger than dorsals. Male with enlarged postanal scales. Olive or brownish above, sides darker; a more or less distinct yellowish dorso-lateral band; frequently a series of transverse blackish spots along each side of the back; a blackish spot, light-edged in front, above axilla; lower surfaces yellowish or dirty white; the male with a large pink blotch on each side of the belly (disappearing in spirits) edged internally with blue.

	millim.		millim.
Total length	158	Fore limb	32
Head	17	Hind limb	51
Width of head	14	Tail	88
Body	53		

Mexico and Guatemala.

a, b, c, d, e–g, h–m. ♂, ♀, & hgr.	Mexico.	
n. ♂.	Jalapa.	
o. ♂.	Hacienda del Hobo, S. Mexico.	P. Geddes, Esq. [P.].
p–q. ♀ & hgr.	Mountains of Huamelula.	
r. ♀.	San Gerónimo.	O. Salvin, Esq. [C.].

22. Sceloporus siniferus.

Sceloporus siniferus, *Cope, Proc. Am. Philos. Soc.* xi. 1869, p. 157, and xxii. 1885, p. 394; *Bocourt, Journ. Zool.* v. 1876, p. 342.
—— humeralis, *Bocourt, Miss. Sc. Mex., Rept.* p. 206, pl. xviii. bis, fig. 3.

Head-shields keeled or striated; two canthal scales; a series of four or five large transverse supraoculars; occipital a little broader than long; parietals much smaller, two on each side; anterior border of ear very slightly denticulated. Lateral cervical fold forming a deep pouch. Dorsal scales much larger than ventrals, strongly keeled, sharply pointed, forming nine or ten parallel longitudinal series; thirty-two to thirty-eight scales between the occipital shield and the base of the tail; seven or eight scales correspond to the length of the shielded part of the head; lateral scales smaller, as large as ventrals, keeled, directed obliquely upwards; ventrals smooth, pointed, not emarginate; forty-four to forty-six scales round the middle of the body. The adpressed hind limb reaches the ear or the eye; tibia longer than the shielded part of the head; the distance between the base of the fifth toe and the extremity of the fourth equals the distance between the end of the snout and the cervical pouch. Four or five (three to seven) femoral pores. Caudal scales as large as, or a little smaller than dorsals. No enlarged postanal scales. Olive-brown above, with coppery gloss, sides darker; a yellowish dorso-lateral line frequently separates the dorsal from the lateral surfaces; a series of oblique blackish cross-lines along each side of the back; lower surfaces uniform yellowish in both sexes.

	millim.		millim.
Total length	153	Fore limb	24
Head	14	Hind limb	41
Width of head	10	Tail	98
Body	41		

Isthmus of Tehuantepec.

a–b. ♀.	Tehuantepec.	
c. ♂.	Oaxaca.	M. Sallé [C.].

23. Sceloporus fulvus *.

Sceloporus fulvus, *Bocourt, Miss. Sc. Mex., Rept.* p. 214, pl. xviii. bis, fig. 8.

Head-shields keeled or striated; one canthal scale; a series of four or five large transverse supraoculars; occipital as long as broad; parietals much smaller, two on each side; anterior border of ear very slightly denticulated. Lateral cervical fold forming a deep pouch. Dorsal scales much larger than ventrals, strongly keeled, sharply pointed, forming eight parallel longitudinal series; thirty to thirty-two scales between the occipital and the base of the tail; seven or eight scales correspond to the length of the shielded part of the head; lateral scales smaller, slightly larger than ventrals, keeled, directed obliquely upwards; ventrals obtusely keeled, pointed, not emarginate; thirty-eight to forty scales round the middle of the body. The adpressed hind limb reaches the eye; tibia longer than the shielded part of the head; the distance between the base of the fifth toe and the extremity of the fourth equals the distance between the end of the snout and the cervical pouch. Three or four (or five) femoral pores. Tail long; caudal scales nearly as large as dorsals. No enlarged postanal scales. Fulvous above, sides dark brown; a yellow dorso-lateral band separating the dorsal from the lateral surfaces; a more or less distinct series of oblique dark-brown spots along each side of the back; lower surfaces uniform yellowish in both sexes.

	millim.		millim.
Total length	149	Fore limb	21
Head	12	Hind limb	35
Width of head	9	Tail	106
Body	31		

Salvador, Nicaragua.

24. Sceloporus squamosus.

Sceloporus squamosus, *Bocourt, Miss. Sc. Mex., Rept.* p. 212, pls. xviii. bis. fig. 7, and xix. fig. 3; *Cope, Proc. Am. Philos. Soc.* xxii. 1885, p. 394.

Differs from *S. siniferus* and *fulvus* by the presence, in the male, of enlarged postanal scales; from the former, by the presence of a single canthal scale; from the latter by the absence of a keel on the ventral scales. Lower surfaces, in both sexes, yellowish, dotted all over with grey.

Guatemala to Costa Rica.

25. Sceloporus utiformis.

Sceloporus utiformis, *Cope, Proc. Ac. Philad.* 1864, p. 177; *Bocourt, Miss. Sc. Mex., Rept.* p. 208, pl. xviii. bis. fig. 6; *Cope, Proc. Am. Philos. Soc.* xxii. 1885, p. 394.

---

* Described from specimens from Nicaragua, in M. Lataste's collection.

Head-shields keeled or striated; two canthal scales; a series of five or six large transverse supraoculars; occipital as long as broad; parietals very small; anterior border of ear very slightly denticulated. Sides of neck strongly plicate, minutely granular. Ten longitudinal rows of large, highly keeled, shortly mucronate dorsal scales, separated by many lateral series of minute flat scales from the smaller entire-edged ventrals; forty scales between the occipital shield and the base of the tail; ten dorsal scales correspond to the length of the head. Tibia longer than the shielded part of the head. Thirteen or fourteen femoral pores. Tail long. General colour blackish above, with numerous indistinct lighter cross bars; or light brown with a double longitudinal series of transverse black spots and a yellow dorso-lateral band; yellowish or greenish inferiorly.

	millim.		millim.
Total length	210	Fore limb	30
Head	15	Hind limb	51
Body	53	Tail	142

Colima, Mexico.

50. PHRYNOSOMA.

Phrynosoma, *Wiegm. Isis*, 1828, p. 367, *and Herp. Mex.* p. 52; *Wagler, Syst. Amph.* p. 145; *Dum. & Bibr.* iv. p. 311; *Fitzing. Syst. Rept.* p. 78; *Gray, Cat. Liz.* p. 227; *Girard, U.S. Explor. Exped., Herp.* p. 388; *Bocourt, Miss. Sc. Mex., Rept.* p. 231.
Batrachosoma, *Fitzing. l. c.* p. 79; *Girard, l. c.* p. 400; *Bocourt, l. c.* p. 239.
Tropidogaster (*non D. & B.*), *Fitzing. l. c.*
Anota, *Hallow. Proc. Ac. Philad.* 1852, p. 182; *Bocourt, l. c.* p. 230.
Tapaya (*non Fitz.*), *Girard, l. c.* p. 394; *Bocourt, l. c.* p. 220.
Doliosaurus, *Girard, l. c.* p. 407.

Tympanum more or less distinct. Body depressed, without dorsal crest; dorsal lepidosis heterogeneous. Head covered with small scales, bordered posteriorly by osseous spines. A transverse gular fold. Digits with keeled lamellæ inferiorly. Femoral pores. Tail short. Lateral teeth subconical or indistinctly tricuspid; no pterygoid teeth. A large sternal fontanelle. No abdominal ribs.

North America and Mexico.

Synopsis of the Species.

I. Nostrils pierced in the line of the canthus rostralis; gular scales small, equal; a series of enlarged scales, parallel with, but not much larger than the lower labials.

Head-spines very small, shorter than the diameter of the eye .... 1. *douglassii*, p. 240.
Head-spines moderately large, the occipital spines measuring about the diameter of the eye; gular scales perfectly smooth.. 2. *orbiculare*, p. 241.

Head-spines moderately large; gular scales feebly keeled .... 3. *boucardii*, p. 243.

II. Nostril pierced in the line of the canthus rostralis or slightly above; several longitudinal series of enlarged gular scales; a series of very large bony shields below the lower labials.

Occipital spines obliquely directed upwards, like the temporals; ventral scales smooth. 4. *coronatum*, p. 243.
Occipital spines vertical, erect; ventral scales keeled........................... 5. *asio*, p. 244.

III. Nostrils pierced within the canthi rostrales; gular scales small, equal, or with one series of enlarged ones on each side.

A. Tail longer than the head.

1. Head bordered posteriorly by a series of contiguous large spines, of which four are occipital .... 6. *regale*, p. 245.

2. Two occipital spines, larger than the other head-spines, with an interval between their bases and those of the temporals.

Enlarged dorsal tubercles spinose, erect; tympanum naked .... 7. *cornutum*, p. 245.
Enlarged dorsal tubercles flat or feebly raised; tympanum covered with scales; occipital spines as long as the horizontal diameter of the orbit................ 8. *maccallii*, p. 246.
Enlarged dorsal tubercles flat or feebly raised; tympanum covered with scales; occipital spines shorter than the horizontal diameter of the orbit .... 9. *platyrhinus*, p. 247.

3. Two occipital spines, not longer than the longest temporals ...... 10. *modestum*, p. 248.

B. Tail not longer than the head.

Occipital spines a little larger than the temporals .... 11. *braconnieri*, p. 248.
Occipital spines much smaller than the largest temporals .... 12. *taurus*, p. 249.

1. Phrynosoma douglassii.

Phrynosoma douglassii, *Gray, Cat.* p. 227.
Agama douglassii, *Bell, Trans. Linn. Soc.* xvi. 1833, p. 105, pl. x.; *Harlan, Med. & Phys. Res.* p. 141, pl. —. fig. 3.
Phrynosoma douglassii, *Wagl. Syst. Amph.* p. 146; *Wiegm. Herp. Mex.* p. 54; *Holbr. N. Amer. Herp.* ii. p. 101, pl. xiv.; *Girard, in Stansb. Exped. Gr. Salt Lake*, p. 362, pl. vii. figs. 6–9; *Cope, Proc. Ac. Philad.* 1866, p. 302; *Coues, in Wheeler's Rep. Explor. W. 100th Mer.* v. p. 591.
—— orbiculare (*non L.*), *Hallow. in Sitgr. Exped. Zuni & Color. Riv.* p. 125, pls. viii. & ix.

? Tapaya hernandesi, *Girard, U. S. Explor. Exped., Herp.* p. 395; *Bocourt, Miss. Sc. Mex., Rept.* p. 228.
Tapaya ornatissima, *Girard, l. c.* p. 396; *Bocourt, l. c.* p. 227, pl. xi. fig. 6.
—— brevirostris, *Girard, l. c.* p. 397; *Bocourt, l. c.* p. 229.
—— douglassi, *Girard, l. c.* p. 397, pl. xxi. figs. 1–5; *Bocourt, l. c.* p. 226, pl. xi. fig. 5.
Phrynosoma brevirostre, *Cope, Proc. Ac. Philad.* 1866, p. 302.
—— douglassi pygmæa, *Yarrow, Proc. U.S. Nat. Mus.* v. 1882, p. 443.

Head broader than long, with the spines very small; nostril pierced in the line of the canthus rostralis; tympanum naked; the head-spines, which are subequal, not or but slightly larger than the largest spinose scales on the body, and turned upwards; they number on each side one postorbital, one occipital, and three temporals; in very young specimens the spines are not distinguishable. Lower labials terminating in a series of four or five large, compressed, obtuse or pointed scales; a series of enlarged scales, as large as or a little larger than, and parallel to, the lower labials; gular scales equal, smooth; gular fold strong; a dermal thickening, bearing a few small erect spines, on each side between the gular fold and the tympanum. Back and limbs with scattered, large, erect, keeled, spinose scales, which are longer than broad; a regular lateral series of spines; pectoral and ventral scales perfectly smooth. Fifteen to twenty-one femoral pores on each side in the male, twelve to fifteen in the female, the series not joining medially. Male with enlarged postanal scales. Tail twice to twice and a half as long as the head. Yellowish, greyish, or brownish above, with more or less distinct large, dark, light-edged spots forming longitudinal and transverse series; lower surfaces whitish, uniform or scantily dotted with grey.

	millim.		millim.
Total length	94	Fore limb	26
Head	14	Hind limb........	33
Width of head	18	Tail	30
Body	50		

United States.

a–b. ♂ & hgr.	California.	T. Bell, Esq. [P.].	(Types.)
c. ♂.	Kansas.		
d–g. ♀ & hgr.	Western N. America.	J. K. Lord, Esq. [P.].	
h. Several specs.: ♀ & yg.	N. America.		

2. Phrynosoma orbiculare.

Phrynosoma orbiculare, *Gray, Cat.* p. 228.
Lacerta orbicularis, part., *Linn. S. N.* i. p. 265.
Phrynosoma orbiculare, *Wiegm. Isis,* 1828, p. 367; *Wagl. Icon. Amph.* pl. xxiii.; *Gravenh. Nova Acta Ac. Leop.-Carol.* xvi. 1833, p. 911, pl. lxiii.; *Wiegm. Herp. Mex.* p. 53; *Dum. & Bibr.* iv. p. 321; *Girard, in Stansb. Exped. Gr. Salt Lake,* p. 359; *Sumichrast, Arch. Sc. Phys. Math.* xix. 1864, p. 60.
Phrynosoma wiegmanni, *Gray, Zool. Beechey's Voy.* p. 96.

Tapaya orbicularis, *Girard, U. S. Explor. Exped., Herp.* p. 395; *Bocourt, Miss. Sc. Mex., Rept.* p. 221, pl. xi. fig. 1.

Head (spines not included) as broad as long or broader than long; nostril pierced in the line of the canthus rostralis; tympanum naked; the head-spines of moderate size, nearly horizontal or obliquely turned upwards; there are on each side a low post-orbital, an occipital, and three temporals; the occipitals are the largest, and their extremities on a line with or beyond the line of the longest temporal; the length of the occipital spines equals about the diameter of the eye. Lower labials terminating in a series of three or four large, compressed, pointed scales; a series of enlarged scales, as large as or a little larger than, and parallel with, the lower labials; gular scales equal, smooth; gular fold very strong, forming a deep pouch on each side of the neck, which bears two short parallel longitudinal series of erect spines. Back and hind limbs with scattered large spinose keeled scales, which are longer than broad and obliquely directed upwards and backwards; a regular lateral series of spines; pectoral and ventral scales perfectly smooth. Fifteen to twenty-one femoral pores on each side in the male, ten to eighteen in the female, the series sometimes meeting in the middle. Male with enlarged postanal scales. Tail twice to thrice as long as the head. Brownish, yellowish, or reddish above, with one or two series of large blackish spots on each side of the back; these spots sometimes pale-bordered posteriorly; a large black spot on each side of the nape is constant; lower surfaces yellowish white, spotted or marked with black.

	millim.		millim.
Total length......	135	Fore limb	37
Head	20	Hind limb	49
Width of head....	21	Tail	53
Body............	62		

Mexico.

a–c. ♂ ♀.	City of Mexico.	Mr. Doorman [C.].
d. ♂.	Plateau of Mexico, 7500 ft.	P. Geddes, Esq. [P.].
e–g. ♂.	Ciudad.	Hr. Forrer [C.].
h–i. ♀ & hgr.	Mexico.	J. E. Harting, Esq. [P.].
k–l. ♀ & hgr.	Mexico.	Mr. Geale [C.].
m. Ad., stuffed.	Mexico.	W. Ogilby, Esq. [P.].

Var. **cortezii.**

Phrynosoma orbiculare, *Wiegm. Herp. Mex.* pl. viii. f. 1.
Tapaya cortezii, *Bocourt, Miss. Sc. Mex., Rept.* pl. xi. fig. 2.
—— orbicularis, var. A, *Bocourt, l. c.* p. 223.

The occipital spines are not longer than, and do not extend posteriorly quite as far as, the longest temporals. Head much broader than long.

a–b. ♂.	Puebla	
b. ♀.	Orizaba.	
c. ♀.	Jalapa.	Mr. Hoege [C.].

Var. **dugesii.**

Tapaya dugesii, *Bocourt, Miss. Sc. Mex., Rept.* pl. xi. fig. 3.
—— orbicularis, var. B, *Bocourt, l. c.* p. 224.

Spines subequal, the occipitals not extending posteriorly quite as far as the longest temporals. Head nearly as broad as long.

Colima.

3. Phrynosoma boucardii.

Tapaya boucardii, *Bocourt, Miss. Sc. Mex., Rept.* p. 225, pl. xi. fig. 4.

Closely allied to *P. orbiculare.* The occipital spines, which are nearly of the same size, and do not extend posteriorly quite as far as the longest temporals, are directed upwards and outwards. Head much broader than long. Gular scales feebly keeled.

Plateau of Mexico.

4. Phrynosoma coronatum.

Phrynosoma blainvillii, part., *Gray, Cat.* p. 228.
Agama (Phrynosoma) coronata, *Blainv. Nouv. Ann. Mus.* iv. 1835, p. 284, pl. xxv. fig. 1.
Phrynosoma coronatum, *Dum. & Bibr.* iv. p. 318; *Holbr. N. Am. Herp.* ii. p. 97, pl. xiii.; *Girard, in Stansb. Exped. Gr. Salt Lake,* p. 360, pl. viii. figs. 7–12; *Hallow. in Sitgr. Exped. Zuni & Color. Riv.* p. 122.
—— blainvillii, *Gray, Zool. Beechey's Voy.* p. 96, pl. xxix. fig. 1.
—— (Batrachosoma) coronatum, *Fitzing. Syst. Rept.* p. 79.
Batrachosoma coronatum, *Girard, U.S. Explor. Exped., Herp.* p. 400, pl. xx. figs. 10–13; *Bocourt, Miss. Sc. Mex., Rept.* p. 239, pl. xii. fig. 10.

Head broader than long; nostril pierced in the line of the canthus rostralis; tympanum naked; the head-spines large; on each side one postorbital, one occipital, three temporals; a more or less distinct additional small spine between the occipitals; latter as large as or a little larger than the largest temporals, directed obliquely upwards and extending posteriorly as far as or beyond the largest temporals. A series of very large projecting scales below the lower labials; three or four longitudinal series of enlarged, pointed, smooth or very feebly keeled scales on each side of the middle of the throat, the outer series the largest; gular folds several, irregular; two short series of spines on each side of the neck. Back with large, keeled, flat, triangular or trihedral scales, which are mostly as broad as long; two lateral series of spines, upper much stronger than lower; pectoral and ventral scales perfectly smooth. Thirteen to sixteen femoral pores on each side, the series not joining medially. Male with slightly enlarged postanal scales. Tail about twice as long as the head. Brownish above, with large dark brown or reddish spots arranged regularly; a very large dark spot on each side of the nape; lower surfaces whitish, uniform or spotted or marbled with brown.

	millim.		millim.
Total length	125	Fore limb	40
Head	19	Hind limb	52
Width of head	23	Tail	38
Body	68		

California.

a. Yg.	California.	Prof. de Blainville [P.]. (Type of *P. blainvillii.*)
b–c. Yg.	California.	Sir E. Belcher [P.].
d. ♀.	California.	
e. ♂.	Monterey.	J. H. Gurney, Esq. [P.].
f. ♂.	——?	W. G. Lettson, Esq. [P.].
g–i. ♂ ♀.	——?	

5. Phrynosoma asio.

Phrynosoma asio, *Cope, Proc. Ac. Philad.* 1864, p. 178; *Sumichrast, Arch. Sc. Phys. Math.* xlvi. 1873, p. 258.
Batrachosoma asio, *Bocourt, Miss. Sc. Mex., Rept.* p. 241, pl. xvii. fig. 9.
Phrynosoma spinimentum, *Peters, Mon. Berl. Ac.* 1873, p. 742.

Head a little broader than long; nostril somewhat within the extension of the supraciliary ridge; tympanum naked; the head-spines large, subequal in size; on each side one postorbital, one occipital, and two temporals; the occipital spines almost perfectly vertical. A series of very large projecting scales below the lower labials; several longitudinal series of enlarged, subequal, keeled, pointed gular scales separated by granular intervals; gular folds several, irregular; a group of spines on the side of the neck. Back with very large distant tubercles, those on the sides of the vertebral line flat, strongly keeled, and as broad as long; a dorso-lateral series of very large spines; two lateral series of spines, upper largest; pectoral and ventral scales keeled. Six to eight femoral pores on each side. Male without distinctly enlarged postanal scales. Tail about three times as long as the head. Olive-grey above, with large, transverse, oblique, brown, black-edged spots on each side of the back; lower surfaces immaculate.

	millim.		millim.
Total length	189	Fore limb	59
Head	24	Hind limb	73
Width of head	28	Tail	75
Body	90		

Mexico; Lower California.

a–c. ♂ ♀. Tehuantepec.

6. Phrynosoma regale.

Phrynosoma blainvillii, part., *Gray, Cat.* p. 228.
Phrynosoma regale, *Girard, U.S. Explor. Exped., Herp.* p. 406, *and U.S. Mex. Bound. Surv., Rept.* p. 9, pl. xxviii. figs. 1–3; *Bocourt, Miss. Sc. Mex., Rept.* p. 235, pl. xii. fig. 12.

Head a little broader than long; nostrils within the canthi rostrales; tympanum naked; postorbital spine small; head bordered posteriorly by a series of contiguous large spines obliquely directed upwards, viz. two smaller median and two larger outer occipitals and, on each side, four temporals. A series of very large projecting scales below the lower labials, almost continuous with the temporal spines; gular scales very feebly keeled, with a longitudinal series of enlarged ones on each side; two groups of spines on each side of the neck; a double transverse gular fold. Back with very large trihedral tubercles, the largest broader than long; two lateral series of spines, lower very small; pectoral and ventral scales very feebly keeled. Fourteen or fifteen femoral pores on each side, extending only on the proximal third of the thigh, and the two series narrowly separated on the præanal region. Male with enlarged postanal scales. Tail a little over twice the length of the head. Brownish above, with an ill-defined darker band on each side of the back; belly whitish, with a few brownish spots.

	millim.		millim.
Total length	138	Fore limb	42
Head	20	Hind limb	54
Width of head	22	Tail	46
Body	72		

Desert of Gila and Colorado.

a. ♂. [California.]

7. Phrynosoma cornutum.

Phrynosoma cornutum, *Gray, Cat.* p. 229.
Agama cornuta, *Harlan, Journ. Ac. Philad.* iv. 1825, p. 299, pl. xx.
Phrynosoma bufonium, *Wiegm. Isis*, 1828, p. 367.
—— cornutum, *Gray, Griff. A. K.* ix. *Syn.* p. 45; *Holbr. N.-Am. Herp.* ii. p. 87, pl. xi.; *Girard, in Stansb. Exped. Gr. Salt Lake*, p. 360, pl. viii. figs. 1–6; *Hallow. in Sitgr. Exped. Zuni & Color. Riv.* p. 119; *Girard, U.S. Explor. Exped., Herp.* p. 403, pl. xxi. figs. 6–9; *Bocourt, Miss. Sc. Mex., Rept.* p. 236, pl. xii. fig. 9.
—— harlanii, *Wiegm. Herp. Mex.* p. 54; *Dum. & Bibr.* iv. p. 314.
—— (Tropidogaster) cornutum, *Fitzing. Syst. Rept.* p. 79.
—— (——) bufonium, *Fitzing. l. c.*

Head as long as broad, or a little broader than long; nostrils within the canthi rostrales; tympanum naked; five spines on each side of the head, viz. a postorbital, three temporals, and an occipital; a more or less distinct small median occipital; the large occipitals much longer than the temporals, obliquely directed upwards and outwards. A series of very large projecting scales below the lower labials; gular scales keeled, with a longitudinal series of enlarged pointed ones on each side; two or three groups of spines on each side of the neck; gular folds several, irregular. Back with very large erect spinose tubercles, the largest of which form a series of three or four on each side of the vertebral line; two lateral series of spines, upper largest; pectoral and ventral scales more or less distinctly keeled. Femoral pores four to twelve on each side, not extending on the præanal region, sometimes very indistinct. No enlarged postanal scales. Tail once and a half to twice and a half as long as head and body. Greyish or brownish above, with a more or less marked light dorsal streak and dark brown spots at the base of the larger dorsal spines; a large dark brown spot on each side of the nape; two cross streaks between the supraciliary ridges, a band from the eye to the angle of the mouth, and another from the eye to the middle temporal spine, dark brown; lower surfaces yellowish, uniform or with a few small brownish spots.

	millim.		millim.
Total length	125	Fore limb	47
Head	21	Hind limb	59
Width of head	21	Tail	40
Body	64		

United States; Northern Mexico.

a. ♀.	Illinois.	G. B. Salmon, Esq. [P.].
b. ♂.	Arkansas.	Dr. Macauley [P.].
c. ♀.	Kansas.	
d. ♀.	Texas.	Edw. Cross, Esq. [P.].
e. Several sp.: ♂, ♀, & hgr.	Texas.	
f–g. ♀.	N. America.	Lord Ampthill [P.].
h–i, k. ♀ & hgr.	——?	
l. Ad., stuffed.	Mexico.	
m, n. Ad., skeletons.	N. America.	

Var. **planiceps.**

Phrynosoma planiceps, *Hallow. Proc. Ac. Philad.* 1852, p. 178, *and in Sitgr. Exped. Zuni & Color. Riv.* p. 124, pl. vii.; *Bocourt, l. c.* p. 238, pl. xii. fig. 11; *Yarrow, in Wheeler's Report Explor. W. 100th Mer.* p. 579, pl. xxiv. fig. 1.

Hardly distinguishable from *P. cornutum*. Ventral and gular scales smooth.

Texas to Sonora.

8. **Phrynosoma maccallii.**

Anota m'callii, *Hallow. Proc. Ac. Philad.* 1852, p. 182, *and in Sitgr. Exped. Zuni & Color. Riv.* p. 127, pl. x.; *Bocourt, Miss. Sc. Mex., Rept.* p. 230.

Doliosaurus mc'calli, *Girard, U.S. Explor. Exped., Herp.* p. 408, *and U.S. Mex. Bound. Surv., Rept.* p. 9, pl. xxviii. figs. 4–6.
Phrynosoma maccallii, *Cope, Proc. Ac. Philad.* 1866, p. 310.

Head as long as broad; nostrils within the canthi rostrales; tympanum covered with scales; six spines on each side of the head, viz. a postorbital, four temporals, and an occipital, the latter much longer than the longest temporal, quite as long as the horizontal diameter of the orbit, obliquely directed upwards and slightly outwards. A series of very large projecting scales below the lower labials; gular scales smooth, uniform; two short series of small spinose scales on each side of the neck; gular folds several, irregular. Back with irregularly scattered, large, flat, keeled tubercles, which are mostly broader than long; a single lateral series of small spines; pectoral and ventral scales perfectly smooth. Seven femoral pores on each side. Tail nearly twice and a half as long as the head. Greyish above, with rather indistinct darker spots; lower surfaces white, with a few scattered blackish dots.

	millim.		millim.
Total length	108	Fore limb	36
Head	15	Hind limb	46
Width of head	15	Tail	37
Body	56		

Desert of Gila and Colorado.

a. ♀.	—— ?	Smithsonian Institution [P.].

9. Phrynosoma platyrhinus.

Phrynosoma platyrhinos, *Girard, in Stansb. Exped. Gr. Salt Lake,* p. 361, pl. vii. figs. 1–5; *Bocourt, Miss. Sc. Mex., Rept.* p. 232.
Doliosaurus platyrhinos, *Girard, U.S. Explor. Exped., Herp.* p. 409.

Very closely allied to the preceding species, from which it is distinguished by the shorter head-spines. The occipital spines are only a little longer than the longest temporals and shorter than the horizontal diameter of the orbit. Nine or ten femoral pores on each side. Sandy-coloured above, with brown marblings or undulated cross bands on the back, interrupted on the vertebral line; a large brown spot on each side of the nape; lower surfaces white.

	millim.		millim.
Total length	121	Fore limb........	37
Head	17	Hind limb	47
Width of head	17	Tail	41
Body	63		

Western North America.

a. ♀.	Camp 12, Nevada.	Smithsonian Institution [P.].
b–d. ♀ & yg.	Rogne River Valley, Oregon.	Dr. Brown [C.].

10. Phrynosoma modestum.

Phrynosoma modestum, *Girard, in Stansb. Exped. Gr. Salt Lake*, p. 361, pl. vi. figs. 4–8; *Bocourt, Miss. Sc. Mex., Rept.* p. 232.
Doliosaurus modestus, *Girard, U.S. Explor. Exped., Herp.* p. 409.

Nostrils within the canthi rostrales; tympanum naked; head-spines of moderate development, five on each side, viz. a very small postorbital, three temporals, and an occipital: latter not larger than the largest temporal. A series of very large scales below the lower labials; gular scales uniform, smooth, exceedingly small. Enlarged dorsal scales very small and flat; no lateral series of spines; ventrals smooth, larger than in any other species. Series of femoral pores meeting on the præanal region. Tail about twice and a half as long as head and body, tapering very suddenly, perfectly cylindrical beyond the dilated portion of its base. Uniform yellowish brown above, with a patch of black on each side of the nape.

Total length 75 millim.

Texas to Oregon and California.

11. Phrynosoma braconnieri.

Phrynosoma braconnieri, *Bocourt, Miss. Sc. Mex., Rept.* p. 233, pl. xii. fig. 7.

Head broader than long; nostrils pierced within the canthi rostrales; tympanum naked; head-spines small, six on each side, viz. a supraorbital, four temporals, and an occipital; latter a little larger than the others. A series of enlarged scales, a little larger than and parallel with the lower labials; gular scales equal [or with a longitudinal series of slightly enlarged scales on each side], keeled; gular fold strong, forming a deep pouch on each side of the neck; a dermal thickening, with two series of spinose scales, on each side between the gular fold and the tympanum. Back and limbs with scattered large spinose keeled scales which, on the back, are about as broad as long; a lateral series of spines; pectoral and ventral scales keeled. Twelve or thirteen femoral pores on each side, not extending on the præanal region. Male with enlarged postanal scales. Tail shorter than the head, as long as broad at the base. Sandy-coloured above, tinged with red on the head and limbs; a large blackish spot on each side of the nape; belly spotted with blackish.

	millim.		millim.
Total length	48	Fore limb	21
Head	11	Hind limb	27
Width of head	13	Tail	8
Body	29		

Mexico.

a. Hgr. ♂. Oaxaca.

12. Phrynosoma taurus.

Phrynosoma taurus *(Dugès), Bocourt, Miss. Sc. Mex., Rept.* p. 234, pl. xii. fig. 8.

Head broader than long, with the temporal region much produced, the posterior outline of the head forming a strong concave curve; nostrils within the canthi rostrales; tympanum naked; the temporal processes covered with strongly enlarged scales, terminating in a long spine with a smaller detached one on the inner side; on each side a small occipital spine and a postorbital of equal size. A series of large scales below the lower labials; gular scales subequal, keeled, with a more or less distinct longitudinal series of slightly enlarged pointed scales on each side: gular folds several, forming a pouch on each side of the neck; a dermal appendage, bearing large spinose scales, on each side between the gular fold and the head. Back and limbs with scattered large spinose keeled scales; on the back they are as broad as long, or a little longer than broad; a lateral series of spines; pectoral and ventral scales keeled. Eight to eleven femoral pores on each side, not extending on the præanal region. Male with enlarged postanal scales. Tail as long as or shorter than the head. Greyish or yellowish above, with transverse series of large blackish spots; a large black patch on each side of the nape; lower surfaces more or less distinctly spotted with blackish.

	millim.		millim.
Total length	86	Fore limb	40
Head	16	Hind limb	54
Width of head	21	Tail	10
Body	60		

Mexico.

a–f. ♂, ♀, & yg.	Foot of Popocatepetl, Puebla.	M. Boucard [C.].

The following Lizard, type of the genus *Pneustes*, described by F. d'Azara, Quadr. Paraguay, ii. p. 325, has not been rediscovered in the course of the present century:—

PNEUSTES, *Merrem. Tent.* p. 49.

"Palmæ tetradactylæ; Plantæ pentadactylæ; Aures latentes; Gula dilatabilis."

Pneustes prehensilis.

Agama prehensilis, *Daud. Rept.* iii. p. 440.
Pneustes prehensilis, *Merr. l. c.*

Paraguay.

Fam. 7. XENOSAURIDÆ.

Xenosauridæ, *Cope, Proc. Ac. Philad.* 1866, p. 322.
Trachydermi aglyphodonta, part., *Bocourt, Miss. Sc. Mex., Rept.* 1878.
Xenosauridæ, *Boulenger, Ann. & Mag. N. H.* (5) xiv. 1884.

Tongue short, villose, with a narrow, feebly incised, retractile anterior part covered with flat papillæ. Dentition strictly pleurodont; teeth numerous, small, with long cylindrical shafts, the laterals with obtusely bi- or tricuspid crowns: the new teeth originate between the bases of the old ones. Palate toothless. Postorbital and frontosquamosal arches bony; supratemporal fossa not roofed over by bone; præmaxillary, frontal, and parietal single; nasals distinct; palatines and pterygoids widely separated medially, both bordering the infraorbital fossa; a supraorbital bone. Limbs well-developed. Clavicle slender, not dilated proximally; interclavicle T-shaped; sternum without fontanelle. No abdominal ribs. Head covered with small tubercular scales, which coalesce with the skull. Eyelids well developed. Tympanum covered with scales. Body depressed, covered above with granular scales intermixed with tubercles, inferiorly with cross rows of squarish juxtaposed scales; no bony plates. Digits well developed, with sharp curved claws. No femoral pores. Tail round.

The unique genus, of which the preceding description is given, constitutes a connecting link between the *Iguanidæ* and the *Anguidæ*. Its affinity to the former is shown by the T-shaped clavicle, the absence of symmetrical bony shields on the head and of osteodermal plates on the body; to the latter by the structure of the tongue, the teeth not hollowed out at the base, and the widely separated palatine bones.

1. XENOSAURUS.

Xenosaurus, *Peters, Mon. Berl. Ac.* 1861, p. 453; *Bocourt, Miss. Sc. Mex., Rept.* p. 303.

Central America.

1. Xenosaurus grandis.

Cubina grandis, *Gray, Ann. & Mag. N. H.* (2) xviii. 1856, p. 270.
Xenosaurus fasciatus, *Peters, l. c.* p. 454.
—— grandis, *Cope, Proc. Ac. Philad.* 1866, p. 322; *Sumichrast, Arch. Sc. Phys. Math.* 1873, p. 243; *Bocourt, l. c.* pl. xx. F. fig. 1.

Head depressed, subtriangular, covered with rough, subconical or obtusely keeled tubercles, which form three longitudinal rows between the orbits; a slightly enlarged median occipital; a series of three or four transversely dilated scales on the supraocular region; nostril pierced in a single nasal; rostral quadrangular, low, three times as broad as high; ten to thirteen upper and nine or ten lower labials; mental short and broad, subtriangular or subpentagonal; two or three rows of enlarged scales bordering the lower labials; gular-region granular. Two strong transverse gular folds. Body depressed, covered above with irregular minute granules intermixed with small round or oval, conical or obtusely keeled tubercles forming irregular longitudinal series; a more or less distinct fold from axilla to groin. Belly covered with smooth squarish scales forming thirty-six transverse series (from fore arm to præanal region); a transverse series in the middle of the belly contains about twenty scales. Limbs covered above with granules and tubercles as on the body, inferiorly with uniform granules; digits slender, cylindrical, slightly swollen at the articulations, inferiorly with a series of transverse smooth scales. Tail cylindrical, tapering, about as long as head and body, with cross rows of subtubercular scales above, of larger quadrangular flat ones inferiorly. Yellowish-brown, marbled with blackish, with four or five more or less marked transverse light bands on the back, and dark and light annuli on the tail.

	millim.		millim.
Total length	240	Fore limb	45
Head	30	Hind limb........	57
Width of head	24	Tail	120
Body	90		

Southern Mexico.

a–c. Ad.	Cordova.	M. Sallé [C.].	(Types.)
d. Ad.	Oaxaca.	M. Sallé [C.].	

Fam. 8. ZONURIDÆ.

Cordyloidea, part., *Fitzinger, Neue Classif. Rept.* 1826.
Autarchoglossæ, part., *Wagler, Syst. Amph.* 1830.
Chamæsauri, part., Ptychopleuri, part., *Wiegmann, Herp. Mex.* 1834.
Chalcidiens, part., *Duméril & Bibron, Erp. Gén.* v. 1839.
Zonuridæ, part., Chamæsauridæ, *Gray, Cat. Liz.* 1845.
Zonuridæ, part., *Cope, Proc. Am. Assoc. Adv. Sc.* xix. 1871.
Zonuridæ, *Boulenger, Ann. & Mag. N. H.* (5) xiv. 1884.

Tongue short, villose, scarcely protractile, entire or very feebly nicked at the end. Dentition pleurodont; teeth numerous, small, with long cylindrical shafts, hollowed out at the base. Palate toothless. Postorbital and frontosquamosal arches bony; supratemporal fossa roofed over by dermo-ossification; præmaxillary, frontal, and parietal single; nasals distinct; palatines and pterygoids widely separated medially, both bordering the infraorbital fossa; head with dermal bony shields. Clavicle slender, not dilated proximally; interclavicle cruciform; sternum without fontanelle. No abdominal ribs. Head symmetrically shielded. Eyelids well developed. Scales on the body, if not granular, arranged in transverse series. Osteodermal plates present in one genus, but devoid of distinct tubules.

Like the preceding, this family has points of resemblance with the *Iguanidæ* and with the *Anguidæ*. From the former it is distinguished by the cranial dermal ossifications and the cruciform interclavicle; from the latter by the tongue, which, like that of the *Iguanidæ*, is not divisible into an anterior and a posterior part, the cœlodont dentition, and the structure of the bony plates of the body, when present.

Four genera are known, inhabiting South and Tropical Africa, and Madagascar.

Synopsis of the Genera.

I. Limbs well developed.

A. Back with osteodermal plates.

Dorsal scales large . . . . 1. **Zonurus**, p. 252.

B. Back without osteodermal plates.

Dorsal lepidosis heterogeneous . . . . . . 2. **Pseudocordylus**, p. 259.
Dorsal lepidosis uniformly granular . . 3. **Platysaurus**, p. 261.

II. Limbs rudimentary, body serpentiform.

Scales lanceolate, keeled . . . . 4. **Chamæsaura**, p. 263.

1. ZONURUS.

Cordylus, part., *Laur. Syn. Rept.* p. 51.
Zonurus, *Merr. Tent. Syst. Amph.* p. 57; *Gray, Cat. Liz.* p. 47.
Cordylus, *Cuv. Règne Anim.* 2nd ed. ii. p. 33; *Gray, l. c.* p. 46.
Hemicordylus, *Smith, Mag. N. H.* (2) ii. 1838, p. 32; *Gray, l. c.* p. 48.
Zonurus, part., *Dum. & Bibr.* v. p. 344.

Head and body depressed; limbs well developed. Head-shields regular; four parietals; nostril pierced in the nasal. Ear-opening large. Eyelids well developed. Dorsal scales large, bony, forming regular transverse series; ventrals large, square or subtriangular, juxtaposed or imbricate, forming longitudinal and transverse series. No collar fold. Digits slightly keeled inferiorly. Tail spinose. Femoral pores.

South Africa.

Synopsis of the Species.

I. Frontonasal in contact with the rostral; scales of the flanks separated from one another by granular intervals.

Head with large spines posteriorly.... 1. *giganteus*, p. 253.
Head without spines .... 2. *tropidosternum*, p. 254.

II. Frontonasal separated from the rostral.

A. Flanks with scales like the dorsals.

a. No supranasals; lower eyelid opaque.

Nasal swollen, hemispherical, pierced in the centre; temporal spines .... 3. *cataphractus*, p. 255.
Nasal not swollen, pierced posteriorly; no temporal spines .... 4. *cordylus*, p. 256.

b. A supranasal, nasal very small; lower eyelid with a transparent disk.

Anterior gular scales flat, moderate .. 5. *polyzonus*, p. 257.
Anterior gular scales almost granular . 6. *pustulatus*, p. 258.

B. Flanks granular............ 7. *capensis*, p. 258.

1. Zonurus giganteus.

Zonurus derbianus, *Gray, Cat.* pp. 48, 270.
Cordylus giganteus, *Smith, Ill. S. Afr., Rept.* pls. xxxv., xxxvi.

Head longer than broad. Head-shields very rough; frontonasal slightly broader than long, in contact with the rostral, separating the nasals; latter not swollen, the nostril pierced in the postero-inferior corner; præfrontals forming a suture; frontal hexagonal, slightly widened anteriorly; frontoparietals longer than broad; interparietal as large as, and on a line with the anterior parietals, forming a suture with the postfrontals; posterior parietals much larger than anterior; four large, striated occipital spines, the inner shorter and slightly curved inwards, the outer largest, directed obliquely outwards, as long as the three anterior supraorbitals; temporal shields large, keeled; four subtriangular, slightly curved temporal spines on each side, the upper nearly as large as the median occipitals; supraoculars four, anterior largest, second broadest; three supraciliaries; lower eyelid opaque; a small loreal or postnasal and a large præorbital; two infraorbitals; six or seven upper labials, fourth or fifth smallest; rostral six-sided, twice and a half as broad as high. Mental large, pentagonal; five lower labials, fifth largest and keeled; a row of five large shields borders the lower labials, and there is a median chin-shield behind the anterior pair; gular scales moderately large, subtriangular, strongly imbricate, feebly keeled; scales on the collar large, lanceolate, and mucronate; large, sharp, slightly curved spines on the sides of the neck. Dorsal scales large, forming regular transverse series, rough, subquadrangular, with a strong keel ending in a short sharp point; twenty-three to twenty-five transverse rows of scales from occiput to base of tail; the median series composed of twelve scales, not including those on the flanks; latter lanceolate, keeled, spinose, separated by wide granular interspaces; a slight lateral fold. Ventral scales large, strongly imbricate, forming ten or twelve regular longitudinal and twenty-five to twenty-seven transverse series; the median and pectoral scales smooth and subtriangular, the laterals feebly keeled and shortly mucronate. Two or three enlarged præanal scales. Limbs above with large spinose imbricate keeled scales, inferiorly with smooth ones. Eleven femoral pores on each side. Upper and lateral caudal scales very large, strongly spinose, the lateral spines largest and horizontal; lower caudal scales long, narrow, pentagonal, smooth. Yellowish, back clouded with brown; upper surface of head brown.

The young light yellowish, marbled and cross-barred with blackish-brown; the spines not at all developed, and all the scales more strongly imbricate.

	millim.		millim.
Total length	382	Fore limb	78
Head	52	Hind limb	98
Width of head	42	Tail	180
Body	150		

South Africa.

a, b. ♂ & yg.	Interior of S. Africa.	Sir A. Smith [P.].	(Types.)
c–d. ♂ & yg.	S. Africa.	Earl of Derby [P.].	(Types of *Zonurus derbianus.*)

2. Zonurus tropidosternum.

Zonurus tropidosternum, *Cope, Proc. Amer. Philos. Soc.* xi. 1869, p. 169.

Head-shields excessively rugose, with longitudinal striæ; rostral in contact with the frontonasal, which is much longer than wide and separates the nasals, and well separated from the frontal by the præfrontals; the other head-shields as in *Z. cordylus*, except that there are six rows of temporals; upper labials six, the fifth not more elevated than the others; lower labials six, in contact with a row of five shields. Gular scales in twenty-two series from angles of mandible; those of the neck abruptly larger, mucronate, keeled, forming a rudimental collar. Dorsal scales in twenty-four series from nape to opposite femur, all strongly mucronate, keeled and rugose; the median series like the others. Median ventrals nearly smooth, laterals mucronate, keeled. Lateral scales increasing regularly in size from the ventrals, subround, widely separated from each other by minutely granular intervals, strongly mucronate, keeled. Præanal plates small, equal, except two marginals a little longer. Femoral pores seven on each side. Caudal whorls very spinous, the scales not serrate, but striate on the surface. Below and upper lip to ear yellow; above rich brown, with several indistinct blackish cross shades; head above wood-brown.

	millim.		millim.
From snout to vent . .	92	Fore limb	32
From snout to ear . .	44	Hind limb	42

Madagascar.

3. Zonurus cataphractus.

Cordylus cataphractus, *Boie, N. Acta Ac. Leop.-Carol.* xiv. 1828, i. p. 140; *Smith, Ill. S. Afr., Rept.* pl. xxix. & xxx. fig. 9.
Zonurus cordylus, *Schleg. Tijdschr. Nat. Gesch. Phys.* i. 1834, p. 211, pl. vii. fig. 3.
Cordylus nebulosus, *Smith, Mag. N. H.* (2) ii. 1838, p. 31.
Zonurus cataphractus, *Dum. & Bibr.* v. p. 355.

Head triangular, much depressed, as broad as long in the adult; snout very short. Head-shields very rugose; nasals in contact, much swollen, pierced in the centre; frontonasal and præfrontals subequal, on a line, the former pentagonal, its anterior angle wedged in between the nasals, its posterior border forming a suture with the frontal; latter seven-sided, a little wider anteriorly; interparietal inclosed between the two pairs of parietals, of which the posterior is the largest; a row of six keeled occipitals; temporals large, striated and keeled, in three longitudinal series; three temporal spines, upper smallest but sharpest; four supraoculars; three supraciliaries; lower eyelid opaque; loreal very small, or absent; a large præorbital; three infraorbitals; six or seven upper labials, the three posterior largest and keeled; rostral pentagonal, twice and a half as broad as high. Mental pentagonal; six lower labials, fifth largest, four posterior keeled; a row of five large shields bordering the lower labials; four small median chin-shields; gular scales small, rather irregular, subquadrangular, smooth, except those on the sides, which are feebly keeled; four or five curved rows of smooth large scales under the neck; sides of neck with large spines. Dorsal scales large, forming regular transverse series; these scales rough, elongate quadrangular, serrate posteriorly, keeled. On the sides the keel becomes stronger and ends in a strong spine; lateral scales otherwise not distinguished from the dorsals. A strong lateral fold. The dorsal scales form fifteen or sixteen transverse rows from occiput to base of tail; in the middle of the body a series contains twenty scales, counting from the lateral fold. Ventral scales quadrangular, longer than broad, not imbricate, smooth, forming regular transverse series; latter twenty to twenty-two, the largest containing twenty scales. A pair of enlarged posterior præanal scales. Limbs above with large spinose imbricate keeled scales, inferiorly with smooth or feebly keeled ones; thirteen to sixteen femoral pores on each side. Tail with whorls of large, strongly keeled, spinose scales, the spines very large on the upper and lateral surfaces. Yellowish brown, belly clouded with dark brown; gular region dark brown, spotted with light yellow.

	millim.		millim.
Total length	203	Fore limb	37
Head	29	Hind limb........	45
Width of head	28	Tail	100
Body	74		

South Africa.

a–b. ♂ & hgr. Western coast of Cape Colony. Sir A. Smith [P.].

4. Zonurus cordylus.

Zonurus cordylus, *Gray, Cat.* p. 47.
—— cataphractus (*non Boie*), *Gray, l. c.*
Lacerta cordylus, *Linn. S. N.* i. p. 361.
Cordylus verus, *Laur. Syn. Rept.* p. 52.
Stellio cordylus, *Daud. Rept.* iv. p. 8.
Zonurus cordylus, *Merr. Tent.* p. 57.
Cordylus griseus, *Cuv. R. A.* 2nd ed. ii. p. 33; *Smith, Mag. N. H.* (2) ii. p. 31, *and Ill. S. Afr., Rept.* pl. xxviii. figs. 2, 3, & xxx. fig. 8.
—— niger, *Cuv. l. c.*
—— dorsalis, *Cuv. l. c.*
Zonurus vertebralis, *Gray, Ann. N. H.* i. 1838, p. 388.
—— griseus, *Dum. & Bibr.* v. p. 350.

Head much depressed, much longer than broad. Head-shields slightly rugose; nasals not swollen, in contact, pierced in their posterior part; frontonasal hexagonal, broader than long, followed by a pair of præfrontals forming a suture; frontal pentagonal, a little wider anteriorly; interparietal small, quadrangular, in the middle between the two pairs of parietals, of which the posterior is the largest; temporals slightly keeled, in four longitudinal rows; no spines; four supraoculars; three supraciliaries; lower eyelid opaque; a small loreal; a large præorbital; three infraorbitals; five or six upper labials, posterior largest; rostral three to four times as broad as high. Mental rather large, tetragonal or pentagonal; five or six lower labials, posterior largest and keeled, bordered by a row of five large shields, the fourth of which is the largest; gular scales moderate, subquadrangular, imbricate, larger and more rounded on the neck; sides of neck with spinose keeled scales. Dorsal scales large, forming regular transverse series, imbricate, subquadrangular, more or less strongly keeled, shortly mucronate, frequently serrated; on the flanks similar, but more strongly keeled; a distinct lateral fold; twenty-five to twenty-eight cross series of scales from occiput to base of tail, the largest containing sixteen or eighteen scales. Ventral scales smooth, quadrangular, a little broader than long, slightly imbricate, forming ten to fourteen regular longitudinal, and twenty-three to twenty-seven transverse series, those of the outermost series shortly mucronate and frequently feebly keeled. A pair of enlarged posterior præanal scales. Limbs above with large spinose, serrated, imbricate keeled scales, inferiorly with smooth or feebly keeled ones. Femoral pores six to nine on each side. Tail with whorls of large, strongly keeled, spinose, serrated scales, the spines strongest on the sides. Upper parts yellowish, olive, or blackish-brown, uniform or variegated with darker; sometimes a reddish band along each side; lower surfaces yellowish or greenish white.

	millim.		millim.
Total length	174	Fore limb	30
Head	24	Hind limb........	41
Width of head	19	Tail	90
Body	60		

South Africa.

a. ♂.	Damara Land.	
b. ♂.	East London, British Kaffraria.	H. Trevelyan, Esq. [P.].
c. ♂.	Kaffraria.	
d–e. Yg.	Cape Colony.	F. P. M. Weale, Esq. [P.].
f. ♀.	Cape.	Prof. Busk [P.].
g. ♂.	S. Africa.	Earl of Derby [P.].
h. ♂.	S. Africa.	C. Darwin, Esq. [P.].
i–l, m–n, o, p, q, r. ♂, ♀, & hgr.	S. Africa.	
s, t. Ad., stuffed.	S. Africa.	(Types of *Z. vertebralis.*)
u. Ad., stuffed.	S. Africa.	

5. Zonurus polyzonus.

Cordylus polyzonus, *Gray, Cat.* p. 47.
Cordylus polyzonus, *Smith, Mag. N. H.* (2) ii. 1838, p. 34, *and Ill. S. Afr., Rept.* pl. xxviii. fig. 1, and xxx. fig. 7.
Zonurus polyzonus, *Dum. & Bibr.* v. p. 357; *Peters, Mon. Berl. Ac.* 1862, p. 18.

Head and body much depressed. Head longer than broad, with slightly rugose shields. Nasal very small, pierced posteriorly; a larger supranasal forming a suture with its fellow; frontonasal hexagonal, broader than long, followed by a pair of præfrontals; frontal hexagonal or pentagonal, slightly wider anteriorly; interparietal small, quadrangular, in the middle between the two pairs of parietals, the posterior of which is a little larger than the anterior; temporals strongly keeled, in five or six longitudinal rows, upper smallest; a short spine in front of the ear; four supraoculars, anterior largest; three supraciliaries; a loreal, a larger præorbital; three infraorbitals, second descending to the lip, separating the fourth and fifth labials; six upper labials; rostral three to four times as broad as high. Mental pentagonal or tetragonal; six lower labials, posterior keeled, in contact with a row of five large shields; gular scales flat, subquadangular or polygonal, slightly imbricate, the anterior smallest, the others nearly as large as the ventrals; scales under the neck rounded, strongly imbricate; sides of the neck with short erect spinose scales. Dorsal scales relatively small, quadrangular, striated or more or less distinctly keeled, forming regular transverse series, which sometimes alternate on the median line; lateral scales same as dorsals, but strongly keeled and serrate; a distinct lateral fold; thirty-two to forty-four transverse series of scales from occiput to base of tail, the largest containing thirty-two to thirty-eight scales. Ventrals quadrangular, scarcely imbricate, smooth, forming twenty or twenty-two longitudinal and thirty-four to thirty-six transverse series. Præanal scales very irregular. Limbs with keeled imbricate scales, except under the thighs, where they are smooth. Femoral pores ten to sixteen on each side. Tail with whorls of keeled spinose scales, strongest on the sides. Brown or yellowish-brown above, with more or less distinct darker spots; a blackish spot on each side of the neck; lower surfaces yellowish or reddish.

	millim.		millim.
Total length	251	Fore limb	39
Head	26	Hind limb	56
Width of head	22	Tail	145
Body	80		

South Africa.

a–d. ♂, ♀, & hgr.	S. Africa.	Sir A. Smith [P.]. (Types.)
e, f. ♂.	S. Africa.	
g. Yg.	S. Africa.	Earl of Derby [P.].
h–i. ♂.	Orange River.	D. Kannemeyer, Esq. [P.].
k. Skeleton.	S. Africa.	Sir A. Smith [P.].

6. Zonurus pustulatus.

Zonurus pustulatus, *Peters, Mon. Berl. Ac.* 1862, p. 19.

In general structure similar to the preceding, but distinguished by the much smaller anterior median gulars, which are almost granular, by the very large femoral pores (in the male), which are in two or three series, and by the scales of the lower surface of the thighs, which increase in size posteriorly. Two large præanal scales, as in *Z. cordylus.* Dorsal scales in thirty-one or thirty-two cross rows, from occiput to base of tail. Olive-brown, darker spotted, and with scattered yellow spots; lower surfaces brownish yellow.

Total length 195 millim.; tail 113 millim.

Hereroland, S.W. Africa.

7. Zonurus capensis.

Hemicordylus capensis, *Gray, Cat.*, p. 48.
Cordylus (Hemicordylus) capensis, *Smith, Mag. N. H.* (2) ii. p. 32, *and Ill. S. Afr., Rept.* pl. xxvii. fig. 2, pl. xxx. fig. 6.
Zonurus capensis, *Dum. & Bibr.* v. p. 360.

Head much depressed, much longer than broad. Nasals not swollen, in contact, pierced in their posterior part; frontonasal hexagonal, as broad as long, followed by a pair of præfrontals forming a suture; frontal pentagonal, a little wider anteriorly; interparietal small, between the four parietals, touching the fronto-parietals with its anterior point; temporals in three longitudinal rows; no spines; four supraoculars, anterior largest; four supra-ciliaries; a loreal, a præorbital, three infraorbitals. Six lower labials, posterior largest, in contact with a row of five large shields; space between the rami of lower jaw covered with scales of two kinds: those on the central portion quadrangular, rather large, and disposed in irregular rows, those on the sides small and granular; scales of under surface of neck moderately large, ovate, and ir-regularly arranged. Scales of the back large, quadrangular, strongly keeled and arranged in transverse rows, generally ten in each row, the keels forming continuous delicate ridges; the lateral and upper parts of the neck, and the lower portion of sides covered with small, subcircular, prominent scales like small granulations. Ventrals in regular transverse rows, eight in each. Two enlarged posterior præanal scales. Upper surface of limbs covered with imbricate keeled scales, inferiorly with smooth scales or granules. Femoral pores small, closely set, eighteen on each side. Tail surrounded by whorls of hard rugged quadrangular scales, each scale strongly keeled and each keel, especially towards the base of the tail, produced posteriorly into a sharp prominent spine, most developed superiorly at the lateral angles; the keels of the subcaudal scales obtuse, only slightly indicated, and without the spinose prolongations of the upper and lateral parts. Purplish brown above, dull brownish red inferiorly.

Total length 264 millim.; tail 156 millim.

Cape of Good Hope.

2. PSEUDOCORDYLUS.

Pseudocordylus, *Smith, Mag. N. H.* (2) ii. 1838, p. 32; *Gray, Cat. Liz.* p. 48.

Zonurus, part., *Dum. & Bibr.* v. p. 344.

Head and body depressed; limbs well developed. Head-shields regular; four parietals; nostril pierced between the nasal and the first labial. Ear-opening large. Dorsal scales roundish, soft, inter-mixed with granules; ventrals large, square, smooth, juxtaposed, forming longitudinal and transverse series. A collar fold, attached medially. Digits slightly keeled inferiorly. Tail spinose. Femoral pores.

South Africa.

1. Pseudocordylus microlepidotus.

Pseudocordylus microlepidotus, *Gray, Cat.* p. 49.

Cordylus microlepidotus, *Cuv. R. A.* 2nd ed. ii. p. 33; *Guér. Icon. R. A., Rept.* pl. vi. fig. 1; *Smith, Ill. S. Afr., Rept.* pls. xxiv.–xxvi. and xxx. figs. 1–5.

Zonurus microlepidotus, *Gray, Griff. A. K.* ix. *Syn* p. 63; *Schleg. Tijdschr. Nat. Gesch. Phys.* i. 1834, p. 216, pl. vii. fig. 1; *Dum. & Bibr.* v. p. 361.
—— wittii, *Schleg. l. c.*
? Zonurus davyi, *Gray, Ann. N. H.* i. 1838, p. 388.
Cordylus (Pseudocordylus) montanus, *Smith, Mag. N. H.* (2) ii. p. 32.
—— (——) melanotus, *Smith, l. c.*
—— (——) algoensis, *Smith, l. c.*
—— (——) subviridis, *Smith, l. c.*
—— (——) fasciatus, *Smith, l. c., and Ill. S. Afr.* pl. xxvii. fig. 1, & xxx. fig. 5.

Head much depressed, much longer than broad. Head-shields slightly rugose; nasals generally forming a suture; frontonasal hexagonal, broader than long, followed by a pair of præfrontals; frontal hexagonal, broader anteriorly; interparietal small, inclosed between the two pairs of parietals, of which the posterior is the largest; occipital small, elongate subtriangular; four supraoculars, anterior largest; four or five supraciliaries; lower eyelid opaque; a loreal, a larger præorbital; four infraorbitals, second descending to the lip, separating the fourth and fifth, or the fifth and sixth labials; temporals in two or three rows; two projecting compressed scales in front of the ear; six or seven upper labials; rostral pentagonal, nearly three times as broad as high. Mental with straight posterior border; six lower labials, posterior largest and keeled, in contact with five large shields. Gular scales small, elongate anteriorly, granular medially, increasing in size, and imbricate, towards the collar; latter angular, attached in the middle; sides of neck folded. Back covered with small flat roundish subtriangular feebly keeled tubercles forming regular longitudinal and transverse series separated by minute granules; the tubercles becoming smaller and passing into the granules on the flanks; a lateral fold. Ventral scales large, quadrangular, broader than long, juxtaposed, forming twelve or fourteen longitudinal and thirty-three or thirty-four transverse series. Præanal scales rather irregular, generally a posterior pair enlarged. Limbs above with imbricate strongly keeled scales, inferiorly with smooth or granular scales. Femoral pores in two or three rows in the males, the inferior row composed of six to ten pores. Caudal scales smooth or feebly keeled inferiorly, strongly keeled above, and spinose on the sides; on the upper surface a whorl of large scales alternates with a whorl of smaller ones. Brown, blackish, reddish, or olive above, with darker or lighter markings; large black spots, bordered with yellow, on the neck; yellowish inferiorly; throat sometimes black.

	millim.		millim.
Total length	258	Fore limb	43
Head	34	Hind limb	63
Width of head	28	Tail (injured)	120
Body	104		

South Africa.

a. ♂.	S. Africa.	Earl of Derby [P.].
b. ♀.	S. Africa.	Sir A. Smith [P.].
c–d. ♂ & yg.	S. Africa.	Dr. Quain [P.].
e, f. Ad., stuffed.	S. Africa.	
g. Skeleton.	S. Africa.	Sir A. Smith [P.].

3. PLATYSAURUS.

Platysaurus, *Smith, Ill. Zool. S. Afr., Rept.*; *Gray, Cat. Liz.* p. 49; *Peters, Reise n. Mossamb.* iii. p. 52.

Head and body much depressed; limbs well developed. Head-shields regular; four parietals; nostril pierced in the nasal; a post-nasal. Ear-opening large. Dorsal scales granular; ventrals large, square, smooth, juxtaposed, forming regular longitudinal and transverse series. A collar fold. Digits slightly keeled inferiorly. Tail without spines. Femoral pores.

South Africa.

Synopsis of the Species.

20 longitudinal rows of ventral scales; no occipital .... 1. *capensis*, p. 261.
20 longitudinal rows of ventral scales; occipital present .... 2. *guttatus*, p. 262.
16 longitudinal rows of ventral scales .... 3. *torquatus*, p. 262.

1. Platysaurus capensis.

Platysaurus capensis, *Gray, Cat.* p. 49.
Platysaurus capensis, *Smith, l. c.* pl. xl.

Head much longer than broad. Frontonasal hexagonal, as long as broad; frontal trapezoid; interparietal small, lozenge-shaped, in the middle between the two pairs of parietals, of which the posterior is the largest; no occipital; four supraoculars, anterior large, triangular, posterior smallest; four supraciliaries; lower eyelid with a transparent disk; two longitudinal rows of large temporals, upper largest; a loreal; a large præorbital; four infraorbitals, second and third bordering the lip; five labials anterior to the second infraorbital. Mental pentagonal; five lower labials, bordered by a row of five large shields, the fourth of which is much larger than the others; gular scales small, elongate quadrangular anteriorly, with a median series of slightly enlarged ones, granular under the neck, again increasing in size towards the collar; latter edged by eight scales. Dorsal scales small, flat, granular, slightly enlarged on the vertebral line, minute on the flanks. Ventrals square, in twenty longitudinal and forty transverse series. Præanal scales small, the largest hardly larger than the ventrals. Limbs long; the length of the tibia equals the distance from the end of the snout to the lower inferior corner of the ear-opening; digits long and slender. Scales on the limbs smooth, feebly keeled on the tibia, granular on the thighs and under the arms; lower surface of tibia with a series of eight large plates; eighteen femoral pores on each side. Tail depressed, tapering, with annuli of elongate quadrangular, smooth scales. Grey-brown above (in spirits), with three light longitudinal bands commencing on the head; lower surfaces whitish.

	millim.		millim.
Total length	197	Fore limb	33
Head	18	Hind limb	48
Width of head	12	Tail	120
Body	59		

Great Namaqualand.

a. ♀. Great Namaqualand. Sir A. Smith [P.]. (Type.)

2. **Platysaurus guttatus.** (Plate XIV. fig. 1.)

Platysaurus guttatus, *Smith, Ill. S. Afr., Rept., App.* p. 8.

Differs from the preceding in the following points:—Frontonasal much broader than long; frontal hexagonal; a small six-sided occipital, separated from the interparietal, which is pentagonal, by an elongate quadrangular shield; four upper labials anterior to the second infraorbital. Granules on sides of neck much enlarged. Dorsal granules smaller, none enlarged. Limbs, and especially digits, shorter; the length of the tibia equals three fourths the distance from end of snout to anterior lower border of ear-opening; seven large plates under the tibia; scales on upper surface of limbs strongly keeled. Eighteen femoral pores. Caudal scales much larger and keeled, the keels strongest on the sides. Upper surfaces bluish (in spirits), guttate with lighter; upper surface of head with three longitudinal light bands, lost on the body; gular region light blue, marbled with blackish-blue; belly dark blue; tail yellowish (red in life).

	millim.		millim.
Total length	165	Fore limb	30
Head	17	Hind limb	42
Width of head	12	Tail	94
Body	54		

Limpopo River.

a. ♂. Limpopo River. Sir A. Smith [P.]. (Type.)

3. **Platysaurus torquatus.**

Platysaurus capensis (*non Smith*), *Peters, Mon. Berl. Ac.* 1854, p. 616.
—— guttatus (*non Smith*), *Peters, l. c.*
—— torquatus, *Peters, Sitzb. Ges. Naturf. Freunde*, 1879, p. 10, *and Reise n. Mossamb.* iii. p. 52, pl. ix. A.

Most like *P. guttatus*, but with only sixteen longitudinal rows of ventrals. Interparietal nearly as large as the frontal, in contact with an occipital which may be broken up into two or three parts; four or five upper labials anterior to second infraorbital; no enlarged gulars on the median line. Femoral pores seventeen or eighteen in the female, twenty-one in the male. A black gular collar.

Total length 185 millim.; tail 113.

Mossambique.

4. CHAMÆSAURA.

Chamæsaura, part. *Schneid. Hist. Amph.* p. 204.
Monodactylus, *Merrem, Tent. Syst. Amph.* p. 76.
Chamæsaura, *Fitzing. Neue Class. Rept.* p. 18; *Wagler, Syst. Amph.* p. 157; *Dum. & Bibr.* v. p. 438; *Gray, Cat. Liz.* p. 61.
Cricochalcis, *Wiegm. Handb. der Zool.* p. 185, *and Herp. Mex.* p. 11; *Peters, Abh. Berl. Ac.* 1862, p. 166.
Mancus, *Cope, Proc. Ac. Philad.* 1862, p. 339.

Body serpentiform, tail extremely long; limbs rudimentary. Head-shields regular; nostril pierced in the nasal. Ear-opening moderate. Scales on the body and tail uniform, lanceolate, strongly keeled, strongly imbricate, forming transverse series; the median dorsal scales larger than the others. Femoral pores.

South Africa.

Synopsis of the Species.

Both pairs of limbs present, pentadactyle .. 1. *ænea*, p. 263.
Both pairs of limbs present, styliform, undivided.......................... 2. *anguina*, p. 264.
Fore limbs absent .... 3. *macrolepis*, p. 264.

1. Chamæsaura ænea.

Cricochalcis ænea (*Wiegm.*), *Fitzing. Syst. Rept.* p. 21; *Peters, l. c.*
Chamæsaura miodactyla, *Günth. Ann. & Mag. N. H.* (5) vi. 1880, p. 235, fig.

Both pairs of limbs present, with five unequal clawed digits. Head-shields strongly striated; frontonasal in contact with the rostral, separating the nasals, followed by a pair of præfrontals forming a median suture; frontal hexagonal; four subequal parietals, hardly larger than the frontoparietals, the anterior pair separated by an interparietal, the posterior by an occipital; three supraoculars, anterior largest; three supraciliaries, anterior largest; nasal large, pierced in its lower anterior portion; two superposed loreals; a large præorbital; a large infraorbital, bordering the lip, separating the third and fourth, or the fourth and fifth labials; upper labials four or five; rostral five times as broad as high; temples covered with imbricate keeled scales. Four lower labials, bordered by four large shields, the anterior pair of which form a suture. Scales on the body in twenty-eight longitudinal and forty-two transverse series. Three femoral pores on each side. Back with two broad dark brown longitudinal bands separating three narrower yellow bands; lower parts greenish white.

	millim.		millim.
Total length	320	Tail (reproduced) ..	230
Head............	14	Hind limb	14

S. Africa.

a. Ad. Peri Bush, S.E. Africa. H. Trevelyan, Esq. [P.]. (Type of *C. miodactyla*.)

2. Chamæsaura anguina.

Chamæsaura anguina, *Gray, Cat.* p. 61.
Lacerta anguina, *Linn. S. N.* i. p. 371.
Chalcides pinnata, *Laur. Syn. Rept.* p. 64.
Chalcida anguina, *Meyer, Syn. Rept.* p. 31.
Chamæsaura anguina, *Schneid. Hist. Amph.* p. 210; *Dum. & Bibr.* v. p. 441; *Gravenh. Nova Acta Ac. Leop.-Carol.* xxiii. i. p. 308, pl. xxx.
Seps monodactylus, *Daud. Rept.* iv. p. 342, pl. lviii. fig. 1.
Chalcides monodactylus, *Oppel, Ord. Rept.* p. 45.

Both pairs of limbs present, styliform, undivided. Head-shields strongly striated; nasals forming a suture, separating the rostral from the frontonasal; latter between the præfrontals, in contact with the frontal, the superior angle of which is cut off; loreal single or divided longitudinally; the other head-shields as in *C. miodactyla*. Scales on the body in twenty-six longitudinal and thirty-eight transverse series. One or two femoral pores. Back with two broad brown longitudinal bands separating three narrower light bands; lower parts whitish.

	millim.		millim.
Total length	520	Tail	400
Head	16	Hind limb	7

S. Africa.

a. Ad.	Kaffraria.	F. P. M. Weale, Esq. [P.].
b. Ad.	Cape of Good Hope.	Dr. Statham [P.].
c. Ad.	S. Africa.	Sir J. McGregor [P.].
d. Ad.	S. Africa.	

3. Chamæsaura macrolepis.

Mancus macrolepis, *Cope, Proc. Ac. Philad.* 1862, p. 339.

Fore limbs absent; hind limbs styliform, undivided. Head-shields as in *C. anguina*. Scales in twenty-two longitudinal and thirty-seven or thirty-eight transverse series to base of tail. One or two femoral pores on each side. Light brown above, with two longitudinal darker bands, whitish inferiorly.

	millim.		millim.
Total length	525	Tail	420
Head	14	Hind limb........	8

S. Africa.

a–b. Ad.	Port Natal.	Mr. Ayres [C.].
c. Ad.	S. Africa.	

Fam. 9. ANGUIDÆ.

Ophisauroidea, part., Chalcidoidea, part., Anguinoidea, *Fitzinger, Neue Classif. Rept.* 1826.
Autarchoglossæ, part., *Wagler, Syst. Amph.* 1830.
Ptychopleuri, part., Scinci, part., *Wiegmann, Herp. Mex.* 1834.
Chalcidiens, part., Scincoïdiens, part., *Duméril & Bibron, Erp. Gén.* v. 1839.
Zonuridæ, part., Scincidæ, part., *Gray, Cat. Liz.* 1845.
Anguidæ, Gerrhonotidæ, *Cope, Proc. Ac. Philad.* 1864.
Anguidæ, *Boulenger, Ann. & Mag. N. H.* (5) xiv. 1884.

The tongue is composed of two distinct portions: a principal posterior, thick, covered with villiform papillæ; and a small anterior, thin, emarginate, covered with lepidoid imbricate papillæ, extensible and more or less retractile into a sheath formed by a transverse fold at the anterior extremity of the villose portion. This retractility of the distal part of the tongue appears to be constant, though more accentuated in some species than in others *; but it should be borne in mind that when the tongue is fully extended, every trace of the transverse fold or sheath disappears, which accounts for the seeming exceptions presented by spirit-specimens in that condition.

The dentition varies from the strictly pleurodont tubercular teeth of *Ophisaurus apus* and some *Diploglossi*, through the conical teeth of the majority of species, to the curved fangs of *Anguis*. The dentition of the latter has no other equivalent but that of *Heloderma*, to which it bears considerable resemblance; and Leydig has even discovered the presence of a slight groove along the anterior surface of the teeth which appears to be homologous with the anterior groove of the poisonous Lizard's fangs; the teeth of *Anguis* are likewise remarkable for the comparatively feeble attachment to the jaws, as in *Heloderma* and the Snakes. The new teeth do not hollow out the base of the old ones, but originate between them. The palate is either toothless, or teeth may be developed on the pterygoids or even on the palatines and vomers (*Ophisaurus*).

The skull belongs to the normal Lacertilian type. The præmaxillary is single; nasals distinct; frontal single in *Gerrhonotus*, double in the other genera; parietal single; palatines and pterygoids well separated along the medial line; infraorbital fossa bounded by the palatine, pterygoid, transverse bone, and maxillary. Dermal cranial ossifications are present, which roof over the supratemporal fossa. The limbs may be more or less developed, or entirely absent externally, in which case, however, the rudiments of the pectoral and pelvic arches are always present. The clavicle is slender, and the interclavicle, in the species with well-developed limbs, cruciform. Abdominal ribs are absent.

The body is protected by bony plates underlying the scales, which

* The common Slow-worm may be taken as a fair example, showing this structure developed to a moderate degree.

are imbricate and subequal; these plates are provided with a system of fine tubercles, as in the *Scincidæ*, which differ from those of the latter family in being arranged irregularly, or forming radiating or arborescent figures. The head-shields present this peculiarity, that an occipital or azygos posterior shield is constantly present, a character which differentiates the *Anguidæ* from most of the *Scincidæ*.

These Lizards are terrestrial. *Anguis* is ovoviviparous. They are most abundantly represented in Central America (*Gerrhonotus*) and the West Indies (*Diploglossus*); a few species occur in North and South America, two in Europe and the borders of the Mediterranean, and one in the Himalayas and Burma.

Synopsis of the Genera.

I. A lateral fold.

Limbs four, pentadactyle .... 1. **Gerrhonotus**, p. 266.

Limbs absent, or the hind pair rudimentary externally.
2. **Ophisaurus**, p. 279.

II. No lateral fold.

Limbs four, pentadactyle .... 3. **Diploglossus**, p. 284.

Limbs four, tetradactyle............ 4. **Sauresia**, p. 294.

Limbs four, mono- or didactyle...... 5. **Panolopus**, p. 295.

External rudiments of the hind limbs only; lateral teeth conical.
6. **Ophiodes**, p. 296.

No trace of limbs externally; lateral teeth fang-like.
7. **Anguis**, p. 297.

1. GERRHONOTUS.

Gerrhonotus, *Wiegm. Isis*, 1828, p. 379, *and Herp. Mex.* p. 29; *Wagl. Syst. Amph.* p. 158; *Dum. & Bibr.* v. p. 394; *Gray, Cat. Liz.* p. 53; *Cope, Proc. Amer. Philos. Soc.* xvii. 1877, p. 96; *Bocourt, Miss. Sc. Mex., Rept.* p. 317.
Pterogasterus, *Peale & Green, Journ. Ac. Philad.* vi. 1830, p. 234; *Cope, l. c.*
Abronia, *Gray, Ann. N. H.* i. 1838, p. 389, *and Cat. Liz.* p. 53.
Elgaria, *Gray, ll. cc.* p. 390, 54.
Barissia, *Gray, ll. cc.*; *Cope, l. c.* p. 97.
Tropidolepis (*non Cuv.*), *Skilton, Am. Journ. Sci.* (2) vii. 1849, p. 202.
Mesaspis, *Cope, l. c.* p. 96.

A lateral fold. Scales squarish or rhomboidal, forming transverse series. Limbs well developed, pentadactyle. Teeth conical. Pterygoid teeth present or absent.

Western and Southern North America; Central America.

Synopsis of the Species.

I. Dorsal scales larger than ventrals.

A. An azygos præfrontal shield.

1. Nuchal scales in four longitudinal series.

No granular zone in the lateral fold; no fold along the side of the neck ...... 1. *oaxacæ*, p. 268.
Lateral fold enclosing granules; a slight fold along the side of the neck ...... 2. *gramineus*, p. 269.

2. Nuchal scales in six or more longitudinal series.

a. No projecting scales above the ear.

Azygos præfontal completely enclosed between two pairs of præfrontals; 13 or 14 longitudinal series of ventrals .... 3. *deppii*, p. 269.
Azygos præfrontal separated from the loreal by a small canthal shield; ventrals in 12 longitudinal series........ 4. *tæniatus*, p. 270.

b. A series of projecting scales above the ear.

Dorsal scales in 12 or 13 longitudinal series .... 5. *vasconcelosii*, p. 270.

B. No azygos præfrontal.

1. A series of projecting scales above the ear.

Four pairs of shields between the frontal and the rostral.................... 6. *fimbriatus*, p. 271.
Three pairs of shields between the frontal and the rostral.................... 7. *auritus*, p. 271.

2. No projecting scales above the ear.

35 to 40 transverse series of dorsal scales between the head and the base of the tail, and 12 or 14 longitudinal series .. 8. *imbricatus*, p. 272.
27 transverse series of scales .... 9. *rudicollis*, p. 272.
16 longitudinal series of dorsal scales.... 10. *planifrons*, p. 273.

II. Dorsal scales not or but scarcely larger than ventrals.

A. Lateral dorsal scales keeled.

Dorsal scales in 14 or 16 longitudinal series .... 11. *cæruleus*, p. 273.

B. Lateral dorsal scales smooth.

1. An azygos præfrontal, separated from the frontal by a pair of præfrontals.

a. Dorsal scales obtusely keeled.

Six or eight median dorsal series keeled.. 12. *kingii*, p. 274.
Three or four median series keeled...... 18. *monticola*, p. 278.

b. Dorsal scales strongly keeled .. 13. *liocephalus*, p. 275.

c. Dorsal scales smooth.

Dorsal scales in 10 or 12 longitudinal series .... 14. *modestus*, p. 276.
Dorsal scales in 16 or 18 longitudinal series .... 15. *rhombifer*, p. 276.

2. No azygos præfrontal.

Dorsal scales in 14 longitudinal series .. 16. *antauges*, p. 277.

3. An azygos præfrontal, in contact with the frontal.

Dorsal scales in 14 longitudinal series .. 17. *obscurus*, p. 277.
Dorsal scales in 18 or 20 longitudinal series .... 19. *moreletii*, p. 278.

1. Gerrhonotus oaxacæ.

Gerrhonotus (Abronia) gramineus (*non Cope*), *Bocourt, Miss. Sc. Mex., Rept.* p. 330, pl. xxvi. A. fig. 6.
—— oaxacæ, *Günth. Biol. C.-Amer., Rept.* p. 36, pl. xxiv. fig. A.

Head very distinct from neck; head-shields thick and rough in the adult, convex on the temples; a pair of small shields behind the rostral; five præfrontals, the central smallest and completely enclosed between the others; nasal separated from the rostral; a small supranasal; two small superposed postnasals and two loreals, second largest; four small outer supraoculars; frontal in contact with the interparietal; ten upper labials, tenth largest and forming with the lower temporals a regular obliquely ascending series of four shields. Sides of neck covered with large irregular scales, separated from the nuchal shields by a naked space with a few scattered tubercular scales; no fold on the side of the neck. Gular scales larger than ventrals. Lateral fold very weak, without any granular zone. Nuchal scales in four longitudinal series. Dorsal scales in sixteen longitudinal series, the outer small and irregular, much larger than the ventrals, obliquely quadrangular, the eight or ten median series obtusely keeled (almost perfectly smooth in the young); twenty-seven transverse series from the head to the base of the tail. Ventrals in fourteen longitudinal series; three ventral transverse series correspond to two dorsals. The adpressed fore limb reaches the eye; the length of the hind limb equals two thirds the distance from axilla to groin. Tail not twice as long as head and body, round, subquadrangular at the base; upper caudal scales very obtusely keeled. Olive above, greenish-white inferiorly; young whitish with metallic gloss, with irregular transverse dark olive bands on the body, limbs, and tail.

	millim.		millim.
Total length	234	Fore limb	26
Head	23	Hind limb........	34
Width of head	18	Tail	138
Body	73		

Oaxaca.

a, b. ♂ & yg. Oaxaca. (Types.)

2. Gerrhonotus gramineus.

Gerrhonotus gramineus, *Cope, Proc. Ac. Philad.* 1864, p. 179; *Günth. Biol. C.-Am., Rept.* p. 36, pl. xxiv. fig. B.

Very closely allied to the preceding species, from which it differs in the following points:—Sides of neck covered with small equal subgranular scales; a slight fold from the ear to the fore limb. Lateral fold stronger, enclosing granular scales. Dorsal scales in twelve or thirteen longitudinal series, very indistinctly keeled. Ventrals in twelve longitudinal series; two ventral series correspond to one dorsal. Eleven upper labials, eleventh forming with the lower temporals an obliquely ascending series of five shields. Pea-green above, each transverse series of dorsal scales black at the base; lower surface of the head yellowish-white, of body pale green.

Orizaba.

a. ♂, not good state. Orizaba.

3. Gerrhonotus deppii.

Abronia deppii, *Gray, Cat.* p. 53.

Gerrhonotus deppii, *Wiegm. Isis,* 1828, p. 379, *and Herp. Mex.* p. 31, pl. ix. fig. 2; *Dum. & Bibr.* v. p. 398.

Abronia deppii, *Gray, Ann. N. H.* i. 1838, p. 389.

Gerrhonotus (Abronia) deppii, *Bocourt, Miss. Sc. Mex., Rept.* p. 325, pl. xxi. A. fig. 3.

Head very distinct from neck; head-shields smooth; a pair of small shields behind the rostral; five præfrontals, the central a little smaller and completely enclosed between the others; nasal separated from the rostral; a small supranasal; two small superposed postnasals and two loreals, second largest; four small outer supraoculars; frontal touching the interparietal; ten upper labials, last two much larger than the others, and than the temporals. Sides of neck covered with granular scales. Lateral fold feeble, commencing in advance of the arm. Nuchal scales in six longitudinal series. Dorsal scales in ten or eleven longitudinal series (thirteen or fourteen in the middle of the body), oblique, much larger than ventrals, smooth; twenty-seven transverse series. Ventrals in thirteen or fourteen longitudinal series. Limbs short. Tail a little longer than head and body; caudal scales smooth. Dark brown above, with yellow spots forming transverse bands on the body, limbs, and tail; lips and lower surfaces yellowish white.

	millim.		millim.
Total length	208	Fore limb	25
Head	20	Hind limb	28
Width of head	16	Tail	114
Body	74		

Mexico.

4. Gerrhonotus tæniatus.

Abronia tæniata, *Gray, Cat.* p. 53.
Gerrhonotus tæniatus, *Wiegm. Isis*, 1828, p. 380, *and Herp. Mex.* p. 32, pl. iv. fig. 1; *Dum. & Bibr.* v. p. 399.
Abronia tæniatus, *Gray, Ann. N. H.* i. 1838, p. 390.
Gerrhonotus (Abronia) tæniatus, *Bocourt, Miss. Sc. Mex., Rept.* p. 327, pl. xxi. A. figs. 4, 5.

Closely allied to *G. deppii*. A small shield on each side between the anterior loreal and the central præfrontal. Dorsal scales a little smaller, less oblique, obtusely keeled; they form fourteen longitudinal series, and twenty-eight or twenty-nine transverse series. Ventrals in twelve longitudinal series. Limbs a little longer. Light green above, with transverse brown bands on the body and tail; an oblique black band on the temple; lower parts yellow.

	millim.		millim.
Total length	216	Fore limb	24
Head	20	Hind limb	28
Width of head	16	Tail	128
Body	68		

Mexico.

5. Gerrhonotus vasconcelosii.

Gerrhonotus vasconcelosii, *Bocourt, Nouv. Arch. Mus.* vii. 1872, *Bull.* p. 107.
—— (Abronia) vasconcelosii, *Bocourt, Miss. Sc. Mex., Rept.* p. 334, pl. xxi. fig. 3, & xxi A. fig. 8.

Head-shields smooth; a pair of small shields behind the rostral and seven præfrontals, the two anterior in contact, the median separating the two other pairs and touching the frontal; nasal separated from the rostral; a small supranasal; two small superposed postnasals; two loreals, second largest; three small outer supraoculars; ten upper labials; a series of short, obtusely conical, projecting scales above the ear. Sides of neck covered with granular scales. Lateral fold pretty well marked though not very deep, commencing in advance of the arm. Dorsal scales obtusely keeled, of the flanks smooth, in twelve or thirteen longitudinal and twenty-seven transverse series; ventrals smaller, in fourteen longitudinal series. Tail moderately long, a little depressed at the base. Bright green above, with broad angular blackish cross bands; the projecting scales above the ear pale yellow; lower surfaces greenish yellow, throat and breast black-dotted.

	millim.		millim.
Total length	145	Fore limb	15
Head	14	Hind limb........	19
Width of head	10	Tail	91
Body	40		

Argueta, W. Guatemala.

6. Gerrhonotus fimbriatus.

Gerrhonotus (Abronia) auritus (*non Cope*), *Bocourt, Miss. Sc. Mex., Rept.* p. 337, pl. xxi. fig. 2, & xxi A. fig. 7.
Barissia fimbriata, *Cope, Proc. Am. Philos. Soc.* xxii. 1885, p. 171.

Head-shields slightly rugose; four pairs of shields between the rostral and the frontal; nasal separated from the rostral; two small superposed postnasals; two loreals, second largest; five large inner and three small outer supraoculars; nine upper labials; a series of long pointed projecting scales above the ear. Sides of neck with granular scales. Lateral fold pretty well marked though not very deep, commencing in advance of the arm. Dorsal scales obtusely keeled, in fourteen longitudinal and thirty transverse series; ventrals smaller, in fourteen longitudinal series. Tail cylindrical, a little depressed at the base, one third longer than head and body. Greenish grey above, usually with angular brown cross bands edged posteriorly with black; the projecting scales above the ear yellow; lower surfaces yellow, nearly all the ventral scales with a brown dot.

	millim.		millim.
Total length	287	Fore limb	27
Head	22	Hind limb........	35
Width of head	17	Tail	177
Body	88		

Vera Paz, Guatemala.

7. Gerrhonotus auritus.

Gerrhonotus auritus, *Cope, Proc. Ac. Philad.* 1868, p. 306.

Distinguished from the preceding in having three pairs of shields between the rostral and the frontal instead of four. Dorsal scales strongly but finely keeled, in sixteeen longitudinal and twenty-six transverse series. Pea-green, more or less shaded with yellow; the rugosities of the scales everywhere black; blackish-green shades form indistinct V's directed backwards on the middle of the back; eyelids and auricular process yellow; below yellowish, with dirty green shades.

From snout to vent 125 millim.

Vera Paz.

8. Gerrhonotus imbricatus.

Barissia imbricata, *Gray, Cat.* p. 55.
—— lichenigerus, *Gray, l. c.*
Gerrhonotus imbricatus, *Wiegm. Isis,* 1828, p. 381, *and Herp. Mex.* p. 34, pl. x. figs. 2, 5; *Dum. & Bibr.* v. p. 407; *Günth. Biol. C.-Am., Rept.* p. 38.
—— lichenigerus, *Wagl. Icon. Amph.* pl. xxxiv. fig. 2; *Wiegm. Herp. Mex.* p. 35, pl. x. fig. 6; *Dum. & Bibr.* v. p. 408.
Barissia imbricata, *Gray, Ann. N. H.* i. 1838, p. 390, *and Zool. Beechey's Voy., Rept.* p. 96, pl. xxxi. fig. 1.
Gerrhonotus (Barissia) imbricatus, *Bocourt, Miss. Sc. Mex., Rept.* p. 363, pl. xxi. B. figs. 1, 2.

Head-shields swollen; three pairs of shields between the frontal and the rostral; nasal separated from the rostral; a supranasal; a postnasal; a loreal; two outer supraoculars; ten or eleven upper labials. Sides of neck covered with granular scales. Lateral fold strong, extending to the neck. Nuchal scales in eight or ten longitudinal series. Dorsal scales large, subrhomboidal, in twelve or fourteen longitudinal series, the four to eight median strongly keeled, the others obtusely keeled or smooth; thirty-five to forty transverse series; ventrals smaller than dorsals, in ten or twelve longitudinal series. The adpressed fore limb usually reaches the anterior corner of the eye; the hind limb measures two thirds to half the distance from axil to groin. Tail not twice as long as head and body, compressed at the base; upper caudal scales strongly keeled. Olive-brown above, uniform or with dark brown or greenish-white spots or dots sometimes forming longitudinal bands; head frequently punctulated with whitish; lower surfaces greenish or yellowish white.

	millim.		millim.
Total length	265	Fore limb	28
Head	28	Hind limb........	35
Width of head	19	Tail	160
Body	77		

Mexico.

a–c. ♂ ♀.	City of Mexico.	Mr. Doorman [C.].
d–f. ♂ & hgr.	Mexico.	M. Sallé [C.].
g. ♂.	Mexico.	Mr. Geale [C.].
h. ♂.	Mexico.	Dr. J. E. Gray [P.].
i. ♀.	Tehuantepec.	
k. ♂.	——?	

9. Gerrhonotus rudicollis.

Barissia rudicollis, *Gray, Cat.* p. 55.
Gerrhonotus rudicollis, *Wiegm. Isis,* 1828, p. 380, *and Herp. Mex.* p. 33, pl. x. fig. 1; *Dum. & Bibr.* v. p. 410.
Barissia rudicollis, *Gray, Ann. N. H.* i. 1838, p. 390.
Gerrhonotus (Barissia) rudicollis, *Bocourt, Miss. Sc. Mex., Rept.* p. 367, pl. xxi. B. fig. 3.

Closely allied to the preceding. Head larger; nasal in contact with the rostral; nuchal shields more strongly keeled. Only twenty-seven transverse series of scales from the head to the base of the tail. Ventrals in twelve or fourteen longitudinal series. Yellowish-green above, flanks with vertical brown bands.

Mexico.

10. Gerrhonotus planifrons.

Gerrhonotus (Barissia) planifrons, *Bocourt, Miss. Sc. Mex., Rept.* p. 361, pl. xxi. c. fig. 1.

Head-shields quite flat. Dorsal scales in sixteen longitudinal rows, the eight median strongly keeled; ventrals in twelve. Otherwise as in *G. imbricatus.* Purplish-brown above, with darker vertical bands on the sides of the neck and body, edged posteriorly with white; head with whitish dots; lower surfaces yellow, speckled with blackish.

Oaxaca.

11. Gerrhonotus cæruleus.

Gerrhonotus cæruleus, *Gray, Cat.* p. 54.
—— burnettii, *Gray, l. c.*
—— wiegmannii, *Gray, l. c.*

Gerrhonotus cæruleus, *Wiegm. Isis,* 1828, p. 380, *and Herp. Mex.* p. 31; *Bocourt, Miss. Sc. Mex., Rept.* p. 353, pl. xxi. c. fig. 3; *Günth. Biol. C.-Am., Rept.* p. 38, pl. xxv. figs. A–D.
—— burnettii, *Gray, Griff. A. K.* ix. *Syn.* p. 64, *and in Beechey's Voy., Rept.* p. 96, pl. xxxi. fig. 2; *Bocourt, l. c.* p. 356, pl. xxi. c. fig. 4.
Cordylus (Gerrhonotus) multicarinatus, *Blainv. Nouv. Ann. Mus.* iv. 1835, p. 289, pl. xxv. fig. 2.
Gerrhonotus multicarinatus, *Dum. & Bibr.* v. p. 404; *Hallow. Rep. U. S. Explor. Surv. R. R.* x. part iv. 1859, pl. ix. fig. 1; *Bocourt, l. c.* p. 357, pl. xxi. c. fig. 5.
Tropidolepis scincicauda, *Skilton, Am. Journ. Sc.* (2) vii. 1849, pp. 202, 312, pl. —. figs. 1–3.
Elgaria scincicauda, *Baird & Gir. in Stansb. Exped. Gr. Salt Lake,* p. 348, pl. iv. figs. 1–3, *and U.S. Explor. Exped., Herp.* p. 210, pl. xxiii. figs. 1–9.
—— principis, *Baird & Gir. Proc. Ac. Philad.* 1852, p. 175, *and U. S. Explor. Exped., Herp.* p. 214, pl. xxii. figs. 9–16.
—— formosa, *Baird & Gir. ll. cc.* pp. 175, 206, pl. xxiii. figs. 10–17.
—— grandis, *Baird & Gir. ll. cc.* pp. 176, 212, pl. xxii. figs. 1–8.
—— marginata, *Hallow. Proc. Ac. Philad.* 1852, p. 179, *and in Sitgreave's Exped. Zuni & Color. Riv.* p. 114, pl. iii.
Gerrhonotus webbii, *Baird, Proc. Ac. Philad.* 1858, p. 255, *and U.S. Mex. Bound. Surv., Rept.* p. 11, pl. xxv. figs. 1–8.
? Gerrhonotus olivaceus, *Baird, ll. cc.*
Gerrhonotus (Elgaria) grandis, *O'Shaughn. Ann. & Mag. N. H.* (4) xii. 1873, p. 47.
—— (——) scincicauda, *O'Shaughn. l. c.*
—— (——) formosus, *O'Shaughn. l. c.*
—— (——) principis, *O'Shaughn. l. c.* p. 48.

Head-shields smooth; an azygos præfrontal, very variable in form and size, either completely enclosed between the other præfrontals or in contact with the loreals or with the frontal, separated from the rostral by one or two pairs of shields; sometimes a small azygos shield behind the rostral, separating the anterior pair; nasal usually in contact with the rostral; two superposed postnasals, followed by one or two loreals; two or three small outer supraoculars; ten or eleven upper labials. Sides of neck granular. Lateral fold very strong, beginning on the neck. Nuchal shields in eight to twelve longitudinal series, keeled except in very young specimens. Dorsal scales in fourteen or sixteen longitudinal series, sharply keeled, of nearly the same size as the ventrals; forty-five to fifty-two transverse series. Ventrals in twelve longitudinal series. The adpressed fore limb reaches the angle of the mouth or the eye; the hind limb measures half to two thirds the distance from axilla to groin. Tail once and two thirds to twice as long as head and body, round, subquadrangular at the base; upper caudal scales sharply keeled. Brown or olive above, usually with irregular undulated blackish cross bars bordered with small yellowish spots posteriorly; back sometimes irregularly spotted with blackish, and the flanks closely spotted with black and dotted with yellowish; lower surfaces whitish olive, uniform or with longitudinal series of confluent brown spots.

	millim.		millim.
Total length	334	Fore limb	30
Head	29	Hind limb........	35
Width of head	21	Tail	220
Body	85		

Western North America to Texas; Central America.

a–b, c. ♂, ♀, & yg.	San Juan I. Vancouver's I.	H.M.S. 'Plumper.'
d. ♂.	Vancouver's I.	Col. Hawkins [P.].
e–f. ♂.	Near Orca I., Gulf of Georgia.	The Lords of the Admiralty [P.].
g. ♂.	Oregon.	J. K. Lord, Esq. [P.].
h. ♀.	Sumass Lake	J. K. Lord, Esq. [P.].
i–l, m–p. Ad., hgr., & yg.	Monterey.	
q. ♂.	California.	Mr. Bridges [C.].
r–t. ♂.	Pacific coast of N. America.	
u. Ad.	N.W. coast of Central America.	(Type of *G. wiegmanni.*)
v. Hgr.	Cartago, Costa Rica.	
w. ♂.	——?	Haslar Collection. (Type of *G. burnettii.*)

12. Gerrhonotus kingii.

Elgaria kingii, *Gray, Cat.* p. 54.
Elgaria kingii, *Gray, Ann. N. H.* i. 1838, p. 390.
Gerrhonotus multifasciatus, *Dum. & Bibr.* v. p. 401.
Elgaria nobilis, *Baird & Gir. Proc. Ac. Philad.* 1852, p. 129.
Gerrhonotus nobilis, *Baird, U.S. Mex. Bound. Surv., Rept.* p. 11, pl. xxv. figs. 1–8.

Gerrhonotus kingii, *O'Shaughn. Ann. & Mag. N. H.* (4) xii. 1873, p. 46; *Bocourt, Miss. Sc. Mex., Rept.* p. 339, pl. xxi. c. fig. 2.

Head-shields smooth; three large præfrontals, the anterior broader than long and normally separated from the frontal by the posterior pair; two pairs of small shields between the rostral and the odd præfrontal; nasal in contact with the rostral; two superposed postnasals and two loreals; five large inner and two or three small outer supraoculars; ten upper labials. Sides of neck granular. Lateral fold very strong, beginning on the neck. Nuchal shields smooth, in ten or twelve longitudinal series. Dorsal scales in fourteen or sixteen longitudinal series, those of the median series obtusely keeled, the laterals smooth; in size nearly equalling the ventrals; fifty to fifty-three transverse series. Ventrals in twelve longitudinal series. Limbs and tail as in *G. cœruleus*. Brownish above, with dark brown cross bands.

From snout to vent 110 millim.; tail 280 millim.

Mexico; New Mexico.

Dr. Günther regards this as one of the variations of *G. cœruleus* (*l. c.* p. 39).

a. Ad., bad state.	Mexico.	T. Bell, Esq. [P.].	(Type.)
b. Yg.	Deming, New Mexico.	Mr. Forrer [C.].	

13. Gerrhonotus liocephalus.

Gerrhonotus tessellatus, *Gray, Cat.* p. 53.

Gerrhonotus liocephalus, *Wiegm. Isis*, 1828, p. 381; *Bocourt, Miss. Sc. Mex., Rept.* p. 342, pl. xxi. A. figs. 1, 2; *Günth. Biol. C.-Am., Rept.* p. 41.

Scincus ventralis, *Peale & Green, Journ. Ac. Philad.* vi. 1830, p. 233.

Gerrhonotus tessellatus, *Wiegm. Herp. Mex.* p. 32, pl. x. fig. 3; *Dum. & Bibr.* v. p. 405.

? Gerrhonotus infernalis, *Baird, Proc. Ac. Philad.* 1858, p. 255.

Gerrhonotus ophiurus, *Cope, Proc. Ac. Philad.* 1866, p. 321.

—— lemniscatus, *Bocourt, Nouv. Arch. Mus.* vii. 1872, *Bull.* p. 105.

Pterogasterus ventralis, *Cope, Proc. Am. Philos. Soc.* xvii. 1877, p. 96.

—— tessellatus, *Cope, l. c.*

—— ophiurus, *Cope, l. c.*

—— infernalis, *Cope, l. c.*

—— lemniscatus, *Cope, l. c.*

Head-shields smooth; three large præfrontals, forming a triangle; three pairs of small shields between the latter and the rostral, the two front pairs usually separated by a small azygos shield; nasal separated from the rostral; two superposed postnasals; two or three loreals, the posterior sometimes fusing with the canthal; three outer supraoculars; ten to thirteen upper labials. Sides of neck with granular scales. Lateral fold very strong, extending on the neck. Dorsal scales in fourteen or sixteen longitudinal series, the eight or ten median series strongly keeled, the others smooth; forty-nine to fifty-five transverse series. Ventrals nearly as large as dorsals, in twelve longitudinal series. Limbs very variable in length. Tail a little compressed at the base, once and two thirds to twice and a half as long as head and body; upper caudal scales usually much less strongly keeled than dorsals. Brown or brownish olive above, with darker brown spots or with more or less regular cross bands of black and whitish spots; a series of black vertical bars or annuli in the lateral fold; usually a yellowish streak, sometimes bordered with black above, from below the eye to the angle of the mouth; lower surfaces yellowish or greenish white, uniform or speckled with black.

	millim.		millim.
Total length	403	Fore limb	39
Head	29	Hind limb........	46
Width of head	18	Tail	270
Body	104		

Western North America and Mexico.

a. ♂.	Cordova	M. Sallé [C.].
b. Hgr.	Mexico.	M. Sallé [C.].
c, d. ♀.	Mexico.	
e–k. Ad. & hgr.	—— ?	

14. Gerrhonotus modestus.

Pterogasterus modestus, *Cope, Proc. Am. Philos. Soc.* xvii. 1877, p. 97; *Günth. l. c.* p. 42.

Differs from *G. liocephalus* in the absence of the small canthal shield which truncates the lateral angle of the odd præfrontal, and in the absence of keels on the dorsal scales, which are in ten or twelve longitudinal series. Upper caudal scales strongly and obtusely keeled. Colour above, brown; below, olivaceous; sides reddish-brown or maroon, bordered above by a blackish line which separates it from the dorsal colour.

Total length 150 millim.; length to vent 72 millim.

Guatemala?

15. Gerrhonotus rhombifer.

Gerrhonotus rhombifer, *Peters, Mon. Berl. Ac.* 1876, p. 298; *Günth. l. c.* p. 42.

—— (Abronia) rhombifer, *Bocourt, Miss. Sc. Mex., Rept.* p. 332, pl. xxi. B. fig. 4.

Head flat and elongate; head-shields smooth; three pairs of small shields, the two anterior separated by two minute azygos shields, between the rostral and a large azygos præfrontal, which is broader than long and separated from the frontal by a pair of nearly equally large præfrontals; nasal separated from the rostral; two superposed postnasals, upper minute; three loreals, second not reaching the labials; four small outer supraoculars, and an odd shield between; fourteen upper labials; temporal scales small. Sides of neck covered with small square shields. Lateral fold feeble, commencing in advance of the arm. Dorsal scales smooth, scarcely larger than ventrals, forming sixteen to eighteen longitudinal series. Ventrals in ten longitudinal series. Upper surface of head and eight irregular rhomboidal spots on the body, yellowish-white, separated by narrow dark brown interspaces; tail annulate brown and yellowish; belly blackish.

	millim.		millim.
Total length	145	Fore limb	11
Head	12	Hind limb........	13
Width of head	7	Tail	96
Body	37		

Chiriqui.

16. Gerrhonotus antauges.

Barissia antauges, *Cope, Proc. Ac. Philad.* 1866, p. 132.
Gerrhonotus viridiflavus, *Bocourt, Ann. Sc. Nat.* xvii. 1873, art. 2.
—— bocourti, *Peters, Mon. Berl. Ac.* 1876, p. 297.
—— antauges, *Bocourt, Miss. Sc. Mex., Rept.* p. 346, pl. xxi. fig. 4, and xxi. B. fig. 7; *Günth. l. c.* p. 37.

Head-shields smooth; four pairs of shields between the rostral and the frontal, the two anterior very small; the second pair sometimes separated; nasal separated from the rostral; two superposed postnasals; one loreal; two or three outer supraoculars; ten upper labials. Sides of neck covered with granular scales. Lateral fold strong, extending on the neck. Dorsal scales in fourteen longitudinal series, the median smooth or obtusely keeled, the laterals smooth; fifty transverse series. Ventral scales nearly the same size as dorsals, in twelve longitudinal series. Tail subquadrangular at the base; upper caudal scales more strongly keeled than the dorsals. Bronzy green above; head spotted with black; three longitudinal bands on the back; flanks and sides of neck with a black network enclosing greenish-white dots; a whitish streak on the side of the head; lower surfaces dark green with numerous small angular yellowish-white spots, or lighter with small greenish-black spots.

	millim.		millim.
Total length (tail reproduced)	149	Body	54
Head	15	Fore limb	15
Width of head	10	Hind limb........	21

Mexico.

17. Gerrhonotus obscurus.

Gerrhonotus obscurus, *Günth. Biol. C.-Am., Rept.* p. 40, pl. xxv. fig. E.

Head-scales smooth; five præfrontals, central largest and in contact with the frontal; a pair of small shields between the rostral and the præfrontals; nasal in contact with the rostral on one side; a supranasal; two superposed postnasals; one loreal; two outer supraoculars; ten upper labials. Sides of neck granular. Lateral fold strong, extending on the neck. Nuchal scales in eight longitudinal rows. Dorsal scales in fourteen longitudinal rows, the laterals smooth, the others obtusely keeled, and forty-seven transverse rows; ventrals nearly as large as dorsals, in twelve longitudinal rows. The adpressed fore limb reaches the posterior border of the eye; the length of the hind limb equals two thirds the distance from axil to groin. Upper caudal scales more strongly keeled than dorsals. Olive above, with a black vertebral line; sides and lower surfaces black, spotted with greenish white.

	millim.		millim.
Total length (tail reproduced)	134	Body	50
Head	14	Fore limb	15
Width of head	10	Hind limb	21

Mexico.

a. ♂, bad state. Mexico. (Type.)

18. Gerrhonotus monticola.

Gerrhonotus fulvus (*non Boc.*), *Cope, Journ. Ac. Philad.* (2) viii. 1876, p. 118.
—— monticolus, *Cope, Proc. Amer. Philos. Soc.* xvii. 1877, p. 97.

Præfrontal separated from the frontal by a pair of shields. Dorsal scales in fifteen longitudinal series, the three or four median rows keeled, the others smooth; forty-five transverse series. Otherwise as in *G. moreletii*. Colour of upper surface and sides brown, the latter a little darker, and bounded above by a narrow black line; a somewhat irregular row of small black spots down the median dorsal line; below yellowish olive, the scales of the abdomen with black borders, those of the gular and thoracic regions with black centres.

Total length 143 millim.; length to vent 61 millim.

Summit of the Pico Blanco (11,500 feet), Costa Rica.

19. Gerrhonotus moreletii.

Gerrhonotus wiegmanni (*non Gray*), *Salvin, Proc. Zool. Soc.* 1860, p. 454.
—— moreletii, *Bocourt, Nouv. Arch. Mus.* vii. 1872, *Bull.* p. 102, *and Miss. Sc. Mex., Rept.* p. 349, pl. xxi. fig. 1, and xxi. B. fig. 5; *Günth. Biol. C.-Am., Rept.* p. 41, pl. xxiv. fig. C.
—— fulvus, *Bocourt, ll. cc.* p. 104, 352, pl. xxi. B. fig. 6.
Mesaspis moreletii, *Cope, Proc. Amer. Philos. Soc.* xvii. 1877, p. 96.
—— fulvus, *Cope, l. c.*

Head-shields smooth; a large præfrontal, sometimes single and forming a suture with the whole anterior border of the frontal, sometimes divided into three, the median or larger shield forming only a short suture with the frontal; two pairs of small shields between the præfrontal and the rostral; nasal separated from the rostral; a supranasal; two superposed postnasals; one loreal and a large canthal descending between the latter shield and the præocular; three outer supraoculars; ten or eleven upper labials. A rather narrow granular zone along the side of the neck. Lateral fold strong, extending on the neck. Dorsal scales in eighteen or twenty longitudinal series, the six median strongly keeled, the others smooth or indistinctly keeled; fifty-two to fifty-four transverse series; ventrals larger than dorsals, in twelve longitudinal series. The adpressed fore limb reaches the eye; the length of the hind limb equals half or three fifths the distance from axil to groin. Tail subquadrangular at the base; upper caudal scales strongly keeled. Olive-brown above, back uniform or black-spotted, sides spotted with black and greenish white; lower surfaces greenish white, uniform or with scattered blackish dots.

	millim.		millim.
Total length (tail reproduced)	185	Body	76
Head	18	Fore limb	22
Width of head	12	Hind limb.	28

Guatemala; Vera Cruz.

a–b. ♀.	Lanquin, Vera Paz.	O. Salvin, Esq. [C.].
c. ♂.	Dueñas.	O. Salvin, Esq. [C.].
d. ♀.	Rio Chisoy.	O. Salvin, Esq. [C.].

2. OPHISAURUS.

Ophisaurus, *Daud. Rept.* vii. p. 346; *Fitzing. N. Classif. Rept.* p. 20; *Wagler, Syst. Amph.* p. 159; *Wiegm. Herp. Mex.* p. 11; *Dum. & Bibr.* v. p. 421; *Gray, Cat. Liz.* p. 56.
Bipes, *Oppel, Ordn.* p. 43; *Wagl. l. c.*
Pseudopus, *Merrem, Tent.* p. 78; *Fitzing. l. c.*; *Wiegm. l. c.*; *Dum. & Bibr. l. c.* p. 412; *Gray, l. c.*; *Günth. Rept. Brit. Ind.* p. 74.
Hyalinus, *Merr. l. c.* p. 79.
Proctopus, *Fischer, Mém. Soc. Nat. Mosc.* iv. 1830, p. 241.
Dopasia, *Gray, Ann. & Mag. N. H.* (2) xii. 1853, p. 389.
Ophiseps, *Blyth, Journ. As. Soc. Beng.* xxii. 1853, p. 655.
Hyalosaurus, *Günth. Ann. & Mag. N. H.* (4) xii. 1873, p. 351.

A lateral fold. Scales squarish-rhomboidal, forming straight longitudinal and transverse series. Limbs absent externally, or reduced to a rudiment of the hind pair. Teeth with conical or subspherical crowns. Pterygoid teeth; palatine and vomerine teeth present or absent.

South-eastern Europe, North Africa, South-western Asia, Himalayas and Burma, South-eastern North America and Mexico.

Synopsis of the Species.

I. Ear-opening distinct; external rudiments of hind limbs .... 1. *apus*, p. 280.

II. Ear-opening distinct; no rudiments of hind limbs externally.

Ear-opening much larger than the nostril; 16 longitudinal rows of dorsal plates.... 2. *ventralis*, p. 281.
Ear-opening not larger than the nostril; 14 longitudinal rows of dorsal plates; two shields on a line from the nasal to the azygos præfrontal .... 3. *attenuatus*, p. 282.
Ear-opening not larger than the nostril; 14 longitudinal rows of dorsal plates; three shields on a line from the nasal to the azygos præfrontal................ 4. *gracilis*, p. 283.

III. Ear concealed; external rudiments of hind limbs .... 5. *koellikeri*, p. 283.

1. Ophisaurus apus.

Pseudopus pallasii, *Gray, Cat.* p. 55.
Lacerta apus, *Pallas, Reise*, iii. p. 702.
—— apoda, *Pallas, Novi Comment. Ac. Petrop.* xix. 1775, p. 435, pls. ix. & x., *and Zoogr. Ross.-As.* iii. p. 33.
Chalcida apus, *Meyer, Syn. Rept.* p. 31.
Chamæsaura apus, *Schneid. Hist. Amph.* ii. p. 212.
Seps sheltopusik, *Daud. Rept.* iv. p. 351.
Bipes pallasii, *Oppel, Ordn.* p. 43; *Wagl. Icon. Amph.* pl. xiv.
Pseudopus serpentinus, *Merr. Tent.* p. 78; *Bonap. Faun. Ital.*
—— oppellii, *Fitzing. N. Classif.* p. 50.
—— pallasii, *Cuv. R. A.* 2nd ed. ii. p. 69; *Bibr. in Bory de St. Vinc. Expéd. Sc. Morée, Rept.* p. 70, pls. xii. & xiii.; *Duvern. R. A., Rept.* pl. xxiii.; *Dum. & Bibr.* v. p. 417; *Demid. Voy. Russ. Mérid.* iii. p. 340, pl. ii. fig. 2; *Strauch, Erp. Alg.* p. 40.
—— durvillii, *Cuv. l. c.*; *Guér. Icon. R. A., Rept.* pl. xvii. fig. 1; *Bibr. l. c.*
Proctopus pallasii, *Fischer, Mém. Soc. Nat. Mosc.* iv. 1830, p. 241, pls. ii.–iv.
Ophisaurus serpentinus, *Eichw. Zool. Spec.* iii. p. 179.
Pseudopus fischeri, *Ménétr. Cat. Rais.* p. 65.
—— apus, *Schreib. Herp. Eur.* p. 362.

Lateral teeth with subspherical crowns; teeth on the pterygoids (in two rows), palatines, and frequently also on the vomers. Shape, arrangement, and number of the head-shields extremely similar to those of *Anguis fragilis*, and likewise subject to a certain amount of variation; azygos præfrontal narrower than the greatest width of the frontal, usually separated from the latter by two or three præfrontals on a transverse line; interparietal narrower than the parietals, as broad as or narrower than the occipital; two shields on a line from the nasal to the azygos præfrontal; three supraoculars and five supraciliaries. Ear-opening distinct, horizontal, slightly larger than the nostril. Dorsal plates in twelve or fourteen longitudinal series, keeled, the keels much stronger in the young than in the adult; ventrals in ten longitudinal series, strongly keeled in the young, the keels disappearing with age. A small rudiment of extremities on each side of the anal flap. Tail once and a half to twice as long as head and body; upper and lower caudal plates keeled. Brown, lighter inferiorly; young olive-grey, with dark-brown undulated cross bands on the back and vertical bars on the sides of the head.

From snout to vent 45 centim.; tail 65.

South-eastern Europe, South-western Asia, North Africa.

a, b. Adult.	Dalmatia.	
c. Hgr.	Greece.	Rev. W. Hennah [P.].
d, e. Ad. & yg.	Europe.	Sir J. Richardson [P.].
f. Ad., stuffed.	Europe.	
g–h, i–k. Ad.	Xanthus.	Sir C. Fellows [P.].
l. Ad.	Palestine.	Rev. H. B. Tristram [P.].
m. Ad.	Lake of Galilee.	Rev. H. B. Tristram [P.].
n. Yg.	Mount Hermon.	Rev. H. B. Tristram [P.].
o, p. Ad., skeletons.	Europe.	

Var. **ornatus.**

Pseudopus apus, forma ornata, *Boettg. Zool. Anz.* 1881, p. 571, *and Abh. Senck. Ges.* xiii. 1883, p. 118.

Flesh-coloured, more greyish along the back, yellowish inferiorly; head uniform, except a few irregularly scattered black dots; back with about twenty narrow undulated cross bands, formed of blue, black-edged scales. Known from a young specimen.

Casablanca, Morocco.

2. **Ophisaurus ventralis.**

Ophisaurus ventralis, *Gray, Cat.* p. 56.
Anguis ventralis, *Linn. S. N.* i. p. 391.
Chamæsaura ventralis, *Schneid. Hist. Amph.* ii. p. 215.
Ophisaurus ventralis, *Daud. Rept.* vii. p. 352, pl. lxxxviii.; *Duvern. R. A., Rept.* pl. xxiv. fig. 1; *Dum. & Bibr.* v. p. 423; *Holbr. N. Am. Herp.* ii. p. 139, pl. xx.; *Yarrow, Bull. U.S. Nat. Mus.* no. 24, p. 46.
Hyalinus ventralis, *Merr. Tent.* p. 79.
Ophisaurus punctatus, *Cuv. R. A.* 2nd ed. ii. p. 70.
—— striatulus, *Cuv. l. c.*
—— lineatus, *Gray, Ann. N. H.* i. 1838, p. 391.

Lateral teeth conical; teeth on the pterygoids in three to five longitudinal series, forming a broad band; one or two series on the palatines. Azygos præfrontal usually narrower than the greatest width of the frontal, usually separated from the latter by a pair of præfrontals; interparietal narrower than the parietals, as broad as or narrower than the occipital; two shields on a line from the nasal to the azygos præfrontal; four supraoculars and five supraciliaries. Ear-opening distinct, horizontal, considerably larger than the nostril. Dorsal plates in fourteen longitudinal and about one hundred and twenty transverse series; the median dorsal plates obtusely keeled, the laterals as well as the ventrals smooth; latter in ten longitudinal series. No rudiments of limbs externally. Tail about twice as long as the body; upper caudal plates obtusely keeled, lower smooth. Green above, marked with black and yellow, or brownish with lateral dark brown longitudinal bands; lower surfaces yellow.

From snout to vent 28 centim.; tail 51.

Eastern North America; from Mexico (Jalapa) northwards to North Carolina and the southern parts of Illinois.

a. Ad.	N. America.	Dr. R. Harlan [P.].
b. Ad.	N. America.	Lord Ampthill [P.].
c. Ad.	N. America.	

3. **Ophisaurus attenuatus.**

Opheosaurus ventralis attenuatus (*Baird*), *Cope, Bull. U. S. Nat. Mus.* no. 17, 1880, p. 18.

—— ventralis, *Shufeldt, Proc. U. S. Nat. Mus.* iv. 1882, p. 392.

Differs from *O. ventralis* in the following points:—A single longitudinal row of small teeth on the posterior part of the palatines and on the pterygoids, where the row becomes double. Head narrower and more depressed. Ear-opening minute, not larger than the nostril. Dorsal shields in fourteen longitudinal and one hundred transverse series. Pale yellowish-olive, back with three longitudinal dark brown bands, the median of which disappears on the tail; sides of head and of anterior part of body sprinkled yellowish and dark brown.

From snout to vent 20 centim.; tail 51 millim.

The geographical range of this form remains to ascertain. Cope calls it a Western form, *i. e.* apparently restricted to Texas and New Mexico. Whether the form occurring in North Mexico is the present or the preceding is not known.

a. Ad.	N. America.	E. Doubleday, Esq. [P.].

Var. **sulcatus.** Under this name, Cope records a form from Texas, which he provisionally regards as a variety of the present. The carinæ are elevated on the ten median dorsal rows, so as to leave sulci between them; on the posterior part of the body the keels extend to the lateral rows, and on the tail even to the inferior surface. These notes are taken from a halfgrown specimen.

4. **Ophisaurus gracilis.** (PLATE XV. fig. 1.)

Pseudopus gracilis, *Gray, Cat.* p. 56.
Dopasia gracilis, *Gray, Ann. & Mag. N. H.* (2) xii. 1853, p. 389; *Günth. Proc. Zool. Soc.* 1860, p. 172.
Ophiseps tessellatus, *Blyth, Journ. As. Soc. Beng.* xxii. 1853, p. 655.
Pseudopus gracilis, *Günth. Rept. Brit. Ind.* p. 75; *Anders. Proc. Zool. Soc.* 1871, p. 156.

Lateral teeth conical; one or two series of teeth on the pterygoids. Azygos præfrontal narrower than the greatest width of the frontal, usually separated from the latter by two or three præfrontals on a transverse series; interparietal broader than the parietals, as broad as or broader than the occipital; three shields on a line from the nasal to the azygos præfrontal; five supraoculars. Ear-opening distinct, not larger than the nostril. Dorsal plates in fourteen longitudinal and one hundred to one hundred and ten transverse series, keeled; ventrals in ten longitudinal series, smooth. No rudiments of limbs externally. Tail about twice as long as head and body; upper caudal scales keeled, lower less distinctly. Brown above, with a darker lateral band, and frequently with irregular transverse series of bluish, black-edged spots; lower surfaces pale brownish or yellowish.

From snout to vent 12 centim.; tail 25. The largest specimen, with reproduced tail, measures 19 centim. from snout to vent.

Eastern Himalayas, Khasia hills, Eastern Bengal, Burma.

a. Ad.	Khasia hills.	East India Company [P.]. (Type.)
b. Ad.	Khasia hills.	Sir J. Hooker [P.].
c–e. Ad.	Khasia hills.	T. C. Jerdon, Esq. [P.].
f–g. Ad.	Darjeeling.	W. T. Blanford, Esq. [P.].

5. **Ophisaurus koellikeri.** (PLATE XV. fig. 2.)

Hyalosaurus koellikeri, *Günth. Ann. & Mag. N. H.* (4) xii. 1873, p. 351.

Lateral teeth conical; two series of teeth on the pterygoids, one on the palatines. Azygos præfrontal large, quite as broad as the frontal, with the entire anterior border of which it forms a suture; interparietal much broader than the parietals or the occipital; two shields on a line from the nasal to the azygos præfrontal; five supraoculars. Ear concealed. Dorsal plates in fourteen longitudinal and one hundred and twenty transverse series, the median obtusely keeled, the laterals smooth; ventrals in ten longitudinal series, smooth. Rudiments of hind limbs, as in *O. apus*. Upper and lower caudal scales keeled. Brownish above, with a darker lateral band and small dark-brown spots on the middle line of the anterior part of the back; belly yellowish.

From snout to vent 13 centim.

Hab. ——?

a. Hgr.?	——?	(Type.)

3. DIPLOGLOSSUS.

Diploglossus, *Wiegm. Herp. Mex.* p. 36; *Dum. & Bibr.* v. p. 585; *Gray, Cat. Liz.* p. 119; *Bocourt, Miss. Sc. Mex., Rept.* p. 375.
Celestus, *Gray, Ann. N. H.* ii. 1839, p. 288, *and Cat. Liz.* p. 117; *Cope, Proc. Ac. Philad.* 1868, p. 123.
Microlepis, *Gray, ll. cc.* pp. 334 & 116.
Camilia, *Gray, Cat. Liz.* p. 118.
Oneyda, *Gray, l. c.* p. 119.
Siderolamprus, *Cope, Proc. Ac. Philad.* 1860, p. 368.

No lateral fold. Scales roundish-rhomboidal, arranged quincuncially on the back, forming more or less regular vertical series on the flanks. Limbs well developed, pentadactyle. Teeth with conical or subspherical crowns. Palate toothless.

Tropical America.

Synopsis of the Species.

I. Digits terminating in a large compressed sheath, into which the claw may be entirely or nearly entirely retracted.

A. Frontal anteriorly in contact with two or three shields.

37 to 39 scales round the middle of the body; dorsal scales with an obtuse median keel .... 1. *monotropis*, p. 285.
40 to 44 scales round the middle of the body; dorsal scales without a median keel .... 3. *fasciatus*, p. 287.
54 scales round the middle of the body . 4. *millepunctatus*, p. 288.

B. Frontal anteriorly in contact with a single shield.

37 scales round the middle of the body . 2. *bilobatus*, p. 286.

II. No large ungueal sheath; frontal anteriorly in contact with a single shield.

A. Fore limb as least as long as the skull.

1. Two pairs of shields between the rostral and the azygos præfrontal.

a. Frontal not twice as long as broad.

α. Interparietal as large as parietals; caudal scales strongly keeled, the keels forming continuous lines.

36 scales round the body .... 5. *rugosus*, p. 288.

β. Interparietal smaller than parietals; caudal scales not or but feebly keeled.

* Canthus rostralis strong; loreal region concave.

41 or 42 scales round the body ...... 6. *phoxinus*, p. 289.

** Canthus rostralis obtuse.

40 to 42 scales round the body; occipital not larger than interparietal; ear-opening smaller than eye-opening .. 7. *striatus*, p. 289.

46 to 50 scales round the body; occipital larger than interparietal; ear-opening at least as large as eye-opening; tail strongly compressed ...... 8. *occiduus*, p. 290.

50 to 56 scales round the body; occipital not larger than interparietal; ear-opening at least as large as eye-opening .... 9. *hewardii*, p. 291.

49 to 51 scales round the body, dorsals with a cross elevation and marginal depression, making rows of pits .... 10. *impressus*, p. 291.

31 to 33 scales round the body ...... 13. *steindachneri*, p. 293.

b. Frontal twice as long as broad.

46 scales round the body .... 11. *microlepis*, p. 292.

2. A single pair of shields between the rostral and the præfrontal .... 12. *owenii*, p. 292.

B. Fore limb shorter than the skull.

31 scales round the body; parietals separated from the frontals by the frontoparietals .... 14. *sagræ*, p. 293.

37 scales round the body; no frontoparietals .... 15. *pleii*, p. 294.

1. Diploglossus monotropis.

Camilia jamaicensis, *Gray, Cat.* p. 119.
? Scincus monotropis, *Kuhl, Beitr. Zool. u. Vergl. Anat.* p. 128.
Diploglossus monotropis, *Wiegm. Herp. Mex.* p. 36, *and Arch. f. Nat.* 1837, p. 130; *Peters, Mon. Berl. Ac.* 1864, p. 48; *Cope, Proc. Ac. Philad.* 1868, p. 98, *and Journ. Ac. Philad.* (2) viii. 1878, p. 118; *Bocourt, Miss. Sc. Mex., Rept.* p. 391, pl. xxii. fig. 6; *Günth. Biol. Centr.-Am., Rept.*, p. 34, pl. xxiii.
Tiliqua jamaicensis, *Gray, Ann. N. H.* ii. 1839, p. 293.

Lateral teeth distinctly bicuspid. Head scarcely distinct from neck; snout short and broad, canthus rostralis rounded; ear-opening moderate, smaller than the eye-opening; three præfrontals, azygos slightly larger, nearly as broad as long, in contact with the frontal on a very small point, separated from the loreal by one shield and from the rostral by two pairs of shields; frontal not twice as long as broad; parietal on each side separated from the frontal and supraoculars by two shields; nasal in contact with the rostral; a small supranasal, a postnasal; second loreal higher than long; rostral much broader than mental; ten or eleven upper labials, suture between sixth and seventh, or seventh and eighth, below the centre of the eye; four or five chin-shields on each side, first or first two in contact with the lower labials. Body subtetragonal. Thirty-seven or thirty-nine scales round the middle of the body: dorsals somewhat pointed posteriorly, obtusely keeled and, as well as the laterals, finely striated. Limbs rather short, failing to meet when adpressed; digits short; claws nearly entirely concealed in a large compressed sheath formed of a larger supero-lateral and a smaller inferior scale*. Tail slightly compressed posteriorly, scaled like the body. Brown above, with transverse whitish dark-edged narrow bands, which may be broken up into ocelli on the flanks; lower surfaces uniform whitish.

	millim.		millim.
Total length (tail reproduced)	340	Body	154
Head	36	Fore limb	45
Width of head	30	Hind limb	56

Costa Rica; Ecuador.

a. Ad.	Esmeraldas.	Mr. Fraser [C.].
b. Ad.	[Jamaica.]	(Type of *T. jamaicensis.*)

2. Diploglossus bilobatus.

Celestus bilobatus, *O'Shaughn. Ann. & Mag. N. H.* (4) xiii. 1874, p. 257.

Diploglossus (Celestus) bilobatus, *Bocourt, Miss. Sc. Mex., Rept.* p. 382, pl. xxii. A. fig. 1.

Lateral teeth distinctly bicuspid. Head scarcely distinct from neck; snout short; canthus rostralis rounded; ear-opening very small; a large azygos præfrontal, broader than and in contact with the entire anterior border of the frontal, separated from the loreal by a small shield, and from the rostral by two pairs of shields; frontal once and a half as long as broad; parietal on each side separated from the frontal and supraoculars by one larger (fronto-parietal) and two small shields; nasal in contact with the rostral; a small supranasal, a postnasal, two consecutive subequal loreals; rostral much broader than mental; the suture between the sixth and seventh upper labials falls below the centre of the eye; three

* These sheaths are the "claws large, broad, slightly compressed, blunt," mentioned in Gray's diagnosis of the genus *Camilia* (p. 118).

chin-shields on each side, first two in contact with the lower labials. Body elongate, rounded, slightly depressed. Thirty-seven scales round the middle of the body, dorsals apparently striated but keelless (all the epidermic plates are lost). Limbs short, not meeting when adpressed; digits short, the claws sheathed as in the preceding species. Tail slightly compressed. "Colour: sides with several widely-placed vertical brown bands or patches, within which one or two conspicuous white spots; a bluish-white stripe before the axilla."

	millim.		millim.
Total length (end of tail lost)	173	Body	72
Head	15	Fore limb	17
Width of head	11	Hind limb	24

Costa Rica.

a. Ad.? Costa Rica. (Type.)

3. Diploglossus fasciatus.

Diploglossus fasciatus, *Gray, Cat.* p. 120.
Tiliqua fasciata, *Gray, Griff. A. K.* ix. *Syn.* p. 71.
Euprepes fasciata, *Reuss, Mus. Senckenb.* i. p. 51, pl. iii. fig. 2.
Diploglossus fasciatus, *Wiegm. Herp. Mex.* p. 36, *and Arch. f. Nat.* 1837, p. 130; *Bocourt, Miss. Sc. Mex., Rept.* p. 389, pl. xxii. fig. 5.
—— houttuynii, (*Coct.*) *Dum. & Bibr.* v. p. 597.

Lateral teeth with obtuse crowns. Head much depresed; snout truncate; canthus rostralis obtuse; ear-opening smaller than the eye-opening; three præfrontals; azygos a little larger, as broad as long or broader than long, separated from the frontal or in contact on a very small point, separated from the loreal by one shield and from the rostral by two pairs; frontal not once and a half as long as broad; parietal on each side separated from the frontal and supraoculars by two shields; nasal usually separated from the rostral; one or two postnasals; rostral much broader than the mental; the suture between the seventh and eighth, or eighth and ninth upper labials falls below the centre of the eye; four chin-shields on each side, two anterior in contact with the lower labials. Body subtetragonal, slightly depressed. Forty to forty-four scales round the middle of the body; dorsals and laterals not keeled, finely striated. The limbs fail to meet when adpressed; digits rather short, the claws sheathed as in the two preceding species. Tail cylindrical, a little longer than head and body. Upper surfaces with alternating pale greenish and brown darker-edged bands of equal width, the brown bands uniting on the abdomen, which is of the latter colour; the first brown band crosses the nostrils, the second the eyes, the third the ears; seven others between the latter and the tail, which is regularly annulate; limbs brown.

	millim.		millim.
Total length	355	Fore limb	35
Head	31	Hind limb	47
Width of head	24	Tail	200
Body	124		

Brazil.

a. Ad.	Tijuco River.	R. Bennett, Esq. [P.].
b. Hgr.	Bahia.	Dr. O. Wucherer [C.].
c. Hgr.	Pernambuco.	J. P. Smith, Esq. [P.].
d. Yg.	Brazil.	Lady Harvey [P.].
e. Yg.	Brazil.	

4. Diploglossus millepunctatus.

Diploglossus millepunctatus, *O'Shaughn. Ann. & Mag. N. H.* (4) xiii. 1874, p. 301; *Bocourt, Miss. Sc. Mex., Rept.* p. 388, pl. xxii. A. fig. 5.

Lateral teeth with obtuse crowns. Head rather strongly depressed; snout narrow, cheeks swollen; canthus rostralis rounded; ear-opening smaller than the eye-opening; three præfrontals, azygos largest, longer than broad, touching the frontal on a very small point, separated from the rostral by two pairs of shields; frontal once and a half as long as broad; parietal on each side separated from the frontal and supraoculars by two shields; nasal in contact with the rostral; a postnasal and three consecutive loreals; rostral broader than the mental; the suture between the seventh and eighth upper labials falls below the centre of the eye; four chin-shields on each side, two anterior in contact with the lower labials. Body subquadrangular. Fifty-four scales round the middle of the body; dorsals smooth or very feebly striated, keelless; laterals much smaller than dorsals and ventrals. Limbs moderate, overlapping when adpressed; claws partly retractile into a sheath similar to that of the preceding species. Tail compressed. Upper parts brown, closely dotted with whitish; belly brown, each scale edged with whitish.

	millim.		millim.
Total length (end of tail lost)	385	Body	185
Head	50	Fore limb	65
Width of head	37	Hind limb	84

Hab. —— ?

a. Ad.	"North-west coast of America."	Lieut. Wood [P.]. (Type.)

5. Diploglossus rugosus.

Celestus rugosus, *Cope, Proc. Am. Phil. Soc.* xviii. 1879, p. 272.

Head flat and rather elongate, its scuta generally similar to those of *D. occiduus*; interparietal as large as parietals; five large chin-shields on each side, separated from the lower labials. General form slender; limbs quite weak, not meeting when adpressed. Scales in thirty-six longitudinal rows; each with a strong median keel, and weaker ones on each side of it; the median keels are strong and continuous from the nape, becoming stronger posteriorly, especially on the tail, whose superior and lateral surfaces are thus thrown into gutters; the keels form oblique lines over the sides; they are strong on the hinder and weaker on the anterior limbs. Ground-colour grey, nape and back crossed by seventeen brown cross-bands, which are nearly in contact medially and taper to disappearance on the upper part of the sides; their dorsal portions sometimes confluent longitudinally; below white, with a few scales here and there brown. From snout to vent 90 millim.

S. Domingo.

6. Diploglossus phoxinus.

Celestus phoxinus, *Cope, Proc. Ac. Philad.* 1868, p. 125.

A fusiform species, the body rather stout and flattened, with the outlines tapering gradually to end of snout and tail; dorso-lateral angle strong on scapular region. Head flattened, with strong canthus rostralis and concave loreal region; head-shields as in *D. occiduus.* Forty-one or forty-two scales round the body, dorsals strongly striate but not keeled. The limbs, when pressed to the sides, fail to meet. Tail slightly compressed. Above light grey; sides from orbit to groin dark brown, with regular vertical brown bars, which are margined behind by a close series of light spots; two series of small brown spots on each side of the dorsal region, the median stronger on the nape, all vanishing behind; limbs with brown light-edged cross-bars; below immaculate. From snout to vent 83 millim.

S. Domingo.

7. Diploglossus striatus. (Plate XVI. fig. 1.)

Celestus occiduus, part., *Gray, Cat.* p. 117.
—— striatus, *Gray, l. c.*
Celestus striatus, *Gray, Ann. N. H.* ii. 1839, p. 288.
Diploglossus cliftii, *Dum. & Bibr.* v. p. 590.
? Diploglossus stenurus, *Cope, Proc. Ac. Philad.* 1862, p. 188.
? Celestus stenurus, *Cope, Proc. Ac. Philad.* 1868, p. 124.
? Celestus weinlandii, *Cope, l. c.*
? Celestus badius, *Cope, l. c.*
Diploglossus (Celestus) occiduus, part., *Bocourt, Miss. Sc. Mex., Rept.* p. 385, pl. xxii. fig. 2.

Although very closely allied to *D. occiduus*, this species may be distinguished by the following characters:—Snout more depressed; ear-opening considerably smaller; occipital not larger than interparietal; the difference in the length of the third and fourth toes very slight; forty or forty-two scales round the middle of the body.

The adpressed limbs meet in specimen *a*, not in the two others*. Olive-brown, with transverse series of more or less confluent brown spots; the brown spots more crowded on the sides of the neck.

	millim.		millim.
Total length	295	Fore limb	38
Head	31	Hind limb	52
Width of head	23	Tail	140
Body	124		

Jamaica; San Domingo?

a. Ad.	W. Indies.	(Type.)
b. Ad.?	Jamaica.	R. Heward, Esq. [P.].
c. Ad.?	—— ?	

8. Diploglossus occiduus.

Celestus occiduus, part., *Gray, Cat.* p. 117.
—— macrolepis, *Gray, l. c.* p. 118.
Sloane, Jam. ii. p. 334, pl. 273. fig. 9.
Lacerta occidua, *Shaw, Zool.* iii. p. 288.
Scincus gallivasp, *Daud. Rept.* iv. p. 239.
—— fossor, *Merr. Tent.* p. 74.
Tiliqua occidua, *Gray, Ann. N. H.* ii. 1839, p. 292.
Diploglossus shawii, *Dum. & Bibr.* v. p. 590.
Celestus occiduus, *Cope, Proc. Ac. Philad.* 1868, p. 124.
Diploglossus (Celestus) occiduus, part., *Bocourt, Miss. Sc. Mex., Rept.* p. 385.

Lateral teeth with subspherical crowns in the adult. Ear-opening at least as large as the eye-opening; a large azygos præfrontal, broader than and in contact with the entire anterior border of the frontal, in contact on each side with the largest loreal, and separated from the rostral by two pairs of shields; frontal not more than once and a half as long as broad; parietal on each side separated from the frontal and supraoculars by two or three shields; occipital larger than interparietal; nasal separated from rostral; a postnasal and two or three loreals, second largest; rostral broader than the mental; the suture between the seventh and eighth or eighth and ninth upper labials falls below the centre of the eye; four or five large chin-shields on each side, first or first two in contact with the lower labials. Body moderately elongate, roundish-subquadrangular. Forty-six to fifty scales round the middle of the body; dorsals finely striated and with a more or less distinct obtuse keel. Limbs nearly meeting when adpressed; digits rather short; fourth toe considerably longer than third; claws exposed. Tail strongly compressed; caudal scales not keeled. Brownish above, with dark brown spots or cross bands.

---

* Probably a sexual difference.

	millim.		millim.
Total length (tail reproduced)	560	Body	250
Head	70	Fore limb	83
Width of head	65	Hind limb	98

Jamaica.

a. Ad.	Jamaica.	J. Weeks, Esq. [P.].
b. Ad.	Jamaica.	
c. Ad.	W. Indies.	(Type of *C. macrolepis.*)
d. Skeleton.	W. Indies.	Sir A. Smith [P.].

9. Diploglossus hewardii. (PLATE XVII.)

Celestus hewardii, *Gray, Cat.* p. 118.
Diploglossus (Celestus) variegatus, *Peters, Mon. Berl. Ac.* 1874, p. 370.
—— (——) occiduus, part., *Bocourt, Miss. Sc. Mex., Rept.* p. 385, pl. xxii. A. fig. 6.

Though very closely allied to the preceding, this species differs by a much smaller size (a gravid female measuring only 120 millim. from snout to vent), a smaller occipital shield, a cylindrical or feebly compressed tail, and smaller scales; these form fifty to fifty-six longitudinal series. The adpressed limbs meet in the male, but not in the female and young. Olive-grey above, spotted or irregularly cross-barred with brown above, or brown, spotted with olive-grey; frequently a light streak along the middle of the nape; throat frequently marbled with brown.

	millim.		millim.
Total length (tail reproduced)	360	Body	140
Head	40	Fore limb	43
Width of head	28	Hind limb	52

Jamaica.

a. Ad.	Jamaica.	R. Heward, Esq. [P.].	(Types.)
b. Ad.	Jamaica.		(Types.)
c. Ad., bad state.	W. Indies.	Dr. Mantell.	(Types.)
d–e. Ad., bad state.	St. Elizabeth, Jamaica.		
f. Yg.	Jamaica.		
g–i. Ad.	W. Indies.		

10. Diploglossus impressus.

Celestus impressus, *Cope, Proc. Ac. Philad.* 1868, p. 127.

Body elongate, cylindrical; limbs short. Head abruptly widened at the temples; ear large; snout short and flat; loreal region grooved; head-shields as in *D. occiduus.* Forty-nine to fifty-one scales round the body, dorsals with a cross elevation and marginal depression, making rows of pits. Tail cylindrical. Colour above olivaceous, below yellowish; back and sides crossed by about eighteen narrow brown bars, which are three times broken and alternate on each side the median line.

	millim.		millim.
Total length (tail reproduced)	204	Body	73
Head, to ear	22	Fore limb	25
		Hind limb	33

Jamaica.

11. Diploglossus microlepis. (Plate XVI. fig. 2.)

Microlepis undulata, *Gray, Cat.* p. 116.
Tiliqua microlepis, *Gray, Griff. A. K.* ix. *Syn.* p. 71.
Microlepis undulata, *Gray, Ann. N. H.* ii. 1839, p. 334.

Lateral teeth bicuspid. Head slightly distinct from neck; canthus rostralis rounded; ear-opening a little smaller than the eye-opening; a large azygos subtriangular præfrontal, nearly as broad as long, broader than and in contact with the entire anterior border of the frontal, in contact with the second loreal on each side, separated from the rostral by two pairs of shields; frontal twice as long as broad; parietal on each side separated from the frontal and supraoculars by two shields; nasal in contact with the rostral; a postnasal; rostral much broader than mental; the suture between the seventh and eighth upper labials falls below the centre of the eye; four chin-shields on each side, none in contact with the lower labials. Body rather short, depressed. Forty-six scales round the middle of the body; dorsals obtusely keeled, and probably finely striated (epidermic scales lost). Limbs moderate, overlapping when adpressed; digits moderate, claws exposed. Tail slender, compressed, longer than head and body. Pale brown, with transverse wavy darker bands.

	millim.		millim.
Total length	167	Fore limb	21
Head	17	Hind limb	27
Width of head	11	Tail	93
Body	57		

Hab. —— ?

a. Ad.? —— ? (Type.)

12. Diploglossus owenii.

Diploglossus owenii, *Gray, Cat.* p. 119.
Diploglossus owenii, *Dum. & Bibr.* v. p. 594.
Celestus owenii, *Cope, Proc. Ac. Philad.* 1868, p. 124.

Only one pair of shields between the azygos præfrontal and the rostral; head-shields otherwise much as in *D. occiduus* and allies. Dorsal scales with an obtuse median keel, forming about twelve longitudinal series. Body much elongate; limbs very short, the fore limb as long as the head. Tail compressed. Fulvous above, with narrow longitudinal reddish-grey bands.

From snout to vent 64 millim.

Hab. —— ?

13. Diploglossus steindachneri.

Diploglossus steindachneri, *Cope, Proc. Ac. Philad.* 1864, p. 179; *Bocourt, Miss. Sc. Mex., Rept.* p. 383, pl. xxii. fig. 3; *Günth. Biol. C.-Am., Rept.* p. 34, pl. xxii. fig. A.
—— chalybæus, *Cope, Proc. Ac. Philad.* 1866, p. 321.
Celestus steindachneri, *Cope, Proc. Ac. Philad.* 1868, p. 123.
—— chalybæus, *Cope, l. c., and Proc. Amer. Philos. Soc.* xxii. 1885, p. 170.

Lateral teeth bicuspid. Head scarcely distinct from neck; snout short, canthus rostralis obtuse; ear-opening small; a large azygos præfrontal, broader than and in contact with the entire anterior border of the frontal, separated from the loreals by a shield and from the rostral by two pairs of shields; frontal once and a half as long as broad; parietal on each side separated from the frontal and supraoculars by three shields; nasal in contact with the rostral; two superposed postnasals; three consecutive loreals; rostral a little narrower than the mental; the suture between the eighth and ninth upper labials falls below the centre of the eye; four chin-shields on each side, first in contact with the lower labials. Body elongate, rounded, a little depressed. Thirty-one or thirty-three scales round the middle of the body; dorsals finely striated, not keeled. Limbs short, not meeting when adpressed; digits moderate, claws exposed. Tail a little compressed in its posterior half, the scales on which are keeled. Olive-brown above, some of the median dorsal scales with short dark brown shaft; sides tessellated with small dark brown and greenish spots; lower surfaces pale olive. Spec. *b* entirely of a dark olive-brown.

	millim.		millim.
Total length (tail reproduced)	200	Body	96
Head	19	Fore limb	24
Width of head	15	Hind limb	31

Mexico; Guatemala.

a. Ad.	Vera Paz, Low Forest.	O. Salvin, Esq. [C.].
b. Ad.	Mexico.	

Siderolamprus enneagrammus, Cope, Proc. Ac. Philad. 1860, p. 368, is perhaps nothing but the young of this species.

14. Diploglossus sagræ.

Diploglossus sagræ, *Gray, Cat.* p. 119.
Diploglossus sagræ, *Cocteau, in R. de la Sagra, Hist. Cuba, Rept.* p. 180, pl. xx.; *Dum. & Bibr.* v. p. 603; *Bocourt, Miss. Sc. Mex., Rept.* p. 378, pl. xxii. fig. 1.
Celestus sagræ, *Cope, Proc. Ac. Philad.* 1868, p. 123.

Head scarcely distinct from neck; snout short; ear-opening very small; an azygos præfrontal, broader than long, slightly broader than and in contact with the entire anterior border of the frontal, separated from the loreal by one shield, from the rostral by two pairs of shields; frontal slightly longer than broad; parietal on each side separated from the frontal and supraoculars by two shields; nasal in contact with the rostral; a postnasal, two consecutive loreals; rostral much broader than mental; fifth upper labial below the eye; no infraorbitals; three chin-shields on each side, first two in contact with the lower labials. Body elongate, subcylindrical. Thirty-one scales round the middle of the body; dorsals finely striated, not keeled. Limbs very short, anterior not measuring more than the width of the head; digits short. Grey-brown above, with metallic gloss; a lateral dark brown band, from the snout to the tail; lower parts yellowish white.

	millim.		millim.
Total length	201	Fore limb	8
Head, to ear	11	Hind limb	15
Width of head	8	Tail	114
Body	76		

Cuba.

15. Diploglossus pleii.

Diploglossus pleii, *Gray, Cat.* p. 119.
Diploglossus pleii, *Dum. & Bibr.* v. p. 605; *Bocourt, Miss. Sc. Mex., Rept.* p. 381, pl. xxii. fig. 4.
Celestus pleii, *Cope, Proc. Ac. Philad.* 1868, p. 123.
—— degener, *Cope, l. c.*

Closely allied to *D. sagræ*. Distinguished by the following characters:—Ear-opening somewhat larger; frontal broader than long; no frontoparietals; the parietals in contact with the frontal; infraorbitals present. Thirty-seven scales round the middle of the body. Limbs shorter still. Reddish brown above, the sides sharply delimited by a dark brown streak and dotted with brown; lower surfaces pale yellowish.

	millim.		millim.
Total length (tail reproduced)	150	Body	75
Head, to ear	13	Fore limb	10
Width of head	11	Hind limb	15

Martinique; Porto Rico.

4. SAURESIA.

Sauresia, *Gray, Ann. & Mag. N. H.* (2) x. 1852, p. 281; *Cope, Proc. Ac. Philad.* 1864, p. 228; *Bocourt, Miss. Sc. Mex., Rept.* p. 455.
Embryopus, *Weinland, Abh. Senck. Ges.* iv. 1862, p. 132.

Limbs tetradactyle; otherwise like *Diploglossus*.
West Indies.

1. Sauresia sepoides. (PLATE XIV. fig. 2.)

Sauresia sepsoides, *Gray, l. c.* p. 282; *Bocourt, l. c.* pl. xxii. A. fig. 5.
Embryopus habichii, *Weinl. l. c.* pl. v. fig. 1.

Head small, not distinct from neck; ear-opening very small; an azygos præfrontal, broader than long, in contact with the entire anterior border of the frontal, separated from the rostral by two pairs of shields; frontal a little longer than broad; parietal on each side separated from the frontal and supraorbitals by three shields; three consecutive shields between the nasal and the præocular, the second in contact with the azygos præfrontal; infraorbital wedged in between fifth and sixth upper labials; four large chin-shields on each side, first in contact with lower labials. Body elongate, subcylindrical; thirty-four scales round the body, dorsals and laterals finely striated. Limbs very short, the front ones much shorter than the head; digits very short, claws retractile. Tail long and nearly as thick as the body. Brown above, the sides closely spotted with darker and separated from the dorsal region by a dark brown line, which extends forwards to the canthus rostralis; labials spotted with dark brown; lower surface yellowish.

	millim.		millim.
Total length (tail reproduced)	132	Body	48
Head	10	Fore limb	6
Width of head	6·5	Hind limb	10

S. Domingo.

a. Ad.	S. Domingo.	M. Sallé [C.].	(Type.)

5. PANOLOPUS.

Panolopus, *Cope, Pr. Ac. Philad.* 1861, p. 494.

No lateral fold. Scaling of the back quincuncial. Supranasals, nasals, first upper labial, and rostral confluent into a large shield in which the nostril is pierced. Four limbs; anterior without digits, posterior with a rudimentary one on the inner border. Teeth obtuse.

West Indies.

1. Panolopus costatus.

Panolopus costatus, *Cope, l. c.*

Form elongate; body fusiform, tetragonal. Anterior extremity consisting of brachium and antebrachium, as long as the commissure of the mouth; it is terminated on its inner face by a tubercle. Posterior extremity as long as the distance from the rictus to the axilla, consisting of femur, tibia, and tarsus; sole tuberculous, with two callosities, that on its external border much the larger; a tubercle-like spur on its inner. Auricular opening present, small.

A large nine-sided præfrontal between the frontal and the rostral; posterior border of latter shield chevron-shaped; parietals small, longer than broad, separated from the frontal by the frontoparietals; seven upper labials. Scales in forty-four longitudinal rows, the dorsals striated. Tail very elongate. Colour above light brown, with numerous small dark brown spots on the dorsal region; sides dark brown, darker anteriorly, where a bandlike arrangement is assumed as far as the orbit; from the ear-opening posteriorly, this is crossed by numerous blackish-brown vertical bars, with irregular light posterior borders; chin whitish, abdomen yellowish.

	millim.		millim.
From snout to vent.	94	Fore limb	13
Head, to ear	17	Hind limb	19

Near Jeremie, S. Domingo.

6. OPHIODES.

Ophiodes, *Wagler, Isis*, 1828, p. 740, *and Syst. Amph.* p. 159; *Dum. & Bibr.* v. p. 788; *Gray, Cat. Liz.* pp. 99, 120; *Bocourt, Miss. Sc. Mex., Rept.* p. 457.
Pygodactylus (*non Merr.*), *Wagl. ll. cc.* pp. 741 & 160.

No lateral fold. Scales roundish-rhomboidal, arranged quincuncially on the back, forming vertical series on the sides. No fore limbs; styliform rudiments of hind limbs. Teeth conical. Palate toothless.

South America.

1. Ophiodes striatus.

Ophiodes striatus, part., *Gray, Cat.* p. 99.
Pygopus striatus, *Spix, Spec. Nov. Lac. Bras.* p. 25, pl. xxviii. fig. 1.
—— cariococca, *Spix, l. c.* p. 26, fig. 2.
Ophiodes striatus, *Wagl. Isis*, 1828, p. 740; *Dum. & Bibr.* v. p. 789; *Gravenh. Nova Acta Ac. Leop.-Carol.* xxiii. 1851, i. p. 379, pl. xliv.; *Girard, U. S. Explor. Exped., Herp.* p. 225; *Cope, Proc. Ac. Philad.* 1862, p. 350; *Bocourt, Miss. Sc. Mex., Rept.* p. 458, pl. xxii. G. fig. 4.
Pygodactylus gronovii (*non Daud.*), *Wagl. l. c.* p. 741.

A large azygos præfrontal (abnormally divided), separated from the rostral by two pairs of shields; frontal large, much longer than broad; interparietal as broad as, or slightly broader than the parietals, longer than the occipital; nasal in contact with the rostral; three or four enlarged shields on each side behind the postmental, only the first or the first two in contact with the lower labials. Ear-opening almost or quite hidden. Scales smooth, the dorsals finely striated; twenty-five or twenty-seven round the middle of the body. Rudimentary hind limbs styliform. Tail about once and a half the length of the body. Brownish or olive-grey above, with a variable, but even, number of dark brown longitudinal lines; upper lip and temple with alternate dark brown and whitish vertical bars.

From snout to vent 185 millim.; tail 265. A larger specimen, with reproduced tail, measures 230 millim. from snout to vent.

South America, from the Guianas to Buenos Ayres.

a. Ad.	Brazil.	
b. Ad.	Para.	
c. Ad.	Garanhuns, Pernambuco.	W. A. Forbes, Esq. [P.].
d. Ad.	Bahia.	Dr. O. Wucherer [C.].
e. Ad.	Tijuco.	R. Bennett, Esq. [P.].
f–g. Ad. & hgr.	Rio Janeiro.	Mrs. Fry [P.].
h, i. Ad. & yg.	Rio Grande do Sul.	Dr. v. Ihering [C.].
k. Yg.	Corrientes, Buenos Ayres.	

2. Ophiodes vertebralis.

Ophiodes striatus, part., *Gray, Cat.* p. 99.
Ophiodes vertebralis, *Bocourt, Miss. Sc. Mex., Rept.* p. 459, pl. xxii. G. fig. 3.

Very closely allied to the preceding, from which it differs in the following points:—Frontal considerably broader than the parietals; three large chin-shields on each side in contact with the lower labials. Twenty-nine or thirty-one scales round the middle of the body. No vertical dark bars on the sides of the head; a narrow dark vertebral line; a broad and a narrow brown streak on each side, separated by a yellowish one.

Southern Brazil and Uruguay.

a. Ad.	Brazil.	C. Darwin, Esq. [P.].
b. Hgr.	Brazil.	T. Bell, Esq. [P.].

7. ANGUIS.

Anguis, *Linn. et omn. auct.*
Siguana, *Gray, Griff. A. K.* ix. *Syn.* p. 74.
Dorfia, *Gray, Ann. N. H.* ii. 1839, p. 334.
Otophis (*Fitzing.*), *Demid. Voy. Russ. Mér.* iii. p. 341.

No lateral fold. Scales roundish, arranged quincuncially on the back, forming vertical series on the sides. No limbs. Teeth fang-like. Palate toothless.

Europe, Western Asia, Algeria.

1. Anguis fragilis.

Anguis fragilis, *Gray, Cat.* p. 100.
Anguis? punctata, *Gray, l. c.*
Anguis fragilis, *Linn. S. N.* i. p. 392; *Laur. Syn. Rept.* p. 68, pl. v. fig. 2; *Daud. Rept.* vii. p. 327, pl. lxxxvii. fig. 2; *Krynicki, Bull. Soc. Nat. Mosc.* 1837, p. 51; *Bonap. Faun. Ital.*; *Bell, Brit. Rept.* p. 39; *Dum. & Bibr.* v. p. 792; *De Betta, Erp. Veron.* p. 160

Strauch, Erp. Alg. p. 46; *Leydig, Deutschl. Saur.* p. 242; *Fatio, Vert. Suisse*, iii. p. 103; *De Betta, Faun. Ital., Rett. Anf.* p. 33; *Schreib. Herp. Eur.* p. 341; *Lataste, Herp. Gir.* p. 99.
Anguis eryx, *Linn. l. c.*; *Daud. l. c.* p. 337.
—— lineata, *Laur. l. c.* p. 68; *Krynicki, l. c.* p. 54.
—— clivica, *Laur. l. c.* p. 69.
Eryx clivicus, *Daud. l. c.* p. 281.
Anguis cinereus, *Risso, Hist. Nat. Eur. Mér.* iii. p. 89.
—— bicolor, *Risso, l. c.*
Siguana ottonis, *Gray, Griff. A. K.* ix. *Syn.* p. 74.
Anguis besseri, *Andrzej. Nouv. Mém. Soc. Nat. Mosc.* ii. 1832, p. 338, pls. xxii. fig. 7, and xxiv.
—— incerta, *Krynicki, l. c.* p. 52, pl. i.
Dorfia punctata, *Gray, Ann. N. H.* ii. 1839, p. 334.
Otophis eryx, var. colchica, *Demid. Voy. Russ. Mér.* iii. p. 341, pl. v.
Anguis orientalis, *Anders. Proc. Zool. Soc.* 1872, p. 376; *Blanf. Zool. E. Persia*, p. 394.

Frontal large, its anterior angle wedged in between a pair of præfrontals; an azygos præfrontal in front of the latter, separated from the nasal by two small shields; one or two small azygos shields behind the very small rostral; nasal very small, separated from the latter by one shield; labials and loreals small, numerous; interparietal as long as and broader than the parietals and in contact with a small occipital; a small frontoparietal between the supraoculars and the interparietal, and a larger shield between the supraoculars and the parietal; supraoculars five or six. Ear-opening minute, usually hidden. Scales smooth, median dorsal and ventrals broad, hexagonal; twenty-four to twenty-eight scales round the middle of the body. Tail at least as long as the body. Young silvery above, with a black vertebral line; sides and lower surfaces blackish. In the adult these markings often disappear, or may be replaced by series of dark dots, and the colour of the upper surfaces becomes brown or bronzy. A variety (*colchica*), in which the ear-opening appears to be normally distinct, is distinguished by having scattered pale blue spots on the back.

From snout to vent 195 millim; tail 230.

Europe, Western Asia, Algeria.

a, b–d. Ad. & hgr.	Sweden.	C. J. A. Thuden, Esq. [P.].
e. Ad.	Bohuslan.	Dr. A. W. Malm [P.].
f–l. Ad., hgr. & yg.	Aberdeen.	Dr. Brine [P.].
m. Yg.	Ringwood.	F. Bond, Esq. [P.].
n–o. Ad.	Düsseldorf.	W. F. Kirby, Esq. [P.].
p. Ad.	Jura.	H. Ling Roth, Esq. [P.].
q. Ad.	France.	
r. Ad.	Santiago, Galicia.	M. V. L. Seoane [P.].
s, t–u. Ad., hgr. & yg.	Oporto.	E. Allen, Esq. [P.].
v–w, x–y. Ad. & yg.	Italy.	J. J. Bianconi [P.].
z. Ad.	——?	
a. Skeleton.	——?	

Fam. 10. ANNIELLIDÆ.

Aniellidæ, *Cope, Proc. Ac. Philad.* 1864.
Aniellidæ, *Boulenger, Ann. & Mag. N. H.* (5) xiv. 1884.

Tongue villose, smooth and bifid anteriorly*. Teeth large, few, fang-like, with short swollen base. Palate toothless. Skull approaching the Ophidian type, *i. e.* lacking the præsphenoidal vacuity and consequently the interorbital septum, and the bones which constitute the brain-case firmly united; no columella cranii, no squamosal; postorbital arch ligamentous; pterygoids in contact with sphenoid; an infraorbital fossa; præmaxillary single; nasal and frontal divided; parietal single; præ- and postorbital in contact, separating the frontal from the border of the orbit. Body vermiform†. Eyes and ears concealed. Scales soft, cycloid-hexagonal, strongly imbricate.

A single genus, which appears to be a strongly degraded form of the *Anguidæ*, similarly as the *Anelytropidæ* and *Dibamidæ* are to the *Scincidæ*.

1. ANNIELLA.

Anniella, *Gray, Ann. & Mag. N. H.* (2) x. 1852, p. 440; *Cope, Proc. Ac. Philad.* 1864, p. 230; *Bocourt, Miss. Sc. Mex., Rept.* p. 460.

California.

1. Anniella pulchra.

Anniella pulchra, *Gray, l. c., and Zool. Herald,* p. 154, pl. xxviii.; *Bocourt, l. c.* pl. xxii. G. fig. 2.

Head much depressed, snout cuneiform; eyes concealed under several small scales resembling lids; rostral large, in contact with the nasal, separated from the frontal by a pair of præfrontals; frontal large, heptagonal, broader than long, its posterior border forming a suture with a nearly equally large interparietal, the posterior border of which is emarginate to receive a small occipital; parietals very small; a larger anterior and two smaller posterior supraoculars; nostril pierced in a large nasal resembling an anterior labial; five upper labials, first very small, situated below the nasal, on the lower surface of the snout; second in contact with the præfrontal, second with a loreal, third with the oculars, fourth small; mental pentagonal, followed by a pair of large postmentals; six lower labials. Scales smooth, hexagonal, twenty-four (or twenty-

---

* Resembling strikingly that of *Heloderma*.
† Whether rudiments of the shoulder and hip-girdles exist, is at present unknown.

six) round the middle of the body, No enlarged anal scales. Tail ending obtusely, when intact measuring one third (or two fifths) of the total length. Silvery, the scales edged with brown; a narrow brown vertebral line.

From snout to vent 130 millim; tail 60.

California.

a. Ad. California. Sir J. Richardson [P.]. (Type.)

2. Anniella nigra.

Anniella nigra, *Fischer, Abh. nat. Ver. Hamb.* ix. 1885, p. —.*

Twenty-eight scales round the body. The three median anal scales twice as long as those preceding. Black above, yellowish-white inferiorly. Otherwise as in *A. pulchra*.

San Diego.

Fam. 11. HELODERMATIDÆ.

Thecoglossæ pleurodontes, part., *Wagler, Syst. Amph.* 1830.
Trachydermi, *Wiegmann, Herp. Mex.* 1835.
Varaniens, part., *Duméril & Bibron, Erp. Gén.* iii. 1836.
Helodermidæ, *Gray, Proc. Zool. Soc.* 1837, *and Cat. Liz.* 1845.
Helodermidæ, *Cope, Proc. Ac. Philad.* 1864 & 1866.
Trachydermi glyphodonta, *Bocourt, Miss. Sc. Mex., Rept.* 1878.
Helodermatidæ, *Boulenger, Ann. & Mag. N. H.* (5) xiv. 1884.

Tongue villose, the anterior portion smooth, bifid, protractile. Dentition resembling that of the Ophidia, the teeth being fang-like, with slightly swollen base, rather loosely attached to the inner edge of the jaws; the teeth grooved in front and behind (*Heloderma*); teeth on the pterygoid and palatine bones. Præmaxillary single; nasals and frontals double; præ- and postfrontals in contact, separating the frontal from the orbit; a strong postorbital arch; no postfrontosquamosal arch; parietal single; pterygoids and palatines widely separated; infraorbital fossa bounded by the pterygoid, palatine, transverse bone, and maxillary. Head with bony tubercles. Eyelids well developed. Limbs well developed; clavicle slender; interclavicle without transverse processes; sternum divided longitudinally. Skin of upper surfaces tubercular, some of the larger tubercles ossifying in the adult, of lower surfaces forming transverse series of squarish plates. No femoral or præanal pores.

The genus *Heloderma*, the only poisonous Lizard at present known, is very closely related to the *Anguidæ*, but well deserves to rank as a distinct family, on account of certain cranial peculiarities and the shape of the interclavicle. Its range is restricted to the western parts of Central America, from the Isthmus of Tehuantepec to as far north as New Mexico.

* Quoted from an advanced separate copy.

1. HELODERMA.

Trachyderma (*non Latr.*), *Wiegm. Isis*, 1829, p. 421.
Heloderma, *Wiegm. l. c.* p. 624; *Wagl. Syst. Amph.* p. 164; *Wiegm. Herp. Mex.* p. 23; *Dum. & Bibr.* iii. p. 499; *Gray, Cat.* p. 14; *Bocourt, Miss. Sc. Mex., Rept.* p. 296.

ANATOMICAL AND PHYSIOLOGICAL: *Troschel, Arch. f. Nat.* 1853, p. 294, pls. xiii. & xiv.; *Sumichrast, Arch. Sc. Phys.-math.* xix. 1864, p. 45; *Kaup, Arch. f. Nat.* 1865, p. 33, pl. iii.; *Cope, Proc. Ac. Philad.* 1869, p. 5; *Gervais, Journ. Zool.* 1873, p. 453, pl. xii.; *Sumichrast, C. R. Ac. Sc.* lxxx. 1875, p. 676; *Bocourt, l. c.* p. 299, pl. 20 G; *Sumichr. Bull. Soc. Zool. France*, 1880, p. 178; *Fischer, Verh. Ver. Hamb.* v. 1882, p. 2, pl. iii.; *Shufeldt, Amer. Nat.* 1882, p. 907; *Bouleng. Proc. Zool. Soc.* 1882, p. 631; *Fayrer, eod. loc.* p. 632; *Mitchell & Reichert, Med. News*, 1883, *and Science*, i. 1883, p. 372.

Ear exposed. A gular fold. Digits with a series of transverse lamellæ inferiorly. Upper surfaces uniformly tubercular; abdominal scales flat, juxtaposed. Labial shields present.
Mexico.

1. Heloderma horridum.

Heloderma horridum, *Gray, Cat.* p. 14.
Hernandez, Hist. Nov. Hisp. p. 315.
Trachyderma horridum, *Wiegm. Isis*, 1829, p. 421.
Heloderma horridum, *Wiegm. Isis*, 1829, p. 624, *and Herp. Mex.* p. 24, pl. i.; *Dum. & Bibr.* p. 499, pl. xxxvi.; *Bocourt, Miss. Sc. Mex.* p. 297, pl. xx. E. figs. 1–12; *Günth. Biol. C.-Am., Rept.* pl. xxvi.

Head depressed; snout short, broadly rounded; lips thick, arcuate, the lower much swollen; nostril large; eye small, with very thick lids; ear-opening elliptical, vertical, slightly oblique; chin-groove strong; gular fold well marked. Body elongate, rather depressed. Limbs moderate; claws strong, of the fingers much longer than of the toes. Tail cylindrical, tapering, about as long as the body. Upper surface of head covered with large, convex, irregular, bony tubercles; rostral broad, pentagonal; nostril between four plates, the largest inferiorly; a pair of large internasals between the two pairs of anterior nasals; ten or eleven upper and about twelve lower labials; mental small, subtrapezoid; chin with large polygonal plates; throat with round or oval convex tubercles. Body and limbs covered above with subequal round convex tubercles separated by a granulate skin: about thirty-five tubercles across the middle of the body. Abdominal scales flat, subquadrangular, longer than broad, in about fifty-five series between the gular fold and the præanal region, and about twenty across the middle of the belly. Caudal scales subequal, tubercular above, flat and subquadrangular inferiorly, arranged in rings. Blackish brown above, spotted with yellow, the latter colour sometimes forming regular rings round the tail.

	millim.		millim.
Total length	597	Fore limb	103
Head	57	Hind limb	114
Width of head	46	Tail	270
Body	270		

Western Mexico.

a. Adult.	Presidio.	Herr A. Forrer [C.].
b. Ad., stuffed.	Presidio.	Herr A. Forrer [C.].
c. Yg.	Tapana, Tehuantepec.	F. Sumichrast, Esq. [C.].
d. Ad.	Mexico.	J. Marshall, Esq. [P.].
e. Ad., stuffed.	Mexico.	Sir J. Richardson [P.].
f, g. Ad., stuffed.	Mexico.	

2. Heloderma suspectum.

Heloderma horridum *(non Wiegm.)*, *Baird, U.S. Mex. Bound. Surv.* ii. *Rept.* pl. xxvi.; *Cope, Proc. Ac. Philad.* 1866, p. 303.
—— suspectum, *Cope, Proc. Ac. Philad.* 1869, p. 5.

Closely allied to the preceding. The digits and the tail are shorter, and the tubercles on the back are separated by wider granular interspaces. Yellowish or orange, with blackish network on the back and cross bands on the tail.

	millim.		millim.
Total length	488	Fore limb	95
Head	58	Hind limb	97
Width of head	50	Tail	148
Body	282		

Arizona and New Mexico.

a. Ad.	Arizona.	Smithsonian Institution [P.].

Whether the following genus is to be placed here, or constitutes a distinct allied family (*Lanthanotidæ*, Steindachner), must remain doubtful until its anatomical characters are known. Its dentition was originally stated to be the same as in *Heloderma*, but this has been subsequently corrected by Steindachner.

LANTHANOTUS.

Lanthanotus, *Steindachner, Denkschr. Ak. Wien,* xxxviii. 1878, p. 95, *and Sitzb. Ak. Wien,* lxxviii. i. 1879, p. 399.

Ear concealed. No gular fold. Digits granular inferiorly. Upper surface of body and tail tubercular, with rows of large wart-like tubercles, each of which bears a horny keeled scale; abdominal scales flat, feebly imbricate. No labial shields.

Borneo.

1. Lanthanotus borneensis.

Lanthanotus borneensis, *Steind. l. c.* pl. ii.

Head depressed, oval, with broad round snout; eyes very small;

nostrils small, oval, obliquely directed upwards and backwards, nearer the end of the snout than the eye. Body elongate. Limbs short, with short, thick, feebly compressed digits; claws short and not much compressed. Tail roundish, feebly depressed at the base, tapering gradually, as long as head and body. Head covered with small tubercular scales, larger on the snout and forehead, largest of all on the temples. Back and upper surface of tail with regular longitudinal rows of large wart-like tubercles, each bearing a horny keeled scale, which does not entirely cover the protuberance; these tubercles form six series on the back; the space between them is covered by juxtaposed small flat tubercles. The lower surface of the head with small scales like those on the snout and forehead, but more convex, gradually increasing in size towards the neck; a slight symphiseal groove on the chin. The abdominal scales slightly imbricate, flat, becoming rather pointed towards the end of the body, passing gradually into the tubercles of the sides. The scales on the the lower surface of the tail like those on the belly, but square and not imbricate, forming regular rows. Reddish brown above; lower surfaces, with the exception of the head, yellow, with brown band-like marblings.

	millim.		millim.
Head and body	20	Fore limb	3
Tail	22	Hind limb........	4

Sarawak.

Fam. 12. VARANIDÆ.

Lacertiens, part., *Cuvier, Règne Anim.* ii. 1817.
Tupinambidæ, part., *Gray, Ann. Phil.* (2) x. 1825.
Ameivoidea, part., *Fitzinger, Neue Classif. Rept.* 1826.
Uaranidæ, *Gray, Phil. Mag.* (2) ii. 1827.
Thecoglossæ pleurodontes, part., *Wagler, Syst. Amph.* 1830.
Monitores, *Wiegmann, Herp. Mex.* 1834.
Varaniens, part., *Duméril & Bibron, Erp. Gén.* iii. 1836.
Monitoridæ, *Gray, Ann. N. H.* i. 1838, *and Cat. Liz.* 1845.
Varanidæ, *Cope, Proc. Ac. Philad.* 1864.
Varanidæ, *Boulenger, Ann. & Mag. N. H.* (5) xiv. 1884.

Tongue smooth, very long and slender, bifid, retractile into a sheath at the base. Teeth large, dilated at the base, which is fixed to the inner side of the jaws; palate toothless. Præmaxillary single, narrowed and much prolonged posteriorly; nasal bones coalesced, narrow; two frontals; a single parietal; a supraorbital bone; postorbital arch incomplete; a bony postfrontosquamosal arch; pterygoids and palatines widely separated; infraorbital fossa bounded by the pterygoid, palatine, and transverse bone, the maxillary being excluded. No dermal cranial ossifications; head covered with small polygonal scales. Eyelids well developed; ear-opening distinct. Limbs well developed; clavicle slender; interclavicle anchor-shaped. Dorsal scales roundish, juxtaposed, surrounded by rings of minute granules; ventral scales squarish, arranged in cross rows. No femoral or præanal pores. Tail very long.

A single genus, confined to the Old World and Australia, and forming a perfectly isolated group.

1. VARANUS.

Tupinambus, part., *Daud. Rept.* iii. p. 5.
Monitors proprement dits, *Cuv. R. A.* ii. p. 24.
Varanus, *Merrem. Tent. Syst. Amph.* p. 58; *Dum. & Bibr.* iii. p. 467; *Gray, Cat.* p. 9; *Günth. Rept. Brit. Ind.* p. 64.
Psammosaurus, *Fitzing. N. Classif. Rept.* p. 50; *Wagl. Syst. Amph.* p. 165; *Gray, Ann. N. H.* i. 1838, p. 392, *and Cat.* p. 7.
Dracæna (*non Merr.*), *Gray, Phil. Mag.* (2) ii. 1827, p. 53.
Hydrosaurus, *Wagl. l. c.* p. 164; *Gray, ll. cc.* pp. 393, 8; *Günth. l. c.* p. 67.
Polydædalus, *Wagl. l. c.*
Monitor, *Gray, ll. cc.* pp. 392, 8; *Schleg. Abbild. Amph.* p. 65; *Peters, Mon. Berl. Ac.* 1870, p. 106.
Empagusia, *Gray, ll. cc.* pp. 393, 9.
Odatria, *Gray, ll. cc.* pp. 394, 7.
Regenia, *Gray, Cat.* p. 8.

Africa; Southern Asia; Australia.

Synopsis of the Species.

I. Nostril an oblique slit; tail round or slightly compressed posteriorly .... 1. *griseus*, p. 306.

II. Nostril an oblique slit; tail compressed, keeled above.

A. Abdominal scales smooth.

1. No transversely enlarged supraocular scales.

Nostril three times more distant from the end of the snout than from the orbit; scales small .... 2. *albigularis*, p. 307.
Nostril three times more distant from the end of the snout than from the orbit; scales large, the nuchals larger than the occipitals .... 3. *ocellatus*, p. 308.
Nostril twice more distant from the end of the snout than from the orbit; scales large .... 4. *exanthematicus*, [p. 308.
Nostril a little nearer the orbit than the end of the snout; scales small ...... 6. *bengalensis*, p. 310.

2. Median supraoculars more or less enlarged transversely.

Nostril a little nearer the end of the snout than the orbit; median supraoculars slightly dilated.................... 5. *flavescens*, p. 309.
Nostril twice more distant from the end of the snout than from the orbit .... 7. *nebulosus*, p. 311.

B. Abdominal scales keeled.

1. Nostril equally distant from the orbit and the end of the snout . . 8. *grayi*, p. 312.

2. Nostril much nearer the orbit than the end of the snout.

Snout moderate; dorsal scales large; nuchals not keeled . . . . 9. *dumerilii*, p. 312.
Snout very long, pointed; nuchal scales extremely large and strongly keeled . . 10. *rudicollis*, p. 313.

3. Nostril much nearer the end of the snout than the orbit. . . . . 11. *salvadorii*, p. 314.

III. Nostril round or oval; tail compressed, keeled above.

A. A transversely dilated median series of supraocular scales.

1. Nostril oval, at least twice as far from the orbit as from the end of the snout.

a. More than 80 transverse rows of abdominal scales from gular fold to groin.

α. Nuchal scales not larger than dorsals . . . . 12. *salvator*, p. 314.

β. Nuchal scales larger than dorsals.

Nuchal scales larger than occipitals . . . . 13. *nuchalis*, p. 315.
Nuchal scales smaller than occipitals . . . . 14. *cumingii*, p. 316.

b. Less than 80 rows of scales from gular fold to groin . . . . 15. *togianus*, p. 316.

2. Nostril round.

Nostril considerably nearer the end of the snout than the orbit; nuchal scales very small, keeled, separated by very wide interspaces . . . . 16. *kalabeck*, p. 316.
Nostril a little nearer the end of the snout than the orbit . . . . 17. *indicus*, p. 316.

B. Supraocular scales subequal.

1. Nostril a little nearer the orbit than the end of the snout . . . . 18. *niloticus*, p. 317.

2. Nostril nearer the end of the snout than the orbit.

Nostril twice as far from the orbit as from the end of the snout; scales on upper surface of head larger than temporals. . 19. *varius*, p. 319.
Nostril more than twice as far from the orbit as from the end of the snout; head-scales very small. . . . . 20. *giganteus*, p. 320.
Nostril a little nearer the end of the snout than the orbit; head-scales very small. 21. *gouldii*, p. 320.

IV. Nostril round; tail round or slightly compressed posteriorly.

A. A median series of transversely dilated supraoculars.

Nuchal scales not keeled .... 22. *prasinus*, p. 321.
Nuchal scales strongly keeled .... 23. *kordensis*, p. 322.

B. Supraocular scales subequal.

1. Dorsal scales much longer than broad.

Upper caudal scales not spinose; nostril slightly nearer the end of the snout than the orbit; 85 to 95 rows of ventral scales from collar-fold to groin .. 24. *punctatus*, p. 322.
Caudal scales not spinose; nostril equally distant from the orbit and the end of the snout; 65 to 75 rows of ventral scales .... 25. *timorensis*, p. 323.
Upper and lateral caudal scales strongly spinose .... 26. *acanthurus*, p. 324.

2. Dorsal scales nearly as broad as long .... 27. *caudolineatus*, p. 324.

1. **Varanus griseus.**

Psammosaurus scincus, *Gray, Cat.* p. 7.
Psammosaurus? caspicus, *Gray, l. c.*
Tupinambis griseus, *Daud. Rept.* viii. p. 352.
—— arenarius, *Geoffr. Descr. Egypte, Rept.* p. 123, pl. iii. fig. 2, pl. vi. figs. 14–15.
Varanus scincus, *Merr. Tent.* p. 59.
Psammosaurus griseus, *Fitzing. N. Class. Rept.* p. 50; *Eichw. Nouv. Mém. Soc. Nat. Mosc.* ix. 1851, p. 416.
—— caspius, *Eichw. Zool. Spec.* iii. p. 190, *and Faun. Casp.-Cauc.* p. 48, pls. vii.–ix.
—— scincus, *Wagl. Syst. Amph.* p. 165; *Jerdon, Proc. As. Soc. Beng.* 1870, p. 70.
Monitor scincus, *Gray, Griff. A. K.* ix. *Syn.* p. 27; *Schleg. Abbild.* p. 69.
Varanus arenarius, *Dum. & Bibr.* iii. p. 471; *Guichen. Expl. Sc. Algér., Zool.* v. p. 6; *De Filippi, Viag. Pers.* p. 352.
—— ornatus (*non Daud., non Gray*), *Carlleyle, Journ. As. Soc. Beng.* xxxviii. 1869, p. 192.
Monitor griseus, *Peters, Mon. Berl. Ac.* 1870, p. 109.

Teeth acute, compressed. Snout depressed at the end, measuring the distance from the anterior border of the orbit to the anterior border of the ear; canthus rostralis distinct. Nostril an oblique slit, about four times as distant from the end of the snout as from the orbit. Digits rather short. Tail round or slightly compressed. Scales of head, including supraoculars, very small, granular, subequal. Scales on upper surfaces small, smooth or feebly keeled; those on the sides of the neck generally conical in adult specimens. Abdominal scales smooth, in one hundred and ten to one hundred and twenty-five transverse series (counted from collar-fold to groin). Caudal scales more or less distinctly keeled; no caudal crest. Greyish yellow, sometimes with more or less distinct brown cross bars on the back and tail and brown streaks along the sides of the neck. Young with round yellow spots and dark brown cross bars and temporal and cervical streaks; a few brown lines across the snout and short vertical brown lines on the lower lip.

	centim.		centim.
Total length	127	Fore limb	19
Head	9	Hind limb	21
Neck	9	Tail	71
Body	38		

North Africa; South-western Asia, from Arabia and the Caspian Sea to North-western India.

a, b. Ad. & yg.	N. Africa.	Sir J. G. Wilkinson [P.].
c, d. Ad., stuffed.	N. Africa.	
e, f. Ad., skeletons.	Algeria.	
g. Hgr., stuffed.	Desert de Souf, Algeria.	
h. Ad., stuffed.	Egypt.	W. K. Loftus, Esq. [P.].
i, k. Ad., stuffed.	Egypt.	Sir J. G. Wilkinson [P.].
l. Hgr., stuffed.	Banks of the Nile.	
m. Ad., stuffed.	Sarakhe, 1000 feet, Caspian Sea.	Capt. Napier [C.].
n–o. Hgr.	Ghainak, Persia.	
p. Yg.	Konah Takhli, S. Persia.	
q. Ad., stuffed.	Kokand.	
r. Ad.	Agra.	
s. Ad.	Kurrachee.	Secretary of State for India [P.].

2. Varanus albigularis.

Regenia albogularis, *Gray, Cat.* p. 8.
Tupinambis albigularis, *Daud. Rept.* iii. p. 72, pl. xxxii.
Monitor (Psammosaurus) albogularis, *Gray, Griff. A. K.* ix. *Syn.* p. 28.
Varanus albogularis, *Dum. & Bibr.* iii. p. 495; *Smith, Ill. S. Afr., Rept.* pl. ii.
Monitor exanthematicus, var. C, *Schleg. Abbild.* p. 71.
—— albogularis, *Peters, Mon. Berl. Ac.* 1870, p. 109, *and Reise Mossamb.* iii. p. 24, pl. iv. fig. 3.

Teeth obtusely conical. Snout obtuse, convex, measuring the distance from the anterior border of the orbit to the anterior border of the ear; canthus rostralis obtuse. Nostril an oblique slit, three times as distant from the end of the snout as from the orbit. Digits short, the length of the fourth toe, measured from its articulation with the tarsus to the base of the claw, equalling about that of the femur. Tail feebly compressed, round in its anterior third, keeled above. Scales of head small, subequal. Scales on upper surfaces small, oval, smooth; nuchal scales not larger than occipitals. Abdominal scales smooth, in eighty-five to one hundred transverse rows. Caudal scales keeled; the caudal keel with a very low, doubly-toothed crest. Greyish-brown above, with large round yellow dark-edged spots arranged in transverse series on the back; a dark temporal streak, extending along the side of the neck; tail with alternate brown and yellowish annuli; lower surfaces yellowish.

	centim.		centim.
Total length	122	Fore limb	17
Head	9	Hind limb........	19
Neck	9	Tail	67
Body	37		

South and South-east Africa.

a, b, c. Ad., stuffed.	S. Africa.	S. African Museum.
d. Hgr., stuffed.	Port Natal.	
e. Hgr.	Lake Nyassa.	
f. Anatomical preparation.	(*Cf.* Günther, Proc. Zool. Soc. 1861, p. 113.)	

3. Varanus ocellatus.

Varanus ocellatus, *Rüppell, Reise N. Afr., Rept.* p. 21, pl. vi.
Monitor ocellatus, *Peters, Mon. Berl. Ac.* 1870, p. 109.

Distinguished from the preceding by the larger dorsal and nuchal scales, the latter being much larger than the occipitals. Abdominal scales in seventy transverse rows. Yellowish-brown above, with more or less distinct, yellowish, brown-edged round spots; no dark streak on the side of the neck.

Abyssinia; Cordofan.

a. Ad.	Anseba, Abyssinia.	W. T. Blanford, Esq. [C.].
b. Ad., stuffed.	Cordofan.	Dr. Rüppell.

4. Varanus exanthematicus.

Regenia ocellatus, *Gray, Cat.* p. 9.
Lacerta exanthematica, *Bosc, Act. Soc. Hist. Nat. Paris,* 1792, p. 25, pl. v. fig. 3.
Tupinambis exanthematicus, *Daud. Rept.* iii. p. 80.
Varanus exanthematicus, *Merr. Tent.* p. 60.
—— ocellatus (*non Rüpp.*), *Dum. & Bibr.* iii p. 496.
Monitor exanthematicus, var. A, *Schleg. Abbild.* p. 70.
—— exanthematicus, *Peters, Mon. Berl. Ac.* 1870, p. 109.

Differs from the two preceding species in the position of the nostrils; these are more anterior, their distance from the orbit being at least half their distance from the end of the snout. Scales as in

V. ocellatus. Sixty to seventy transverse rows of abdominal scales. Markings as in *V. albigularis.*

West Africa.

a. Ad., stuffed.	W. Africa.	
b. Hgr., stuffed.	Senegal.	
c. Yg.	Gambia.	Capt. Moloney [P.].
d. Hgr.	Cape (?).	
e. Ad.	——?	
f. Anatomical preparation.	(*Cf.* Günther, Proc. Zool. Soc. 1861, p. 60.)	

5. Varanus flavescens.

Empagusia flavescens, *Gray, Cat.* p. 9.
Monitor flavescens, *Gray, Griff. A. K.* ix. *Syn.* p. 25, *and Ind. Zool.* pl. —.
Varanus picquotii, *Dum. & Bibr.* iii. p. 485.
Empagusia flavescens, *Gray, Ann. N. H.* i. 1838, p. 393.
Monitor exanthematicus, var. B, *Schleg. Abbild.* p. 71.
Varanus flavescens, *Günth. Rept. Brit. Ind.* p. 65, pl. ix. fig. A.

Teeth subconical, scarcely compressed. Snout short, convex, measuring a little less than the distance from the anterior border of the orbit to the anterior border of the ear; canthus rostralis distinct. Nostril an oblique slit, a little nearer the end of the snout than the orbit. Digits short, the length of the fourth toe, measured from its articulation with the tarsus to the base of the claw, not exceeding the length of the femur. Tail feebly compressed, keeled above. Scales of head small, subequal; the median series of supraocular scales slightly dilated transversely. Scales on upper surfaces moderate, oval, keeled. Abdominal scales smooth, in sixty-five to seventy-five transverse rows. Caudal scales keeled; the caudal keel with a very low, doubly-toothed crest. Olive or yellowish brown above, with irregular darker markings which are generally confluent into broad cross bars; a blackish temporal streak; lower surfaces yellowish, with rather indistinct brown cross bars, which are most distinct on the throat. Young dark brown above, with yellow spots confluent into cross bars; lower surfaces yellow, with dark brown cross bars.

	centim.		centim.
Total length	92	Fore limb	105
Head	7	Hind limb	13
Neck	8	Tail	51
Body	26		

Northern India; Burma; Malay Peninsula.

a, b. Ad., stuffed.	Nepal.	B. H. Hodgson, Esq. [P.].
c, d. Ad., stuffed.	India.	
e. Yg.	India.	Gen. Hardwicke [P.].
f. Hgr.	India.	

6. Varanus bengalensis.

Uaranus heraldicus, *Gray, Cat.* p. 9.
—— lunatus, *Gray, l. c.* p. 10.
Monitor dracæna, *Gray, l. c.* p. 11.
? Lacerta dracæna, *Linn. S. N.* i. p. 360; *Shaw, Zool.* iii. p. 218, pl. lxvii.
Tupinambis bengalensis, *Daud. Rept.* iii. p. 67.
Varanus guttatus, *Merr. Tent.* p. 58; *Lesson, in Bélang. Voy. Ind. Or., Rept.* p. 308.
—— punctatus, *Merr. l. c.* p. 59; *Less. l. c.* p. 309.
Tupinambis cepedianus, *Kuhl, Beitr. z. Zool.* p. 124.
Monitor gemmatus, *Guérin, Icon. R. A., Rept.* pl. iii. fig. 1.
—— heraldicus, *Gray, Griff. A. K.* ix. *Syn.* p. 25.
—— bengalensis, *Gray, l. c.* p. 26.
Varanus bengalensis, *Dum. & Bibr.* iii. p. 480.
Monitor cepedianus, *Schleg. Abbild.* p. 74.
? Varanus olivaceus, *Hallow. Proc. Ac. Philad.* 1856, p. 150.
Varanus bibronii, *Blyth, Journ. As. Soc. Beng.* xi. 1842, p. 869.
—— dracæna, *Günth. Rept. Brit. Ind.* p. 65, pl. ix. fig. B; *Blanford, Journ. As. Soc. Beng.* xlviii. 1879, p. 120.
—— lunatus, *Günth. l. c.* p. 66, pl. ix. fig. C; *Blanf. l. c.* pp. 111, 120.

Teeth acute, compressed. Snout high, very convex at the end, measuring the distance from the anterior corner of the eye to the ear; canthus rostralis well marked. Nostril an oblique slit, a little nearer the orbit than the tip of the snout. Ear-opening oblique. Digits strong, moderately elongate. Tail compressed, keeled above. Scales of head small, subequal; supraocular scales smaller, subequal. Scales of upper surface of body and limbs small, oval, tectiform on the hinder part of the back. Abdominal scales smooth, in ninety to one hundred and ten transverse rows. Caudal scales keeled; the caudal keel with a very low, doubly-toothed crest. Upper surfaces yellowish, brownish, or olive, with more or less numerous blackish dots; a more or less distinct dark streak on the temple; lower surfaces yellowish, uniform or dotted with blackish; these dots most numerous on the throat. Young with numerous whitish ocelli frequently alternating with blackish transverse bands, which may persist after the ocelli have disappeared (*V. lunatus*).

	centim.		centim.
Total length	171	Fore limb	23
Head	10	Hind limb	29
Neck	14	Tail	98
Body	49		

India.

a, b, c. Ad., stuffed.	Nepal.	B. H. Hodgson, Esq. [P.].
d–e. Yg.	N. India.	East India Comp. [P.].
f. Hgr.	Kurrachee	Kurrachee Mus. [P.].
g–h. Ad. & yg	Kurrachee.	Secretary of State for India [P.].
i–k. Hgr. & yg.	Ajmere.	W. T. Blanford, Esq. [P.].
l. Hgr.	Agra.	Col. Beddome [C.].

m. Hgr., stuffed.	Sultanpore.	
n–p. Yg.	Benares.	Dr. Sayer [P.].
q. Hgr.	Deccan.	Col. Sykes [P.].
r. Hgr.	Godavery Valley.	W. T. Blanford, Esq. [P.].
s. Yg.	Matheran.	Dr. Leith [C.].
t. Yg.	Goozerat.	Dr. Leith [C.].
u, *v*. Yg.	Anamallays.	Col. Beddome [C.].
w. Yg.	Ceylon.	Dr. Kelaart.
x. Yg.	Ceylon.	Sir E. Tennant [P.].
y, *z*. Ad. & yg.	India.	East India Comp. [P.].
α. Yg.	India.	W. Masters, Esq. [P].
β, *γ*, *δ*, *ε*, *ζ*. Yg.	India.	
η, *θ*, *ι*, *κ*, *λ*, *μ*, *ν*. Ad., hgr. & yg., stuffed.	India.	
ξ. Hgr., stuffed.	India.	(Type of *V. lunatus*.)
o. Hgr., stuffed.	—— ?	Dr. Shortt.

π. Copulatory claspers. (*Cf.* J. E. Gray, Ann. & Mag. N. H. (4) vii. p. 283.)

7. Varanus nebulosus.

Uaranus nebulosus, *Gray, Cat.* p. 10.
Monitor nebulosus, *Gray, Griff. A. K.* ix. *Syn.* p. 27.
Varanus nebulosus, *Dum. & Bibr.* iii. p. 483, pl. xxxv. figs. 2, 3; *Günth. Rept. Brit. Ind.* p. 66, pl. ix. fig. D.
Monitor nebulatus, *Schleg. Abbild.* p. 75.

Teeth acute, compressed. Snout rather elevated, pointed, convex, measuring the distance from the anterior corner of the eye to the posterior border of the ear; canthus rostralis distinct. Nostril an oblique slit, twice nearer the orbit than the end of the snout. Ear-opening oblique. Digits moderately elongate. Tail compressed, keeled above. Scales of head small, subequal; supraocular region with a series of four to seven transversely dilated scales. Scales on upper surfaces small, oval, obtusely keeled, those on anterior part of neck larger, subcircular, not keeled. Abdominal scales smooth, in about eighty transverse rows. Caudal scales keeled; the caudal keel with a very low, doubly-toothed crest. Greenish or brownish olive, irregularly marbled and dotted with darker and lighter; chin and throat with transverse blackish bands or marbled with blackish. The young have numerous yellowish ocelli on the back, limbs, and base of tail, and the whole of the lower surfaces marbled with blackish; nape with two posteriorly directed, chevron-shaped blackish bands, the anterior proceeding from the eyes.

	centim.		centim.
Total length	95	Fore limb	115
Head	6	Hind limb	14
Neck	8	Tail	60
Body	21		

Bengal; Burma; Siam.

a. Hgr.	N. India.	East India Comp. [P.].
b. Yg.	India.	

c. Ad., stuffed	India.	
d. Hgr.	Pegu.	W. Theobald Esq. [P.].
e. Ad.	——?	

8. Varanus grayi.

Uaranus ornatus, (*non Daud.*) *Gray, Cat.* p. 10.

Snout pointed, depressed at the end, slightly shorter than the distance from the anterior corner of the eye to the anterior border of the ear; canthus rostralis distinct. Nostril oblique, equally distant from the orbit and the end of the snout. Digits long. Tail strongly compressed, keeled above. Scales of head moderate, larger than those on the temples, those between the orbits largest; no transversely dilated supraocular scales. The nine anterior rows of nuchal scales nearly as large as the head-scales, as broad as long, smooth; the following become gradually smaller and elongate; dorsal scales small, oval, strongly keeled. Scales on upper face of limbs strongly keeled. Abdominal scales keeled, in a hundred transverse rows. Caudal scales keeled; the caudal keel with a very low, doubly-toothed crest. Greenish anteriorly, with broad blackish bands across the upper face of the neck and back, these bands darkest at their borders; hinder part of back, tail and hind limbs olive-brown, dotted with greenish; a few blackish spots on the upper face of the head, and a blackish temporal streak, commencing from the eye.

	centim.		centim.
Total length	54·6	Fore limb	7·9
Head	4·6	Hind limb........	10·5
Neck	5	Tail	32
Body	13		

Philippines.

a. Yg.	Philippines.	H. Cuming, Esq. [C.].	(Type.)

9. Varanus dumerilii.

Uaranus dumerilii, *Gray, Cat.* p. 10.
Monitor dumerilii, (*Müller*) *Schleg. Abbild.* p. 78; *Müller & Schleg. Verh. Naturl. Gesch. Nederl. ov. Bezitt., Rept.* p. 44, pl. vi.
Varanus macrolepis, *Blanford, Journ. As. Soc. Beng.* l. 1881, p. 239, pl. xvi.

Teeth acute, compressed. Snout depressed at the end, measuring a little more than the distance from the anterior border of the orbit to the anterior border of the ear; canthus rostralis not well marked. Nostril oblique, about twice nearer the orbit than the tip of the snout. Digits moderate. Tail strongly compressed, keeled above. Scales of head moderate, subequal; those in the middle of the supra-ocular region slightly enlarged transversely. Scales on upper surface of neck very large, about as broad as long, flat, the hindermost slightly keeled; dorsal scales large, oval, keeled; scales above the limbs keeled. Abdominal scales slightly keeled, in seventy-five to eighty-five transverse rows. Caudal scales keeled, the lateral sometimes intermixed with larger ones; the caudal keel with a very low, doubly-toothed crest. Light brown above; a dark temporal streak, from eye to ear, generally confluent with a U-shaped dark marking on the neck; back with very broad dark transverse bars, broader than the interspaces between them; limbs dark brown, spotted with yellow; more or less distinct vertical dark bars on the lips; belly yellowish, uniform or with transverse dark bands.

	centim.		centim.
Total length	81	Fore limb	12
Head	6	Hind limb	15
Neck	8	Tail (injured)	44
Body	23		

Borneo; Sumatra; Tenasserim.

a. Ad.	Borneo.	
b. Ad.	Borneo.	H. Low, Esq. [P.].
c. Ad.	Sumatra.	Hr. C. Bock [C.].
d. Hgr.	——?	Dr. Bleeker.

10. Varanus rudicollis.

Uaranus rudicollis, *Gray, Cat.* p. 10.

Teeth acute, compressed. Snout pointed, very long, considerably longer than the distance from the anterior corner of the eye and the ear; canthus rostralis swollen. Nostril an oblique slit, its distance from the tip of the snout twice and a half that from the orbit. Digits elongate. Tail compressed, keeled above. Scales of head moderate, subequal; supraocular scales with a posterior central series of three to six transversely dilated scales. Scales on upper surface of neck extremely large and prominent, strongly keeled, forming ten to twelve longitudinal series; they become gradually smaller as they pass into the dorsal region; latter and limbs with small, strongly-keeled scales. Abdominal scales keeled, in eighty-five transverse series. Caudal scales keeled; the caudal keel with a very low, doubly-toothed crest. Blackish above, neck and anterior part of body yellowish, the former with three black longitudinal streaks, the latter with two broad transverse black bands; hinder part of back and flanks with yellowish ocelli; limbs with small yellowish spots; lower surfaces obscured by black reticulations.

	centim.		centim.
Total length	84	Fore limb	17
Head	9	Hind limb	20
Neck	11	Tail (injured)	40
Body	24		

Philippines; Borneo.

a. Ad., stuffed.	Philippines.	H. Cuming, Esq. [C.]. (Type.)
b. Ad., stuffed.	Sarawak.	

11. Varanus salvadorii.

Monitor salvadorii, *Peters & Doria, Ann. Mus. Genov.* xiii. 1878, p. 337, pl. ii. figs. 1 & 2.

Teeth acute, compressed. Snout long, rounded and convex at the end; canthus rostralis rounded. Nostril long, oblique, close to the end of the snout. Tail rounded at the base, feebly compressed and keeled above in the rest of its length. Head-scales small, largest on the snout and between the orbit; supraocular scales irregular. Scales on upper surfaces small, elongate, keeled. Abdominal scales strongly keeled. Caudal scales keeled; caudal keel with a very low, doubly-toothed crest. Black above, dotted and spotted with yellow, the spots arranged in cross bands on the back; tail with irregular yellow transverse bands; lower surfaces yellow, marbled and irregularly banded with black.

	centim.		centim.
Total length	165	Fore limb	19
Head	11	Hind limb........	23·5
Neck and body	39	Tail	115

Northern New Guinea.

12. Varanus salvator.

Hydrosaurus salvator, *Gray, Cat.* p. 13.
Stellio salvator, *Laur. Syn. Rept.* p. 56.
Tupinambis bivittatus, *Kuhl, Beitr. z. Zool.* p. 125.
Monitor elegans, *Gray, Zool. Journ.* iii. 1828, p. 225.
Varanus vittatus, *Lesson, in Bélang. Voy. Ind. Or., Rept.* p. 307.
Hydrosaurus bivittatus, *Wagl. Syst. Amph.* p. 164.
—— marmoratus, *Wiegm. Nova Acta Ac. Leop.-Carol.* xvii. 1835, i. p. 196, pl. xiv.
Varanus bivittatus, *Dum. & Bibr.* iii. p. 487.
—— salvator, *Cantor, Cat. Mal. Rept.* p. 29.
Monitor bivittatus, *Schleg. Abbild.* p. 76, pls. xxi. & xxii. figs. 1, 2; *Müll. & Schleg. Verh. Naturl. Gesch. Nederl. ov. Bezitt., Rept.* p. 38.
Hydrosaurus salvator, *Günth. Rept. Brit. Ind.* p. 67, pl. ix. fig. E.

Teeth acute, compressed. Snout depressed at the tip, long, the distance from that point to the anterior corner of the eye being generally a little greater than the distance from the latter to the anterior border of the ear; canthus rostralis obtuse. Nostril oval, at least twice as far from the orbit as from the tip of the snout. Digits moderate. Tail strongly compressed, keeled above. Scales of head moderate, subequal, larger than those on the temples; supraocular region with a series of four to eight large transversely dilated scales. Scales on upper surfaces small, oval, keeled; abdominal scales keeled, in eighty-five to ninety-five transverse rows. Caudal scales keeled; the caudal ridge with a very low, doubly-toothed crest. Dark brown or blackish above, with yellow spots or ocelli; snout generally lighter, with transverse black bars, most distinct on the lips, and continued across the chin; a black temporal band, commencing from the eye and bordered by a yellow band which sometimes extends along the side of the neck; lower surfaces yellow. The markings most accentuated in the young.

	centim.		centim.
Total length	220	Fore limb	30
Head	16	Hind limb	34
Neck	23	Tail (injured)	142
Body	62		

East Indies to Cape York.

a. Hgr.	China.	J. Lindsay, Esq. [P.].
b. Hgr., stuffed.	Nepal.	B. H. Hodgson, Esq. [P.].
c. Ad., stuffed.	Ceylon.	
d. Ad., stuffed.	Tenasserim.	Dr. Packman [P.].
e. Yg.	Siam.	C. Bowring, Esq. [P.].
f. Ad., stuffed.	Sumatra.	Sir S. Raffles [P.].
g–h. Ad., skins.	Sumatra.	H. O. Forbes, Esq. [C.].
i. Hgr.	Java.	
k–l. Yg.	Java.	G. Lyon, Esq. [P].
m. Hgr.	Borneo.	
n–o, *p*. Yg.	Borneo.	L. L. Dillwyn, Esq. [P.].
q, *r*. Ad., stuffed.	Cataguan, Philippines.	
s. Hgr.	Celebes.	Leyden Museum.
t. Ad.	Manado.	Dr. A. B. Meyer [C.].
u. Hgr.	Somerset, Cape York.	'Challenger' Exped.
v, *w*, *x*. Hgr., stuffed.	—— ?	
y. Yg.	—— ?	
z. Skeleton.	—— ?	

13. Varanus nuchalis.

Hydrosaurus nuchalis, *Günth. Proc. Zool. Soc.* 1872, p. 145, pl. viii., and 1873, p. 165.
—— marmoratus, (*non Wiegm.*) *Günth. l. c.* 1873, p. 165.

Agrees in every respect with *V. salvator*, except in having the nuchal scales very large, larger than the scales on the head, distant from one another, frequently intermixed with smaller ones; these large scales are flat and arranged in eight to twelve longitudinal series. The colour agrees with that of *V. salvator*, but the young have the belly reticulated with blackish; this reticulation becomes less distinct in the adult.

Philippines.

a. Ad., skin.	Philippines.	H. J. Veitch, Esq. [P.]. (Type.)
b. Yg.	Philippines.	Hr. Salmin [C.].
c. Hgr.	Philippines.	A. Everett, Esq. [C.].
d–e, *f*. Ad. & hgr.	Negros.	Dr. A. B. Meyer [C.].
g–h. Ad. & yg.	S. Negros.	A. Everett, Esq. [C.].
i. Ad.	Luzon.	Dr. A. B. Meyer [C.].
k, *l*. Ad., stuffed.	Cataguan.	H. Cuming, Esq. [C.].
m. Ad., stuffed.	—— ?	Zoological Society.

14. Varanus cumingii.

Varanus cumingi, *Martin, Proc. Zool. Soc.* 1838, p. 69.
Hydrosaurus cumingii, *Günth. Proc. Zool. Soc.* 1872, p. 145, pl. vii.

Also a very close ally of *V. salvator*. Distinguished in having the nuchal scales larger than the dorsal, and the scales on the occiput larger, less numerous. The colour and markings of the young much as in *V. salvator*, but the belly transversely barred with black. In the adult, the ground colour of the head and back is generally yellowish; more or less distinct transverse dark bands are seen on the head, but they do not extend down to the lips; a well-marked black temporal band; the neck is marbled, and the back transversely barred with black; limbs black, yellow-spotted; tail with black annuli, which are interrupted inferiorly; lower surfaces yellowish, belly with traces of the cross bands so distinct in the young.

Philippines.

a. Ad., skin.	Mindanao.	H. Cuming, Esq. [C.]. (Type.)
b–c. Ad. & yg.	Placer, N.E. Mindanao.	A. Everett, Esq. [C.].
d–e. Yg.	Butuan.	A. Everett, Esq. [C.].
f. Ad.	Zamboanga.	A. Everett, Esq. [C.].
g. Ad.	——?	A. Everett, Esq. [C.].

15. Varanus togianus.

Monitor (Hydrosaurus) togianus, *Peters, Mon. Berl. Ac.* 1872, p. 582.

Differs from *V. salvator* in the larger scales, there being only seventy-four transverse rows of abdominals, and the greater longitudinal development of the supraocular scales. Black, yellow-dotted, without a trace of transverse bands or ocelli, even on the tail; lower surfaces yellow, with transverse black bands on the throat and breast, becoming indistinct on the belly.

Togian Islands, Celebes.

16. Varanus kalabeck.

Monitor kalabeck, *Lesson, Voy. Coquille, Zool.* ii. p. 52; *Peters & Doria, Ann. Mus. Genov.* xiii. 1878, p. 332, pl. i. fig. 3.

Closely allied to *V. indicus*. Snout longer; nostril relatively nearer the tip of the snout. Head-scales rather smaller. Nuchal scales very small, keeled, separated by very wide interspaces.

Waigiou Island.

17. Varanus indicus.

Monitor chlorostigma, *Gray, Cat.* p. 12.
Tupinambis indicus, *Daud. Rept.* iii. p. 46, pl. xxx.
Monitor douarrha, *Lesson, Voy. Coquille, Zool.* ii. p. 53.
Varanus chlorostigma (*Cuv.*), *Dum. & Bibr.* iii. p. 489.
Monitor chlorostigma, *Schleg. Abbild.* p. 75, pl. xxii. fig. 6; *Müll. & Schleg. Verh. Naturl. Gesch. ov. Bezitt., Rept.* p. 40; *Doria, Ann. Mus. Genov.* vi. p. 331.

Monitor doreanus, *Meyer, Mon. Berl. Ac.* 1874, p. 130.
—— indicus, *Peters, Mon. Berl. Ac.* 1876, p. 531; *Peters & Doria, Ann. Mus. Genov.* xiii. 1878, p. 330, pl. i. figs. 1, 2.

Teeth acute, compressed. Snout depressed at the tip, long, the distance from its end to the anterior corner of the eye equalling the distance from the latter point to the anterior border of the ear; canthus rostralis obtuse. Nostril round, a little nearer the tip of the snout than the eye. Digits rather long. Tail strongly compressed, keeled above. Scales of head rather large, much larger on snout and forehead than on temples; supraocular region with a series of four to six large transversely dilated scales. Scales on upper surface of body and limbs small, oval, keeled; abdominal scales smooth, in ninety to one hundred and ten transverse rows. Caudal scales keeled; the caudal ridge with a very low, doubly-toothed crest. Olive-brown or blackish above, dotted all over with yellow; lower surfaces uniform yellowish. Young: spots larger, sometimes forming ocelli; belly frequently with anastomosing blackish lines.

	centim.		centim.
Total length	135	Fore limb	17
Head	9	Hind limb	22
Neck	12	Tail	82
Body	32		

Celebes, Moluccas, New Guinea, Polynesia, North Australia.

a. Hgr., skin.	Celebes.	
b, c. Ad. & yg.	N. Ceram.	
d. Ad., skin.	Timor Laut.	H. O. Forbes, Esq. [C.].
e. Hgr.	N. Guinea.	Rev. S. Macfarlane [C.].
f. Ad., stuffed.	Thaitoro Bay, N. Guinea.	— Comree, Esq. [C.].
g–h. Ad. & hgr.	Murray Island.	Rev. S. Macfarlane [C.].
i, k–l. Ad. & hgr.	Cornwallis Island.	Rev. S. Macfarlane [C.].
m. Ad., stuffed.	Cape York.	Hr. Dämel [C.].
n–q. Hgr. & yg.	Duke of York Island.	Rev. G. Brown [C.]..
r–s. Yg.	Solomon Islands.	G. Krefft, Esq.
t–u. Ad. & yg.	Shortland Islands.	H. B. Guppy, Esq. [P.].
v. Yg.	San Christoval.	E. Brenchley, Esq. [P.].
w. Yg.	Santa Anna Island.	H. B. Guppy. Esq. [P.].
x–y. Ad.	Ke Dolau, Ki Islands.	'Challenger' Exped.
z. Ad.	Wild Islands, Admiralty.	'Challenger' Exped.
α. Ad., stuffed.	—— ?	Paris Museum.
β. Yg.	—— ?	Sir E. Belcher [P.].

18. Varanus niloticus.

Monitor niloticus, *Gray, Cat.* p. 12.
Hasselq. Reise nach Paläst. p. 361; *Geoffr. Descr. Egypte, Rept.* p. 121, pl. iii. fig. 1.
Lacerta nilotica, *Linn. S. N.* i. p. 369; *Forsk. Descr. An. Ægypt.* p. 13.
Stellio saurus, *Laur. Syn. Rept.* p. 56.
Lacerta capensis, *Sparrmann, Resa t. G. H.* p. 749.

Scincus niloticus, *Schneid. Hist. Amph.* ii. p. 195.
Tupinambis elegans, part., *Daud. Rept.* iii. p. 37.
—— stellatus, *Daud. l. c.* p. 59, pl. xxxi.
—— ornatus, *Daud. l. c.* viii. p. 353, *and Ann. Mus.* ii. p. 240, pl. xlviii.
Varanus elegans, *Merr. Tent.* p. 58.
—— ornatus, *Merr. l. c.* p. 59.
Polydædalus niloticus, *Wagl. Syst. Amph.* p. 164.
—— capensis, *Wagl. l. c.*
Monitor pulcher, *Leach, in Griff. A. K.* ix. pl. —.
—— ornatus, *Gray, Griff. A. K.* ix. *Syn.* p. 27.
Varanus niloticus, *Dum. & Bibr.* iii. p. 476; *Steindachn. Sitzb. Ak. Wien,* lxii. i. 1870, p. 330.
—— capensis, *Wiegm. Arch. f. Nat.* 1837, p. 228.
Monitor niloticus, *Gray, Ann. N. H.* i. p. 293; *Duvern. R. A., Rept.* pl. x. bis. fig. 1; *Peters, Mon. Berl. Ac.* 1870, p. 108.
—— elegans, *Schleg. Abbild.* p. 75.
—— saurus, *Peters, l. c.* p. 109, *and Reise n. Mossamb.* iii. p. 23, pl. iv. fig. 2.

Anterior teeth acute, the others becoming rounded-off in adult specimens. Snout depressed at the tip, the distance from its extremity to the anterior corner of the eye equalling the distance from the latter point to the anterior border of the ear; canthus rostralis well marked. Nostril round, a little nearer the eye than the tip of the snout. Digits strong, moderately elongate. Tail strongly compressed, keeled above. Scales of head, including supraoculars, moderate, subequal, larger than those on temples. Scales of upper surface of body and limbs small, oval, tectiform; abdominal scales smooth, in seventy-five to one hundred transverse rows. Caudal scales keeled; the caudal keel with a very low, doubly-toothed crest. Colour of adult brownish or greenish grey above, with darker reticulation and more or less distinct yellowish ocelli on the back and limbs; yellowish beneath, with more or less distinct blackish cross bands. Young black above; head with yellow cross lines, neck with posteriorly directed, chevron-shaped yellow lines; lips with vertical black and yellow bars; back with transverse series of yellow spots or ocelli; limbs black above, with round yellow spots; yellow beneath, reticulated with black; lower surface of head and body yellow, with transverse black bands; tail with alternating black and yellow bars.

	centim.		centim.
Total length	165	Fore limb	19
Head	9	Hind limb	25
Neck	14	Tail	98
Body	44		

Africa, with exception of the north-western part.

a, b, c. Ad., hgr., & yg., stuffed.	N. Africa.	
d. Ad., stuffed.	Egypt.	J. Burton, Esq. [P.].
e. Ad., stuffed.	Egypt.	Sir J. G. Wilkinson [P.].
f, g. Ad. & yg.	Zambesi.	Sir J. Kirk [C.].
h. Yg.	S. Africa.	

i, *k*. Yg., stuffed.	Cape of Good Hope.	
l–*m*. Yg.	Condo, Cuanza.	Dr. Welwitsch [P.].
n. Yg.	Fernando Po.	
o. Yg.	Niger.	Dr. W. Stanger [P.].
p, *q*. Yg.	Ashantee.	T. E. Bowditch, Esq. [P.].
r–*s*. Yg.	Sierra Leone.	
t. Ad., stuffed.	Senegal.	
u. Yg.	Senegal.	
v. Ad., stuffed.	W. Africa.	Mr. Fraser [C.].
w. Yg.	W. Africa.	W. Raddon, Esq. [P.].
x. Hgr.	W. Africa.	
y. Ad.	—— ?	Lord Lilford [P.].
z. Hgr., stuffed.	—— ?	Col. Baikie [C.].
α–*γ*. Yg.	—— ?	
δ, *ε*. Ad., skeletons.	—— ?	

ζ. Anatomical preparation. (*Cf.* Günther, Proc. Zool. Soc. 1861, p. 60.)

19. Varanus varius.

Hydrosaurus varius, *Gray, Cat.* p. 12.
Lacerta varia, *Shaw, in White's Voy. N. S. Wales*, p. 246, pl. iii. fig. 2, *and Zool. Miscell.* iii. pl. lxxxiii.
Tupinambis variegatus, *Daud. Rept.* iii. p. 76.
Varanus varius, *Merr. Tent.* p. 58; *Dum. & Bibr.* iii. p. 491.
Hydrosaurus variegatus, *Wagl. Syst. Amph.* p. 164.
Monitor varius, *Gray, Griff. A. K.* ix. *Syn.* p. 25; *Schleg. Abbild.* p. 78.
Varanus (Hydrosaurus) mustelinus, *De Borre, Bull. Ac. Belg.* (2) xxix. 1870, p. 125.

Teeth acute, compressed. Snout depressed at the end, measuring the distance from the anterior border of the orbit to the ear; canthus rostralis obtuse. Nostril suboval, twice nearer the tip of the snout than the orbit. Digits long. Tail compressed, keeled above. Scales of head small, larger than those on the temples; supraocular scales equal, very small, granular. Scales on upper surfaces small, oval, tectiform. Abdominal scales feebly keeled, in one hundred and twenty to one hundred and thirty transverse series. Caudal scales keeled; the caudal keel with a very low, doubly-toothed crest. Upper surfaces black, with yellow punctulations arranged in transverse bands on the back and lunate bands on the neck; limbs with large yellow spots or annuli; lower surfaces yellow or greenish, with transverse black bands; tail alternately black and yellow in its posterior half.

	centim.		centim.
Total length	148	Fore limb	19
Head	9	Hind limb	25
Neck	13	Tail	93
Body	33		

Australia.

a. Ad.	Gayndah.	Godeffroy Museum.
b. Ad.	Australia.	
c, *d*. Ad., stuffed.	Australia.	J. Gould, Esq. [P.].
e, *f*. Ad., stuffed.	Australia.	

Var. **belli.**

Hydrosaurus bellii, *Gray, Cat.* p. 13.
Varanus bellii, *Dum. & Bibr.* iii. p. 493, pl. xxxv.

Black, with a few very broad yellowish cross bands, generally black-dotted; belly uniform yellowish.

a. Ad., stuffed. Australia.

20. Varanus giganteus.

Hydrosaurus giganteus, *Gray, Cat.* p. 13.
Hydrosaurus giganteus, *Gray, Zool. Erebus & Terror, Rept.*, p. 3, pl. iv.

Teeth acute, compressed. Snout very long, depressed at the tip; the distance from that point to the anterior corner of the eye greater than the distance from the latter point to the anterior border of the ear; canthus rostralis sharp. Nostril oval, more than twice nearer the tip of the snout than the eye. Digits moderate. Tail compressed, keeled above. Scales of head very small, about the size of those on the temples; supraocular scales much smaller still, granular. Scales of upper surfaces small, oval, keeled. Abdominal scales smooth, in one hundred and fifty or one hundred and fifty-five transverse rows. Caudal scales keeled; the caudal keel with a very low, doubly-toothed crest. Brown or blackish above, with large light round spots or ocelli arranged in transverse series on the back and tail; limbs with smaller round spots; sides of neck and throat with large blackish reticulations on a light ground; belly immaculate.

	centim.		centim.
Total length	206	Fore limb	27
Head	14	Hind limb	35
Neck	22	Tail	117
Body	53		

Australia.

a. Ad., stuffed. N. coast of Australia. Capt. Stokes [P.]. (Type.)
b. Ad., stuffed. Australia.

21. Varanus gouldii.

Monitor gouldii, *Gray, Cat.* p. 12.
Hydrosaurus gouldii, *Gray, Ann. N. H.* i. 1838, p. 394, *and in Grey's Trav. Austr.* ii. p. 422.
Monitor gouldii, *Schleg. Abbild.* p. 78; *Gray, Zool. Erebus & Terror, Rept.* pl. iii.; *Peters & Doria, Ann. Mus. Genov.* xiii. 1878, pl. i. fig. 4.
Varanus gouldii, *A. Dum. Cat. Méth. Rept.* p. 52.

Teeth acute, compressed. Snout depressed at the tip, long, the distance from its end to the anterior corner of the eye equalling the distance from the latter point to the anterior border of the ear; canthus rostralis sharp. Nostril round, nearer the tip of the snout than the orbit. Digits strong, moderately elongate. Tail strongly compressed, keeled above. Scales of head, including supraoculars, subequal, very small, not larger than those on the temples. Scales of upper surface of body and limbs small, oval, tectiform. Abdominal scales smooth, in one hundred and twenty-five to one hundred and forty transverse rows. Caudal scales keeled; the caudal keel with a very low, doubly-toothed crest. Brown above, with more or less distinct round yellow spots or ocelli on the back and limbs and yellow annuli round the tail; temple with two yellow streaks, separated by a black band; these streaks extending more or less distinctly along the sides of the neck; lower surfaces yellowish, uniform or with small blackish spots. Young with the markings much accentuated.

	centim.		centim.
Total length	130	Fore limb	20
Head	8	Hind limb	22
Neck	13	Tail	75
Body	34		

Australia; New Guinea.

a. Ad.	Thursday Island, Torres Straits.	Dr. Coppinger [C.]. (H.M.S. 'Alert.')
b, c, d, e. Hgr., stuffed.	N.W. Australia.	J. Gould, Esq. [C.].
f. Ad., stuffed.	Port Essington.	J. Gould, Esq. [C.].
g, h, i. Ad. & hgr., stuffed.	Port Essington.	Capt. Chambers [P.].
k–l. Hgr. & yg.	N. Australia.	J. R. Elsey, Esq. [P.].
m. Ad., stuffed.	N.W. Australia.	Mr. Duboulay [C.].
n. Ad., stuffed.	Dirk Hartog Island, Shark's Bay.	Lords of the Admiralty [P.].
o. Hgr.	Gayndah.	Godeffroy Museum.
p. Yg.	Queensland.	
q. Yg.	W. Australia.	
r. Ad., stuffed.	Australia.	Sir T. L. Mitchell [P.].
s. Yg.	Australia.	
t. Ad.	——?	W. Theobald, Esq. [P.].
u. Yg.	——?	

22. Varanus prasinus.

Hydrosaurus prasinus, *Gray, Cat.* p. 13.
? Monitor viridis, *Gray, Griff. A. K.* ix. *Syn.* p. 26.
Monitor prasinus (*Müll.*), *Schleg. Abbild.* p. 78, pl. xxii. fig. 5; *Müll. & Schleg. Verh. Naturl. Gesch. Nederl. ov. Bezitt., Rept.* p. 42, pl. v.; *Peters & Doria, Ann. Mus. Genov.* xiii. 1878, p. 335, pl. ii. fig. 3.
Varanus prasinus, *A. Dum. Arch. Mus.* viii. p. 499; *Sauvage, Bull. Soc. Philom.* (7) ii. 1878, p. 33.
Odatria prasina, *Günth. Ann. & Mag. N. H.* (5) iii. 1879, p. 84.

Teeth acute, compressed. Snout depressed at the end, measuring the distance from the anterior border of the orbit to the ear; canthus rostralis indistinct. Nostril round or oval, a little nearer the end of the snout than the eye. Digits rather long. Tail round, slightly compressed posteriorly, not keeled. Head-scales large, flat, the largest being those between the orbits; a central series of three or four transversely dilated supraocular scales. Scales on upper surfaces large, oval, keeled on posterior part of back. Abdominal scales keeled, in eighty to ninety transverse rows. Cuadal scales keeled. Upper parts bright green, with a black network sometimes forming transverse bars on the back; lower surfaces yellowish, unspotted.

	centim.		centim.
Total length......	68·8	Fore limb........	9·3
Head	4·5	Hind limb	11·5
Neck	4·3	Tail	44
Body............	16		

New Guinea, and islands of Torres Straits.

a. Ad.	New Guinea.	Dr. Coppinger [C.]. (H.M.S. 'Alert.')
b. Ad.	Islands of Torres Straits.	Rev. S. Macfarlane [C.].
c–d. Ad. & yg.	Cornwallis Island.	Rev. S. Macfarlane [C.].
e–h. Ad.	Murray Island.	Rev. S. Macfarlane [C.].

23. **Varanus kordensis.**

Monitor kordensis, *Meyer, Mon. Berl. Ac.* 1874, p. 131; *Peters & Doria, Ann. Mus. Genov.* xiii. 1878, p. 335, pl. ii. fig. 4.
—— beccarii, *Doria, Ann. Mus. Genov.* vi. 1874, p. 331, pl. xi. fig. *a*.

Distinguished from the preceding in having the scales of the neck and back smaller, more elongate, and very strongly keeled. Specimen *a* is dark olive above, with a black network; specimen *b* is entirely black, lighter beneath.

Western New Guinea.

a. Ad.	Kordo, Mysore.	Marquis G. Doria [P.]. (One of the types.)
b. Ad.	New Guinea, South of Huon Gulf.	— Comree, Esq. [C.].

24. **Varanus punctatus.**

Odatria punctata, *Gray, Cat.* p. 7.
Odatria punctata, *Gray, Ann. N. H.* i. 1838, p. 394, *and Zool. Erebus & Terror, Rept.* pl. i.
Monitor tristis, *Schleg. Abbild.* p. 73.
Varanus punctatus, *A. Dum. Cat. Méth. Rept.* p. 49.
Monitor punctatus, *Peters & Doria, Ann. Mus. Genov.* xiii. 1878, p. 338.

Teeth acute, compressed. Snout depressed at the end, measuring less than the distance from the anterior border of the orbit to the ear; canthus rostralis sharp. Nostril round, slightly nearer the end of the snout than the orbit. Digits moderate. Tail round, sometimes compressed posteriorly, not keeled. Head-scales very small, equal; supraocular scales smaller still, granular. Scales of upper surfaces small, elongate, keeled. Abdominal scales smooth, in eighty-five to ninety-five transverse rows. Caudal scales strongly keeled, the keels forming small spines. Male with a tuft of spines on each side of the base of the tail. Olive above, with narrow black reticulated lines leaving large hexagonal spots; limbs blackish, yellowish spotted; tail, at least in its posterior portion, uniform black; lower surface of body with or without more or less distinct transverse blackish bands.

	centim.		centim.
Total length......	72·5	Fore limb	8
Head	4·5	Hind limb	11
Neck	7	Tail	44
Body............	17		

North and West Australia.

a. Ad.	Torres Straits.	J. B. Jukes, Esq. [P.].
b. Ad., stuffed.	Port Essington.	J. McGillivray, Esq. [C.].
c. Hgr., stuffed.	Port Essington.	J. Gould, Esq. [C.].
d-e, f. Ad. & yg.	North Austral. Exped.	Dr. J. R. Elsey [C.].
g. Ad., stuffed..	N.W. Australia.	J. Gould, Esq. [C.].
h. Yg.	Perth, W. Australia.	Mr. Duboulay [C.].
i. Yg.	W. Australia.	
k, l. Ad. & yg.	Australia.	
m, n. Ad. & hgr.	—— ?	Haslar Collection.
o. Hgr.	—— ?	

25. Varanus timorensis.

Odatria timorensis, *Gray, Cat.* p. 8.
Monitor timorensis, *Gray, Griff. A. K.* ix. *Synops.* p. 26; *Schleg. Abbild.* p. 74.
Varanus timoriensis, *Dum. & Bibr.* iii. p. 473.
—— (Odatria) semiremex, *Peters, Mon. Berl. Ac.* 1869, p. 65.

Very closely allied to the preceding, from which it is distinguished in having the nostril exactly the same distance from orbit and end of snout, and fewer transverse rows of abdominal scales, viz. sixty-five to seventy-five. Male without tufts of spines near the vent. Back yellowish or light olive with black network, or blackish closely dotted with yellowish; head and limbs blackish, dotted or spotted with yellowish; a black temporal streak is generally very distinct; upper face of tail generally with well-defined black annuli; lower surfaces whitish; throat with blackish dots.

Timor; North Australia.

a. Ad.	Timor.	Leyden Museum.
b-c. Ad.	Murray Island.	Rev. S. Macfarlane [C.].
d. Ad.	Thursday Island, Torres Straits.	Dr. Coppinger [C.]. (H.M.S. 'Alert.')

e. Hgr.	Prince-of-Wales' Island, Torres Straits.	Dr. Coppinger [C.]. (H.M.S. 'Alert.')
f. Yg.	Cape York.	Hr. Dämel [C.].
g. Ad.	Port Essington.	Earl of Derby [P.].
h. Yg.	Port Darwin.	R. G. S. Buckland, Esq. [C.].
i. Ad.	—— ?	Earl of Derby [P.].

26. Varanus acanthurus.

Odatria ocellata, *Gray, Cat.* p. 8.
Odatria ocellata, *Gray, Zool. Erebus & Terror*, pl. ii.
Monitor ocellatus, *Peters & Doria, Ann. Mus. Genov.* xiii. 1878, p. 338.

Teeth acute, compressed. Snout depressed at the end, measuring less than the distance from the anterior border of the orbit to the ear; canthus rostralis well marked. Nostril round, slightly nearer the end of the snout than the orbit. Digits moderate. Tail round, depressed at the base, not keeled. Head-scales very small, equal; supraocular scales smaller still, granular. Scales of upper surfaces small, elongate, tectiform. Abdominal scales smooth, in sixty-five to seventy-five transverse series. Caudal scales strongly keeled, the keels on the sides and upper surface forming large spines. Blackish, with large yellow rings; limbs and tail yellow-spotted; a black and yellow temporal streak; neck sometimes with black and yellow longitudinal streaks; lower surfaces yellowish, the throat dotted with blackish.

	millim.		millim.
Total length	665	Fore limb	65
Head	35	Hind limb........	90
Neck	45	Tail	450
Body	135		

Northern and Western Australia.

a. Ad.	N.W. Australia.	(Type.)
b. Ad.	N. Australia.	J. R. Elsey, Esq. [C.].
c. Ad., skin.	N. Australia.	J. R. Elsey, Esq. [C.].
d. Yg.	Nicol Bay.	Mr. Duboulay [C.].

28. Varanus caudolineatus. (Plate XVIII.)

Snout very short, depressed at the end, measuring much less than the distance from the anterior border of the orbit to the ear; canthus rostralis not very distinct. Nostril round, equally distant from the end of the snout and the orbit. Digits moderate. Tail round, depressed at the base, not keeled, not much longer than head and body. Head-scales, including supraoculars, small, subequal. Scales of upper surfaces small, perfectly round, simply convex. Abdominal scales smooth, in seventy transverse rows. Upper and lateral caudal scales strongly keeled, forming small spines. Yellowish-grey above, with round dark brown spots; a dark brown temporal streak; tail with four longitudinal dark brown streaks; lower surfaces pale yellow, the throat dotted with brown.

	millim.		millim.
Total length	198	Fore limb	25
Head	18	Hind limb	33
Neck	17	Tail	107
Body	56		

N.W. Australia.

a. Yg.	Champion Bay.	Mr. Duboulay [C.].

Fam. 13. XANTUSIIDÆ.

Xantusiidæ, *Baird, Proc. Ac. Philad.* 1858, p. 254.
Lacertidæ, part., *Cope, Proc. Ac. Philad.* 1864.
Trachydermi aglyphodonta, part., *Bocourt, Miss. Sc. Mex., Rept.* 1878.
Lepidophymiens, *Sumichrast, Bull. Soc. Zool. France,* 1880, p. 178.
Xantusiidæ, *Cope, Proc. Ac. Philad.* 1883, p. 30.
Xantusiidæ, *Boulenger, Ann. & Mag. N. H.* (5) xiv. 1884.
Lepidophymidæ, *Günther, Biol. C.-Am., Rept.* 1885.

Tongue short, scarcely extensible, indistinctly nicked anteriorly, with oblique overlapping plicæ converging towards the median line, and scale-like imbricate papillæ on the distal extremity. Dentition strictly pleurodont; teeth numerous, the laterals with obtusely tricuspid crowns. Palate toothless, with a large dermal flap overlapping its fellow on the other side. Postorbital and frontosquamosal arches bony, strong, the supratemporal fossa roofed over by the cranial bones; præmaxillary single; nasals, frontals, and parietals two; palatines in contact with each other anteriorly, and with the maxillary and os transversum, there being no infraorbital fossa. Limbs well developed. Clavicle dilated, loop-shaped proximally; interclavicle T-shaped; sternum without fontanelle. No abdominal ribs. Head covered with symmetrical shields, without dermal ossifications. Body granular or tubercular above, plated inferiorly. No moveable eyelids; pupil vertically elongate. Ear exposed. Femoral pores.

Synopsis of the Genera.

Dorsal lepidosis heterogeneous; no supraoculars; two frontals, forming a longitudinal suture between the præfrontal and the interparietal .... 1. **Lepidophyma**, p. 326.

Dorsal scales uniformly granular; a series of supraoculars on each side, separated by a pair of frontals forming a longitudinal suture .... 2. **Xantusia**, p. 327.

Dorsal scales uniform; a series of supraoculars on each side, separated by a large frontal .... 3. **Cricosaura**, p. 329.

1. LEPIDOPHYMA.

Lepidophyma, *A. Dum. Cat. Méth. Rept.* p. 137; *Bocourt, Miss. Sc. Mex., Rept.* p. 306; *Cope, Proc. Ac. Philad.* 1883, p. 30.
Poriodogaster (*Smith*), *Gray, Proc. Zool. Soc.* 1863, p. 154.

No supraoculars; two frontals, forming a longitudinal suture; frontoparietals large, separating the interparietal from the temporals. Dorsal scales granular, intermixed with large tubercles. No plates on the gular fold.

Central America.

1. Lepidophyma flavomaculatum.

Lepidophyma flavomaculatum, *A. Dum. l. c.* p. 138, *and Rev. & Mag. Zool.* 1852, p. 409, pl. xvii.; *Sumichrast, Arch. Sc. Phys. Math.* 1873, p. 251; *Bocourt, Miss. Sc. Mex., Rept.* p. 306, pl. xx. F. fig. 2.
Poriodogaster grayii, (*Smith*) *Gray, l. c.* pl. xxi.; *Peters, Proc. Zool. Soc.* 1874, p. 307.
Lepidophyma smithii, *Bocourt, Journ. Zool.* v. 1876, p. 342, *and l. c.* p. 309, pl. xx. F. fig. 3, and xx. G. fig. 2.
Lepidophyma, sp., *F. Müller, Verh. Naturf. Ges. Basel*, vi. 1878, p. 389, pls. i. & ii.

Head pyramidal; snout pointed, about twice as long as the diameter of the orbit, which nearly equals the vertical diameter of the tympanum. Rostral pentagonal, scarcely visible from above, separated from the large hexagonal or octagonal frontonasal by the pair of anterior nasals which form a short suture; the frontonasal is in contact on each side with the posterior nasal and the first loreal, posteriorly with the præfrontals; latter three, median abnormally missing, sometimes very small and enclosed between the laterals and the frontals, sometimes larger than the laterals and in contact with the frontonasal, laterals extending over the canthus rostralis to the second loreal; frontals large, extending to the supraciliary edge, forming a longitudinal suture between the azygos præfrontal and the interparietal; latter hexagonal, elongate, separating the pair of frontoparietals, which are much broader and a little shorter; a pair of parietals, the largest head-shields; two large temporals on each side, anterior largest and in contact with the frontoparietal and the parietal; nostril pierced between two nasals and the first labial; two loreals, second largest; two postoculars; seven or eight upper labials, sixth largest; mental large, triangular; four large lower labials, the two anterior forming a suture with their fellows on the median line. Throat irregularly folded; one or two transverse gular folds; gular scales uniform, granular. Body depressed; nuchal and dorsal scales irregular, granular, intermixed with trihedral or subconical tubercles forming a regular longitudinal series on each side of the vertebral line, and more or less regular transverse series on the sides; the size of these tubercles very variable. Ventral scales quadrangular, forming ten longitudinal and about thirty transverse series. Four or more enlarged præanal scales. Limbs well developed, the hind ones not reaching the axil when extended, scaled like the body; digits slender, compressed, bent at the articulations. Three (♀) to sixteen (♂) femoral pores on each side. Tail cylindrical, tapering, covered with rings of smaller and larger keeled scales; when intact, its length a little exceeds that of head and body. Brown above, with darker brown and yellowish spots on the body and limbs; lips with vertical dark brown bars; lower surfaces yellowish or brownish white.

	millim.		millim.
Total length	235	Fore limb	30
Head	23	Hind limb........	43
Width of head	16	Tail	135
Body	77		

Panama; Guatemala; South-western Mexico.

a. ♀.	Isthmus of Tehuantepec.	Brussels Museum [E.].
b–e. Hgr.	Mexico.	M. Sallé [C.].
f. Yg.	Mexico.	
g. ♂.	——?	Sir A. Smith [P.]. (Type of *Poriodogaster grayii.*)

2. XANTUSIA.

Xantusia, *Baird, Proc. Ac. Philad.* 1858, p. 255; *Cope, Proc. Ac. Philad.* 1883, p. 30.

A series of small supraoculars, separating the supraciliaries from the frontals; latter two, forming a longitudinal suture; frontoparietals large, separating the interparietal from the temporals. Dorsal scales uniformly granular. A series of slightly enlarged scales on the gular fold.

California.

1. Xantusia vigilis.

Xantusia vigilis, *Baird, l. c.*

Three folds on the throat, the anterior connecting the ears inferiorly and encircling the head. The adpressed hind limb reaches the first gular fold; its length is contained about twice and three fourths in head and body. Colour above dark brownish yellow, varied with blackish spots on single scales ; a yellowish line on each side of the neck, with two others on the nape, making four parallel ones ; underparts whitish. Young vermiculated with yellowish on a brown ground.

Head and body about 50 millim. long.

Fort Tejon, California.

2. Xantusia riversiana.

Xantusia riversiana, *Cope, l. c.* p. 29.

Eye rather large, its diameter more than half the length of the snout ; vertical diameter of the ear-opening a little less. Nostril situated in a small scute at the junction of the sutures which separate the internasal, rostral, first labial, and loreal scuta ; three loreals, increasing in size posteriorly ; frontonasal nearly square ; azygos præfrontal hexagonal, broader than long ; interparietal as large as each frontoparietal ; a single large temporal bounds the frontoparietal and parietal, and it is followed by two small scuta which are in contact with the parietal ; eight upper labials, fifth largest ; posterior labials small, separated by nine rows of still smaller scales from the large temporal. Scales of the gular region flat and hexagonal, one row on the gular fold a little larger, and equal to the anterior gulars. Scales of the dorsal and lateral regions rather coarsely and uniformly granular, not keeled ; ventrals quadrate, in sixteen longitudinal and thirty-two transverse series. Præanal scales in three transverse rows, the anterior two of four scales, with the median pair in both much enlarged, and the posterior row of six scales. Hind limb remarkably short ; it reaches the elbow of the adpressed fore limb ; the first digits of both extremities very short ; scales on the anterior aspects of the fore-leg and femur larger than the others, those of the tibia small, and those of the posterior face of the femur still smaller. Femoral pores twelve on each side. Tail compressed, with a flat inferior surface ; scales of equal size, in whorls of equal width, convex but not keeled. Colour light brown, with dark umber-brown spots on the superior surface ; dark brown bands cross the snout and form a wide **U** on the back of the head ; sides of head with rather large brown spots ; inferior surfaces with minute brown spots which are least numerous on the middle line ; tail with irregular pale spots.

	millim.		millim.
Total length	118	Fore limb	17
Head	16	Hind limb........	23
Width of head	11·5	Tail	58
Body	44		

California.

3. CRICOSAURA.

Cricosaura, *Gundl. & Peters, Mon. Berl. Ac.* 1863, p. 362; *Bocourt, Miss. Sc. Mex., Rept.* p. 312.

A series of supraoculars; a single frontal; frontoparietals very small; interparietal in contact with the temporals. Dorsal scales uniformly granular. A series of plates on the gular fold, forming a collar.

Cuba.

1. Cricosaura typica.

Cricosaura typica, *Gundl. & Peters, l. c.* pl. —. figs. 1–6; *Bocourt, l. c.* p. 313, pl. XX. E. figs. 14–18.

Rostral pentagonal, scarcely visible from above, separated from the frontonasals by the anterior nasals, which form a short suture; a pair of frontonasals and a pair of præfrontals, subequal in size; frontal heptagonal, the largest head-shield; a series of four supraoculars; a large heptagonal interparietal, not much smaller than, and forming a broad suture with, the frontal; a small frontoparietal on each side, between the supraoculars, the frontal, the interparietal, and a large temporal, which forms a suture with the interparietal and the parietal; latter one pair, smaller than the interparietal; nostril pierced between two nasals and the first labial; two loreals, second largest; temple covered with small flat squarish scales; seven upper labials; mental trapezoid, followed by a heptagonal postmental, and three large chin-shields on each side, the first pair forming a suture; lower labials not larger than upper. A strong fold separates the head from the neck; the gular scales anterior to the fold, squarish, those between the latter and the collar roundish. Nuchal and dorsal scales equal, granular, forming transverse series; ventrals large, square, forming eight longitudinal and twenty-six transverse series. Four squarish præanals, posterior pair largest. Limbs weak; digits short, inferiorly with a double series of tubercular scales. Eight femoral pores on each side. Tail cylindrical, tapering to a fine point, a little longer than head and body; caudal scales squarish, verticillate, largest inferiorly. Light brown above, yellowish inferiorly.

	millim.		millim.
Total length	81	Body	29
Head	8	Hind limb........	13
Width of head	5	Tail	44

Cuba.

Fam. 14. TEIIDÆ.

Lacertiens, part., Scincoïdiens, part., *Cuvier, Règne Anim.* ii. 1817.
Tupinambidæ, part., Chalcididæ, part., *Gray, Ann. Phil.* (2) x. 1825.
Cordyloidea, part., Tachydromoidea, part., Chalcidoidea, Ameivoidea, part., *Fitzinger, Neue Classif. Rept.* 1826.
Teiidæ, *Gray, Phil. Mag.* (2) ii. 1827.
Autarchoglossæ acrodonta et pleurodonta, part., *Wagler, Syst. Amph.* 1830.
Ameivæ, Lacertæ, part., Chamæsauri, part., Ptychopleuri, part., Gymnophthalmi, part., *Wiegmann, Herp. Mex.* 1834.
Lacertiens pléodontes, Chalcidiens, part., Scincoïdiens, part., *Duméril & Bibron, Erp. Gén.* v. 1839.
Teidæ, Chalcidæ, Anadiadæ, Chirocolidæ, Cercosauridæ, Gymnophthalmidæ, part., *Gray, Cat. Liz.* 1845.
Argaliadæ, *Gray, Ann. & Mag. N. H.* xviii. 1846, p. 67.
Ecpleopoda, *Tschudi, Arch. f. Nat.* 1847.
Riamidæ, *Gray, Proc. Zool. Soc.* 1858, p. 445.
Cercosauri, *Peters, Abh. Berl. Ac.* 1862.
Teidæ, Chalcididæ, Ecpleopidæ, *Cope, Proc. Ac. Philad.* 1864.
Teidæ, Chalcididæ, *Cope, Proc. Am. Assoc. Adv. Sc.* xix. 1871.
Teiidæ, *Boulenger, Ann. & Mag. N. H.* (5) xiv. 1884.

The tongue is flat, more or less elongate, ending in two long smooth points, the greater part of its surface covered with rhomboidal, imbricate, scale-like papillæ; in *Alopoglossa*, these scale-like papillæ are replaced by oblique plicæ, as in *Xantusiidæ* and the Lacertoid genus *Tachydromus*. In a few genera the tongue is particularly long and narrow at the base, which is retractile into a sheath; in the others the tongue is bicuspid posteriorly, the whole organ, when the distal points are close together, being arrow-headed.

The teeth vary considerably, but are constantly to be distinguished from those of the analogous family *Lacertidæ* in not being hollow at the base, the new teeth developing in small sockets at the base of the old ones. In some of the higher forms (*Tupinambis*, *Dracæna*, *Teius*), the lateral teeth are inserted almost on the parapet of the jaws, so that, in fact, they might be termed *acrodont*; this dentition is, like that of the "pleurodont" Amphisbænoids, truly intermediate between the acrodont and the pleurodont. In the other forms the teeth are attached more distinctly to the inner side of the jaws and there is a basal shaft. The præmaxillary teeth are constantly conical; the laterals may be conical, bicuspid, tricuspid, obtuse or molar-like (adult *Tupinambis*), or enormous oval crushers (*Dracæna*); the bicuspid teeth may be either compressed longitudinally, or perpendicularly to the jaws (*Dicrodon*, *Teius*). Pterygoid teeth are but seldom present, and if so, but feebly developed.

In the most highly developed forms the skull presents the typical Lacertilian or Cionocranian structure; but, as we approach the Chalcidine forms, the skull becomes more depressed, the vacuity between the ante- and postorbital portions, and consequently the interorbital septum, become much reduced, the arches weaker, and, at last, we have in *Ophiognomon* a skull that approaches the Amphisbænian type: the pterygoids are largely in contact with the sphenoid, the arches are very weak, the columella is so reduced as to be hardly distinguishable, and the whole skull forms an almost continuous solid mass.

The præmaxillary bone is single, the nasals double, the frontal and the parietal single; a small supraorbital bone exists in some Ameivas; the palatines are in contact anteriorly; the maxillary is excluded from the infraorbital fossa, which is bounded by the palatine, pterygoid, and transverse bone, as in *Varanus*. The head-shields are always free from the cranial ossification, a character which separates sharply this family from the Old-World *Lacertidæ*.

Limbs, or their rudiments, are present; the degradation begins with *Proctoporus*, which has very short limbs, but with five well developed clawed digits; the limbs are still more reduced in size in *Scolecosaurus*, and the digital formula is incomplete, although claws are still present; in *Cophias* and *Ophiognomon*, digits are either entirely lost or reduced to bud-like rudiments; and in a species of the latter genus, the pelvic limbs have altogether disappeared. The clavicle is dilated and perforated proximally; however, Cope mentions *Tretioscincus* and a species of *Scolecosaurus* as forming exceptions, having a simple clavicle; the interclavicle is cruciform.

This family comprises a large number of species, all of which are inhabitants of the New World, which show great variety of form and scaling. The *Teiidæ* with Lacertoid or Varanoid appearance pass, through *Crocodilurus* and *Neusticurus*, into the Cercosaurs (to which the latter is nearer on account of its separated nasal plates, although hitherto placed with the true Teioids). We have then several diverging series, the two principal being:—towards *Heterodactylus* through *Arthrosaura*, and towards *Cophias* through *Proctoporus*. On the other hand, *Cophias* resembles *Heterodactylus* in the position of the nostril, and the latter genus leads us through *Perodactylus* to the Skink-like genera such as *Iphisa* and *Gymnophthalmus*, which all agree with *Heterodactylus* in the rudimentary condition (or absence) of the inner finger. And, at last, the vermiform Chalcidines by their scaling and cranial structure approach the Amphisbænoids.

Synopsis of the Genera.

I. Anterior nasal plates not separated by the fronto-nasal*; limbs well developed.

A. Scaly portion of the tongue not much widened, and not or but feebly emarginate posteriorly, frequently retractile into a basal sheath.

1. Ventral plates small, forming more than 20 longitudinal series.

* Separated by several small scales in *Callopistes*; sometimes separated in *Teius*.

Tail not or but feebly compressed; dorsal scales small, uniform.
1. **Tupinambis**, p. 334.

Tail strongly compressed, with a double longitudinal keel above; dorsal scales intermixed with large keeled tubercles.
2. **Dracæna**, p. 337.

2. Ventral plates large, forming less than 20 longitudinal series.

Ventral plates keeled; femoral pores.
3. **Centropyx**, p. 339.

Ventral plates keeled; no pores .. 4. **Monoplocus**, p. 343.

Ventral plates smooth .... 5. **Ameiva**, p. 343.

B. Scaly portion of the tongue arrow-headed, bifid and not retractile posteriorly.

1. Tail rounded.

a. Teeth longitudinally compressed.

Head-shields large, regular; ventral shields in less than 20 longitudinal rows .... 6. **Cnemidophorus**, p. 360.

Head-shields small, numerous; ventral shields in more than 20 longitudinal rows .... 7. **Callopistes**, p. 375.

b. Teeth transversely compressed.

Toes five...................... 8. **Dicrodon**, p. 377.

Toes four .... 9. **Teius**, p. 378.

2. Tail strongly compressed, bicarinate.

Dorsal scales small, uniform ...... 10. **Crocodilurus**, p. 380.

II. Nasal plates widely separated by one or two frontonasals; fingers and toes five, all clawed.

A. Præfrontals present.

1. Tail compressed, with a double denticulated crest.

Dorsal lepidosis composed of small scales intermixed with large keeled ones.................. 11. **Neusticurus**, p. 381.

2. Tail round or cyclo-tetragonal.

α. Lingual papillæ, oblique plicæ.

Dorsal scales large, pointed, strongly keeled.
12. **Alopoglossus**, p. 383.

b. Lingual papillæ normal, scale-like, imbricate.

α. Dorsal scales strongly keeled.

* Dorsal scales arranged in regular transverse and oblique series.

Ventrals like the dorsals, keeled .. 13. **Leposoma**, p. 386.

Ventrals quadrate, smooth; no collar.
14. **Loxopholis**, p. 387.

Ventrals quadrate, smooth; a collar. 15. **Pantodactylus**, p. 388.

** Dorsal scales forming transverse series only.

Dorsal and lateral scales equal .... 16. **Arthrosaura**, p. 389.

Lateral scales small, irregular .... 17. **Prionodactylus**, p. 390.

*** Dorsal scales forming regular transverse and longitudinal series.

Lateral scales small, irregular .... 18. **Cercosaura**, p. 395.

β. Dorsal scales smooth or feebly keeled.

* Dorsal scales not larger than ventrals.

Scales not forming uninterrupted series round the body; a continuous series of femoral and præanal pores.
19. **Placosoma**, p. 396.

Scales subequal, forming uninterrupted series round the body.
20. **Anadia**, p. 398.

Scales not forming uninterrupted annuli, the laterals being much smaller; dorsals hexagonal; collar-fold feeble; males without femoral pores .... 21. **Ecpleopus**, p. 401.

Scales not forming uninterrupted annuli, the laterals being much smaller; collar-fold strong; males with femoral pores.
23. **Euspondylus**, p. 404.

** Dorsal scales larger than ventrals.

Scales quadrangular, smooth; collar-fold strong; both sexes with femoral pores .... 24. **Argalia**, p. 408.

B. No præfrontals.

Dorsal scales hexagonal, striated, imbricate; collar-fold feeble.
22. **Pholidobolus**, p. 403.

Dorsal scales keeled, juxtaposed, separated from ventrals by a wide zone of small scales; collar-fold strong.
25. **Oreosaurus**, p. 409.

Dorsal scales smooth or striated, juxtaposed, separated from ventrals by a fold; collar-fold strong .... 26. **Proctoporus**, p. 412.

III. Nostril pierced between the nasal and the first labial; no ear-opening.

A. Digits rudimentary or absent.

Digits distinct, clawed; a pair of præfrontals and a frontonasal.
27. **Scolecosaurus**, p. 416.

Digits, if distinct, clawless; a frontonasal; no præfrontals.
28. **Cophias**, p. 417.

No frontonasal, no præfrontals.... 29. **Ophiognomon**, p. 420.

B. Digits well developed; inner finger rudimentary.

Dorsal scales narrow, hexagonal-lanceolate, keeled.
30. **Heterodactylus**, p. 422.

IV. Nasal plates widely separated by a frontonasal; ear exposed; inner finger, if distinct, clawless.

A. Eyelids developed.

Dorsal scales narrow, hexagonal-lanceolate, keeled.
31. **Perodactylus**, p. 423.

All the scales rounded, imbricate, two median dorsal and two ventral series strongly enlarged transversely.
32. **Iphisa**, p. 424.

All the scales rounded, imbricate, subequal.
33. **Tretioscincus**, p. 425.

B. No eyelids; scales cycloid, quincuncial.

No præfrontals; frontoparietals present.
34. **Micrablepharus**, p. 426.

Præfrontals present; no frontoparietals.
35. **Gymnophthalmus**, p. 427.

1. TUPINAMBIS.

Tupinambis, part., *Daud. Rept.* iii. p. 6.
Sauvegardes, part., *Cuv. R. A.* ii. p. 26.
Teius, part., *Merr. Tent. Syst. Amph.* p. 60.
Monitor, *Fitzing. N. Classif. Rept.* p. 21.
Exypnestes, *Kaup, Isis,* 1826, p. 87.
Podinema, *Wagl. Syst. Amph.* p. 153; *Wiegm. Herp. Mex.* p. 8.
Ctenodon, *Wagl. l. c.*; *Wiegm. l. c.*

Teguixin, *Gray, Griff. A. K.* ix., *Syn.* p. 29.
Teius, *Gray, Ann. N. H.* i. 1838, p. 276, *and Cat. Liz.* p. 16.
Salvator, *Dum. & Bibr.* v. p. 78.

Tongue long, of nearly equal width throughout, sheathed at the base. Lateral teeth longitudinally compressed and tricuspid in the young, the crowns obtuse in the adult. Head with large regular shields; anterior nasal in contact with its fellow; nostril pierced in the suture between the two nasals. Eyelids developed. Ear exposed. Limbs well developed; digits five. Dorsal scales small, juxtaposed, equal; ventral plates small, elongate-quadrangular, smooth, forming regular series. A double collar-fold. Femoral pores. Tail rounded.

South America.

Synopsis of the Species.

Two loreal plates; dorsal scales oval, convex; 46 to 50 ventral scales across the middle of the belly .... 1. *rufescens*, p. 335.
Two loreal plates; dorsal scales oval, flat; 36 to 40 ventral scales across the middle of the belly .... 2. *teguixin*, p. 335.
One loreal plate; dorsal scales regularly hexagonal, flat; about 30 ventral scales across the middle of the belly ...... 3. *nigropunctatus*, p. 337.

1. Tupinambis rufescens.

Tejus rufescens, *Günth. Proc. Zool. Soc.* 1871, p. 541, *and Trans. Zool. Soc.* xi. p. 221, pl. xlv.

Very similar to *T. teguixin*, but with all the scales considerably smaller, this being most conspicuous on the temporal, dorsal, and gular regions; the granules of the back more convex; a double series of enlarged scales above the temple. A single anterior undivided chin-shield. Ventral scales in thirty-two to thirty-four transverse and forty-six to fifty longitudinal series. Seven or eight femoral pores on each side. Posterior part of the tail scarcely compressed. Reddish yellow, above with blackish-brown transverse bands; lower parts with more or less indistinct irregular transverse dark spots.

From snout to vent 40 centim.; tail 60 centim.

Mendoza.

a, b. ♂.	Mendoza.	Zoological Society.	(Types.)

2. Tupinambis teguixin.

Teius teguexim, *Gray, Cat.* p. 16.
Lacerta teguixin, *Linn. S. N.* i. p. 368; *Shaw, Zool.* iii. p. 235, pls. lxxiii., lxxiv.; *Duvern. R. A., Rept.* pl. xi. fig. 1.

Seps marmoratus, *Laur. Syn. Rept.* p. 59.
Tupinambis monitor, *Daud. Rept.* iii. p. 20; *Wied, Beitr. Nat. Bras.* i. p. 155, *and Abbild.*; *Spix, Spec. Nov. Lacert. Bras.* p. 19, pl. xix.
Monitor meriani, *Blainv. Bull. Soc. Philom.* 1816, p. 111.
Teius monitor, *Terr. Tent.* p. 61.
Monitor teguixin, *Fitz. N. Classif.* p. 51.
Podinema teguixin, *Wagl. Syst. Amph.* p. 153.
Teius teguixin, *Gray, Ann. N. H.* i. p. 276; *Girard, U. S. Explor. Exped., Herp.* p. 202, pl. xviii. figs. 1–7.
Salvator merianæ, *Dum. & Bibr.* v. p. 85; *Guichen. in Casteln. Amér. Sud, Rept.* p. 30.

Nasals forming a short suture behind the posterior angle of the rostral; one large internasal; a pair of smaller præfrontals; a frontal, nearly as large as the internasal; a pair of small fronto-parietals; an interparietal and a pair of parietals, these three shields small, subequal, on a line; occiput with small polygonal scales; four supraoculars, second largest; two large loreals; eight to ten upper labials, separated from the orbit by a row of infraorbitals; temples with small granular scales. Mental broad, truncate posteriorly; eight or nine lower labials, the first constantly, and sometimes the three following, in contact with the chin-shields, the others separated from the latter by small shields; chin-shields large, the first or the first two single, the others in pairs, the first pair forming a suture; gular scales flat, granular between the chin-shields and in the folds of the neck, larger and hexagonal on the remaining parts. Body covered above with small oval flat granular scales arranged in regular transverse series; these granules becoming smaller still on the flanks. Ventral plates at least twice as long as broad, in thirty to thirty-five regular transverse * and thirty-six to forty longitudinal series. Præanal scales small, like the gulars. Limbs granular above, the fore limb anteriorly and the hind-limb inferiorly with enlarged flat scales, a few of which, on the inner edge of the tibia, are transversely dilated. Femoral pores small, about twenty on each side. Tail subcylindrical, slightly compressed towards the end, with small keeled scales arranged in regular transverse series, two series on the upper surface corresponding to one on the lower. Olive above, marbled and transversely barred with black, and with more or less distinct rows of lighter spots; yellowish inferiorly, with interrupted black cross bands; the sutures of the plates on the upper surface of the head black.

	centim.		centim.
Total length	92	Fore limb	14
Head	10	Hind limb........	21
Width of head	0·6	Tail	55
Body	27		

South America, from the Guianas to Uruguay; West Indies.

* Counted from a line drawn at the posterior border of the arms.

a, b, c. ♂, ♀, & hgr.	Brazil.	
d. Hgr.	Brazil.	Sir E. Belcher [P.].
e. Yg.	Rio Janeiro.	Mrs. Fry [P.].
f–g. Yg.	Rio Grande do Sul.	Dr. v. Ihering [C.].
h, i, k, l. Ad. & hgr., stuffed.	S. America.	
m. ♂, skeleton.	S. America.	

3. Tupinambis nigropunctatus.

Teius nigropunctatus, *Gray, Cat.* p. 16.
Tupinambis nigropunctatus, *Spix, Spec. Nov. Lacert. Bras.* p. 18, pl. xx.
Ctenodon nigropunctatus, *Wagl. Syst. Amph.* p. 153.
Salvator nigropunctatus, *Dum. & Bibr.* v. p. 20; *Guichen. in Casteln. Amér. Sud, Rept.* p. 31.
Podinema nigropunctatum, *Peters, Mon. Berl. Ac.* 1877, p. 414.

Differs from *T. teguixin* in the following points:—A single loreal. A few large plates bordering the temple superiorly. Constantly a single undivided anterior chin-shield. All the scales larger, especially those on the gular and præanal regions; the dorsal scales hexagonal; about thirty longitudinal and as many transverse series of ventral plates. About ten femoral pores on each side. Coloration rather similar to that of *T. teguixin*, but no cross bands on the belly, which is uniform or black-spotted; head with large black spots above.

Guianas; Brazil; Eastern Peru.

a. ♂.	Demerara.	Col. Sabine [P.].
b–c. Yg.	Demerara.	
d. Skin.	Better Hope, Brit. Guiana.	Rev. W. Turner [C.].
e. Yg.	Berbice.	
f. Hgr.	Brazil.	Lord Stuart [P.].
g. Yg.	Brazil.	T. Bell, Esq. [P.].
h. Yg.	Sarayacu, Peru.	Messrs. Veitch [P.].
i. Yg.	Pampa del Sacramento.	Messrs. Veitch [P.].
k, l, m, n–o, p–u. ♂, hgr., & yg.	S. America.	
v. Ad., stuffed.	S. America.	

2. DRACÆNA.

Dracæna, *Daud. Rept.* ii. p. 421.
Teius, part., *Merr. Tent. Syst. Amph.* p. 60.
Thorictis, *Wagler, Syst. Amph.* p. 153; *Wiegm. Herp. Mex.* p. 8; *Dum. & Bibr.* v. p. 51.
Ada, part., *Gray, Griff. A. K.* ix. *Syn.* p. 28.
Ada, *Gray, Cat. Liz.* p. 25.

Tongue long, not distinctly sheathed, slightly widened, but not emarginate posteriorly. Lateral teeth very large, molar-like, with rounded crowns. Head with large regular shields; anterior nasal in contact with its fellow; nostril pierced in the suture between two nasals. Eyelids developed. Ear exposed. Limbs well developed, each with five digits, the toes with a feebly denticulated fringe laterally, inferiorly with small squarish scales. Dorsal lepidosis composed of small scales intermixed with very large keeled tubercles; ventral plates small, elongate subquadrangular, feebly keeled, forming regular series. A double collar fold. Femoral pores. Tail strongly compressed, with a double denticulated keel superiorly.

Guianas and Basin of the Amazon.

1. Dracæna guianensis.

Ada guianensis, *Gray, Cat.* p. 25.
Dracæna guianensis, *Daud. Rept.* ii. p. 423, pl. xxviii.
Teius crocodilinus, *Merr. Tent.* p. 62.
Thorictis guianensis, *Wagl. Syst. Amph.* p. 153.
Teius (Ada) crocodilus, *Gray, Griff. A. K.* ix. *Syn.* p. 28.
Thorictes dracæna, *Dum. & Bibr.* v. p. 56; *Duvern. R. A., Rept.* pl. x. bis. fig. 2.

Head large, not depressed, subpyramidal. Nostrils opening on the upper surface of the snout, on the canthus rostralis; anterior nasals forming a long suture. Upper head shields: a pair of nasofrontals, three smaller præfrontals on a line, followed by several irregular small shields, a large frontal, a pair of frontoparietals, three subequal parietals on a line, four or five supraoculars; occiput with numerous small shields; a row of three or four large supratemporals; two or three loreals and several small frenoorbitals; a row of infraorbitals; labials large, eight or nine upper and eleven to thirteen lower; chin-shields, an anterior and nine or ten pairs, all except the anterior separated from the labials by smaller shields, the outer row of which is the largest; gular and mesoptychial scales flat, subhexagonal. Dorsal scales small, irregular, intermixed with irregularly arranged large oval strongly keeled tubercles. Abdominal scales small, narrow, feebly keeled, in thirty-four or thirty-six transverse and about forty longitudinal series. Præanal scales small, irregular. Scales on the limbs small. Four or five small pores on each side in front of the præanal region and two or three under each thigh. Caudal scales narrow, feebly keeled, forming annuli; the dorsal tubercles are gradually modified into a strong double crest along the upper surface of the tail. Olive-brown above, flanks with more or less distinct lighter spots; the sutures of the labials and chin-shields black; lower surface yellowish, marbled with blackish.

	centim.		centim.
Total length	83	Fore limb	11·5
Head	8·5	Hind limb	15
Width of head	5·5	Tail	50
Body	24·5		

Guianas and Basin of the Amazon.

a. Ad., stuffed.	Para.	
b. ♂.	Santarem.	H. W. Bates, Esq. [C.].
c. ♀.	Upper Amazon.	
d. Hgr.	Cashiboya.	Messrs. Veitch [P.].

3. CENTROPYX.

Kentropyx, *Spix, Spec. Nov. Lacert. Bras.* p. 21.
Pseudoameiva, *Fitzing. N. Class. Rept.* p. 21.
Trachygaster, *Wagl. Syst. Amph.* p. 154.
Centropyx, *Wiegm. Herp. Mex.* p. 9; *Dum. & Bibr.* v. p. 146; *Gray, Cat.* p. 23.
Acanthopyga, *Gray, l. c.*

Tongue long, its scaly portion slightly widened and emarginate posteriorly, retractile at the base. Lateral teeth compressed longitudinally, tricuspid. Head with large regular shields; anterior nasal in contact with its fellow; nostril in the suture between two nasals. Eyelids developed. Ear exposed. Limbs well developed, each with five digits, the borders of which are serrated. Dorsal scales small or large, juxtaposed or imbricate, usually keeled; ventral plates large, imbricate, keeled, forming regular series. A strong collar fold, with a row of large scales. Femoral pores. Tail cyclo-tetragonal.

South America.

Habit lacertoid. Head shields: a frontonasal, two præfrontals, a frontal, two frontoparietals, two parietals, an interparietal, two occipitals, three or four supraoculars, four or five supraciliaries; a rostral, an anterior and a posterior nasal, a loreal, three or four infraorbitals, six or seven upper labials; a mental, five or six lower labials; an anterior chin-shield and five or six pairs, the first pair forming a suture. Caudal scales large, imbricate, strongly keeled.

Synopsis of the Species.

I. Median dorsal scales much larger than laterals, strongly imbricate, keeled.

Enlarged dorsal scales very large, with sharp keels forming fourteen straight longitudinal lines .... 1. *striatus*, p. 340.
Enlarged dorsal scales smaller, the keels not continuous and directed towards the median line of the back .... 2. *intermedius*, p. 340.

II. Median dorsal scales a little larger than laterals, hexagonal, juxtaposed, keeled.

The femoral scales which touch the row of femoral pores anteriorly hardly as large as the median gulars .... 3. *calcaratus*, p. 341.
These scales larger than the median gulars. 4. *pelviceps*, p. 342.

III. Dorsal and gular scales all smooth. 5. *altamazonicus*, p. 342.

1. Centropyx striatus.

Acanthopyga striata, part., *Gray, Cat.* p. 23.
Lacerta striata, *Daud. Rept.* iii. p. 247.
Pseudoameiva striata, *Fitzing. N. Class. Rept.* p. 51.
Trachygaster striatus, *Wagl. Syst. Amph.* p. 154.
Teius (Centropyx) striata, *Gray, Griff. A. K.* ix. *Syn.* p. 31.
Centropyx striatus, part., *Dum. & Bibr.* v. p. 151.
—— decodon, *Cope, Proc. Philad. Ac.* 1861, p. 495.
—— renggerii, *Peters, Mon. Berl. Ac.* 1869, p. 63.

Posterior head-shields keeled. Gular scales rhomboidal, imbricate, strongly keeled, gradually enlarged towards the middle; the border of the collar strongly dentated, formed of twelve plates. Median dorsal scales very large, strongly imbricate, with sharp keels forming fourteen straight longitudinal lines; scales on the nape and on the sides small, subimbricate, keeled. Ventral plates in fourteen longitudinal and thirty-two transverse series. Six longitudinal series of enlarged keeled brachial scales and five of antebrachials, the keels of third and fourth of former continuous with the keels of fourth and fifth of latter. Ten rows of femoral plates and about twelve of tibials. Femoral pores seven. Male with two large præanal spines on each side. Olive; a few black spots on each side of the anterior part of the body.

	millim.		millim.
Total length	307	Fore limb	34
Head............	23	Hind limb........	61
Width of head	15	Tail	215
Body	69		

Northern South America.

a. ♂. —— ?

2. Centropyx intermedius.

Acanthopyga striata, part., *Gray, Cat.* p. 23.
Centropyx calcaratus, *Gray, l. c.*
Merrem, Ann. Wetterauisch. Gesellsch. i. 1809, p. 2, pl. i.
Lacerta striata (*non Daud.*), *Merr. Tent.* p. 65.
Teius (Centropyx) intermedius (*Schleg.*), *Gray, Griff. A. K.* ix. *Syn.* p. 31.
Centropyx striatus, part., *Dum. & Bibr.* v. p. 151; *Troschel, in Schomburgk, Reise Brit. Guian.* iii. p. 651.
—— intermedius, *Cope, Proc. Ac. Philad.* 1861, p. 496.
—— borckiana, *Peters, Mon. Berl. Ac.* 1869, p. 64.

Distinguished from the preceding by the much smaller dorsal scales, the keels of which are not continuous, but directed towards the median line of the back. Ventral plates in fourteen or sixteen longitudinal and thirty-three to thirty-five transverse series. Femoral pores eleven to fifteen. Digital denticulations stronger. Olive; a light line on each side, beginning from the eye, bordered above and beneath by a broad black band or a series of black spots; a second, more or less distinct, light lateral line, beginning from the ear; lower surfaces greenish white.

	millim.		millim.
Total length	344	Fore limb	37
Head	24	Hind limb	68
Width of head	15	Tail	245
Body	75		

Guianas, Brazil.

a. ♀.	Demerara.	
b. ♀.	British Guiana.	
c, d. ♀.	S. America.	T. Bell Esq. [P.].
e–h, i, k. ♀ & hgr.	S. America.	

The fact that all the specimens examined by various authors as well as by myself are females, makes it possible that *C. intermedius* is the female of *C. striatus*, as already suggested by Troschel. However, this is by no means certain, and the characters which separate the two forms are not of a kind known to vary according to sexes in other genera of lizards.

3. Centropyx calcaratus.

Kentropyx calcaratus, *Spix, Spec. nov. Lacert. Bras.* p. 21, pl. xxii. fig. 2.
Lacerta striata (*non Daud.*), *Wied, Abbild.*
Trachygaster calcaratus, *Wagl. Syst. Amph.* p. 154.
Teius (Centropyx) calcaratus, *Gray, Griff. A. K.* ix. *Syn.* p. 31.
Centropyx vittatus, *Wiegm. Herp. Mex.* p. 26.
—— calcaratus, *Dum. & Bibr.* v. p. 149; *Cope, Proc. Ac. Philad.* 1861, p. 495; *Peters, Mon. Berl. Ac.* 1869, p. 64, *and* 1877, p. 412.

Interorbital and occipital region concave in full-grown specimens, bordered by a prominent ridge. Gular scales small, subhexagonal, keeled, juxtaposed or slightly imbricate, slightly enlarged on the middle of the throat; the edge of the collar not, or but very slightly, denticulated; the last row of mesoptychial plates composed of fifteen or sixteen plates, sometimes even separated from the fold by granules. Median dorsal scales only a little enlarged, subhexagonal, juxtaposed, keeled; scales on the flanks very small, almost granular. Median temporal scales minute. Ventral plates in fourteen or sixteen longitudinal and thirty-three to thirty-five transverse series. Præanal scales smooth in the males, strongly keeled in the females; the former with two large præanal spines on each side. Inferior femoral scales very small, those of the lower series hardly as large as the median gulars. Femoral pores seventeen to nineteen. Olive above; a light line on each side of the body, commencing from the eye, bordered inferiorly, and sometimes also superiorly, by a row of black spots, which may be confluent into a band; females with a broad median yellowish band on the head, which is generally lost on the anterior part of the back; females sometimes with a second light line along the side; males with light spots on the flanks; lower surfaces light olive.

	♂. millim.	♀. millim.
Total length	291	253
Head	26	22
Width of head	16	12
Body	65	61
Fore limb	35	34
Hind limb	67	65
Tail	200	170

South America.

a-b. ♂ ♀.	Demerara Falls.	
c. ♀.	Cayaria, E. Peru.	Messrs. Veitch [P.].
d, e. ♂ ♀.	Sarayacu, Peru.	Messrs. Veitch [P.].

4. Centropyx pelviceps.

Centropyx pelviceps, *Cope, Proc. Ac. Philad.* 1868, p. 98.
—— dorsalis, *O'Shaughn. Proc. Zool. Soc.* 1881, p. 228.

Very closely allied to the preceding. The occipital concavity is stronger; the temporal granules and lower femoral scales larger; femoral pores nineteen to twenty-two. Præanal scales in both sexes keeled. Dark olive above, with a pale light median band commencing on the snout and gradually disappearing on the back; this band bordered on each side by a broad festooned black band breaking up into spots on the hind part of the back; lower surfaces pale green.

	♂. millim.	♀. millim.
Total length	343	348
Head	31	31
Width of head	18	18
Body	82	87
Fore limb	45	50
Hind limb	88	91
Tail	230	230

Ecuador; N. Peru.

a–b. ♀.	Canelos, Ecuador.	Mr. Buckley [C.].
c. d. ♂ ♀.	Peruvian Amazon.	Mr. E. Bartlett, [C.].
f. Yg. ♂.	(Stated to be the type of *Monoplocus dorsalis.*)	

5. Centropyx altamazonicus.

Centropyx altamazonicus, *Cope, Journ. Ac. Philad.* (2) viii. 1876, p. 162.

Dorsal scales very small, hexagonal, smooth, but little larger than the lateral. Ventrals keeled, in sixteen longitudinal and thirty-five transverse rows. Two rows of keeled collar scales, the second largest. Gular scales small, equal, smooth. Bluish, with a median light band from the middle of the back to the end of the nose, covering the muzzle and front as far as the frontoparietal scuta; a light band from orbit to near femur and another from tympanum to groin, the intervening space crossed by vertical black spots and bars; lower surfaces green.

This species is established on a single, young, specimen measuring 40 millim. from snout to vent.

Moyobamba, Peru.

4. MONOPLOCUS.

Monoplocus, *Günth. Proc. Zool. Soc.* 1859, p. 404.

"Tongue elongate, free, not sheathed, terminating in two very fine points. Palatine teeth none; the posterior teeth in the jaws bi- or tricuspid. Tympanum distinct. Throat with a single fold. Scales of the back exceedingly small, those of the sides granular: gular and ventral shields keeled. Tail rounded, covered with keeled and verticillated scales of moderate size. Femoral pores none."

1. Monoplocus dorsalis.

Monoplocus dorsalis, *Günth. l. c.*

This species has been identified with *Centropyx pelviceps* (*cf.* O'Shaughnessy, P. Z. S. 1881, p. 228), from the examination of a specimen labelled as the type of *Monoplocus dorsalis*. Although that specimen agrees pretty well in size, proportions, and coloration with Günther's detailed description, it differs in several points which certainly cannot be attributed to difference in the mode of describing or to oversight:—1. The tongue is distinctly sheathed at the base; 2. Femoral pores are as distinct as can be; 3. There are fourteen instead of eleven rows of ventral shields: 4. The dorsal scales are not smooth and imbricate, but strongly keeled and juxtaposed. There is therefore no doubt that a young specimen of *Centropyx pelviceps* has been substituted in the bottle containing the original *Monoplocus dorsalis*, which has disappeared from the collection.

W. Ecuador.

5. AMEIVA.

Ameiva, *Cuv. Règne Anim.* ii. p. 27; *Fitzing. N. Classif. Rept.* p. 21; *Dum. & Bibr.* v. p. 92; *Gray, Cat.* p. 17; *Cope, Proc. Ac. Philad.* 1862, p. 61; *Bocourt, Miss. Sc. Mex., Rept.* p. 250.
Teius, part., *Merr. Tent. Syst. Amph.* p. 60.
Cnemidophorus, part., *Wagler, Syst. Amph.* p. 154; *Wiegm. Herp. Mex.* p. 9.
Holcosus, *Cope, l. c.* p. 60.

Tongue long, narrow, more or less retractile at the base. Lateral teeth compressed longitudinally, bi- or tricuspid. Head with large regular shields; anterior nasal in contact with its fellow; nostril in the suture between the two nasals or anterior to that suture. Eyelids developed. Ear exposed. Limbs well developed, each with five digits. Dorsal scales small; ventral plates large, subquadrangular, smooth, forming regular series. A double collar-fold. Femoral pores. Tail cyclotetragonal.

Tropical America.

Synopsis of the Species.

I. Six longitudinal rows of ventral plates; head-shields small, numerous.

Head-shields strongly keeled; mesoptychium* with scarcely enlarged scales . 1. *bridgesii*, p. 345.
Head-shields not keeled; a row of enlarged shields on the mesoptychium . . 2. *septemlineata*, p. 346.

II. Eight rows of ventral plates; caudal scales straight, keeled.

Outer row of ventrals considerably smaller than the others . . . . 3. *festiva*, p. 347.
Outer row of ventrals as large as the others; one row of large brachial shields . . . . 4. *undulata*, p. 347.
Several rows of brachials; male with a group of spines on each side of the præanal region . . . . 5. *edracantha*, p. 349.

III. Eight or ten rows of ventral plates; caudal scales oblique, smooth or feebly keeled above, smooth on the sides of the tail.

Upper caudal scales perfectly smooth . . 6. *lineolata*, p. 349.
Upper caudal scales keeled; eight rows of ventrals . . . . 7. *tæniura*, p. 350.
Upper caudal scales keeled; ten rows of ventrals . . . . 8. *polops*, p. 350.

IV. Ten or twelve rows of ventral plates; caudal scales straight, keeled; femoral pores less than 30 on each side.

A. A broad band of enlarged gular scales across the throat.

Second and third supraoculars separated from the frontal by granules . . . . 9. *bifrontata*, p. 351.
Outer toe, laid forwards, not extending as far as inner . . . . 10. *surinamensis*, p. 352.
Outer toe extending as far as inner . . . . 11. *pleii*, p. 354.

B. A small central group of enlarged gulars.

Four supraoculars . . . . 12. *riisii*, p. 354.
Three supraoculars . . . . 13. *vittipunctata*, p. 355.

---

* The space between the two collar-folds.

C. All the gular scales minutely granular.

1. Males without heel spurs.

Four supraoculars; enlarged mesoptychial scales in three series.......... 14. *chrysolæma*, p. 355.
Three supraoculars; enlarged mesoptychial scales in six series .... 15. *thoracica*, p. 356.

2. Males with heel-spurs.

Brachial plates continuous with antebrachials; femoral pores less than 20. 16. *auberi*, p. 356.
Brachial plates not continuous with antebrachials; femoral pores more than 20 17. *dorsalis*, p. 357.

V. Twelve to sixteen rows of ventral plates; femoral pores at least 30.

Ventral plates in 12–14 rows; uniform black above .... 18. *corvina*, p. 358.
Ventral plates in 14–16 rows; eight to ten supraciliaries .... 19. *punctata*, p. 359.

1. Ameiva bridgesii.

Ameiva sexscutata, *Günth. Proc. Zool. Soc.* 1860, p. 234 (*nec Günth.* 1859); *Bocourt, Miss. Sc. Mex., Rept.* pl. XX. A. fig. 11.
Holcosus bridgesii, *Cope, Proc. Ac. Philad.* 1868, p. 306.

Nostril between the two nasals; upper head-shields numerous, strongly keeled, arranged somewhat as in some *Anoles*; three or four longitudinal rows of small shields on the interorbital region: occipitals numerous, very small, irregular; supraoculars two, bordered by granules, except the anterior internally; five supraciliaries, second very long; a second small loreal, or frenoorbital; seven or eight upper labials; five lower labials, third longest; chin-shields, one anterior and four or five pairs; no strongly enlarged gular scales, the granules of that region slightly keeled; mesoptychium with numerous slightly enlarged scales, which are a little larger than the median gulars. Dorsal granules minute, distinctly keeled. Ventral plates in six longitudinal series; twenty-five transverse series, counting from the granules of the collar. Three large, rounded anal plates, anterior largest, arranged in a triangle. Brachial plates small, irregular; a row of large antebrachials; one or two rows of postbrachials. Three rows of large femoral plates, those of the median row smallest; two rows of large tibial plates. Femoral pores twenty-three or twenty-four. Digits strongly serrated. Caudal scales keeled, very strongly above and on the sides. Greenish-olive; a black band along each side, beginning at the end of the snout and passing through the eye, bordered above and below by a more or less marked light streak; sometimes a light vertebral streak.

	millim.		millim.
Total length	440	From end of snout to vent	120
Head	30	Fore limb	48
Width of head	22	Hind limb	102
From end of snout to fore limb	51	Tail	320

Ecuador.

a–b. ♂.	Guayaquil.	Mr. Fraser [C.].

2. Ameiva septemlineata.

Ameiva septemlineata, *A. Dum. Cat. Méth. Rept.* p. 112; *Bocourt, Miss. Sc. Mex., Rept.* pl. xx. A. fig. 4.
—— sexscutata, *Günth. Proc. Zool. Soc.* 1859, p. 402.
Holcosus septemlineatus, *Cope, Proc. Ac. Philad.* 1862, p. 60.
—— sexscutatus, *Cope, l. c.*

Nostril between the two nasals; interorbital region with two or three longitudinal rows of plates; occipital region with numerous small irregular shields; supraoculars two, bordered by granules, except usually the anterior on the inner side; four or five supraciliaries, second (or first) very long; loreal undivided; eight upper labials. Four or five lower labials, third very long; chin-shields, one anterior and five pairs; no strongly enlarged median gular scales; a row of large plates on the mesoptychium. Dorsal granules minute, slightly keeled. Ventral plates in six longitudinal series; twenty-five to twenty-eight transverse series, commencing from the granules of the collar. Three large, rounded anal plates, anterior largest, arranged in a triangle. Brachial and antebrachial plates large, in one row, confluent; a row of large postbrachials. Three rows of large femoral plates, those of the median row smallest; two rows of large tibial plates. Femoral pores eighteen to twenty. Digits rather strongly serrated. Caudal scales keeled, the keels very strong, almost spinose on a series on each side of the anterior part of the tail. Greenish olive; a whitish dorsal streak from the snout to the middle of the tail; two other whitish streaks on each side, separated by a black band.

	millim.		millim.
Total length	267	From end of snout to vent	87
Head	22	Fore limb	35
Width of head	14·5	Hind limb........	65
From end of snout to fore limb	43	Tail	180

Ecuador.

a. ♂.	Andes of W. Ecuador.	Mr. Fraser [C.]. (Type of *Ameiva sexscutata.*)
b–c. ♀ and hgr.	Tanti, 2000 ft.	Edw. Whymper, Esq. [C.].

3. Ameiva festiva.

Cnemidophorus festivus, *Lichtenst. Nomencl. Rept. Mus. Berol.* p. 13.
Ameiva eutropia, *Cope, Proc. Ac. Philad.* 1862, p. 62.
—— edwardsii, *Bocourt, Ann. Sc. Nat.* xvii. 1873, art. 17.
—— festiva, *Bocourt, Miss. Sc. Mex., Rept.* p. 260, pls. xx. A. fig. 10, xx. D. fig. 3, and xx. fig. 2; *Günth. Biol. C.-Amer., Rept.* p. 24.

Nostril between the two nasals; upper head-shields large, normal; three large occipitals; three (or four) supraoculars; five supraciliaries, second very long; loreal undivided; seven or eight upper labials. Six or seven lower labials, third longest; chin-shields, one anterior and four or five pairs; a few very large median gulars; mesoptychium with a row of large shields. Dorsal granules minute, becoming indistinctly keeled towards the posterior part of the body. Ventral plates in eight longitudinal series, those of the outer series considerably smaller than the others; twenty-eight transverse series, counting from the granules of the collar. Three large, rounded anal plates, anterior largest, arranged in a triangle. One row of large brachials and one of antebrachials (the latter double near the articulation of the arm), continuous with each other; one row of rather large postbrachials. Three rows of large femoral plates, those of the median row smallest; two rows of large tibial plates. Eighteen to twenty femoral pores. Digits strongly serrated. Caudal scales keeled, very strongly above and on the sides. Olive-green above, with a more or less distinct festooned light green band along the middle of the back, commencing on the head and extending on the tail, bordered laterally with black; a deep black lateral band, beginning from the tip of the snout, passing through the eye, bordered above and below by a light line; lower surfaces greenish.

	millim.		millim.
Total length (tail injured)	302	From end of snout to vent	112
Head	29	Fore limb	44
Width of head	19	Hind limb	95
From end of snout to fore limb	49		

Central America, Guatemala to Panama.

a–b. ♂, & hgr. Vera Paz (Lower Forests). O. Salvin, Esq. [C.].

4. Ameiva undulata.

Ameiva undulatus, *Gray, Cat.* p. 20.
Cnemidophorus undulatus, *Wiegm. Herp. Mex.* p. 27.
Ameiva undulata, *A. Dum. Cat. Méth. Rept.* p. 113; *Cope, Proc. Ac. Philad.* 1862, p. 62; *Bocourt, Miss. Sc. Mex., Rept.* p. 254, pl. xx. A. fig. 7, and xx. fig. 1; *Günth. Biol. C.-Am., Rept.* p. 22.

Cnemidophorus quadrilineatus, *Hallow. Proc. Ac. Philad.* 1860, p. 483.
Ameiva pulchra, *Hallow. l. c.*; *Cope, l. c.*
—— quadrilineata, *Cope, l. c.* p. 61; *Bocourt, l. c.* p. 259, pl. xx. A. fig. 9; *Günth. l. c.*
—— gabbiana, *Cope, Journ. Ac. Philad.* (2) viii. 1876, p. 117, pl. xxviii. fig. 3; *Günth. l. c.*

Nostril between the two nasals; three large occipitals; three supraoculars, sometimes a small posterior fourth; five or six supraciliaries, second very long; generally a second small loreal inferoposteriorly; seven to nine upper labials. Five or six lower labials, third and fourth longest; chin-shields, one anterior and five or six pairs; generally several much enlarged median gulars, their number, size, and arrangement being very variable; mesoptychium with one or two rows of large shields. Dorsal granules minute, becoming slightly keeled towards the posterior part of the body. Ventral plates in eight longitudinal series, those of the outer series quite as large as those next to it; twenty-nine to thirty-three transverse series, counting from the granules of the collar. Præanal plates very variable, generally small and in two longitudinal rows. One row of large brachials and two of antebrachials, the outer of latter continuous with the former. Four rows of femoral, and two or three of tibial plates. Sixteen to twenty-three femoral pores in the males, and fourteen to seventeen in the females. Digits moderately serrated. Caudal scales keeled. Brown or olive above, more or less spotted with black; flanks black-marbled; males frequently with a lateral series of large light spots; females with two whitish lines along each side of the body, continued on the tail; lower surfaces greenish white.

	♂. millim.	♀. millim.
Total length	332	258
Head	27	20
Width of head	17	13
From end of snout to fore limb	43	32
From end of snout to vent	102	83
Fore limb	37	30
Hind limb	82	62
Tail	230	175

Central America; Ecuador.

a. Several spec.: ♀, hgr., & yg.	Hacienda del Hobo, S. Mexico.	P. Geddes, Esq. [P.].
b. Several spec.: ♂, ♀, & hgr.	Dueñas, Guatemala.	O. Salvin, Esq. [C.].
c. Yg.	Guatemala.	F. D. Godman, Esq. [P.].
d. ♂.	Honduras.	
e–g. ♂ ♀.	Neighbourhood of Irazu, Costa Rica.	
h. ♂.	Guayaquil.	Mr. Fraser [C.].
i, k. ♂.	——?	

5. Ameiva edracantha.

Ameiva edracantha, *Bocourt, Ann. Sc. Nat.* xix. 1874, art. 4, *and Miss. Sc. Mex., Rept.* pl. xx. A. fig. 6; *Günth. Biol. C.-Am., Rept.* p. 25.
Cnemidophorus armatulus, *Cope, Journ. Ac. Philad.* (2) viii. 1876, p. 164.
Verticaria hedracantha, *Cope, Proc. Am. Philos. Soc.* xxii. 1885, p. 172.

Nostril pierced between the two nasals; five occipitals on a transverse line, the central largest, the outer pair smallest; fronto-parietals united or distinct; three supraoculars, sometimes a small posterior fourth; five supraciliaries, second longest; loreal undivided; eight or nine upper labials. Five or six lower labials; chin-shields, one anterior and five or six pairs; several irregular enlarged median gulars; mesoptychium with two or three rows of enlarged scales. Dorsal granules considerably larger than in *A. undulata*, smooth. Ventral plates in eight longitudinal series (with an outer rudimentary series), all nearly of the same size, thirty transverse series, counting from the granules of the collar. Two median larger præanal plates; on each side, a group of six or seven spinose plates in the male. Brachium with several rows of enlarged shields passing gradually into the postbrachials, those of the anterior row largest, continuous with the antebrachials, which are in two rows. Three or four rows of femoral and two of tibial shields. Thirteen femoral pores. Digits feebly serrated. Caudal scales keeled. Olive-brown above, back with transverse black lines; five yellowish longitudinal lines, the median widest, commencing from the snout; sides of body, and limbs, marbled with blackish; lower surfaces yellowish white.

	♂. millim.	♀. millim.
Total length	151	148
Head	13	13
Width of head	8	8·5
From end of snout to fore limb	19	20
From end of snout to vent	50	54
Fore limb	19	19
Hind limb	35	35
Tail	101	94

Ecuador; Northern Peru.

a–b. ♂ ♀.	Guayaquil.	Mr. Fraser [C.].

6. Ameiva lineolata.

Ameiva lineolata, *Gray, Cat.* p. 20.
Ameiva lineolata, *Dum. & Bibr.* v. p. 119; *Bocourt, Miss. Sc. Mex., Rept.* pl. xx. A. fig. 5.

Nostril pierced in the anterior nasal only; five occipitals; four supraoculars; loreal undivided. Median gular and mesoptychial scales scarcely enlarged. Dorsal granules rather large. Ventral plates in eight longitudinal rows. Three enlarged præanal plates, in a triangle. Brachium with several rows of enlarged plates, those of the anterior row largest and continuous with the antebrachials, which are in two rows. Five rows of femoral and two of tibial shields. Fifteen femoral pores. Caudal scales all rhomboidal and smooth, oblique on the upper surface of the tail. Black above, with nine longitudinal pure white lines; limbs with white vermiculations; lower surfaces pure white.

	millim.		millim.
Total length	186	Hind limb........	30
Head	9	Tail	130
Fore limb	20		

San Domingo.

7. Ameiva tæniura.

Ameiva tæniura, *Cope, Proc. Ac. Philad.* 1862, p. 63.

Nostril in the anterior nasal; five occipitals; supraoculars three; supraciliaries five, second longest. Plates of the mesoptychium numerous. One series of brachial plates, two of antebrachial, scarcely continuous; postbrachials small. Four series of large femoral plates, bounded by smaller ones anteriorly and posteriorly; three tibial rows, two upon the inferior face of the limb, the external composed of seven plates, the third very large. Femoral pores fifteen. Abdominal plates in eight longitudinal series. Two transverse marginal and three longitudinal median præanal plates. Lateral caudal plates smooth, superior keeled. Above brown; a narrow yellowish line extends from the supraciliary margin to a nearer or more distant point upon the tail; this is bordered above by a black band, three times its width, which is sometimes faintly margined above with yellowish; sides black as far as a yellowish line which extends from the superior border of the ear to the groin, and thence with increased width for some distance upon the tail; on the latter region it is bounded below by a black band; beneath yellowish white.

Total length 255 millim.; head and body 72 millim.

San Domingo.

8. Ameiva polops.

Ameiva polops, *Cope, Proc. Ac. Philad.* 1862, p. 66.
—— örstedii, *Reinh. & Lütk. Vidensk. Meddel.* 1862, p. 232; *Bocourt, Miss. Sc. Mex., Rept.* pl. XX. B. fig. 4.

Nostril between the two nasals; five occipitals; four supraoculars, posterior small; seven or eight supraciliaries; loreal undivided; seven upper labials. Five lower labials; chin-shields, one anterior and six pairs; median gular scales feebly enlarged; mesoptychium with numerous small shields. Dorsal granules flat, smooth, considerably enlarged on the middle zone of the back. Ventral plates in ten longitudinal rows, those of the external row small; thirty-two transverse rows; several irregular, enlarged præanal plates. A row of large brachials, hardly continuous with a row of large antebrachials; a few large postbrachials. Four series of femoral plates; two series of tibials, inner small. Seventeen or eighteen femoral pores. Scales on the upper surface of the tail oblique, those on the sides smooth, the others feebly keeled. Olive-grey above, the median dorsal zone uniform; on each side three white longitudinal lines, the interspaces between them blackish brown; limbs reticulated with blackish; a white, black-edged line along the hinder side of the femur and tibia and the anterior side of the latter; lower surfaces greenish white.

	millim.		millim.
Total length	172	From end of snout to vent	62
Head	16	Fore limb	22
Width of head	9	Hind limb........	45
From end of snout to fore limb	23	Tail	110

St. Croix, W. Indies.

a. ♀. St. Croix. Copenhagen Museum [P.]. (As typical of *A. örstedii.*)

9. Ameiva bifrontata.

Ameiva bifrontata, *Cope, Proc. Ac. Philad.* 1862, p. 67.
? Cnemidophorus divisus, *Fischer, Verh. Naturw. Ver. Hamb.* (2) iii. 1879, p. 99, pl. v.

Nostril between the two nasals; five occipitals; frontal divided transversely, with slight longitudinal keels; three (or four) supra-oculars, the two (or three) posterior entirely surrounded by granules; six supraciliaries; loreal undivided; seven upper labials; six lower labials; chin-shields, one anterior and seven pairs; gular granules enlarged on a band across the throat; mesoptychial scales about the size of the largest granules. Dorsal granules small, smooth. Ventral plates in ten longitudinal rows, those of the external rows small; transverse rows, thirty-four. Three enlarged præanal plates, forming a triangle. One row of large brachial plates sub-continuous with the antebrachials, which are in two rows; one row of dilated postbrachials. Femoral plates numerous; tibials in three rows. Digits moderately serrated; outer toe not reaching quite so far as inner. Femoral pores fifteen. Caudal scales keeled. Green above; two light streaks on each side of the body, the upper continued on the tail; two alternating series of roundish black spots along the back, and another between each pair of light streaks; hind limbs whitish-spotted; lower surfaces yellowish.

	millim.		millim.
Total length	286	From end of snout to vent	116
Head	28	Fore limb	39
Width of head	17	Hind limb........	77
From end of snout to fore limb	42	Tail (reproduced) ..	170

St Thomas, W. Indies; Colombia; Venezuela.

a. ♂. Venezuela.

10. Ameiva surinamensis.

Ameiva surinamensis, *Gray, Cat.* p. 18.
—— guttata, *Gray, l. c.*
Lacerta ameiva, *Linn. S. N.* i. p. 362.
Seps surinamensis, *Laur. Syn. Rept.* p. 59.
—— zeylanicus, *Laur. l. c.*
Lacerta litterata, *Daud. Rept.* iii. p. 106.
—— graphica, *Daud. l. c.* p. 112.
—— gutturosa, *Daud. l. c.* p. 119.
Tejus ameiva, *Merr. Tent.* p. 61; *Wied. Abbild. Nat. Bras.*; *Spix, Spec. Nov. Lacert. Bras.* p. 21, pl. xxiii.
—— lateristriga, *Spix, l. c.* p. 22, pl. xxiv. fig. 1.
—— tritæniatus, *Spix, l. c.* p. 22, pl. xxiv. fig. 2.
Ameiva vulgaris, *Lichtenst. Verz. Doubl. Mus. Berl.* p. 91; *Dum. & Bibr.* v. p. 100; *Guichen. in Casteln. Amér. Sud, Rept.* p. 32; *Reinh. & Lütk. Vidensk. Meddel.* 1862, p. 230.
—— maculata, *Gray, Ann. N. H.* i. 1838, p. 277.
Cnemidophorus præsignis, *Baird & Gir. Proc. Ac. Philad.* 1852, p. 129; *Girard, in Gillis, U.S. Nav. Astron. Exped.* ii. p. 227, pl. xxxviii.
Ameiva læta, *Cope, Proc. Ac. Philad.* 1862, p. 65.
—— præsignis, *Cope, l. c.* p. 67; *Bocourt, Miss. Sc. Mex., Rept.* p. 265, pl. xx. B. fig. 9.
—— surinamensis, *Cope, l. c.* p. 67; *Bocourt, l. c.* fig. 10; *Cope, Proc. Amer. Philos. Soc.* xviii. 1879, p. 276; *Günth. Biol. C.-Am., Rept.* p. 22, pl. xix.
—— petersii, *Cope, Proc. Ac. Philad.* 1868, p. 99.
—— pleurotænia, *Peters, Mon. Berl. Ac.* 1871, pp. 398, 652.
Cnemidophorus maculatus, *Fischer, Verh. Naturw. Ver. Hamb.* (2) iii. 1879, p. 95, pl. iv.

Nostril between the two nasals; five occipitals; four supraoculars; six or seven supraciliaries; loreal undivided; six to eight upper labials. Five or six lower labials; chin-shields, one anterior and four to six pairs; gular granules enlarged on a band across the throat; mesoptychium with slightly enlarged scales, about the size of the largest gulars. Dorsal granules small, smooth. Ventral plates in ten or twelve longitudinal rows, the outer small; thirty to thirty-three transverse rows. Præanal plates irregular, in two rows or forming a triangle. Two or three rows of brachial plates, those of the anterior row largest, and generally continuous with the antebrachials; latter composed of one row of very large plates, and one or two of much smaller ones; postbrachials small. Six to eight rows of femoral plates and three or four of tibials. Femoral pores fifteen to twenty-three. Digits feebly serrated; fifth toe short, not extending as far as inner. Caudal scales feebly keeled, especially inferiorly. The coloration varies considerably. The young are olive-brown with darker markings or white dots, and have constantly a black, white-edged band on each side, commencing from the eye and extending on the tail. With age, these bands disappear generally, but they are retained by some female specimens. Usually, the upper surfaces are greenish, more or less spotted or reticulated with black, especially on the sides, where there are round white spots which, in males, may be confluent into vertical bars. In a few specimens, there are five more or less distinct longitudinal light dark-edged lines on the body. Lower surfaces greenish-white, black-spotted on the sides.

	♂. millim.	♀. millim.
Total length	525	380
Head	49	32
Width of head	29	20
From end of snout to fore limb	71	49
From end of snout to vent	165	129
Fore limb	62	44
Hind limb	126	84
Tail	360	251

South America to Nicaragua.

a, b. ♂ ♀.	Costa Rica.	Mr. Geale [C.].
c. ♂.	Sabanna-larga, Colombia.	Dr. J. G. Fischer [E.].
	(One of the types of *Cnemidophorus maculatus.*)	
d. ♂.	British Guiana.	
e. ♂.	Better Hope, British Guiana.	Rev. W. Turner [C.].
f–i. Hgr. & yg.	Demerara Falls.	
k. ♂.	Demerara.	D. Hancock, Esq. [P.].
l. ♂.	Demerara.	(Type of *A. guttata.*)
m. ♂.	Surinam.	
n. ♀.	Brazil.	(Type of *A. guttata.*)
o–p, q. ♂ & yg.	Para.	R. Graham, Esq. [P.].
r. Yg.	Para.	
s. Yg.	Caripe, Para.	J. R. Smith, Esq. [P.].
t–v. ♂ ♀.	Pernambuco.	Mrs. J. P. G. Smith [P.]. (Types of *A. guttata.*)
v–x. ♂.	Bahia.	
y–β. ♀ & hgr.	Pebas.	H. W. Bates, Esq. [C.].
γ. ♀.	Huallaga River.	
δ. ♂.	Yurimaguas.	
ε–ζ, η–λ. ♂ ♀.	Sarayacu, Peru.	Messrs. Veitch [P.].
μ–ν. Yg.	Puerto del Mairo.	Messrs. Veitch [P.].
ξ–ο. ♀.	Moyobamba.	Mr. A. H. Roff [C.].
π. ♂.	Montevideo.	C. Darwin, Esq. [C.].
ρ, σ. ♂ ♀.	——?	(Types of *A. guttata.*)
τ. ♀.	——?	L. L. Dilwyn, Esq. [P.].

11. Ameiva pleii.

Ameiva scutata, *Gray, Cat.* p. 19.
—— plei, *Gray, l. c.*
Ameiva plei, *Dum. & Bibr.* v. p. 114; *Reinh. & Lütk. Vidensk. Meddel.* 1862, p. 239; *Bocourt, Miss. Sc. Mex., Rept.* pl. xx. B. fig. 6.
? Ameiva analifera, *Cope, Proc. Amer. Phil. Soc.* xi. 1869, p. 158.

Closely allied to *A. surinamensis*, from which it may be distinguished by the following characters:—Nostril pierced in the posterior border of the anterior nasal; fifth toe relatively longer, reaching quite as far as the inner; mesoptychial shields a little larger; one row of brachials and two of antebrachials, widely separated by several rows of scales; three rows of tibials, outer very large, inner very small; a regular transverse row of eight or ten posterior præanal plates, decreasing gradually in size from the centre; femoral pores very numerous, twenty-eight (twenty-four according to Duméril and Bibron). The colour is described as rubiginous olive, with light spots on the loins, limbs, and tail.

From snout to vent 107 millim.

Martinique.

a. ♀, bad state. —— ? (Type of *A. scutata.*)

12. Ameiva riisii.

Ameiva riisei, *Reinh. & Lütk. Vidensk. Meddel.* 1862, p. 232; *Bocourt, Miss. Sc. Mex., Rept.* pl. xx. B. fig. 3.
—— plei (*non D. & B.*), *Cope, Proc. Ac. Philad.* 1862, p. 65.

Nostril in the posterior border of the anterior nasal; five occipitals; four supraoculars; seven or eight supraciliaries; loreal undivided; six or seven upper labials. Five or six lower labials; chin-shields, one anterior and four or five pairs; a few slightly enlarged gular scales forming a small group on the middle of the throat; mesoptychium with several rows of enlarged scales, which are larger than the median gulars. Dorsal granules small, smooth. Ventral plates in ten or twelve longitudinal rows; thirty-four to thirty-six transverse rows. Three enlarged median præanal plates, forming a triangle. A single row of humeral plates, more or less distinctly continuous with the antebrachials, which are in two rows; postbrachials small. Six or seven rows of femoral shields; two or three rows of tibial shields, the outer extremely large, the inner, if present, very small and irregular. Femoral pores thirteen to seventeen. Digits feebly serrated, fifth extending quite as far as inner. Caudal scales keeled. Olive-brown above; a series of black spots on each side of the body; flanks, hind limbs, tail, and sometimes hinder part of back, with numerous round white spots; lower surfaces greenish-white.

	millim.		millim.
Total length	420	From end of snout to vent............	150
Head	38	Fore limb	55
Width of head......	25	Hind limb	102
From end of snout to fore limb	59	Tail	270

St. Thomas, West Indies.

a-b. ♂.	St. Thomas.	Hr. Riise [C.]
c. ♂.	St. Thomas.	R. Swift, Esq. [P.].
d. Hgr.	St. Thomas.	Copenhagen Museum [P.].
e. Hgr.	St. Thomas.	H.M.S. 'Challenger.'
f. ♂.	—— ?	

13. Ameiva vittipunctata.

Ameiva vittipunctata, *Cope, Proc. Ac. Philad.* 1871, p. 220.
? Cnemidophorus affinis, *Fischer, Oster Progr. Ak. Gymn. Hamb.* 1883, p. 1, pl. —. figs. 1, 2.

Three supraoculars; seven supraciliaries; seven chin-shields, separated by granules and scales from the labials throughout; no plates inside the chin-shields behind; a few gulars a little enlarged in the middle of the area; edge of gular fold widely granular. Twelve series of ventral plates. Præanals, a marginal series with small central, with larger ones bounding it, and three short rows of diamond-shaped scuta in front of them. Brachials rather small, one row larger; postbrachials similar, two rows enlarged; antebrachials two narrow, one transverse row, well separated from brachials. Femoral pores eighteen. Olive above, with three pale lines on each side of the median line; a light lateral band from axilla to groin, separated from outer dorsal line by a broad black band, which is marked by a median series of light dots; sides and limbs white-spotted; tail brown-spotted, black below; belly and throat uniform light olive.

	millim.		millim.
Total length......	273	From end of snout to ear	21
From end of snout to vent	88	From end of snout to fore limb	30

San Domingo.

14. Ameiva chrysolæma.

Ameiva plei, part., *Dum. & Bibr.* v. p. 114.
—— chrysolæma, *Cope, Proc. Ac. Philad.* 1868, p. 127; *Bocourt, Miss. Sc. Mex., Rept.* pl. xx. B. fig. 5.

Nostril in the posterior border of the anterior nasal; five occipitals; four supraoculars; loreal undivided; six upper labials; seven lower labials; median and lateral gular granules equal, not enlarged; mesoptychium with three larger series of small hexagonal scales. Twelve rows of ventral plates. Brachials rather small, in seven rows above, and two below, not separated from antebrachials by granules, but joining an area of small scales in seven rows above, and ending in one row of broad and one of narrow antebrachials below; postbrachials small, three rows larger; tibial plates in four rows, two on under face, the outer of seven plates, of which the third and fourth are very large. Outer toe just exceeding inner. Femoral pores twenty. Olive with numerous series of white spots, sometimes indistinct on dorsal region; gular fold black, throat yellow; belly green and yellow.

	millim.		millim.
Total length	477	From end of snout to edge of collar	48
From end of snout to vent............	138	Fore limb	54
		Hind limb	95

San Domingo.

15. Ameiva thoracica.

Ameiva thoracica, *Cope, Proc. Ac. Philad.* 1862, p. 64.

Nostril in the anterior nasal; three supraoculars; five supraciliaries, second longest; median gular scales minute, those of the mesoptychium larger, in six rows. Ten longitudinal rows of ventral plates. Præanals in a series of three or four large marginal, one or two large median, longitudinally arranged. Three (two small, one large) antebrachial series, scarcely continuous with brachial row; postbrachials small, in three short rows; femoral plates numerous; external tibial plates seven, very wide, second, third, and fourth largest, third broad, fourth narrow, transverse; a slight tendency to acumination in the heel-scales of some specimens. External toe extending beyond extremity of internal. Femoral pores eighteen. Two indistinct light bands—the superior from the supercilium, the interior from the superior margin of the auricular opening—enclose a black band, which is continued some distance upon the base of the tail, with its inferior light border; beneath greenish or yellowish white, the pectoral and gular regions more or less black. The female differs in having the brown of the back lighter and marked with a narrow median line. The black upon the throat is also sometimes wanting.

	♂. millim.	♀. millim.
Total length	380	233
From end of snout to vent	115	77

New Providence Island, Bahamas.

16. Ameiva auberi.

Ameiva trilineata, ***Gray, Cat.*** **p. 19.**
—— auberi, ***Gray, l. c.*** **p. 20.**
Ameiva auberi, ***Cocteau, in R. de la Sagra, Hist. Cuba, Rept.*** **p. 74,**

pl. vi.; *Dum. & Bibr.* v. p. 111; *Cope, Proc. Ac. Philad.* 1862, p. 68; *Bocourt, Miss. Sc. Mex., Rept.* pl. xx. B. fig. 2.
Ameiva trilineata, *Gray, Ann. N. H.* v. 1840, p. 114; *Cope, l. c.* p. 69.

Nostril in the posterior part of the anterior nasal; five occipitals; three or four supraoculars; five or six supraciliaries; loreal undivided; seven upper labials; six or seven lower labials; chin-shields, one anterior and six or seven pairs, separated from the labials by several rows of granules; median gular scales not enlarged; mesoptychium with three or four rows of feebly enlarged scales. Dorsal granules rather large, smooth. Ventral plates in ten or twelve longitudinal rows; thirty-four to thirty-six transverse rows, counting from the granules of the collar. Præanal plates, two or three anterior median and three or four posterior on a line. One row of brachials and two of antebrachials, continuous; postbrachials large. Six or seven rows of femoral plates; three rows of tibials, the inner small and irregular, the outer extremely large. Heel with a few more or less prominent spinose scales. Femoral pores eleven to sixteen. Digits moderately serrated; fifth toe extending further than inner. Caudal scales keeled. Olive above, with three or five more or less distinct light longitudinal streaks, and a lateral band of blackish marblings or reticulations; lower surfaces yellowish, marbled with olive.

	♂. millim.	♀. millim.
Total length	312	190
Head	20	15
Width of head	13	9
From end of snout to fore limb	33	21
From end of snout to vent	87	60
Fore limb	32	22
Hind limb	68	48
Tail	225	130

Cuba.

a–b, c. ♂ ♀.	Cuba.	W. S. MacLeay, Esq. [P.]. (Types of *A. trilineata.*)
d. ♀.	—— ?	

17. Ameiva dorsalis.

Ameiva dorsalis, *Gray, Cat.* p. 19.
Ameiva dorsalis, *Gray, Ann. N. H.* i. 1838, p. 277; *Gosse, Proc. Zool. Soc.* 1848, p. 24; *Cope, Proc. Ac. Philad.* 1862, p. 69; *Bocourt, Miss. Sc. Mex., Rept.* pl. xx. B. fig. 1.
—— sloanei, *Dum. & Bibr.* v. p. 107.

Nostril in the posterior part of the anterior nasal; five or seven occipitals; three or four supraoculars; six or seven supraciliaries; loreal frequently divided; seven or eight upper labials; seven to nine lower labials; chin shields, one anterior and six to eight pairs, separated from the labials by several rows of granules; median gular scales not enlarged; mesoptychium with three or four rows of feebly enlarged scales. Dorsal granules small, smooth. Ventral plates in ten or twelve longitudinal rows; thirty-six to thirty-eight transverse rows, counting from the granules of the collar. Large præanal plates, generally three, arranged in a triangle. Brachials generally small and in several rows, widely separated from the ante-brachials, which are in two or three rows; postbrachials large. Six or seven rows of femoral plates, gradually passing into granules; two or three rows of tibials, the inner, if present, small and irregular, the outer extremely large. Heel with more or less prominent spinose scales. Femoral pores twenty-one to twenty-seven. Digits moderately serrated; fifth toe extending further than inner. Caudal scales keeled. Olive-grey above, blackish on the back; a median light green vitta begins from the occiput where it is linear, and gradually widens to the tail; sides with light spots, sometimes confluent into two longitudinal lines on each side of the median band; lower surfaces greenish white; the sides of the belly, in males, olive, with round lighter spots.

	♂. millim.	♀. millim.
Total length	342	260
Head	28	23
Width of head	17	12·5
From end of snout to fore limb	45	33
From end of snout to vent	112	90
Fore limb	41	35
Hind limb	83	73
Tail	230	170

Jamaica.

a. ♂.	Jamaica.	R. Heward, Esq. [P.].
b. Hgr.	Jamaica.	G. E. Dobson, Esq. [E.].
c–g. ♂, ♀, & yg.	Bluefields, Jamaica.	P. H. Gosse, Esq. [C.].
h, i, k, l. ♂, ♀, & hgr.	—— ?	
m. Skeleton.	Jamaica.	Sir A. Smith [P.].

18. Ameiva corvina.

Ameiva corvina, *Cope, Proc. Ac. Philad.* 1861, p. 312, and 1862, p. 68; *Reinh. & Lütk. Vidensk. Meddel.* 1862, p. 242.

Nostril in the posterior part of the anterior nasal; five occipitals; four supraoculars; six or seven supraciliaries; loreal undivided; six or seven upper labials; five or six lower labials; chin-shields, one anterior and five pairs; median gulars slightly enlarged, forming a wide band across the throat; enlarged scales of the mesoptychium about the size of the larger gulars. Dorsal granules small, smooth. Ventral plates in twelve or fourteen longitudinal rows, those of the outer row very small; thirty-four to thirty-six transverse rows, counting from the collar. Præanal plates irregular, with three larger median ones. No regular brachial plates; one row of large transverse antebrachials and one or two rows of small ones; postbrachials very small. Femoral plates numerous; three rows of tibial shields, outer very large. Femoral pores thirty to thirty-three (or thirty-six, *Cope*). Caudal scales keeled. General colour black; lower surfaces glaucous green.

	millim.		millim.
Total length	320	From end of snout to vent	110
Head	25	Fore limb	40
Width of head	15	Hind limb	70
From end of snout to fore limb	41	Tail	210

Sombrero and Anguilla, West Indies.

a. ♀.	Sombrero.	F. G. Sawkins, Esq. [P.].
b–c. ♀.	Anguilla.	W. J. Cooper, Esq. [P.].

19. **Ameiva punctata.**

Ameiva major, *Gray, Cat.* p. 17.
—— punctata, *Gray, l. c.*
Ameiva punctata, *Gray, Ann. N. H.* i. 1838, p. 277.
—— undulata, *Gray, l. c.*
—— major, *Dum. & Bibr.* v. p. 117, pl. lii.; *Bocourt, Miss. Sc. Mex., Rept.* p. 246, pl. xx. B. figs. 7 & 8.
—— erythrops, *Cope, Proc. Ac. Philad.* 1871, p. 221.

Nostril in the posterior part of the anterior nasal; five occipitals; four supraoculars; eight to ten supraciliaries; loreal undivided; seven upper labials; five or six lower labials; chin-shields, one anterior and six pairs; median gulars slightly enlarged; enlarged scales on the mesoptychium a little larger than the largest gulars. Dorsal granules small, smooth. Ventral plates in fourteen or sixteen longitudinal series, those of the outer row very small; thirty-five or thirty-six transverse rows, counting from the collar. Præanal plates small. No regular brachial plates; one row of large transverse antebrachials and two rows of small ones; postbrachials very small. Femoral plates numerous; three or four rows of tibial shields, outer very large. Digits feebly serrated, outer toe extending a little further than inner. Femoral pores thirty-four to thirty-seven. Caudal scales keeled. Back olive, with black reticulations or wavy cross lines; flanks and belly blackish green; head lighter, reddish above, whitish beneath. (Young with longitudinal lighter bands on the sides of the body.)

	♂. millim.	♀. millim.
Total length	418	350
Head	33	27
Width of head	21	16
From end of snout to fore limb	56	45
From end of snout to vent	138	120
Fore limb	52	43
Hind limb	96	80
Tail	280	230

Guianas; Trinidad; St. Eustatius.

a. ♂.	Demerara.	Col. Sabine [P.]. (Type.)
b. ♀.	—— ?	(Type of *A. undulata*.)

6. CNEMIDOPHORUS *.

Ameiva, part., *Fitzing. N. Class. Rept.* p. 21.
Cnemidophorus, part., *Wagler, Syst. Amph.* p. 154; *Wiegm. Herp. Mex.* p. 9.
Cnemidophorus, *Dum. & Bibr.* v. p. 123; *Gray, Cat.* p. 20; *Bocourt, Miss. Sc. Mex., Rept.* p. 268.
Verticaria, *Cope, Proc. Amer. Philos. Soc.* xi. 1869, p. 158.

Tongue long, narrow, bilobate and not retractile at the base. Lateral teeth compressed longitudinally, bi- or tricuspid. Head with large regular shields; anterior nasal in contact with its fellows; nostril in the suture between the two nasals, or anterior to that suture. Eyelids developed. Ear exposed. Limbs well developed, pentadactyle. Dorsal scales small; ventral plates large, subquadrangular, smooth, forming regular series. A double collar-fold. Femoral pores. Tail cyclo-tetragonal.

America.

Synopsis of the Species.

I. Dorsal scales equal.

A. Nostril between the two nasals; males with a spine on each side of the præanal region.

1. 10 or 12 longitudinal rows of ventral plates.

Brachial shields small .... 1. *murinus*, p. 361.

2. 8 longitudinal rows of ventral plates.

An anterior row of large brachials; antebrachials, outer row largest and continuous to the hand .... 2. *espeuti*, p. 362.

An anterior row of large brachials; antebrachials, inner row largest and continuous to the hand .... 3. *lemniscatus*, p. 363.

Brachials small, in two rows; mesoptychial scales hardly larger than those of the chin .... 4. *nigricolor*, p. 364.

B. Nostril anterior to nasal suture.

1. 8 longitudinal rows of ventral plates.

a. More than 12 femoral pores on each side.

---

* *Cnemidophorus perplexus*, Baird & Gir. Proc. Ac. Philad. 1852, p. 128.—South-western United States.

C. inornatus, Hallow. eod. loc. 1858, p. 255.—N. Leon.

C. decemlineatus, Hallow. eod. loc. 1860, p. 482.—Nicaragua.

a. Frontoparietals distinct.

* Supraoculars normally four.

Three large præanals; caudal scales strongly oblique, the keels distinctly subdiagonal, their posterior border forming an angle .... 5. *sexlineatus*, p. 364.
Three large præanals; caudal scales scarcely oblique .... 6. *mariarum*, p. 368.
Præanals in 4 transverse series, the posterior two containing 6 .... 7. *maximus*, p. 369.

** Supraoculars normally three.

Posterior row of mesoptychial scales separated from the edge of the collar by 2 or 3 rows of granules; femoral plates in 8 to 12 rows, tibials in 3 .... 8. *guttatus*, p. 370.
No granules between the mesoptychial scales and the border of the collar; femoral plates in 10 rows .... 9. *immutabilis*, p. 370.
Posterior row of mesoptychial scales separated from the edge of the collar by a row of granules; femoral plates in 6 or 7 rows, tibials in 2 .... 10. *deppii*, p. 371.

β. Frontoparietals united .. 11. *hyperythrus*, p. 371.

b. Not more than 12 femoral pores on each side.

Inner edge of supraoculars bordered by a row of granules .... 12. *ocellifer*, p. 372.
No granules between the frontal and the supraoculars .... 13. *multilineatus*, p. 373.

2. 10 or 12 longitudinal rows of ventral plates.

No granules between the frontal and the supraoculars; dorsal scales granular .. 14. *lacertoides*, p. 373.
Inner edge of supracular bordered by a row of granules; dorsal scales flat, hexagonal .... 15. *longicauda*, p. 374.

II. Scales on the hind part of the back larger, rhomboidal, keeled, imbricate .... 16. *heterolepis*, p. 375.

1. Cnemidophorus murinus.

Cnemidophorus murinus, *Gray, Cat.* p. 21.
Seps murinus, *Laur. Syn. Rept.* p. 63.
Lacerta ameiva, part., *Daud. Rept.* iii. p. 98.
Ameiva murina, *Fitzing. N. Class.* p. 51.
Cnemidophorus murinus, *Dum. & Bibr.* v. p. 126; *Bocourt, Miss. Sc. Mex., Rept.* pl. xx. c. fig. 1.

Head moderate. Nostril between the two nasals. Parietals five, narrow; four supraoculars; five or six supraciliaries; no frenoorbital: median gular scales gradually enlarged; mesoptychium with three or four rows of enlarged scales, which are a little larger than the largest gulars, separated from the free edge of the collar by four or five rows of small granules. Dorsal scales minutely granular, smooth. Ventral plates in ten or twelve longitudinal and thirty-eight to forty transverse series *. Three large præanals, forming a triangle. Brachial and postbrachial plates small, without strongly enlarged series; two or three rows of antebrachials; forearm entirely granular inferiorly. Femoral plates in ten to twelve rows, one of which is large; tibials in three or four rows, outer very large. Femoral pores twenty-nine to thirty-five. Male with a spine on each side of the vent. Caudal scales oblique, with straight keels, not distinctly pointed posteriorly. Olive above, sides with large round bluish-white spots; hinder side of thigh with a whitish longitudinal band; lower surface of head and body brownish.

	millim.		millim.
Total length	435	From end of snout to fore limb	52
Head	32	From end of snout to vent	145
Width of head	21	Fore limb	54
From end of snout to interparietal plate	22	Hind limb	98
Width at posterior corners of supra-ocular region....	14	Tail	290

Guianas; Trinidad.

a, *b*. ♂. ——?

2. **Cnemidophorus espeuti.** (Plate XIX.)

Head moderate. Nostril between the two nasals. Parietals five; four supraoculars; five supraciliaries; no frenoorbital; anterior gular scales subequal, without enlarged medians; mesoptychial scales gradually increasing in size, those of the last row large, separated from the free edge of the collar by one or two rows of granules. Dorsal scales small, granular, smooth. Ventral plates in eight longitudinal and thirty-two transverse series. Three large præanal scales forming a triangle. An anterior row of large brachials, followed by several smaller ones, passing into the postbrachials; antebrachials large, continuous with the brachials, in two rows, the outer largest and continued to the hand; forearm entirely granular inferiorly. Femorals in six or seven rows, one of which is large; tibials in two or three rows, the outer very large and composed of six plates, the other small and irregular. Femoral pores nineteen. Caudal scales oblique, with strong straight keels,

* Counting from the collar.

slightly pointed posteriorly. Olive above, with a broad lighter dark-edged median band from the occiput to the base of the tail; a few rather indistinct round light spots on the flanks; temples and sides of neck lapis-blue; lower surfaces pale green, bluish on the sides and on the throat.

	millim.		millim.
Total length	320	From end of snout to fore limb	35
Head	24	From end of snout to vent	100
Width of head	15	Fore limb	38
From end of snout to interparietal plate	16·5	Hind limb	68
Width at posterior corners of supra-ocular region....	10	Tail	220

Old Providence Island, Colombia.

a. ♀. Old Providence Island. W. B. Espeut, Esq. [P.].

3. Cnemidophorus lemniscatus.

Cnemidophorus lemniscatus, *Gray, Cat.* p. 21.
—— scutata, *Gray, l. c.*
Lacerta lemniscata, *Daud. Rept.* iii. p. 175, pl. xxxvi. fig. 1.
—— cœruleocephala, *Wolf, Abbild. u. Beschr. merkw. nat. Gegenst.* ii. p. 43, pl. xii.
? Teius cyanomelas, *Wied, Abbild.*
Ameiva lineata, *Gray, Ann. N. H.* i. 1838, p. 278.
Cnemidophorus lemniscatus, *Dum. & Bibr.* v. p. 128; *Bocourt, Miss. Sc. Mex., Rept.* pl. xx. c. fig. 2.

Head short. Nostril between the two nasals. Parietals five; supraoculars four; five or six supraciliaries; no frenoorbital; anterior gular scales subequal, without enlarged medians; mesoptychium with three or four rows of enlarged scales, separated from the free edge of the collar by one or two rows of granules. Dorsal scales small, granular, smooth. Ventral plates in eight longitudinal and thirty to thirty-three transverse series. Generally three large præanal scales forming a triangle. An anterior row of large brachials, followed by several smaller ones, continuous with the postbrachials; antebrachials more or less continuous with the brachials, in two rows, the inner largest, the outer frequently extending only half down the inner; forearm entirely granular inferiorly. Femorals in six to eight rows, one of which is large; tibials in three rows, outer very large. Femoral pores eighteen to twenty-four. Male with a spine on each side of the vent. Caudal scales oblique, with straight keels, shortly pointed posteriorly. Olive above, with eight or nine whitish longitudinal lines on the body in the females, the space between the two upper on each side black or blackish; limbs with round whitish spots, and sometimes a band of the same colour along the hinder side of the thighs; in males the four or five median lines alone persist, and the flanks are spotted with whitish; lower surfaces greenish white.

	♂. millim.	♀. millim.
Total length	298	216
Head	23	17
Width of head	16	10
From end of snout to interparietal plate	16	12
Width at posterior corners of supra-ocular regions	10	7·5
From end of snout to fore limb	31	25
From end of snout to vent	88	71
Fore limb	32	24
Hind limb	64	46
Tail	210	145

Tropical America, east of the Andes.

a. ♂.	Colombia.	
b. Hgr.	Caracas.	
c–d. ♂ & yg.	Venezuela.	
e. ♂.	Trinidad.	C. Taylor, Esq. [P.].
f. Yg.	Berbice.	
g. ♀.	Para.	
h. Several specs., ♂, ♀, & yg.	Barakela.	Mr. Rippon [C.].
i. ♀.	Santarem.	Mr. Wickham [C.].
k–m, ♀ & yg.	S. America.	
n–o. ♂.	S. America.	(Types of *C. scutata*.)

4. Cnemidophorus nigricolor.

Cnemidophorus nigricolor, *Peters, Sitzb. Ges. Nat. Fr.* 1873, p. 76; *Bocourt, Miss. Sc. Mex., Rept.* pl. xx. c. fig. 3.

Distinguished from *C. lemniscatus* in the following characters:—Mesoptychial scales hardly larger than those of the chin; brachial plates very small, in two rows. Uniform black or blackish brown, with lighter brownish lines on the nape, and spo ts on the outer sie of the forearm, thigh, and leg, and marblings on the chin.

Los Roques, group of small islands north of La Guayra, Venezuela.

5. Cnemidophorus sexlineatus.

Cnemidophorus sexlineatus, *Gray, Cat.* p. 21.

Lacerta sexlineata, *Linn. S. N.* i. p. 364; *Daud. Rept.* iii. p. 183; *Harl. Journ. Ac. Philad.* vi. 1829, p. 18.

Ameiva sexlineata, *Holbr. N. Amer. Herp.* p. 63, pl. vi. & 2nd ed. ii. p. 109, pl. xv.

Cnemidophorus sexlineatus, *Dum. & Bibr.* v. p. 131; *Cope, Proc. Ac. Philad.* 1866, p. 303; *Bocourt, Miss. Sc. Mex., Rept.* p. 273, pl. xx. c. fig. 11; *Günth. Biol. C.-Am., Rept.* p. 25.

—— gularis, *Baird & Gir. Proc. Ac. Philad.* 1852, p. 128, *and U.S. Mex. Bound. Surv., Rept.* pl. xxxiv. figs. 1–6.

Cnemidophorus guttatus (*non Wiegm.*), *Hallow. Proc. Ac. Philad.* 1854, p. 192.
—— octolineatus, *Hallow. eod. loc.* 1858, p. 255.
—— sexlineatus, *var.* gularis, part., *Bocourt, l. c.* p. 278.

Head rather short. Nostril anterior to nasal suture. Three or five parietals; four (exceptionally three) supraoculars, anterior sometimes divided into two; five or six supraciliaries; seldom a freno-orbital; median gular scales gradually enlarged; mesoptychium with three or four rows of enlarged scales, the posterior of which are larger than the largest gulars, and generally form the edge of the collar. Dorsal scales rather large, granular, smooth. Ventral plates in eight longitudinal and thirty-one to thirty-five transverse series. Generally three large præanals, forming a triangle. Four rows of brachials, anterior largest, continuous with the postbrachials; two or three rows of antebrachials, outer largest; the scales on the outer side of the inferior surface of the forearm more or less enlarged, sometimes forming a band of much dilated transverse plates. Six to eight rows of femorals, one of which is large; tibials in two or three rows, outer largest. Femoral pores fourteen to nineteen. Males without anal spines. Caudal scales oblique, with straight keels, obtusely pointed posteriorly. Olive or brownish grey, with three or four white (yellow) lines on each side separated by blackish interspaces; sometimes series of round light spots on the latter; limbs with dark reticulations; a more or less distinct white band along the hinder side of the thighs; lower surfaces whitish; belly blue in breeding males.

	♂. millim.	♀. millim.
Total length	248	232
Head	19	17
Width of head	12	10
From end of snout to interparietal plate	14	12
Width at posterior corners of supra-ocular regions	7·5	7
From end of snout to fore limb	30	25
From end of snout to vent	78	72
Fore limb	27	25
Hind limb	53	51
Tail	170	160

Southern North America.

a–b. Hgr.	N. America.	F. Doubleday, Esq. [P.].
c, d, e–g. ♂, ♀, & hgr.	N. America.	
h, i. ♂ ♀.	Texas.	
k. ♀.	Kansas.	
l. ♀.	California.	
m–n. ♂ & hgr.	Mexico.	

Var. **mexicanus.**

Cnemidophorus mexicanus, *Peters, Mon. Berl. Ac.* 1869, p. 62.
? Cnemidophorus communis, *Cope, Proc. Amer. Philos. Soc.* xvii. 1877, p. 95.
Cnemidophorus sexlineatus, *var.* mexicanus, *Bocourt, Miss. Sc. Mex., Rept.* p. 281, pl. xx. c. fig. 7.

Size larger. Head rather more elongate. Brown above, with three white lines on each side, the interspaces between them blackish and, in males, white spotted; in the latter the longitudinal lines more or less broken up into spots; lower surfaces white, spotted with black in the males.

	♂. millim.	♀. millim.
Total length	303	254
Head	22	17
Width of head	13·5	10·5
From end of snout to interparietal plate	17	13
Width at posterior corners of supra-ocular region	9	7
From end of snout to fore limb	33	25
From end of snout to vent	93	74
Fore limb	28	26
Hind limb	63	55
Tail	210	180

Mexico.

a–f. ♂, ♀, & hgr. Presidio, W. Mexico. Hr. A. Forrer [C.].

Var. **angusticeps.**

Cnemidophorus angusticeps, *Cope, Proc. Amer. Philos. Soc.* xvii. 1877, p. 95.

Differs from *C. sexlineatus* in the very narrow parietal plates. The light bands are much wider than the interspaces, and send off lateral processes, which give the dark ground-colour a very broken character; the colour of the bands is an olive-green. The adult male of about the size of *C. guttatus*.

Yucatan.

Var. **costatus.**

Cnemidophorus costatus, *Cope, Proc. Amer. Philos. Soc.* xvii. 1877, p. 95.

Resembles the last two in general characters, but differs in the shorter head, and strikingly in the coloration. Brown, with black cross bands on sides, which join across the middle line on the lumbar region; sacral region and femora white spotted.

Mexico.

Var. **bocourtii.**

Cnemidophorus sexlineatus, var. gularis, part., *Bocourt, Miss. Sc. Mex., Rept.* p. 278.

Head short. Three parietals; four supraoculars; six supraciliaries; frenoorbital present or absent; a band of enlarged gular scales across the throat; mesoptychium with three rows of enlarged scales, the posterior composed of nine to twelve, forming the border of the collar. Dorsal granules larger than in *C. sexlineatus*, smooth. Ventrals in eight longitudinal and thirty-five transverse series. Three large præanals, forming a triangle. Four rows of brachials, anterior largest, continuous with the postbrachials; two or three rows of antebrachials, outer largest; enlarged scales under the forearm. Six or seven rows of femorals, one of which is large: tibials in three rows, outer largest. Femoral pores sixteen. Caudal scales oblique, with straight keels, obtusely pointed posteriorly. Olive above, with round yellow spots on the back, hind limbs, and base of tail; sometimes traces of six darker longitudinal lines on the back, not interrupting the yellow spots; throat yellowish; belly whitish, in both sexes spotted with blackish blue, which markings may nearly entirely cover the ground-colour.

	♂. millim.	♀. millim.
Total length	310	207
Head	24	21
Width of head	17	13
From end of snout to interparietal plate	16	15
Width at posterior corners of supraorbital regions	9·5	9
From end of snout to fore limb	36	32
From end of snout to vent	100	87
Fore limb	34	30
Hind limb	64	61
Tail	210	120 *

California; Mexico.

a–b. ♂ ♀.	California.
c. ♂.	Mexico.

Var. **tessellatus.**

Cnemidophorus sackii, *Gray, Cat.* p. 22.
Ameiva tessellata, *Say, in Long's Exped. Rocky Mount.* ii. p. 50.
Cnemidophorus sackii, *Wiegm. Herp. Mex.* p. 29; *Peters, Mon. Berl. Ac.* 1869, p. 63.
—— tigris, *Baird & Gir. Proc. Ac. Philad.* 1852, p. 69, *and in Stansb. Exped. Great Salt Lake*, p. 338, pl. ii., *and U.S. Mex. Bound. Surv.* ii., *Rept.* p. 10, pl. xxxiii.

* Tail reproduced.

Cnemidophorus marmoratus, *Baird & Gir. Proc. Ac. Philad.* 1852, p. 128.
—— gracilis, *Baird & Gir. l. c.* p. 128, *and U.S. Mex. Bound. Surv.* ii., *Rept.* pl. xxxiv. figs. 7–14.
—— undulatus, *Hallow. Proc. Ac. Philad.* 1854, p. 94, *and Rep. U.S. Explor. Surv. R. R.* x. part iv. 1859, p. 8, pl. ix. fig. 2.
—— tessellatus, *Baird, Rept. U.S. Explor. Surv. R. R.* x. pt. iv. p. 18.
Ameiva sackii, *Cope, Proc. Ac. Philad.* 1862, p. 23.
Cnemidophorus melanostethus, *Cope, op. cit.* 1863, p. 104.
—— sexlineatus, var. sackii, *Bocourt, Miss. Sc. Mex., Rept.* p. 276, pl. xx. c. fig. 6.
—— sexlineatus, var. grahamii, *Boc. l. c.* p. 277, pl. xx. fig. 8.
—— sexlineatus, var. tessellata, *Günth. Biol. C.-Am., Rept.* p. 26.

Head moderate. Three parietals; four supraoculars; seven supraciliaries; no frenoorbital; median gular scales gradually enlarged; mesoptychium with three or four rows of enlarged scales, the posterior largest, scarcely or not larger than the largest gulars; the mesoptychial scales forming the border of the collar or separated by one row of granules. Dorsal granules small, smooth. Ventral plates in eight longitudinal rows, those of the outer row largest of all; thirty-one to thirty-three transverse rows. Three large præanals, forming a triangle. Four or five rows of brachials, anterior largest, continuous with the postbrachials; antebrachials continuous with brachials, in three rows, the outer of which is the largest, the median the smallest; slightly enlarged scales under the forearm. Eight to ten rows of femorals, one of which is large; tibials in three rows, outer largest. Femoral pores nineteen to twenty-two. Caudal scales broad, very oblique, with straight keels, slightly pointed posteriorly. Olive-grey above, spotted or marbled with black, with or without more or less defined light longitudinal lines; lower surfaces yellowish, the belly spotted with black.

	millim.		millim.
Total length	450	From end of snout to fore limb	57
Head	38	From end of snout to vent	140
Width of head	25	Fore limb	50
From end of snout to interparietal plate	27	Hind limb	93
Width at posterior corners of supraocular regions	14	Tail	310

South-western North America.

a, b. ♂. California.

6. Cnemidophorus mariarum.

? Cnemidophorus grahamii, *Baird & Gir. Proc. Ac. Philad.* 1852, p. 128, and *U.S. Mex. Bound. Surv., Rept.* pl. xxxii. figs. 1–6.
Cnemidophorus mariarum, *Günth. Biol. C.-Am., Rept.* p. 28, pl. xx.

Head narrow. Nostril anterior to nasal suture. Three parietals; four supraoculars; seven supraciliaries; a frenoorbital; median gular scales gradually enlarged; mesoptychium with three or four rows of enlarged scales, the posterior largest, a little larger than the largest gulars; the mesoptychial scales forming the border of the collar. Dorsal granules extremely minute, smooth. Ventral plates in eight longitudinal and thirty-six to thirty-eight transverse rows. Three large præanals, forming a triangle. Four or five rows of brachials, anterior largest, continuous with the postbrachials; antebrachials continuous with brachials, in three rows, the outer largest; enlarged scales under the fore arm. Eight to ten rows of femorals one of which is large; tibials in two or three rows, outer largest. Femoral pores nineteen to twenty-two. Males without anal spines. Caudal scales slightly oblique, feebly keeled, not pointed posteriorly. Olive or reddish brown above, the sides largely spotted or marbled with black, the spots sometimes forming vertical bars on the flanks; young with six light longitudinal lines, as in *C. sexlineatus*, which gradually disappear; hind limbs with round light spots; belly blackish blue in males.

	millim.		millim.
Total length	383	From end of snout to fore limb	46
Head	30	From end of snout to vent	123
Width of head	17	Fore limb	45
From end of snout to interparietal plate	22	Hind limb	86
Width at posterior corners of supraocular regions	11	Tail	260

Tres Marias Islands.

a–e. ♂, hgr., & yg. Tres Marias Islands. Hr. A. Forrer [C.].

7. Cnemidophorus maximus.

Cnemidophorus maximus, *Cope, Proc. Ac. Philad.* 1863, p. 104.

Nostril anterior to nasal suture. Four supraoculars; anterior gulars abruptly larger, their median largest; plates of the collar graduating into the posterior gulars, the marginal largest, the series concave anteriorly in the middle. Dorsal scales minute. Ventral plates in eight rows. Præanals in four transverse series, the posterior two containing six. Brachials in seven rows; antebrachials in four; femorals in seven; tibials in four. Above olive-brown, with three brown bands on each side, which are as wide as the intervals between them, and are so broken by spots of the ground-colour as to resemble series of confluent brown variegations; posterior extremities coarsely reticulate with the same colour; superior surface of tail and gular region blackish, varied; abdominal shields black-tipped.

From end of snout to gular fold 4 centim.; from gular fold to vent 8; from vent to end of tail 22.

Cape St. Lucas, California.

8. Cnemidophorus guttatus.

Cnemidophorus guttatus, *Gray, Cat.* p. 22.
Cnemidophorus guttatus, *Wiegm. Herp. Mex.* p. 29; *Peters, Mon. Berl. Ac.* 1869, p. 63; *Bocourt, Miss. Sc. Mex., Rept.* p. 285, pl. xx. c. fig. 4; *Günth. Biol. C.-Am., Rept.* p. 28, pl. xxi.
—— sexlineatus, part., *Dum. & Bibr.* v. p. 131; *A. Dum. Cat. Méth. Rept.* p. 116.
Ameiva guttata, *Cope, Proc. Ac. Philad,* 1862, p. 63.
Cnemidophorus microlepidopus, *Cope, Proc. Am. Philos. Soc.* xvii. 1877, p. 93.
—— unicolor, *Cope, l. c.*

Head large. Nostril anterior to nasal suture. Three parietals; three supraoculars; six supraciliaries; a frenoocular usually present; median gulars enlarged; mesoptychial scales in three rows of subequal size, a little larger than the largest gulars, separated from the free edge of the collar by two or three rows of granules. Dorsal granules minute, smooth. Ventral plates in eight longitudinal and thirty-three transverse series. Three large præanals, forming a triangle. Brachial plates in two or three rows, anterior largest and continuous with the antebrachials; a few small postbrachials; antebrachials in two rows, outer largest; lower surface of forearm uniformly granular. Femoral plates in eight to twelve rows; tibials in three, outer largest. Femoral pores twenty to twenty-four. Male without anal spines. Caudal scales slightly oblique, keeled, not distinctly pointed posteriorly. Olive above, the back spotted with yellowish; a blackish, above light-edged, band along each side; belly olive, with lighter spots on the sides; mesoptychium plumbeous. Young uniformly olive, or with traces of six or eight narrow lighter longitudinal lines; a whitish line on each side of the tail, continuous with a similar line along the hinder side of the thighs.

	millim.		millim.
Total length	420	From end of snout to fore limb	54
Head	37	From end of snout to vent	130
Width of head..........	22	Fore limb	53
From end of snout to interparietal plate	25	Hind limb............	92
Width at posterior corners of supraocular regions..	12	Tail	290

Central America.

a. ♂. Vera Cruz.
b–e. Hgr. & yg. Huamelula.

9. Cnemidophorus immutabilis.

Cnemidophorus immutabilis, *Cope, Proc. Amer. Philos. Soc.* xvii. 1877, p. 93.

Differs from *C. guttatus* in the absence of granules between the mesoptychial scales and the border of the collar. Adult with eight longitudinal bands, the inferior lateral brown band extending to the orbit; femora pale spotted; throat pale, breast plumbeous.

West Tehuantepec.

10. Cnemidophorus deppii.

Cnemidophorus deppii, *Gray, Cat.* p. 22.

Cnemidophorus deppii, *Wiegm. Herp. Mex.* p. 28; *Peters, Mon. Berl. Ac.* 1869, p. 63; *Bocourt, Miss. Sc. Mex., Rept.* p. 281, pl. xx. fig. 3, and xx. c. fig. 5; *Günth. Biol. C.-Am., Rept.* p. 27.

? Cnemidophorus lineatissimus, *Cope, Proc. Amer. Philos. Soc.* xvii. 1877, p. 94.

? Cnemidophorus lativittis, *Cope, l. c.*

Head short. Nostril anterior to nasal suture. Parietals three three supraoculars (seldom four); six supraciliaries; no frenoorbital; median gulars enlarged; mesoptychial scales in three or four rows of subequal size, scarcely larger than the largest gulars, separated from the free edge of the collar by a row of granules. Dorsal granules small, smooth. Ventral plates in eight longitudinal and thirty-five transverse series. Three large præanals, forming a triangle. Brachial plates in three rows, continuous with the postbrachials, anterior largest and continuous with the antebrachials; latter in two rows, outer largest; lower surface of fore arm uniformly granular. Femoral plates in six or seven rows, one of which is large; tibial plates in two rows, outer largest. Femoral pores seventeen. Male without anal spines. Caudal scales oblique, with straight keels, slightly pointed posteriorly. Olive above, with nine longitudinal white lines; a white line along the hinder side of the thigh; lower surfaces whitish.

	millim.		millim.
Total length	216	From end of snout to fore limb	28
Head	19	From end of snout to vent	76
Width of head	11	Fore limb	28
From end of snout to interparietal plate	13	Hind limb	54
Width at posterior corners of supraocular regions	8	Tail	140

Central America; Venezuela.

a. ♂. Caracas.

11. Cnemidophorus hyperythrus.

Cnemidophorus hyperythrus, *Cope. Proc. Ac. Philad.* 1863, p. 103.

Verticaria hyperythra, *Cope, Proc. Amer. Philos. Soc.* xi. 1869, p. 158.

Head rather narrow, snout long. Nostril anterior to nasal suture. Frontoparietal plates united; supraoculars four, rarely three; posterior gular scales small, abruptly separated from the anterior, which are large, the median twice the size of the surrounding ones; three rows of scales on the gular fold, those of the posterior row largest. Ventral plates in eight series, sometimes with additional small external plates. Three larger præanals. Brachials in four series, antebrachials in three. Above brown or blackish; two yellow bands on each side, which extend some distance on the tail, with a median dark stripe between; two paler median dorsal bands which diverge on the nape, and are frequently united for most of their length.

From end of snout to collar 18 millim.; from collar to vent 40; vent to end of tail 140.

Cape St. Lucas, California.

12. Cnemidophorus ocellifer.

Teius ocellifer, *Spix, Spec. Nov. Lacert. Bras.* p. 23, pl. xxv.
Cnemidophorus hygomi, *Reinh. & Lütk. Vidensk. Meddel.* 1861, p. 231; *Bocourt, Miss. Sc. Mex., Rept.* pl. xx. c. fig. 12.
—— ocellifer, *Peters, Mon. Berl. Ac.* 1877, p. 412, 414.

Snout acutely pointed. Nostril anterior to nasal suture. Parietals five; supraoculars four, the posterior three separated from the shields of the vertex by a row of granules; five supraciliaries; no frenoorbital; all the scales in front of the anterior gular fold enlarged, subequal; mesoptychium with two or three rows of shields, those of the posterior row largest, considerably larger than the largest gulars. Dorsal scales granular, smooth. Ventral plates in eight longitudinal and twenty-seven or twenty-eight transverse series. Three large præanals, forming a triangle. Brachials in two or three rows, anterior largest and continuous with the antebrachials; postbrachials large; antebrachials in two rows, outer largest; lower surface of fore arm uniformly granular. Femoral plates in five or six rows, one of which is large; tibial plates in two rows, outer largest. Femoral pores nine to eleven. Male without anal spines. Caudal scales oblique, with straight keels, shortly pointed posteriorly. Females and young brown or olive above, with six narrow light longitudinal lines along the body; the interspace between the three light lines on each side blackish or with alternating black and light spots; in the adult males the light lines disappear, and sometimes also the dark markings, and there is a series of light spots along each side of the back; lower surfaces white.

	♂. millim.	♀. millim.
Total length	198	189
Head	25	16
Width of head	15	10
From end of snout to interparietal plate	16	12
Width at posterior corners of supra-ocular regions	9	7
From end of snout to fore limb	35	24
From end of snout to vent	85	64

	♂. millim.	♀. millim.
Fore limb	31	22
Hind limb	53	40
Tail	113*	125

Brazil.

a. ♀.	Brazil.	Copenhagen Museum [P.]. (As typical of *C. hygomi.*)
b. ♂.	Brazil.	Prof. C. Machado [P.].
c–e, f–g. ♂, ♀, & yg.	Pernambuco.	W. A. Forbes, Esq. [P.].
h. ♂.	Pernambuco.	

13. Cnemidophorus multilineatus.

Cnemidophorus multilineatus, *Philippi, Arch. f. Nat.* 1869, p. 41.

Nostril anterior to nasal suture. Three parietals; three supraoculars. Ventral plates in eight longitudinal and thirty-three transverse series. Brachial plates in three rows, antebrachials in two. Femoral pores twelve. Caudal scales keeled, obtusely pointed posteriorly. Grey-brown above, with ten longitudinal whitish lines, the lowermost twice as wide as the others, which are as wide as the dark interspaces between them; lower surfaces white.

	millim.		millim.
Head to ear-opening	11	Tail	98
From snout to vent	45		

Mendoza.

14. Cnemidophorus lacertoides.

Cnemidophorus lacertinoides, *Gray, Cat.* p. 22.
Cnemidophorus lacertoides, *Dum. & Bibr.* v. p. 134; *Bocourt, Miss. Sc. Mex., Rept.* pl. xx. c. fig. 10.
—— grandensis, *Cope, Proc. Amer. Philos. Soc.* xi. 1869, p. 158.

Head very short. Nostril anterior to nasal suture. Parietals three; supraoculars three or four; supraciliaries five; no frenoorbital; median gular scales gradually enlarged; mesoptychium with two or three rows of shields, the largest considerably larger than the median gulars, the posterior row separated from the edge of the collar by a row of granules. Dorsal granules rather large, smooth. Ventral plates in ten longitudinal and thirty-three to thirty-five transverse series. Præanals irregular, sometimes three larger median ones forming a triangle. Brachials in two or three rows, anterior largest and continuous with the antebrachials; postbrachials rather large, more or less continuous with brachials; antebrachials in two rows, outer largest; lower surface of fore arm uniformly granular. Femoral plates in five or six rows, one of which is large; tibials in three rows, outer largest. Femoral pores ten to twelve. Male

* Tail reproduced.

without anal spines. Caudal scales slightly oblique, keeled, shortly pointed posteriorly. Olive above, with longitudinal series of black spots and two narrow white lines along each side, the upper, commencing from the supraciliaries, constant, the lower frequently broken up into spots; flanks and limbs black and white spotted; lower surfaces yellowish white; throat with a few black dots; ventral plates sometimes black-tipped in the males.

	♂. millim.	♀. millim.
Total length	174	132
Head	15	13
Width of head	9·5	9
From end of snout to interparietal plate	10	9
Width at posterior corners of supra-ocular regions	7	6·5
From end of snout to fore limb	23	21
From end of snout to vent	64	62
Fore limb	22	19
Hind limb	34	31
Tail	110	70*

Uruguay; Southern Brazil.

a. Several specs., ♂ ♀. Montevideo. C. Darwin, Esq. [P.].
b. ♂. Montevideo.

15. Cnemidophorus longicauda.

Cnemidophorus longicauda, *Gray, Cat.* p. 21.
Ameiva longicauda, *Bell, Zool. Beagle, Rept.* p. 28, pl. xv. fig. 1.
Cnemidophorus longicauda, *Bocourt, Miss. Sc. Mex., Rept.* pl. xx. c. fig. 13.

Head moderate, snout pointed. Nostril in the middle of the lower part of the anterior nasal. Parietals three; supraoculars four, the posterior three separated from the plates of the vertex by a row of granules; supraciliaries six; no frenoorbital; anterior gular scales largest, subequal; mesoptychial scales very small, in three or four rows. Dorsal scales rather large, flat, hexagonal, juxtaposed, smooth. Ventral plates in ten or twelve longitudinal and thirty-three or thirty-four transverse series. Præanal plates irregular or with two larger medians. Brachials in three rows, continuous with the postbrachials, anterior largest; antebrachials in two or three rows, outer largest. Femorals in five or six series, one of which is large; tibials in three rows, outer largest. Femoral pores twelve to fourteen. Males without anal spines. Caudal scales broad, very oblique, with straight keels, pointed posteriorly. Olive above, with nine white longitudinal lines, the median ones close together, the outer more widely separated; lower surfaces white.

---

* Tail reproduced.

	millim.		millim.
Head	13	From end of snout to fore limb	19
Width of head	8	From end of snout to vent	55
From end of snout to interparietal plate	9	Fore limb	19
Width at posterior corners of supraocular regions	5·5	Hind limb	34

Northern Patagonia.

a. ♂.	Bahia Blanca.	C. Darwin, Esq. [C.]. (Type.)
b. ♂.	—— ?	Haslar Collection.

16. Cnemidophorus heterolepis.

Cnemidophorus heterolepis, *Tschudi, Faun. Per., Herp.* p. 40.
Verticaria heterolepis, *Cope, Proc. Amer. Philos. Soc.* xi. 1869, p. 158.

Dorsal scales smooth and granular on the nape and anterior part of the back, then becoming gradually larger, rhomboidal, keeled and imbricate. Ventral plates in ten rows. Præanal plates irregular, numerous. Frontoparietals united. Femoral pores seventeen to nineteen. Olive-green above; young with several longitudinal dark bands; throat dirty white, collar and breast shiny black; belly bright bluish green.

Eastern Slope of Andes, Peru.

7. CALLOPISTES.

Callopistes, *Gravenh. N. Acta Ac. Leop.-Carol.* xviii. 1838, ii. p. 743; *Gray, Cat.* p. 17.
Aporomera, *Dum. & Bibr.* v. p. 69.
Tejovaranus, *Steindachn. Denkschr. Ak. Wien,* xxxviii. 1878, p. 93.

Tongue long, narrow, bilobate and retractile at the base. Teeth compressed longitudinally, bi- or tricuspid. Head with very small shields; anterior nasal separated from its fellow; nostril pierced in the suture between the two nasals. Eyelids developed. Ear exposed. Limbs well developed, pentadactyle. Dorsal scales small, juxtaposed, equal; ventral plates rather small, elongate-quadrangular, smooth, forming regular series. A double collar-fold. No femoral pores. Tail cyclotetragonal.

Peru, Chili.

1. Callopistes maculatus.

Callopistes celestis (*non D'Orb.*), *Gray, Cat.* p. 17.
Callopistes maculatus, *Gravenh. l. c.* p. 744, pl. lv. fig. 1.
Aporomera ornata, *Dum. & Bibr.* v. p. 76; *Guichen. in Gay's Chile, Zool.* ii. p. 58, pl. iii. fig. 1; *Philippi, Reise Atacama,* p. 168.
Ameiva oculata, *D'Orbigny, Voy. Amér. Mér., Rept.* p. 9, pl. v. figs. 6–9.
Aporomera ocellata, *Guichen. l. c.* p. 61.

In habit somewhat similar to *Lacerta ocellata*. Upper surface of snout with numerous small polygonal plates, there being seven to nine on a line between the frontal and the rostral; vertex with much larger plates, a frontal, a pair of fronto-parietals, and an interparietal being constantly well developed, the former largest, hexagonal; supraocular region with a row of five or six transversely enlarged plates bordered by granules; only one row of granules between the frontal and the supraocular plates; occipital region with small irregular scales; rostral pentagonal, with truncate upper angle, separated from the anterior nasal by one plate; one anterior and two (or one) posterior nasals; three loreals, higher than broad, anterior smallest, third largest, separated from the labials by one or two rows of small plates; a series of infraorbitals, separated from the labials by one row of small plates; upper labials small, twelve or thirteen; temporal scales granular. Mental broadly trapezoid, its posterior edge in contact with an anterior, single, chin-shield, which is followed by three or four pairs of others forming (at least the two anterior pairs) a suture; eleven or twelve lower labials, the first or the two first forming a suture with the chin-shields, the others separated from the latter by granules; gular region with flat granules. Dorsal scales small, flat, round, granular, separated by minute granules; on the flanks the scales are smaller still. Ventral plates square, much longer than broad, in forty to forty-five transverse* and twenty-six or twenty-eight longitudinal series. Præanal scales small, numerous, subequal. Fore limb anteriorly, and hind limb inferiorly, with enlarged flat scales, but without regular plates. Tail with rings of small keeled scales. Males with a small projecting conical tubercle on each side behind the vent. Olive or bluish above (in spirits); back with four longitudinal series of large black spots, white-edged laterally; head, limbs, flanks, and tail, variegated with black; lower surfaces whitish, uniform or with small black spots.

	♀. millim.	♂. millim.
Total length	410	307
Head	39	26
Width of head	30	16
Body	131	86
Fore limb	58	43
Hind limb	97	70
Tail	250	195

Chili.

a, b, c, d, e–f. ♂ ♀. Chili.

2. Callopistes flavipunctatus.

Callopistes flavipunctatus, *Gray, Cat.* p. 17.
Aporomera flavipunctata, *Dum. & Bibr.* v. p. 72, pl. 51.
Tejovaranus branickii, *Steind. Denkschr. Ak. Wien*, xxxviii. p. 93, pl. i., *and Sitzb. Ak. Wien*, lxxviii. 1. 1879, p. 399.

* Counted from the gular fold to the groin.

More slender, resembling a *Varanus*. Vertex with numerous small plates, similar to those on the snout; a few slightly enlarged supraocular plates, separated from the plates on the vertex by two or three rows of granules; loreal plates small, numerous, irregular; two rows of small scales between the labials and the infraorbitals. Dorsal scales granular, oval. Blackish-brown above, dotted with yellow; lower surfaces yellow, the throat marbled, and the belly more or less irregularly spotted with brown.

	millim.		millim.
Total length	925	Fore limb	100
Head	70	Hind limb........	172
Body	180	Tail	630

Peru.

8. DICRODON.

Dicrodon, *Dum. & Bibr.* v. p. 137; *Gray, Cat.* p. 23.

Tongue moderately elongate, not retractile, widened and emarginate posteriorly. Lateral teeth compressed transversely, bicuspid. Head with large regular shields; nostril pierced in the anterior nasal. Eyelids developed. Ear exposed. Limbs well developed, each with five digits. Dorsal scales small; ventral plates large. subquadrangular, smooth, forming regular series. A double collar-fold. Femoral pores. Tail cyclotetragonal.

Peru.

1. Dicrodon guttulatum.

Dicrodon guttulatum, *Gray, Cat.* p. 23.
Dicrodon guttulatum, *Dum. & Bibr.* v. p. 138.

A single frontoparietal shield, followed by numerous small shields; the anterior head-shields as in the *Ameivæ*; eight upper and six lower labials; chin-shields one anterior and three large pairs, which are followed by smaller ones; gular scales hexagonal, smooth, smallest between the chin-shields; mesoptychial scales imbricate, hardly larger than the largest gulars. A row of large transversely dilated brachial plates, followed by other smaller oneses a row of similar large antebrachials. Six series of femoral plates; three rows of tibial plates, outer largest. Femoral pores ten. Several large, angular, imbricate præanal plates. Caudal scales elongate, quadrangular, keeled. Olive above, the back and sides with round whitish or yellowish spots; on each side two feebly marked yellowish longitudinal lines, the upper extending from the supraciliaries to the tail, the lower from the shoulder to the hind limb; a yellow, black-edged band along the hinder side of the thigh; belly and lower surface of limbs and tail white; throat, chest, and sides of the belly bluish grey.

	millim.		millim.
Total length	450	Fore limb	45
Head	29	Hind limb	85
Body	101	Tail	320

Peru.

2. Dicrodon calliscelis.

Dicrodon calliscelis, *Cope, Journ. Ac. Philad.* (2) viii. 1876, p. 163.

The inner cusp of each tooth smaller than the outer. Nostril in the anterior nasal close to the posterior and inferior suture. Frontal narrow behind, angulate in front, divided transversely by a suture at the penultimate supraocular scute; frontoparietals very small, separated by the small interparietal, which is in contact with the frontal; two small suboval parietals on each side, and ten or twelve large scales surrounding them behind; four supraoculars, the last two separated from the frontal by granules; five upper and five lower labials; six pairs of large chin-shields, the last three smaller and separated by two plates from the labials, the anterior pair in contact; gular scales small, smooth, a wide band of smaller scales separating them from those of the collar; margin of collar granular; its middle with five transverse rows of unequal transverse scales. Dorsal scales commencing as granulations on the interscapular region, and gradually increasing in size until they are similar in size to the large ones which cover the tail in whorls; like the latter they are keeled, the keels forming continuous lines; there are thirteen series at the groin, and eighteen one third the length behind the axillæ; nuchal scales and those of the sides of the neck and body granular. Ventral plates in eight longitudinal rows. Præanals in three longitudinal series of alternating rhombic plates. Posterior face of arm, postero-superior faces of fore arm, superior and posterior faces of femur, and superior face of tibia granular; other faces of limbs scutate. Femoral pores eighteen. Olive above, becoming yellow posteriorly and on the tail; a dark line bounding the dorsal scales on the side, and a row of small blackish spots along the middle of the granular lateral band; femur yellow behind, with two parallel black bands; tail with light bluish spots on the sides; lower surfaces yellowish olive.

	millim.		millim.
Total length	343	From snout to vent .	99
Head (to ear)	22	Hind limb	73
Width of head	14		

Pacasmayo, coast of Northern Peru.

9. TEIUS.

Teius, part., *Merrem. Tent. Syst. Amph.* p. 60.
Teius, *Fitzing. N. Classif. Rept.* p. 21.
Acrantus, *Wagler, Syst. Amph.* p. 154; *Wiegm. Herp. Mex.* p. 8; *Dum. & Bibr.* v. p. 141; *Gray, Cat.* p. 23.

Tongue moderately elongate, not retractile, widened and emarginate posteriorly. Lateral teeth compressed transversely, bicuspid. Head with large regular shields; nostril pierced in the anterior nasal. Eyelids developed. Ear exposed. Limbs well developed, anterior with five, posterior with four digits, the fifth being rudimentary. Dorsal scales small; ventral plates large, subquadrangular, smooth, forming regular series. A double collar fold. Femoral pores. Tail cyclotetragonal.

South-eastern South America.

1. Teius teyou.

Acrantus teyou, *Gray, Cat.* p. 23.
Lacerta teyou, *Daud. Rept.* iii. p. 195.
Teius viridis, *Merr. Tent.* p. 60.
Ameiva teiu, *Lichtenst. Verz. Doubl.* p. 91.
Teius teyou, *Fitzing. N. Class.* p. 51.
Acrantus viridis, *Wagl. Syst. Amph.* p. 154; *Dum. & Bibr.* v. p. 143; *Hensel, Arch. f. Nat.* 1868, p. 347.
Ameiva cœlestis, *D'Orb. Voy. Amér. Mér., Rept.* p. 2, pl. v. figs. 1–5.
Dicrodon cœlestis, *Peters, Mon. Berl. Ac.* 1869, pp. 64, 433, & 720.

Habit of *Lacerta viridis*. Upper head-shields: a frontonasal, a pair of præfrontals, a frontal, a pair of frontoparietals, three small parietals; four supraoculars, the three posterior forming a disk entirely surrounded by granules; occipital region covered with small scales; rostral frequently forming a suture with the frontonasal; a postnasal, a loreal, a præorbital, three infraorbitals; six or seven upper and as many lower labials; chin-shields, one anterior and five to seven pairs, the posterior separated from the labials by another row of shields; a broad transverse band of slightly enlarged gular scales; enlarged mesoptychial scales varying considerably in size, in two or three rows. Dorsal scales granular, smooth. Ventral plates in eight or ten longitudinal, and thirty-three to thirty-six transverse series. Several irregular præanal plates. Two rows of brachial plates, outer small, inner continuous with outer series of antebrachials; latter in two or three rows, outer very large. Femorals in seven or eight rows, one of which is much enlarged; tibials in three or four rows, outer largest. Femoral pores fifteen to twenty-one. Caudal scales narrow, four-sided, strongly keeled above, feebly beneath. Green above, the sides sometimes brownish; a series of regular transverse black spots on each side of the back; two yellow black-edged lateral lines on each side, the upper commencing from above the ear, the lower extending from the shoulder to the hind limb; flanks black-spotted; a yellow, black-edged band along the hinder side of the thighs; lower surfaces yellow.

	♂. millim.	♀. millim.
Total length	290	248
Head	24	24
Width of head	17	15
Body	76	94
Fore limb	33	34
Hind limb	63	65
Tail	190	130*

Southern Brazil, Paraguay, Uruguay, Argentine Republic.

* Tail reproduced.

a. ♀.	Rio Grande do Sul.	Dr. v. Ihering [C.].
b–c. ♂.	Dept. of Soriano, Uruguay.	R. Havers, Esq. [P.].
d–i. ♂, ♀, hgr., & yg.	Uruguay.	
k. Yg.	Montevideo.	C. Darwin, Esq. [P.].
l–n. Hgr., ♂, ♀.	Montevideo.	Dr. Cunningham [P.].
o. ♂.	S. America.	B. Bynoe, Esq. [P.].
p. ♂.	—— ?	C. Darwin, Esq. [P.].
q. ♀, skeleton.	S. America.	Berlin Museum.

10. CROCODILURUS.

Tupinambis, part., *Daud. Rept.* iii. p. 5.
Crocodilurus, *Spix, Spec. Nov. Lacert. Bras.* p. 19; *Wagler, Syst. Amph.* p. 153; *Wiegm. Herp. Mex.* p. 8; *Dum. & Bibr.* v. p. 42; *Gray, Cat.* p. 25.
Custa, *Fleming, Philos. Zool.* ii. p. 274.
Ada, part., *Gray, Griff. A. K.* ix. *Syn.* p. 28.

Tongue moderately elongate, not retractile, widened and emarginate posteriorly. Lateral teeth compressed longitudinally, bi- or tricuspid. Head with large regular shields; nostril pierced in the suture between two nasals. Eyelids developed. Ear exposed. Limbs well developed, each with five digits, the toes with a denticulated fringe laterally, inferiorly with small squarish scales. Dorsal scales small, equal; ventral plates large, elongate, subquadrangular, smooth, forming regular series. A collar fold. Femoral pores. Tail strongly compressed, with a double denticulated keel superiorly.

Guianas, Brazil.

1. Crocodilurus lacertinus.

Crocodilurus lacertinus, *Gray, Cat.* p. 25.
Tupinambis lacertinus, *Daud. Rept.* iii. p. 85.
Crocodilurus amazonicus, *Spix, Spec. Nov.* p. 19, pl. xxi.; *Wagl. Icon. Amph.* pl. xv.
—— ocellatus, *Spix, l. c.* p. 20, pl. xxii. fig. 1.
Teius (Ada) bicarinata (*non L.*), *Gray, Griff. A. K.* ix. *Syn.* p. 29.
Crocodilurus lacertinus, *Dum. & Bibr.* v. p. 46.

Habit of a *Varanus*. Upper head-shields: a fronto-nasal, a pair of præfrontals, a frontal, a pair of frontoparietals, a large interparietal, a pair of parietals and a pair of occipitals; one or two small shields between the interparietals and occipitals; four large, and a small posterior, supraoculars; two large supratemporals on each side; a loreal, a præorbital, six or seven infraorbitals; eight upper and seven or eight lower labials; chin-shields, one anterior and six pairs, the posterior separated from the labials by another row of large shields; gular scales small between the chin-shields, large and subhexagonal on the throat; the scales in front of the collar-fold larger. Dorsal scales small, elongate, oval, slightly keeled or tectiform. Abdominal plates with the posterior border rounded, in twenty-four longitudinal and thirty-seven transverse series. Præanal plates in three transverse series, the median posterior pair largest. Femoral pores very small, rather indistinct, six to ten. Caudal scales elongate, keeled, straight, in regular cross rows, the upper with pointed posterior edge, the laterals and inferior rounded posteriorly. Brown above, dotted with black; digits with black annuli; lower surfaces yellow, with scattered blackish spots.

	millim.		millim.
Total length	510	Fore limb	49
Head	35	Hind limb	78
Width of head	20	Tail	360
Body	125		

Guianas; Brazil.

a. ♂.	Amazons.	
b. ♀.	——?	Sir A. Smith [P.].

11. NEUSTICURUS.

Neusticurus, *Dum. & Bibr.* v. p. 61.
Custa (*non Flem.*), *Gray, Cat. Liz.* p. 25.

Tongue moderately elongate, arrow-headed. Lateral teeth compressed, bi- or tricuspid. Head with large regular shields; frontonasal separating the nasals; præfrontals and frontoparietals present; nostril pierced in the middle of a single nasal. Eyelids developed, lower with a transparent disk composed of one or several scales. Ear exposed. Limbs well developed, pentadactyle. Dorsal lepidosis composed of small scales intermixed with large keeled ones forming longitudinal series; ventral plates large, squarish, forming regular longitudinal and transverse series. A collar fold. Males with femoral pores, females with a few præanal pores only. Tail compressed, with a double denticulated keel superiorly.

South America.

1. Neusticurus bicarinatus.

Custa bicarinata, *Gray, Cat.* p. 25.
Lacerta bicarinata, *Linn. S. N.* i. p. 361.
Dracæna bicarinata, *Guér. Icon. R. A., Rept.* pl. iii. fig. 2.
Neusticurus bicarinatus, *Dum. & Bibr.* v. p. 64.

Snout pointed; canthus rostralis very strong; the diameter of the ear-opening about half that of the eye. Transparent disk of lower eyelid composed of several pieces. Head-shields: a pair of frontonasals, a pair of præfrontals, a small azygos shield at the intersection of the four latter; a frontal, slightly grooved medially; a pair of frontoparietals and an interparietal, of subequal size, the latter further back than the former; occipital region with small irregular shields; four supraoculars, anterior more or less divided; a loreal; a row of six infraorbitals; a row of granules between the loreal and anterior infraorbital and the labials; latter, six or seven upper and four lower; chin-shields, one anterior and four pairs. Gular scales rather large, median smallest, smooth or indistinctly keeled; a row of six or seven large collar-plates. Smaller dorsal scales irregular, slightly imbricate, feebly keeled; the larger scales strongly keeled, also subimbricate, forming six longitudinal series on the back, the two outer on each side in contact; the two median series may be less distinct and interrupted. Abdominal plates rounded posteriorly, in about ten longitudinal and twenty-six transverse series; the plates on the six median longitudinal rows smooth, the others keeled and passing gradually into the scaling of the flanks; the plates of the two median rows narrower. Three large præanal plates, one antero-median and two postero-lateral; a small plate between the latter. Upper limb-scales imbricate, keeled. Females with three or four præanal pores on each side; [males with thirty femoral pores on each side]. Caudal scales forming regular annuli, the lateral strongly keeled; the large dorsal scales continued to the upper surface of the tail, where they form first four and then two denticulated crests. Olive-brown above, spotted with dark brown, with more or less distinct whitish ocelli on the flanks; three vertical dark brown bars on the upper eyelid and supercilium; lower surfaces whitish, throat and belly immaculate.

	millim.		millim.
Total length	270	From end of snout to vent	90
Head	22	Fore limb	30
Width of head	12	Hind limb	43
From end of snout to fore limb	38	Tail	180

Brazil.

a, b. ♀.	S. America.
c. Yg.	Interior of Brazil.

2. Neusticurus ecpleopus.

Custa bicarinata (*non L.*), *Günth. Proc. Zool. Soc.* 1859, p. 404.
Neusticurus ecpleopus, *Cope, Journ. Ac. Philad.* (2) viii. 1876, p. 161; *O'Shaughn. Ann. & Mag. N. H.* (5) iv. 1879, p. 295.

Compared with the preceding, this species has a shorter and more obtuse snout; ear-opening much larger, almost as large as the eye. Transparent disk of lower eyelid undivided. Frontonasals frequently united; no azygos shield between the latter and the præfrontals; interparietal much larger than parietals; no granules between the loreal and labials; five or six upper and four or five lower labials; anterior gular scales large, roundish. Nape granular, with about eight longitudinal series of roundish, keeled or subconical tubercles; enlarged dorsal scales more tubercular, with more elevated keels; they form four equidistant longitudinal series on the back; flanks with regular vertical folds, which support mingled small and large scales. Abdominal plates perfectly square, in eight longitudinal and twenty transverse series; plates of the outer series smaller, keeled, of the median series as large as the others. Anterior large præanal frequently divided longitudinally. Hind limbs above granular, with large keeled tubercles. Males with fifteen to nineteen femoral pores; females with two or three præanal pores on each side. Brown above, darker spotted; flanks with more or less distinct lighter spots; generally a large black ocellus with white centre at axilla; lower surfaces whitish, usually irregularly brown-dotted.

	millim.		millim.
Total length	169	From end of snout to vent	61
Head	15	Fore limb	20
Width of head	10	Hind limb	31
From end of snout to fore limb	24	Tail	108

Ecuador; Upper Amazons.

a. ♂.	Amazons.	
b. ♂.	Brazil.	
c. Hgr.	Pallatanga, Ecuador.	Mr. Buckley [C.].
d. ♀.	W. Ecuador.	Mr. Fraser [C.].
e. Hgr.	——?	

12. ALOPOGLOSSUS.

Tongue moderately elongate, arrow-headed, without imbricate scale-like papillæ, these being replaced by oblique plicæ converging towards the median line*. Lateral teeth compressed, bi- or tricuspid. Head with large regular shields; frontonasal separating the nasals; præfrontals and frontoparietals present; nostril pierced in the suture between two nasals*. Eyelids developed, lower with a transparent disk composed of several scales. Ear exposed. Limbs well developed, pentadactyle. Dorsal and lateral scales subequal, large, pointed, imbricate, arranged in transverse and oblique rows; ventral plates large, pointed, imbricate, smooth or feebly keeled, arranged in regular longitudinal and transverse rows. A rather indistinct collar-fold. Tail subcylindrical. Males with femoral pores, females without.

Ecuador; Peru.

1. **Alopoglossus copii.** (Plate XX. fig. 1.)

Leposoma carinicaudatum (*non Cope*), *O'Shaughn. Proc. Zool. Soc.* 1881, pp. 233, 234.

Head short, snout very obtuse, almost truncate; body rather elongate. Frontonasal broad; two præfrontals, forming a short suture; a frontal; a pair of frontoparietals; an interparietal and two parietals, these three shields large, equal in size; no occipitals; four supraoculars; a small oblique loreal, resting on the postnasal and a frenoorbital; second infraorbital very long; temporal scales keeled; six or seven upper labials, third very long; five lower

* *A. carinicaudatus* not examined.

labials; chin-shields very large, one anterior and three pairs, the first two pairs in contact: gular scales equal, lanceolate-triangular, strongly keeled and imbricate, arranged in oblique rows. Dorsal scales very large, those on the middle of the back largest and their exposed portion as broad as long, all very strongly imbricate, leaf-shaped, ending in a sharp point, and very strongly keeled; on the hinder part of the back the keels become continuous, forming ridges which become most conspicuous on the tail. The ventral scales similar to the dorsals, but very feebly keeled and arranged in longitudinal and transverse series. Eighteen to twenty scales round the middle of the body; twenty-one to twenty-four scales from occiput to sacrum inclusive; twenty-five or twenty-six from chin-shields to præanals. Latter, four anterior and four posterior, all keeled, the postero-median pair largest. Limbs with large imbricate keeled scales. Males with sixteen to eighteen femoral pores on each side. Tail with large keeled imbricate scales, largest on the upper surface, where the keels form four strong, longitudinal, slightly denticulated ridges. Brown above, yellowish on the upper surface of the head and nape; back indistinctly variegated with darker and lighter; tail with yellowish dark-edged spots, having a tendency to form rings; lower surfaces yellowish white; labials and chin-shields brown-spotted; each ventral shield with a more or less marked brown spot.

	♂. millim.	♀. millim.
Total length	152	197
Head	13	16
Width of head	9·5	11
From end of snout to fore limb	23	28
From end of snout to vent	56	74
Fore limb	19	23
Hind limb	30	37
Tail	96	123

Ecuador.

a–b. ♂ ♀.	Pallatanga.	Mr. Buckley [C.].
c–d. ♂ ♀.	Canelos.	Mr. Buckley [C.].

2. Alopoglossus carinicaudatus.

Lepidosoma carinicaudatum, *Cope, Journ. Ac. Philad.* (2) viii. 1876, p. 160.

This species is probably closely allied to the preceding, but, judging from Cope's description, must be considered distinct. The limbs are shorter; the temporal scales are smooth, at least anteriorly; twenty-one scales round the body; an oval median anterior præanal, instead of a pair; eleven femoral pores on each side. The coloration is very different:—Colour above cinnamon-brown, below yellow; sides, from ear to an indefinite point on the tail, black, inclosing a narrow yellow band from axilla to groin; black on scapular region, pale-bordered above ; from nape to tail a median series of small black spots ; head brown ; throat yellow.

	millim.		millim.
Total length (tail reproduced)	115	From end of snout to axilla	21
Length of head to ear	11	From end of snout to vent	50
Width of head	8·5	Hind limb	23

Upper Amazons.

3. Alopoglossus buckleyi.

Leposoma buckleyi, *O'Shaughn. Proc. Zool. Soc.* 1881, p. 233, pl. xxii. fig. 2*.

Differs from *A. copii* in the following points :—Suture between the præfrontals extremely short ; parietals and interparietal with longitudinal ridges ; all three pairs of chin-shields in contact ; gular, ventral, and præanal scales smooth ; dorsal scales much narrower ; twenty-nine scales round the body ; thirty-one scales from occiput to sacrum inclusively, and twenty-eight from gulars to præanals. Femoral pores fourteen. Caudal keels weaker, not forming such regular ridges. Upper surfaces pale brown, with a longitudinal row of black spots on middle of back ; a light lateral stripe from the supraorbital angle, along the parietal border and extreme edge of the dorsal surface, to the tail ; sides of body black, variegated with blue or yellowish spots on the neck ; entire mental region yellowish, without spots ; a dark coloration predominates over the whole inferior surface of the throat, of the body, limbs, and tail, the scales being blackish at their root or for half their length, and yellowish at the tips.

	millim.		millim.
Total length (tail reproduced)	115	From end of snout to vent	51
Head	12	Fore limb	16
Width of head	8	Hind limb	23
From end of snout to fore limb	21		

Ecuador.

a. ♂. Canelos. Mr. Buckley [C.]. (Type.)

---

* The description is incorrect, owing to the imperfect state of preservation of the unique specimen. The epidermis has come off in many places, and the shape of the scales has changed ; this explains the sudden modification of the pectoral into the ventral scales, the former being intact, the latter deprived of the epidermis.

13. LEPOSOMA.

Leposoma, *Spix, Spec. Nov. Lacert. Bras.* p. 24; *Peters, Mon. Berl. Ac.* 1880, p. 217.
Lepidosoma, *Wagl. Syst. Amph.* p. 157; *Tschudi, Arch. f. Nat.* 1847, p. 44; *Peters, Abh. Berl. Ac.* 1862, p. 189.
Lepisoma, part., *Gray, Cat. Liz.* p. 60.

Tongue moderately elongate, arrow-headed. Lateral teeth compressed, bi- or tricuspid. Head with large regular shields; frontonasal separating the nasals; præfrontals and frontoparietals present; nostril pierced in the middle of a single nasal. Eyelids developed, lower with a transparent disk composed of several scales. Ear exposed. Limbs well developed, pentadactyle. Scales of the body, above and beneath, subequal, lanceolate-subhexagonal, strongly keeled, imbricate, arranged in transverse and oblique rows. No collar-fold. Tail subcylindrical.

South America.

1. Leposoma scincoides.

Lepisoma scincoides, part., *Gray, Cat.* p. 60.
Leposoma scincoides, *Spix, l. c.* pl. xxvii. fig. 2.
Lepidosoma scincoides, *Tschudi, l. c.* p. 55; *Peters, l. c.* p. 190, pl. ii. fig. 1.

Head and body moderate. A broad frontonasal, with concave posterior border; a pair of small præfrontals; a frontal; a pair of frontoparietals; two large parietals and a large interparietal; no occipitals; four supraoculars; all the upper head-shields rough, longitudinally striated; no loreal; a very small upper and a larger lower frenoorbital; a row of very small infraorbitals; temporal scales small, keeled; six upper and five lower labials; chin-shields very large, one anterior and three pairs, the two first pairs forming a suture; gular scales similar to the ventrals, but smaller; no trace of a collar. Body covered with uniform scales, which are rather narrower on the belly; twenty-three or twenty-four scales round the middle of the body; thirty-four scales from the occiput to the base of the tail, and thirty-three from the third pair of chin-shields to the vent. Four enlarged præanals. Limbs with imbricate keeled scales. Caudal scales like those of the body. Yellowish brown above, with three or four rows of small black spots on the back; lower surfaces yellowish.

	millim.		millim.
Total length	99	From end of snout	
Head	9	to vent	37
Width of head	4	Fore limb	9·5
From end of snout		Hind limb........	15
to fore limb	13	Tail	62

Amazons.

2. Leposoma dispar.

Leposoma dispar, *Peters, Mon. Berl. Ac.* 1880, pp. 217, 309, pl. —. fig. 2.

Closely allied to the preceding; differing chiefly in the following points:—Frontonasal obtusely angular posteriorly; interparietal shorter and broader, much larger than parietals; four pairs of chin-shields, the three anterior in contact; a complete row of smaller scales across the throat, from one ear to the other. Scales distinctly broader; abdominals rounded or truncate posteriorly, mucronate. Dark brown above, black-dotted; on each side of the back a light line, commencing from the supraocular region; sides of head with oblique black lines; lower surfaces yellowish white.

	millim.		millim.
Total length	104	Fore limb	12
To base of tail	37	Hind limb........	17

Caceres, Colombia.

14. LOXOPHOLIS.

Loxopholis, *Cope, Proc. Ac. Philad.* 1868, p. 305.

Head with large regular shields; frontonasal separating the nasals; præfrontals and frontoparietals present. Eyelids developed, lower with a transparent disk. Ear exposed. Limbs well developed, each with five digits. Scales imbricate, arranged in oblique rows or quincuncially, the exposed portion triangular, strongly keeled; lateral and gular scales like the dorsals; ventrals broad, quadrate, smooth. No collar. Tail cylindrical.

Colombia.

1. Loxopholis rugiceps.

Loxopholis rugiceps, *Cope, l. c.*

Upper head-shields rugose, with longitudinal striæ; two loreals or præoculars between nasal and eye; four supraoculars; frontonasal as long as broad; præfrontals largely in contact; upper labials five, long and narrow, third and fourth under orbit, with a narrow series of scales between; five long and narrow lower labials; chin-shields large, one anterior and five pairs, the first two pairs in contact. Twenty rows of lance-triangular scales on back and both sides, and four rows of smooth abdominal scales, which are broader than long. Six large præanal plates, five reaching the margin, the two outer narrow, the median short. Scales of the limbs large, keeled. Caudal scales like those of the back of *Pantodactylus*; keels stronger below than above. Above yellowish brown, with a narrow blackish band on each side from the upper margin of the ear to near the end of the tail; a median pair of bands appear near the rump, and continue half the length of the tail, but are very indistinct; the upper and lower labials with a continuous transverse black band through the middle of each; lower surfaces yellow.

	millim.		millim.
Total length (tail injured)	112	Fore limb..........	10
From end of snout to vent	39	Hind limb	14
From end of snout to axilla	15		

Magdalena River, Colombia.

15. PANTODACTYLUS*.

Pantodactylus, *Dum. & Bibr.* v. p. 428; *Tschudi, Arch. f. Nat.* 1847, p. 47.
Cercosaura, part., *Gray, Cat. Liz.* p. 60.
Lepisoma, part., *Gray, l. c.*
Pantodactylus, part., *Peters, Abh. Berl. Ac.* 1862, p. 182.

Tongue moderately elongate, arrow-headed. Lateral teeth compressed, bi- or tricuspid. Head with large regular shields; frontonasal separating the nasals; præfrontals and frontoparietals present; nostril pierced in the middle of a single nasal. Eyelids developed, lower with a transparent disk, which is undivided. Ear exposed. Limbs well developed, each with five digits. Dorsal and lateral scales subequal, large, lanceolate, imbricate, arranged quincuncially in transverse and oblique rows; ventral plates large, subquadrangular, smooth, arranged in regular longitudinal and transverse series. A collar fold. Tail cylindrical. Femoral pores.

South-eastern South America.

1. Pantodactylus schreibersii.

Cercosaura schreibersii, *Gray, Cat.* p. 60.
Lepisoma scincoides, part., *Gray, l. c.*
Cercosaura schreibersii (*Fitz.*), *Wiegm. Herp. Mex.* p. 10.
Pantodactylus dorbignyi, *Dum. & Bibr.* p. 431; *Tschudi, l. c.* p. 49.
Cercosaura (Pantodactylus) schreibersii, *Peters, l. c.* pl. i. fig. 4.
Pantodactylus bivittatus, *Cope, Proc. Ac. Philad.* 1863, p. 103.

Head small, body moderately elongate. A broad frontonasal; a pair of præfrontals, forming a short suture; a frontal; a pair of frontoparietals; a pair of parietals separated by a large interparietal; a pair of small occipitals, separated by one or two smaller median occipitals; three supraoculars; a loreal; a frenoorbital; a row of very small infraorbitals; temples with large shields above, with small ones inferiorly; six or seven upper and four or five lower labials; chin-shields, one anterior and four pairs, the two first pairs forming a suture; anterior gulars large; a double median row of transversely enlarged gulars; a pair of broad collar shields. Dorsal scales large,

* *Pantodactylus concolor*, Tschudi, Arch. f. Nat. 1847, p. 48.—Northern Brazil.

strongly imbricate, strongly keeled; twenty-four to twenty-six scales round the middle of the body, including ventrals; thirty to thirty-four scales from occiput to sacrum inclusively. Ventrals large, in four large longitudinal series, passing gradually into the dorsal scales. Two pairs of large præanals, posterior largest. Limbs with large smooth shields; lower surface of arm and hinder side of thighs granulate; digits slightly, but distinctly, toothed inferiorly. Four femoral pores on each side in the males, one or two in the females. Tail, when intact, extremely long, gradually tapering to a fine point; the scales similar to those of the body. Olive or brownish, above, sides darker; a light streak on each side, commencing above the ear; a more or less distinct dark vertebral line; lower surfaces whitish, more or less dotted with grey.

	millim.		millim.
Total length	151	From end of snout to vent..............	39
Head	9	Fore limb	11
Width of head	6	Hind limb	16
From end of snout to fore limb	15	Tail	112

Southern Brazil, Uruguay, Buenos Ayres.

a–c. ♂, ♀, & yg.	Rio Grande do Sul.	Dr. v. Ihering [C.].
d–g. ♀ & yg.	S. Lorenzo, Rio Grande do Sul.	Dr. v. Ihering [C.].
h. ♀.	Brazil.	
i. ♀.	Montevideo.	
k–l. ♀ & hgr.	Dept. of Soriano, Uruguay.	R. Havers, Esq. [P.].
m–p. ♀ & yg.	Uruguay.	

16. ARTHROSAURA.

Tongue moderately elongate, arrow-headed. Lateral teeth compressed, bi- or tricuspid. Head with large regular shields; frontonasal separating the nasals; præfrontals and frontoparietals present; nostril pierced in the middle of a single nasal. Eyelids developed, lower with a transparent disk composed of several scales. Ear exposed. Limbs well developed, pentadactyle. Dorsal and lateral scales equal, hexagonal-lanceolate, keeled, imbricate, arranged in regular transverse series; ventral plates large, subquadrangular, rounded and overlapping posteriorly, smooth, arranged in regular longitudinal and transverse series. A collar fold. Tail cylindrical. Female without femoral pores.

Ecuador.

This genus forms a connecting link between *Prionodactylus* and *Heterodactylus.*

1. Arthrosaura reticulata.

Cercosaura (Pantodactylus) reticulata, *O'Shaughn. Proc. Zool. Soc.* 1881, p. 230, pl. xxii. fig. 1.

Head depressed, snout short, obtuse; body rather elongate. A broad frontonasal; two præfrontals, forming a short suture; a frontal; two frontoparietals; a pair of large pariteals separated by an equally long, narrower interparietal; no occipital; three large supraoculars and a minute posterior fourth; a loreal and a freno-orbital; a row of small infraorbitals; temples with small irregular shields; seven upper and six lower labials; chin-shields very large, three anterior pairs forming a suture; a double longitudinal row of large transverse gulars; collar-shields five, elongate. Dorsal scales narrow, exactly similar to those of *Heterodactylus*, forming regular transverse series which are not at all interrupted on the median line of the back, and are not separated from the ventrals by any smaller scales; thirty-nine scales round the middle of the body, including ventrals; twenty-eight scales from occiput to sacrum inclusively. Ventral plates longer than broad, strongly rounded and overlapping posteriorly, also very similar to those of *Heterodactylus*; they form eight or ten longitudinal and eighteen transverse series. Four principal præanals, two median, with their points touching, and two lateral. Fore limb with large shields, granulate under the arm; hind limb with large shields anteriorly and inferiorly; hinder side of thighs granulate; upper side of tibia with very small keeled scales. Tail with a distinct groove along the side; the scales narrow, forming regular imbricate verticilles, the upper and lateral keeled and sharply pointed, the lower smooth and obtusely pointed posteriorly. Light reddish brown above, reticulated with blackish and with longitudinal series of small light spots; head variegated with blackish; lower surfaces cream-coloured, lips and throat black-spotted.

	millim.		millim.
Total length	118	From end of snout to vent	43
Head	11	Fore limb	12
Width of head........	6·5	Hind limb	19
From end of snout to fore limb.............	17	Tail..................	75

Ecuador.

a. ♀. Canelos. Mr. Buckley [C.]. (Type.)

17. PRIONODACTYLUS.

Pantodactylus, part., *Peters, Abh. Berl. Ac.* 1862, p. 182.
Prionodactylus, *O'Shaughn. Proc. Zool. Soc.* 1881, p. 231.

Tongue moderately elongate, arrow-headed. Lateral teeth compressed, bi- or tricuspid. Head with large regular shields; frontonasal separating the nasals, præfrontals and frontoparietals present; nostril pierced in the middle of a single nasal, or in the suture between two nasals. Eyelids developed, lower with a transparent disk composed of several scales. Ear exposed. Limbs well developed, pentadactyle. Dorsal scales large, hexagonal-lanceolate, strongly keeled, imbricate, arranged in regular transverse series; lateral scales small; ventral plates large, quadrangular, smooth, arranged in regular longitudinal and transverse series. A collar fold. Tail cylindrical. Femoral pores.

South America.

Synopsis of the Species.

I. Nostril pierced in the middle of a single nasal.

Ventrals in six longitudinal rows, outer row smaller; twenty-nine scales, including ventrals, round the middle of the body........................ 1. *argulus*, p. 391.

Ventrals in six equal longitudinal rows; thirty-three or thirty-five scales round the middle of the body .... 2. *oshaughnessyi*, p. 392.

Ventrals in four longitudinal rows; twenty-seven scales round the middle of the body...................... 3. *quadrilineatus*, p. 393.

Ventrals in six longitudinal rows; forty-three scales round the middle of the body; digits distinctly toothed inferiorly 4. *manicatus*, p. 393.

II. Nostril pierced in the suture between two plates .... 5. *vertebralis*, p. 394.

1. Prionodactylus argulus.

Cercosaura (Pantodactylus) argulus, *Peters, l. c.* p. 184, pl. i. fig. 3.

Snout short; body rather short. The ordinary head-shields; frontonasal longitudinally bisected; præfrontals very small, not forming a suture, separated by the anterior angle of the frontal; parietals and interparietal subequal in size; three occipitals, median smallest; four supraoculars; a loreal and a frenoorbital; temples shielded; six upper and as many lower labials; chin-shields very large, one anterior and four pairs, the first two forming a suture, the two others separated by one or two rows of granules; two longitudinal rows of large transverse gular shields; five collar shields. Dorsal scales strongly keeled, very narrow; lateral scales very small, keeled; twenty-nine scales, including the ventrals, round the middle of the body; thirty-one scales from the occiput to the base of the tail. Ventrals large, in six longitudinal rows, the outer row smaller, and twenty transverse rows. A pair of large præanal shields, in front of which is another, much smaller, pair. Limbs with large shields; hinder side of thighs granulate. Seven femoral pores on each side (♂). Tail above and on the sides with scales similar to the dorsals, inferiorly with a double row of large smooth scales. Brown above, with five longitudinal rows of distant black dots; a black streak on each side, from the nostril to the ear, passing through the eye; two black lines on the side of the body, the lower bordered inferiorly with white; between these two black lines and above the arm two black ocelli with white centre; lower surfaces yellowish.

	millim.		millim.
Total length	162	From end of snout to vent	39
Head	11	Fore limb	14
Width of head	6	Hind limb............	18
From end of snout to fore limb	15·5	Tail	123

Sta Fé de Bogota.

2. **Prionodactylus oshaughnessyi.** (Plate XXI. fig. 1.)

Cercosaura (Pantodactylus) argulus (*non Peters*), *O'Shaughn. Proc. Zool. Soc.* 1881, p. 229.

Snout rather longer than in *P. argulus*. Frontonasal longitudinally bisected; præfrontals forming a suture (in one specimen with a small azygos shield separating them); the other head-shields as in *P. argulus*. Large shields between the posterior chin-shields; two more or less regular longitudinal rows of large transverse gular shields; five collar-shields. Dorsal and lateral scales as in *P. argulus*, but the former not so narrow, about twice as long as wide; thirty-three or thirty-five scales, including the ventrals, round the middle of the body; thirty-eight to forty scales from the occiput to the base of the tail. Ventrals large, in six longitudinal rows, the outer quite as large as the others, rounded; eighteen or nineteen transverse rows. A pair of large præanal shields, in front of which is another, much smaller pair; the large præanals may each be longitudinally divided. Femoral pores nine in the male, six in the female. Tail as in the preceding. Brown above; a black vertebral line; a dark lateral band, well marked on the sides of the head and neck, less distinct on the body; below this a whitish streak; male with a series of eight or nine strongly marked black ocelli with white centre on each side of the body; these ocelli very indistinct in the female; lower surfaces cream-coloured, each shield with a blackish central dot.

	millim. ♀.	millim. ♂.
Total length	103	142
Head	13	12
Width of head	7	6
From end of snout to fore limb .	18	18
From end of snout to vent	45	47
Fore limb	16	15
Hind limb	21	21
Tail	58	95

Ecuador.

a–c. ♀ ♂.	Canelos.	Mr. Buckley [C.].
d. ♀.	Pallatanga.	Mr. Buckley [C.].

3. Prionodactylus quadrilineatus.

Cercosaura (Pantodactylus) quadrilineata, *Boettger, Ber. Senck. Ges.* 1876, p. 141, pl. —.

Allied to the two preceding. Head-shields as in *P. oshaughnessyi*, except in the presence of a pair of large postoccipitals, which are quite as large as the occipitals; interparietal much larger than the parietals. Collar quite indistinct. Dorsal scales very narrow, as in *P. argulus*; twenty-seven scales, including the ventrals, round the middle of the body, and thirty scales from the occiput to the base of the tail. Ventral plates in four longitudinal series. Præanal plates five, equally large, a pair anteriorly, three on a line posteriorly. Nine femoral pores on each side (♂). Olive-brown above; a whitish streak on each side of the body, beginning from above the ear, and another from axilla to groin; lower surfaces whitish, throat, sides of belly and base of tail scantily dotted with black.

	millim.		millim.
Total length	152	From end of snout to vent	40·5
Head	10	Fore limb	13·5
Width of head	6	Hind limb	17·5
From end of snout to fore limb	16·5	Tail	111·5

S. Paul, Brazil.

4. Prionodactylus manicatus.

Cercosaura (Prionodactylus) manicata, *O'Shaughn. Proc. Zool. Soc.* 1881, p. 231, pl. xxii. fig. 3.

Head depressed, body moderately elongate. Head-shields normal; præfrontals forming a short suture; interparietals a little larger than parietals; median occipital small; temples with large shields above, with small keeled scales inferiorly; loreal large; six or seven upper and five or six lower labials; chin-shields, one anterior and four pairs, the two first pairs forming a suture, the two others widely separated by large shields; a double longitudinal row of transversely enlarged gulars; collar-shields six, the two median extremely large. Dorsal scales strongly keeled, narrow; lateral scales small, keeled, imbricate; forty-three scales, including the ventrals, round the middle of the body; thirty-five to thirty-eight scales from the occiput to the base of the tail. Ventrals large, in six or eight rows, on the median rows a little smaller than on the next, those of the outer rows rounded; eighteen or nineteen transverse rows. Præanals five, a small anterior pair and three large posterior, the median very narrow. Digits distinctly toothed inferiorly. Limbs with large shields; lower surface of arm and hinder side of thighs granulate; upper surface of hind limbs with imbricate keeled scales. Male with thirteen or fourteen femoral pores on each side, female with twelve to fourteen. Upper and lateral caudal scales elongate hexagonal, keeled; lower smooth, the two median rows larger; all forming regular annuli. Dark greenish above; the sides black, bordered inferiorly by a white streak from the rostral to the hind limb; fore limb anteriorly, and the three inner digits, white; a white spot on the fore part of the arm, near the body; hind limbs paler brown, with faint-coloured ocelli; lower surfaces whitish, the belly more or less spotted with bluish; upper surface of tail with a light zigzag band.

	♂ millim.	♀ millim.
Total length	138	191
Head	14	17
Width of head	9	11
From end of snout to fore limb	24	28
From end of snout to vent	54	71
Fore limb	21	25
Hind limb	32	38
Tail	84	120

Ecuador.

a. ♂.	Pallatanga.	Mr. Buckley [C.].	(Types.)
b–d. ♀.	Canelos.		

5. **Prionodactylus vertebralis.** (Plate XXI. fig. 2.)

Cercosaura (Pantodactylus) vertebralis, *O'Shaughn. Ann. & Mag. N. H.* (5) iv. 1879, p. 298.

Snout very short, body moderate. Præfrontals forming a short suture; interparietals longer than broad; parietals broader than long; a small median occipital; two large transverse postoccipitals; three supraoculars, anterior largest; anterior supraciliary very large; nostril in the suture between two nasals, the posterior of which is the largest; two small loreals, the upper resting on the lower and on the posterior nasal; a small frenoorbital; a row of small infraorbitals; temples with large shields above; six upper and five lower labials; chin-shields, one anterior and four pairs, the two first of which form a suture, the two others widely separated by large gulars; a double longitudinal series of large transverse gulars; collar-shields six. Dorsal scales elongate hexagonal, strongly keeled, scarcely imbricate, not mucronate; lateral scales small, keeled; thirty-seven scales round the body, including ventrals; thirty scales from occiput to sacrum inclusively. Ventral plates quadrangular, in six longitudinal rows, median and outer rows smaller, and twenty transverse rows. Four large præanals, in two pairs, a small lateral on each side, and a small triangular one wedged in between the large posterior pair. Limbs with large shields; lower surface of arms and hinder side of thighs granulate. Six femoral pores on each side. Upper and lateral caudal scales like the dorsals, lower larger and smooth. Dark olive above; a broad bluish black-edged band extends from the tip of the snout along the middle of the back and the upper surface of the tail; a narrow whitish line along the upper labials to the arm-pit; a series of very small white ocelli surrounded by black rings along the side of the body, and a few on the limbs; a larger ocellus at the junction of the fore-arm and the chest; lower surfaces lead-coloured, sprinkled with blackish.

	millim.		millim.
Total length	98	From end of snout to vent	33
Head	9	Fore limb	12
Width of head	5·5	Hind limb	18
From end of snout to fore limb	14	Tail	65

Ecuador.

a. Hgr. Intac. Mr. Buckley [C.]. (Type.)

18. CERCOSAURA.

Cercosaura, *Wagler, Syst. Amph.* p. 158; *Wiegm. Herp. Mex.* p. 10; *Peters, Abh. Berl. Ac.* 1862, p. 174.
Emminia, *Gray, Cat. Liz.* p. 24.
Cercosaura, part., *Gray, l. c.* p. 60.

Tongue moderately elongate, arrow-headed. Lateral teeth compressed, bi- or tricuspid. Head with large regular shields; frontonasal separating the nasals; præfrontals and frontoparietals present; nostril pierced in the middle of a single nasal. Eyelids developed, lower with a transparent disk composed of several scales. Ear exposed. Limbs well developed, pentadactyle. Dorsal scales large, subquadrangular, slightly imbricate, strongly keeled, arranged in regular transverse and longitudinal series; lateral scales small; ventral plates large, quadrangular, smooth, arranged in regular longitudinal and transverse series. An indistinct collar fold. Femoral pores few. Tail cyclotetragonal.

Brazil.

1. Cercosaura ocellata.

Emminia olivacea, *Gray, Cat.* p. 24.
Cercosaura ocellata, *Gray, l. c.* p. 60.
Cercosaura ocellata, *Wagl. l. c.*; *Peters, l. c.* p. 175, pl. i. fig. 1.
Emminia olivacea, *Gray, Proc. Zool. Soc.* 1851, pl. vi. fig. 2.
Cercosaura humilis, *Peters, l. c.* p. 180, pl. i. fig. 2.
—— olivacea, *O'Shaughn. Proc. Zool. Soc.* 1881, p. 228.

Head depressed, body moderately elongate. Ear-opening large. The ordinary head-shields; interparietal large, longer than the parietals; two large lateral and a small median occipital; four supraoculars; a loreal and a frenoorbital; temples with large shields superiorly, scaly inferiorly; six upper and four or five lower labials; chin-shields very large, one anterior and four pairs, the first two pairs in contact on the median line, the two others separated by a row of granules; two longitudinal rows of large imbricated transverse gular shields; three collar-shields. Dorsal scales very large, slightly rounded posteriorly, imbricate, a little oblique, with strong keels forming straight continuous lines along the back; they form eight longitudinal series, and, including the nuchal scales, which are more irregular and feebly keeled, thirty transverse series; lateral scales much smaller, feebly keeled, forming rather irrregular transverse series. Ventral plates large, in six longitudinal and twenty transverse series, the outer series smallest. Præanal plates normally four, a small anterior pair and a very large posterior; these shields are sometimes more divided. Fore limbs with large smooth shields; hind limbs inferiorly with very large smooth shields; scales on upper face of tibia keeled; posterior femoral scales granular. Five femoral pores in the male, one or two in the female. Tail long, gradually tapering to a fine point, covered with quadrangular scales forming regular annuli; those on the upper and lateral surfaces longer than broad and keeled, those on the lower surface broader than long, smooth, and arranged in two longitudinal series. Olive above; females uniform, males with a lateral series of large black ocelli with white centre *; lower surfaces yellowish white.

	millim.		millim.
Total length	174	From end of snout to vent	54
Head	13	Fore limb	18
Width of head	7	Hind limb	24
From end of snout to fore limb	20	Tail	120

Brazil.

a. ♀. Pernambuco Mrs. J. P. G. Smith [P.]. (Type of *Emminia olivacea.*)

b. ♀. Para.

19. PLACOSOMA.

Placosoma, *Tschudi, Arch. f. Naturg.* 1847, p. 50.
Urosaura, *Peters, Mon. Berl. Ac.* 1870, p. 641.

Tongue moderately elongate, arrow-headed. Lateral teeth compressed, bi- or tricuspid. Head with large regular shields; frontonasal separating the nasals; præfrontals and frontoparietals present; nostril pierced in the middle of a single nasal. Eyelids developed, lower with a transparent disk composed of several scales. Ear exposed. Limbs well developed, pentadactyle. Dorsal scales large, subquadrangular, slightly imbricate, smooth, arranged in

* It seems to me that the presence of these ocelli is a sexual character; the type specimen which has them is a male, whilst the two specimens in the British Museum (*E. olivacea*) as well as Peters's *C. humilis*, which show no trace of such markings, are females. This view is besides fortified by our knowledge of *Prionodactylus oshaughnessyi*, in which species the ocelli are much less distinct in the females.

regular transverse series; lateral scales smaller, feebly keeled; ventral plates large, quadrangular, smooth, arranged in regular longitudinal and transverse series. A collar fold. Tail cylindrical. Male with præanal and femoral pores.

Brazil.

1. Placosoma cordylinum.

Placosoma cordylinum (*Fitz.*), *Tschudi, l. c.* p. 51.
Cercosaura (Urosaura) glabella, *Peters, l. c.* pl. —. fig. 1.

Head depressed, pointed; body rather short. Upper head-shields as in *Cercosaura*, but no median occipital; temples with rather large shields; seven upper labials, only a row of granular scales separating them from the lower eyelid; six lower labials; chin-shields, one anterior and four pairs, the anterior two forming a suture, the others widely separated; gular scales irregular, juxtaposed, largest between the chin-shields and on the median line in front of the collar, where they form a short double longitudinal series; collar-shields six or seven. Dorsal scales large, about equilateral, slightly rounded posteriorly, slightly imbricate, smooth; they form regular transverse series, which are interrupted on the median line of the back, where the series alternate; the shields on each side the dorsal line the largest; on the flanks the scales become gradually smaller and more irregular, forming oblique as well as transverse series, and are rather indistinctly keeled; twenty scales round the body (including ventrals), and thirty-three from occiput to sacrum inclusively. Ventral shields not larger than dorsals, in six longitudinal and twenty-two or twenty-three transverse series. Præanal shields four, two median and two larger laterals. Limbs with large shields; hinder side of thighs granular. Males with an uninterrupted series of femoral and præanal pores; twenty pores altogether. Tail very long, gradually tapering to a fine point, covered with quadrangular scales forming annuli; the scales on the upper and lateral surfaces longer than broad and very slightly keeled, those on the lower surface equilateral, smooth, and forming two longitudinal series. Brownish yellow above, with a brown band along each side, commencing from the snout and passing through the eye; limbs with a few yellow spots; a median series of brown spots on the upper surface of the tail; lower surfaces cream-coloured, finely dotted with brown.

	millim.		millim.
Total length (tail injured)	128	From end of snout to vent	43
Head	11	Fore limb	13
Width of head	6	Hind limb	18
From end of snout to fore limb	17		

Brazil.

a. ♂.	Rio Janeiro.	Mrs. Fry [P.].
b. ♂.	Brazil.	T. Bell, Esq. [P.].

20. ANADIA.

Anadia, *Gray, Cat. Liz.* p. 74.
Euspondylus, part., *Peters, Abh. Berl. Ac.* 1862, p. 206.
Xestosaurus, *Peters, l. c.* p. 216.
Chalcidolepis, *Cope, Journ. Ac. Philad.* (2) viii. 1876, p. 116.

Tongue moderately elongate, arrow-headed. Lateral teeth compressed, tricuspid. Head with large regular shields; frontonasal separating the nasals; præfrontals and frontoparietals present; nostril pierced in the middle of a single nasal, or in the suture between two nasals. Eyelids developed, lower with a more or less transparent disk composed of several scales. Ear exposed. Limbs well developed, pentadactyle. Scales all smooth, forming uninterrupted annuli round the body; dorsals juxtaposed or slightly imbricate, subhexagonal, forming transverse and oblique series; ventrals little larger than dorsals, quadrangular, forming regular longitudinal and transverse series. A collar fold. Tail cylindrical. Males with femoral pores, females without.

Costa Rica to Ecuador.

Synopsis of the Species.

I. Supraoculars four; body and tail very long, limbs feeble.

30 scales round the middle of the body, and 58 from occiput to sacrum inclusively; anterior supraocular smallest .... 1. *ocellata*, p. 398.
33 scales round the middle of the body and 48 from occiput to sacrum; posterior supraocular smallest .... 2. *rhombifera*, p. 399.
23 scales round the middle of the body, and 43 from occiput to sacrum .... 3. *metallica*, p. 400.

II. Two supraoculars .... 4. *bogotensis*, p. 400.

1. Anadia ocellata.

Anadia ocellata, *Gray, Cat.* p. 74.
Anadia ocellata, *Gray, Proc. Zool. Soc.* 1851, pl. vi. fig. 1.
Ecpleopus? (Euspondylus) ocellatus, *Peters, Abh. Berl. Ac.* 1862, p. 212.

Head depressed, snout pointed; body much elongate. Nostril in the suture between two nasals. Frontonasal pentagonal; a pair of præfrontals forming a long suture; frontal pentagonal, narrow posteriorly; a pair of large anterior and a pair of large posterior parietals; a large parietal; several small occipitals, the smallest central; four supraoculars, first smallest, second largest, third separated from supraciliaries by a small supplementary shield; a loreal, a frenoorbital, and a row of infraorbitals; temples with rather large shields; seven or eight upper and six lower labials; chin-shields one anterior and four pairs, the first two and the third anteriorly forming a suture; gular scales subequal, squarish, juxtaposed; collar-scales seven, scarcely enlarged. Scales of the body subequal, the dorsals quadrangular-subhexagonal, slightly longer than broad, the laterals a little smaller and more oval, the ventrals regularly quadrangular; about thirty scales round the middle of the body, and fifty-eight from occiput to sacrum inclusively. Præanal plates small, in three rows, the anterior composed of two, the next of four, and the last of six shields, the latter elongate. Limbs short, covered with large scales, except on the hinder side of the thighs and the lower surface of the arms. Ten femoral pores on each side. Tail long and thick, covered with equal elongate quadrangular smooth scales, arranged in rings. Light greyish, with strong metallic gloss; a darker band on each side, best marked between the eye and the fore limb; three or four black ocelli with blue centre on each side forming a short series, beginning from above the fore limb; a few blackish dots on the labials and on the base of the tail.

	millim.		millim.
Total length	138	From end of snout to vent	48
Head	10	Fore limb........	10·5
Width of head	5·5	Hind limb	14
From end of snout to fore limb	16	Tail	90

Hab. ——?

a. ♂. ——? (Type.)

2. Anadia rhombifera.

Cercosaurus rhombifer, *Günth. Proc. Zool. Soc.* 1859, p. 405, pl. xx. fig. A.
Ecpleopus (Euspondylus) rhombifer, *Peters, l. c.* p. 208.

Closely allied to the preceding; differs in the following points:—Nostril in a single plate; præfrontal suture shorter; the first supraocular is absent, but there is a small posterior fourth; thirty-three scales round the middle of the body, and forty-eight from occiput to sacrum inclusively. Colour as in the preceding, but without ocelli; upper surface of tail with a dark brown zigzag band.

	millim.		millim.
Total length	182	From end of snout to vent	57
Head	12	Fore limb........	14
Width of head	7	Hind limb	18
From end of snout to fore limb	20	Tail	125

Ecuador.

a. ♀. W. Ecuador. Mr. Fraser [C.]. (Type.)

3. Anadia metallica.

Chalcidolepis metallicus, *Cope, Journ. Ac. Philad.* (2) viii. 1876, p. 116, pl. xxiv. fig. 5.
Anadia metallica, *Günth. Biol. C.-Am., Rept.* p. 30.

Head narrowed and acute in front, with produced rostral shield; body much elongate; limbs feeble. Lepidosis much as in the preceding. Nostril in a single nasal; twenty-three scales in an annulus of the body, and forty-three from occiput to sacrum inclusively; two large longitudinal præanal scuta, which embrace a scale between them on the anal border; they are preceded by another large pair, but of reduced size. Colour light grey with red and green metallic reflections; the sides are brown, and the middle of the back darker than a line above the brown of the side; near the light bands a few scales are blackish, forming a row on each side; below dusted with brown; head with deep brown sides and white upper lip; sides of tail brown, with a zigzag upper margin.

	millim.		millim.
Total length	165	Head and body ..	58
Head to tympanum .	11	Fore limb	13
Head to axilla	17	Hind limb	14·5

Aguacate Mountains, Costa Rica.

4. Anadia bogotensis.

Ecpleopus (Xestosaurus) bogotensis, *Peters, l. c.* p. 217, pl. iii. fig. 3.

Head small, snout short, obtusely pointed; body rather elongate. Nostril in a single plate; supraoculars two, subequal; a large interparietal; three subequal occipitals; a small loreal and a frenoorbital; a row of small infraorbitals; temples with large shields; seven or eight upper and six or seven lower labials; chin-shields, one anterior and four pairs, the two first in contact; gular scales roundish subquadrangular, increasing in size towards the collarfold; collar-shields six or seven. Scales of the body subequal, slightly imbricate, squarish, with rounded posterior border, slightly longer than broad; twenty-eight scales round the middle of the body, and forty from occiput to sacrum inclusively. Præanals small, in two or three rows, the posterior composed of six plates. Exposed parts of limbs with large scales. [Males with seven to fourteen femoral pores.] Tail thick, covered with subequal elongate quadrangular scales forming regular annuli. Olive-brown above, with rather indistinct darker markings; lower surfaces lead-colour.

	millim.		millim.
Total length	120	From end of snout	
Head	11	to vent	55
Width of head	7	Fore limb	14
From end of snout		Hind limb	18
to fore limb ...	18	Tail (reproduced) ..	65

Sta Fé de Bogota.

a–b. ♀ & hgr. Sta Fé de Bogota. Prof. Peters [P.]. (Typical specimens.)

21. ECPLEOPUS.

Ecpleopus, *Dum. & Bibr.* v. p. 434; *Tschudi, Arch. f. Naturg.* 1847, p. 47; *Peters, Abh. Berl. Ac.* 1862, p. 195.
Cercosaura, part., *Gray, Cat.* p. 60.
Aspidolæmus, *Peters, l. c.* p. 199.

Tongue moderately elongate, arrow-headed. Lateral teeth compressed, bi- or tricuspid. Head with large regular shields; frontonasal separating the nasals; præfrontals and frontoparietals present; nostril pierced in the middle of a single nasal or in the suture between two nasals. Eyelids developed, lower with a non-transparent disk composed of several scales. Ear exposed. Limbs well developed, pentadactyle. Dorsal scales juxtaposed or but slightly imbricate, elongate subhexagonal, smooth or very slightly keeled, arranged in regular transverse series; lateral scales small; ventral plates large, quadrangular, smooth, arranged in regular longitudinal and transverse series. A slight collar fold. Tail cylindrical. No præanal or femoral pores.

South America.

1. Ecpleopus gaudichaudii.

Cercosaura gaudichaudi, *Gray, Cat.* p. 60.
Ecpleopus gaudichaudii, *Dum. & Bibr.* v. p. 436; *Tschudi, l. c.*; *Guichen. in Casteln. Amér. Sud, Rept.* p. 34, pl. vii. fig. 2; *Peters, l. c.* p. 195.

Nasal single; no loreal; præfrontals forming a suture, separated from the loreal by the anterior supraciliary, which is large, and reaches the frontonasal; frontal twice as long as broad, smaller than the interparietal; latter much longer than broad, as broad as, and a little shorter than, the parietals; three supraoculars. Gular scales small, squarish. Dorsal scales longer than broad, quadrangular, obtusely keeled, in thirty-three transverse series from the occiput to the base of the tail; ventrals scarcely larger than dorsals, smooth. Three nearly equal præanals, on a transverse row. Reddish brown above, with one or two light longitudinal lines on each side; lower surfaces white.

	millim.		millim.
Total length	120	Fore limb........	9
Head	10	Hind limb	12
From end of snout to vent	40		

Brazil.

2. Ecpleopus affinis.

Cercosaurus gaudichaudii (*non D. & B.*), *Günth. Proc. Zool. Soc.* 1859, p. 405.
Ecpleopus (Aspidolæmus) affinis, *Peters, l. c.* p. 199, pl. iii. fig. 1.

Snout short, obtuse; body rather elongate. Two nasals, nostril between; a loreal; præfrontals usually forming a short suture; frontal broad; interparietal a little larger than the latter and than the parietals; three small occipitals, median smallest; three supra-oculars; anterior supraciliary large, not reaching the frontonasal; a row of small infraorbitals; temples scaly; seven or eight upper labials, fourth largest, under the eye: six lower labials; chin-shields, one anterior and four pairs, the two first pairs forming a suture, the others widely separated by the rather larger anterior gulars; a short double longitudinal row of transversely enlarged gulars in front of the collar; latter composed of six to eight scales. Dorsal scales longer than broad, hexagonal, feebly keeled*, separated from the ventrals by very small irregular scales; thirty-nine to forty-one scales round the middle of the body, ventrals included; forty-six to forty-eight scales from occiput to sacrum inclusively. Ventral plates quadrangular, smooth, in eight or ten longitudinal and twenty-seven or twenty-eight transverse rows. Præanals six, two pairs larger; sometimes a small additional plate wedged in between the posterior pair. Limbs with large smooth shields; upper surface of tibia with small keeled scales. Tail long, tapering to a fine point, covered with scales similar to those of the body, the upper and lateral keeled, the lower smooth. Olive-brown above; sides closely spotted with black; a yellow, black-edged line on each side, along the canthus rostralis and supraciliaries, becoming indistinct on the body; a series of small black spots along the middle of the body, more or less confluent into a vertebral line; lower surfaces greenish white, spotted with blackish.

	♂. millim.	♀. millim.
Total length	162	193
Head	13	14
Width of head	8	8
From end of snout to fore limb	20	23
From end of snout to vent	52	65
Fore limb	18	19
Hind limb	27	28
Tail	110	128

Andes of Ecuador.

a, b, c. ♂ ♀.	W. Ecuador.	Mr. Fraser [C.].
d. ♀.	Intac.	Mr. Buckley [C.].

* The scales are perfectly smooth when the epidermis is removed, and such is probably the case with Peters's type specimen.

22. PHOLIDOBOLUS.

Pholidobolus, *Peters, Abh. Berl. Ac.* 1862, p. 195.

Tongue moderately elongate, arrow-headed. Lateral teeth bi- or tricuspid. Head with large regular shields; nasals separated by the frontonasal; no præfrontals; frontoparietals present; nostril pierced in the suture of a divided nasal*. Eyelids developed, lower with a non-transparent disk. Ear exposed. Limbs well developed, pentadactyle. Dorsal scales elongate hexagonal, striated, imbricate, separated from the ventrals by small scales; ventral plates large, smooth, quadrangular, forming regular longitudinal and transverse series. A collar fold. Tail cylindrical. No femoral pores in either sex.

Ecuador.

1. Pholidobolus montium.

Ecpleopus (Pholidobolus) montium, *Peters, l. c.* p. 196, pl. ii. fig. 3.

In habit very similar to *Lacerta vivipara*. Snout short, obtuse. Frontonasal large, subquadrangular or trapezoid, a little broader than long; frontal narrowed posteriorly; frontoparietals forming a rather long suture; parietals short; interparietal as long as, but narrower than, latter; three occipitals, median smallest; two supraoculars; a loreal and a frenoorbital; infraorbitals very small; temples shielded; seven upper and five or six lower labials; chin-shields, one anterior and four pairs, the two first pairs forming a suture; gular scales irregular between the rami of the mandible, and a double longitudinal row of large transverse ones to the collar; collar-shields six to eight. Dorsal scales rather large, regularly hexagonal, each with three feeble keels; lateral scales small, more irregular; about thirty-five scales round the middle of the body, including ventrals, and thirty-three to forty from occiput to sacrum inclusively. Ventral plates large, square, in eight longitudinal and twenty-two to twenty-six transverse series. A pair of large præanals, bordered anteriorly by smaller shields. Limbs with smooth or feebly striated shields; lower surface of arm, hinder side of thighs, and inner side of tibia granulate. Upper caudal scales elongate hexagonal, keeled; lower larger, subquadrangular, smooth; all forming regular annuli. Brown above, sides blackish; a black vertebral line; a light, black-edged streak from the tip of the snout along the canthus rostralis and supraciliaries to the side of the back, where it often gradually disappears; another along the upper lip to the fore arm; two more or less distinct parallel light lines along the flanks; lower surfaces olive or greyish, black-spotted.

* This shield is, however, undivided in one of our specimens; and Peters also gives it as undivided.

	♂ millim.	♀ millim.
Total length	123	155
Head	11	11
Width of head	7·5	8
From end of snout to fore limb	18	20
From end of snout to vent	51	57
Fore limb	16	18
Hind limb	21	25
Tail	72	98

Andes of Ecuador.

a–b. ♂ & yg.	W. Ecuador.	Mr. Fraser [C.].
c. ♂.	Quito.	
d–i. ♂ ♀.	Intac, Ecuador.	Mr. Buckley [C.].
k. ♂.	Pichincha, Ecuador. 11,000 feet.	Edw. Whymper, Esq. [P.].
l. ♀.	Peru?	
m–n. ♀ & yg.	——?	Mme. Pfeiffer [C.].

23. EUSPONDYLUS.

Euspondylus, *Tschudi, Arch. f. Naturg.* 1845, p. 160; *Peters, Abh. Berl. Ac.* 1862, p. 206.

Tongue moderately elongate, arrow-headed. Lateral teeth bi- or tricuspid, Head with large regular shields; frontonasal separating the nasals; præfrontals and frontoparietals present; nostril pierced in a single (or divided) nasal. Eyelids developed, lower with a transparent disk formed of several scales. Ear exposed. Limbs well developed, pentadactyle. Dorsal scales smooth or very feebly keeled, quadrangular, juxtaposed, forming transverse series, separated from the ventrals by small scales; ventral plates large, smooth, quadrangular, forming regular longitudinal and transverse series. A strong collar fold. Tail cylindrical. Males with femoral pores, females without.

Venezuela to Peru.

Synopsis of the Species.

I. Labials separated from orbit by a row of very small infraorbitals.

Dorsal scales much shorter than ventrals; nasal in contact with the frenoorbital .. 1. *strangulatus*, p. 405.

Dorsal scales nearly as long as ventrals; loreal separating the nasal from the frenoorbital .... 2. *maculatus*, p. 406.

II. Infraorbitals large.

19 transverse series of scales from the chin-shields to the edge of the collar, and 42 from occiput to base of tail.......... 3. *acutirostris*, p. 407.

11 transverse series of scales from the chin-shields to the edge of the collar, and 32 from occiput to base of tail.......... 4. *guentheri*, p. 407.

1. Euspondylus strangulatus.

Euspondylus strangulatus, *Cope, Proc. Ac. Philad.* 1868, p. 99.
Ecpleopus (Euspondylus) strangulatus, *O'Shaughn. Proc. Zool. Soc* 1881, p. 236.

Head not depressed, large and distinct from neck, especially in the males; body moderately elongate; limbs long. Interparietal large, longer than parietals; occipitals split up into numerous irregular shields; four supraorbitals, anterior small; nasal, frenoorbital, and loreal of subequal size, the latter resting on the two former, the three plates forming a triangle; a row of infraorbitals; temples granular inferiorly; six upper and five or six lower labials; chin-shields, one anterior and four pairs, the two first pairs forming a suture; gular scales rather small, round, larger near the chin-shields, becoming largest and quadrangular near the collar; collar-shields six or seven. Dorsal scales rather small and rather irregular, a little oblique, the transverse series broken up on the median dorsal line; these scales are subquadrangular and slightly keeled; lateral scales very small, almost granular; forty-five to forty-eight scales round the middle of the body, including ventrals; nuchal scales minute, granular. Ventral shields large, square, in eight longitudinal and twenty-two to twenty-four transverse series. A pair of large præanal plates, bordered anteriorly by a row of small plates. Limbs with smooth shields; lower surface of arm and hinder side of thigh granulate; tibia above with small feebly keeled scales. Male with twenty-six to twenty-nine femoral pores on each side. Caudal scales quadrangular, forming regular annuli, narrow and feebly keeled above, larger and smooth inferiorly. Olive-brown above, with more or less distinct large round darker spots; sides of body with more or less marked light spots or ocelli, the three or four anterior of these ocelli being very strongly marked in the young; sides and lower surface of the head blackish in the males; lips and throat white and spotted with black in the females; belly cream-coloured.

	♂ millim.	♀ millim.
Total length	233	202
Head	25	18
Width of head	17	11
From end of snout to fore limb	40	30
From end of snout to vent	98	78
Fore limb	33	27
Hind limb	50	40
Tail	135	124

Ecuador.

a–d. ♂, ♀, & yg.	Canelos.	Mr. Buckley [C.].
e. ♂.	Sarayacu.	Mr. Buckley [C.].
f–g. ♀.	Pallatanga.	Mr. Buckley [C.].

2. Euspondylus maculatus.

Euspondylus maculatus, *Tschudi, Arch. f. Naturg.* 1845, p. 160, *and Faun. Per., Herp.* p. 42, pl. ii. fig. 1.
Ecpleopus (Euspondylus) maculatus, *Peters, l. c.* p. 206, pl. ii. fig. 4.
—— (Proctoporus) fraseri, *O'Shaughn. Ann. & Mag. N. H.* (5) iv. 1879, p. 296.

Head depressed, moderately large; body moderately elongate. The ordinary head-shields *; parietals short, broad, interparietal narrow; four occipitals; four supraorbitals, anterior smallest; a loreal and a frenoorbital; a row of extremely small infraorbitals; temporal shields large; seven upper and six lower labials; chin-shields, one anterior and four pairs, the two first forming a suture; gular scales subquadrangular, three transverse rows in front of the collar largest; collar-shields nine. Dorsal scales quadrangular, slightly rounded posteriorly, about twice as long as broad, straight, slightly keeled on the posterior part of the body; lateral scales very small, subquadrangular; forty scales round the middle of the body, including ventrals, and thirty-five from occiput to sacrum inclusively. Ventrals larger than dorsals, more equilateral, in ten longitudinal and twenty-two transverse series. Præanal shields in two rows, anterior composed of two large shields, posterior containing five narrower ones. Limbs with the normal scaling. Digits rather elongate. [Male with six femoral pores on each side.] Caudal scales quadrangular, forming regular annuli, upper narrower and slightly keeled. Olive above, with small black spots; a black lateral streak, commencing from the eye, becoming indistinct on the hind part of the body; lips black-spotted; lower surfaces whitish, with indistinct small dark spots.

	millim.		millim.
Total length	135	From end of snout to vent	55
Head	13	Fore limb	16
Width of head	75	Hind limb............	23
From end of snout to fore limb	23	Tail (injured)	80

Northern Peru; Ecuador.

a. ♀.	Guayaquil.	Mr. Fraser [C.].	(Type of *Ecpleopus fraseri.*)

* In the unique specimen in the British Museum the præfrontals are abnormally united with the frontonasal, on which account it has been referred to *Proctoporus*. I have examined a second specimen, a large male, in the Brussels Museum; the head-shields are normal, but the nasal is divided.

3. Euspondylus acutirostris.

Ecpleopus (Euspondylus) acutirostris, *Peters, l.c.* p. 209.

Snout acutely pointed. Scaling apparently very similar to that of *E. guentheri*, but with the following differences:—Nineteen transverse series of scales from the chin-shields to the edge of the collar; nuchal scales smaller, more hexagonal; dorsal scales feebly keeled; forty-two scales from occiput to base of tail, and twenty-seven transverse rows of ventral shields. Ten or eleven femoral pores on each side. Greenish brown above; a white, inferiorly black-edged, line on each side, beginning from above the eye and breaking up into spots on the hinder half of the back; sides of body brown, with a series of seven to nine very small ocelli with white centre; lower surface of head and body greenish bronze.

	millim.		millim.
Total length	157	Fore limb	16
Head	15	Hind limb	21
Width of head	8	Tail	101
From end of snout to vent	56		

Venezuela.

4. Euspondylus guentheri.

Ecpleopus (Euspondylus) guentheri, *O'Shaughn. Proc. Zool. Soc.* 1881, p. 235, pl. xxiii. fig. 1.

Head not distinct from neck, body slender and elongate. Frontonasal longer than broad; interparietal large, longer than parietals, hexagonal; parietals forming an oblique suture with a large supratemporal; a pair of large occipitals; four supraorbitals, anterior smallest; loreal large, no frenoorbital; a series of large infraorbitals; temples with large shields; seven upper and six lower labials; chin-shields, one anterior and four pairs, the two first pairs forming a suture; gular scales large, quadrangular, forming regular transverse series, largest anteriorly; collar-shields eight; eleven transverse series of scales from the chin-shields to the edge of the collar. Dorsal scales large, regular, perfectly smooth, elongate quadrangular, as long as, but narrower than the ventrals; lateral scales very small, suboval; thirty-five scales round the middle of the body, including ventrals, and thirty-two from occiput to sacrum inclusively. Ventral plates in ten or twelve longitudinal and twenty-one transverse series. Two large anterior and five narrow posterior præanals. Limbs with large smooth shields; digits elongate. Male with eight præanal pores on each side. Caudal scales like those of the body. Pale brown above, with eight transverse black bars on the neck and back, some of which are broken up; similar bars and spots on the tail; head above with round black spots; lower surfaces yellowish white, with roundish dark-brown spots, largest on the labial region.

	millim.		millim.
Total length	249	From end of snout to vent	94
Head	22	Fore limb	30
Width of head	12·5	Hind limb	41
From end of snout to fore limb	37	Tail	155

Ecuador.

a. ♂. Sarayacu. Mr. Buckley [C.]. (Type.)

24. ARGALIA.

Argalia, *Gray, Ann. & Mag. N. H.* xviii. 1846, p. 67; *Peters, Abh. Berl. Ac.* 1862, p. 213.

Tongue moderately elongate, arrow-headed. Lateral teeth compressed, bi- or tricuspid. Head with large regular shields; frontonasal separating the nasals; præfrontals and frontoparietals present; nostril pierced in the middle of a single nasal. Eyelids developed, lower with a non-transparent disk. Ear exposed. Limbs well developed, pentadactyle. Scales perfectly smooth, quadrangular, juxtaposed, forming regular transverse series on the back, and transverse and longitudinal series on the belly; the abdominal scales considerably smaller than the dorsals; the lateral scales smaller and irregular. A strong collar fold. Tail cylindrical. Femoral pores in both sexes.

Venezuela and Colombia.

1. Argalia marmorata.

Argalia marmorata, *Gray, l. c.*
—— olivacea, *Gray, Proc. Zool. Soc.* 1847, p. 97, and 1858, pl. xv. fig. 1.
Gerrhonotus pœcilochilus, *Licht. Nomencl. Rept. Mus. Berol.* p. 16.
Ecpleopus (Argalia) olivaceus, *Peters, l. c.*
—— (——) marmoratus, *Peters, l. c.*
—— (——) pœcilochilus, *Peters, l. c.*

Appearance somewhat similar to that of a *Gerrhonotus*. Head depressed, large and distinct from neck in the males. Frontonasal longer than broad, pentagonal; a pair of præfrontals forming a short or very short suture; frontal short; frontoparietals moderate; parietals forming an oblique suture with a large supratemporal; interparietal hexagonal, longer than the parietals; a pair of large occipitals; four supraoculars; a loreal; a row of rather large infraorbitals; temples with large shields; seven or eight upper and six or seven lower labials; chin-shields, one anterior and four pairs, the two first pairs forming a suture; gular scales squarish, subequal, forming transverse rows; collar-shields nine to eleven; fourteen to sixteen transverse rows of scales between the chin-shields and the edge of the collar. Scales quadrangular, at least twice as long as broad on the back, shorter on the belly; lateral scales smaller, sub-oval; thirty to thirty-five scales round the middle of the body, thirty from occiput to sacrum inclusively, and twenty-seven or twenty-eight from collar fold to præanal region. Præanal shields in two rows, four in the first, six in the second. Limbs with large smooth shields; hinder side of thighs with small scales. Nine to twelve femoral pores on each side. Caudal scales elongate quadrangular, smooth, forming regular annuli. Olive-brown above, with or without reddish brown marblings; flanks with equidistant groups of cream-coloured spots; the sutures of the head-shields generally dark brown; lower surfaces yellowish.

	♂. millim.	♀. millim.
Total length	228	212
Head	25	19
Width of head	17	12
From end of snout to fore limb	38	29
From end of snout to vent	93	82
Fore limb	28	21
Hind limb	37	28
Tail	135	130

Colombia; Venezuela.

a–b. ♀.	Colombia.	(Types.)
c–d. ♂.	Venezuela.	(Types of *A. olivacea.*)

25. OREOSAURUS.

Oreosaurus, *Peters, Abh. Berl. Ac.* 1862, p. 201.

Tongue moderately elongate, arrow-headed. Lateral teeth compressed, bi- or tricuspid. Head with large regular shields; frontonasal separating the nasals; no præfrontals; frontoparietals present; nostril pierced in the middle of a single nasal. Eyelids developed, lower with a non-transparent disk. Ear exposed. Limbs well developed, pentadactyle. Dorsal scales elongate quadrangular, feebly keeled, juxtaposed, forming regular transverse rows; lateral scales small; ventrals large, quadrangular, smooth, arranged in longitudinal and transverse series. A strong collar fold. Tail cylindrical. Males with femoral pores, females without.

Colombia, Ecuador, Brazil.

Synopsis of the Species.

I. A loreal; four supraoculars.

Thirty-five scales from occiput to base of tail; male with five or six femoral pores on each side . . . . 1. *oculatus*, p. 410.

Forty-three scales from occiput to base of tail; male with ten to twelve femoral pores on each side . . . . 2. *luctuosus*, p. 410.

II. No loreal.

Four supraoculars .... 3. *striatus*, p. 411.
Three supraoculars........................ 4. *petersii*, p. 411.

1. Oreosaurus oculatus. (Plate XX. fig. 2.)

Ecpleopus oculatus, *O'Shaughn. Ann. & Mag. N. H.* (5) iv. 1879, p. 297.

Head rather elongate; body moderately elongate; limbs well developed. Frontonasal quadrangular, longer than broad; frontal trapezoid, anteriorly as broad as the frontonasal; frontoparietals forming a long suture; parietals rather small; interparietal nearly as large, hexagonal, narrower anteriorly; a pair of occipitals; four supraoculars; a loreal; a row of large infraorbitals; temple shielded; six upper labials, posterior largest; five lower labials; chin-shields, one anterior and four pairs, the two first pairs forming a suture; these are followed by large, irregular hexagonal gulars; the other gulars quadrangular, increasing in size towards the collar-shields, which are largest and eight in number; seven transverse rows of quadrangular gulars, including collar. Dorsal scales quadrangular, keeled, about twice and a half as long as broad; lateral scales much smaller, half as long, also quadrangular and keeled; forty-one scales round the middle of the body, including ventrals; thirty-five scales from occiput to base of tail. Ventrals large, quadrangular, smooth, forming ten or twelve longitudinal and twenty transverse series. Præanals seven, an anterior pair and five shields in the posterior row. Limbs with large smooth shields; tibia above with keeled scales; lower surface of arm and hinder side of thighs granulate. Five or six femoral pores on each side. Caudal scales like those on the body. Brown above, with a lighter, dark-edged band on each side, commencing from the eye; a series of black ocelli with white centre along each side; gular and ventral scales black in the middle, white on the edges, forming black and white longitudinal bands.

	millim.		millim.
Total length (tail broken)	100	From end of snout to vent	63
Head................	15	Fore limb	20
Width of head	9	Hind limb	29
From end of snout to fore limb	27		

Ecuador.

a. ♂. Intac. Mr. Buckley [C.] (Type.)

2. Oreosaurus luctuosus.

Ecpleopus (Oreosaurus) luctuosus, *Peters, l. c.* p. 203.

Apparently closely allied to the preceding. Eight or nine transverse rows of scales between the enlarged anterior gulars and the edge of the collar. Forty-three scales from occiput to base of tail. Ventrals in ten longitudinal and twenty-two transverse series. Ten to twelve femoral pores on each side.

	millim.		millim.
Total length	188	From end of snout to vent	92
Head	23	Fore limb	27
Width of head	11	Hind limb	37
From end of snout to fore limb	34	Tail (reproduced) ..	96

Venezuela.

3. Oreosaurus striatus.

Ecpleopus (Oreosaurus) striatus, *Peters, l. c.* p. 201, pl. iii. fig. 2.

Head short, body and limbs moderate. Frontonasal quadrangular, as long as broad; frontal pentagonal, narrower than the frontonasal; parietals and interparietal subequal in size, the latter hexagonal; a pair of occipitals; four supraoculars; no loreal; a row of large infraorbitals; temple shielded; five or six upper and four or five lower labials; chin-shields, one anterior and three pairs, the two first pairs in contact; nine or ten transverse rows of scales between the chin-shields and the edge of the collar; collar-shields eight. Dorsal scales quadrangular, keeled, about twice as long as broad; lateral scales smaller, convex, subgranular; thirty-five scales round the middle of the body, ventrals included; thirty-five scales from occiput to base of tail. Ventrals large, quadrangular, in eight or ten longitudinal and twenty-three transverse series. [Males with six femoral pores on each side.] Caudal scales like those on the body. Light brown above, with a yellowish band on each side commencing from the supraciliary region, bordered on each side by a dark line; a more or less distinct dark median band; lower surfaces whitish, black-spotted.

	millim.		millim.
Total length	82	From end of snout to vent	33
Head	7·5	Fore limb	9
Width of head	4·5	Hind limb	12
From end of snout to fore limb	12	Tail	49.

Sta Fé de Bogota.

a. Hgr. Bogota.

4. Oreosaurus petersii.

Ecpleopus (Oreosaurus) petersii, *Boettger, Jahresb. Offenb. Ver. Nat.* xvii.–xviii. 1879, p. 9, pl. i. fig. 2.

Head very short; body elongate; limbs short. Interparietal very large, seven-sided; three occipitals, median smallest; three supraoculars; no loreal. Scaling otherwise as in *O. striatus.* Thirty-nine scales from occiput to base of tail, and twenty-six from collar fold to vent. Nine or ten femoral pores on each side. Light brown above, with small blackish dots and marblings; lower surfaces blackish brown.

	millim.		millim.
Total length	113	Width of head	6·5
Head	9	From snout to vent	47

Para.

26. PROCTOPORUS.

Proctoporus, *Tschudi, Arch. f. Naturg.* 1847, p. 53; *Peters, Abh. Berl. Ac.* 1862, p. 220.
Riama, *Gray, Proc. Zool. Soc.* 1858, p. 445.
Emphrassotis, *O'Shaughn. Ann. & Mag. N. H.* (5) iv. 1879, p. 295.

Tongue moderately elongate, arrow-headed. Lateral teeth conical, indistinctly bicuspid. Head with large regular shields; frontonasal separating the nasals; no præfrontals; frontoparietals present; nostril pierced in the middle of a single nasal. Eyelids developed, lower with a non-transparent disk. Ear exposed. Limbs well developed, short, pentadactyle. Dorsal scales elongate-quadrangular, smooth or striated, juxtaposed, forming regular transverse rows, separated from the ventral shields by a fold extending from axilla to groin; ventrals large, quadrangular, smooth, arranged in longitudinal and transverse series. A strong collar fold. Tail cylindrical. Femoral or præanal pores in both sexes.

Andes of Ecuador and Peru.

Synopsis of the Species.

I. Dorsal scales feebly keeled or striated.

Fifty-four scales from occiput to base of tail; a loreal .... 1. *pachyurus*, p. 412.
Thirty-seven to forty-four scales from occiput to base of tail; no loreal.......... 2. *unicolor*, p. 413.

II. Dorsal scales perfectly smooth; thirty-four scales from occiput to base of tail.

Thirty-six scales round the middle of the body .... 3. *simoterus*, p. 414.
Forty-four scales round the middle of the body .... 4. *meleagris*, p. 415.

1. Proctoporus pachyurus.

Proctoporus pachyurus, *Tschudi, l. c.* p. 54, *and Faun. Per., Herp.* p. 43, pl. ii. fig. 2.
Ecpleopus (Proctoporus) pachyurus, *Peters, l. c.* p. 221, pl. ii. fig. 2.

Body elongate, limbs weak. Frontonasal much longer than broad, larger than frontal; frontoparietals forming a long suture; interparietal narrower and a little shorter than parietals, pentagonal, hardly wider posteriorly than anteriorly, once and a half as long as broad; three subequal occipitals; three supraoculars; a loreal and a frenoorbital; a row of very small infraorbitals; temples irregularly shielded; chin-shields, one anterior and four pairs, the first two pairs entirely, the third anteriorly, in contact; gular scales subquadrangular, those in contact with the chin-shields a little larger; about twelve transverse rows of scales between the chin-shields and the edge of the collar; about ten collar-shields. Nuchal scales very small, round; dorsal scales narrow, quadrangular, feebly keeled or striated, the transverse series interrupted on the median dorsal line; a lateral line of granular scales between the dorsals and ventrals; fifty-four scales between the occiput and the base of the tail, fourteen of which are anterior to the fore limbs. Ventral plates quadrangular, in twelve longitudinal and twenty-six transverse series. A pair of anterior præanal shields, and a second row of six narrow elongate shields, with a small triangular one wedged in between the median pair posteriorly. Limbs with smooth shields. Females with two or three femoral pores on each side. Tail thick, covered with scales like those of the body. Dark brown above, bluish grey inferiorly, dotted with darker.

	millim.		millim.
Total length (tail broken)	87	From end of snout to vent	58
Head	12·5	Fore limb	13
Width of head	7	Hind limb	17
From end of snout to fore limb	17·5		

Chanchamayo River, eastern slope of the Peruvian Andes.

2. Proctoporus unicolor.

Riama unicolor, *Gray, Proc. Zool. Soc.* 1858, p. 446, pl. xv. fig. 2.
Proctoporus pachyurus (*non Tsch.*), *Günth. Proc. Zool. Soc.* 1859, p. 89.
Ecpleopus (Proctoporus) unicolor, *Peters, l. c.* p. 223.

This species differs from *P. pachyurus* by the still more elongate body and weaker limbs, the less depressed head, the shorter snout, and in the following points:—Interparietal hexagonal, considerably broader posteriorly than anteriorly, longer than parietals; a median occipital is frequently absent, and, if present, is very small; no loreal; temporal shields larger; only the two anterior pairs of chin-shields in contact; they are followed by several large similar shields, the median pair resembling an additional pair of chin-shields; collar-shields more elongate, eight or nine. Nuchal scales much larger and more quadrangular; thirty-seven to forty-four scales from occiput to sacrum inclusively; thirty-six to thirty-eight scales round the middle of the body; twenty-three to twenty-six transverse rows of ventrals. Præanals five or seven, a small pair anteriorly, three or five elongate shields forming a second row. Males with nine to twelve femoral pores on each side, one or two of which may be termed præanal; only two præanal and one or two femoral pores on each side in the females. Dark brown above, uniform or dotted with black; a more or less distinct short light line or series of light dots on each side on the neck and anterior part of the body; lower surfaces blackish or lead-coloured; throat and lower surface of tail generally variegated with whitish.

	millim.		millim.
Total length	138	From end of snout to vent	63
Head	13	Fore limb	13
Width of head	8·5	Hind limb........	17
From end of snout to fore limb	23	Tail	75

Ecuador.

a. ♂.	W. Ecuador.	Mr. Fraser [C.]. (Type.)
b–c, d–e. ♂.	W. Ecuador.	Mr. Fraser [C.].
f–g. ♂.	Intac.	Mr. Buckley [C.].
h–k. ♂ ♀.	Hacienda de Ollalla, 8490 feet.	Edw. Whymper, Esq. [C.].
l–o. ♂, ♀, & yg.	Chillo, 9000 feet.	Edw. Whymper Esq. [C.].

3. Proctoporus simoterus. (Plate XXII. fig. 1.)

Emphrassotis simoterus, *O'Shaughn. Ann. & Mag. N. H.* (5) iv. 1879, p. 296.

The unique specimen is evidently anomalous in its head-shields, and this is the cause of the absence of frontoparietals and loreal, mentioned in the original description; this is so much the more certain as traces of sutures still indicate the normal position of the missing shields. Head short and thick, snout very short, rounded; body elongate, limbs weak. Frontonasal quadrangular, longer than broad; frontal of nearly the same size; parietals large; interparietal narrow, wedge-shaped, extending posteriorly beyond the parietals; a very large lateral occipital on each side; no median occipital; four supraoculars; a row of rather large infraorbitals, fourth largest, between three labials; temporal shields very large; upper labials five or six, last very large, penultimate very small; four lower labials; chin-shields, one anterior and four pairs, the two first pairs in contact; gular scales large, quadrangular, anterior row very large; eight transverse rows of plates between the chin-shields and the edge of the collar; collar-shields eight, large, elongate. Dorsal scales elongate-quadrangular, perfectly smooth, the transverse series interrupted on the median dorsal line; thirty-four scales from occiput to sacrum inclusively; thirty-six scales round the middle of the body. Ventrals broader, but not longer than dorsals, in twelve longitudinal and twenty-two transverse series. Præanals in two transverse rows, the anterior composed of a pair of quadrangular shields, the posterior of five elongate shields, of which the median and the outer are very narrow. Limbs with large smooth shields.

Six femoral pores on each side. Caudal scales like those of the body. Above light brown, thickly variegated with black; sides black; lower surfaces dark lead-coloured.

	millim.		millim.
Total length (tail reproduced)	105	From end of snout to vent	73
Head	15	Fore limb	15
Width of head	11	Hind limb	19
From end of snout to fore limb	26		

Ecuador.

a. ♂. Intac. Mr. Buckley [C.]. (Type.)

4. Proctoporus meleagris. (Plate XXII. fig. 2.)

Head small, short; body elongate, limbs weak. Frontonasal longer than broad; frontal slightly smaller; frontoparietals moderate; parietals forming an oblique suture with a large supratemporal; interparietal narrow, not widened posteriorly, as long as parietals; three occipitals, median very small; four supraoculars, anterior small; a loreal; a row of large infraorbitals; temporal shields large; six or seven upper and four lower labials; chin-shields, one anterior and three pairs, the two first pairs forming a suture; gulars large, quadrangular, a row in contact with the chin-shields very large and elongate; eight rows of shields between the chin-shields and the edge of the collar; collar-shields eight, large, elongate. Dorsal scales elongate quadrangular, perfectly smooth, the transverse series interrupted on the median dorsal line; thirty-four scales from occiput to sacrum inclusively; forty-four scales round the middle of the body. Ventrals broader, but not longer than dorsals, in twelve longitudinal and twenty-one transverse series. Three præanal shields, median smallest, triangular. Limbs with large smooth shields. Eleven or twelve femoral pores on each side, two of which may be termed præanal. Tail thick, tapering at the end; scales as on the body. Black, upper parts minutely dotted with whitish.

	millim.		millim.
Total length	175	From end of snout to vent	80
Head	15	Fore limb	17
Width of head	10	Hind limb	22
From end of snout to fore limb	29	Tail	95

Ecuador.

a. ♂. W. Ecuador. Mr. Fraser [C.].

27. SCOLECOSAURUS.

Brachypus (*non Swains.*), *Fitzing. N. Classif. Rept.* p. 20; *Wiegm. Herp. Mex.* p. 12; *Gray, Cat. Liz.* p. 57.
Chalcis, part., *Wagler, Syst. Amph.* p. 196.
Chalcides, part., *Dum. & Bibr.* v. p. 450.

Tongue moderately elongate, arrow-headed. Teeth conical. Head with large regular shields; frontonasal separating the nasals; præfrontals present; no frontoparietals; nostril pierced in the suture between the nasal and the first labial. Eyes minute, eyelids developed. No ear-opening. Body and tail elongate, vermiform; limbs rudimentary, the hind pair as much developed as the front pair, with four or three distinct, clawed digits. Scales smooth, forming regular annuli, and also longitudinal series on the belly; the dorsals narrow, hexagonal, overlapping the ventrals, broader, quadrangular, juxtaposed. A collar fold. No femoral or præanal pores (?).

South America.

1. Scolecosaurus cuvieri.

Brachypus cuvieri, *Gray, Cat.* p. 57.
Brachypus cuvieri, *Fitz. l. c.* p. 50.
Chalcis cuvieri, *Wagl. l. c.*
Chalcides cuvieri, *Dum. & Bibr.* p. 453.

Upper head-shields: a pentagonal frontonasal, a pair of præfrontals forming a short suture, a frontal, as broad as long, seven-sided, a pair of large parietals, a narrow quadrangular interparietal of equal length, and four supraoculars. Nasal twice as broad as high; a loreal, half the size of the latter; an infraorbital, resting on the third, fourth, and fifth labials; seven temporals; six upper labials, posterior largest; four lower labials; mental very small; chin-shields, one anterior and three pairs, the first pair forming a suture; collar-shields slightly enlarged, six or eight. Twenty-eight scales round the middle of the body, and forty from occiput to base of tail; ventrals longer than broad, forming six or eight longitudinal and twenty-six or twenty-seven transverse series; a pair of enlarged pectoral shields. Præanal shields four, median posterior smallest. Digits four. Tail long, obtuse at the end, covered with imbricate elongate hexagonal scales, the upper smooth, the lower striated. Reddish brown, lighter beneath; a lighter, dark-edged line along each side of the back.

	millim.		millim.
Total length	124	From end of snout to vent	47
Head	7	Fore limb	5
Width of head	4·5	Hind limb........	6
From end of snout to fore limb	14	Tail	77

South America.

a–b. Adult.	W. Indies (?).	T. Bell, Esq. [P.].
c, d. Adult.	S. America.	

2. Scolecosaurus pallidiceps.

Brachypus pallidiceps, *Cope, Proc. Ac. Philad.* 1862, p. 356.

Fingers four, toes three. Scales in thirty-four series from axilla to groin, and twenty-three longitudinal. Frontonasal longer than broad, hexagonal; five temporals; two supraoculars; frontal much longer than broad. Otherwise like *S. cuvieri*. General colour plumbeous, the scales margined with black; head cinereous; a light brown band extends along each side of the back, separated from its fellow by four rows of scales. Total length 153 millim.

Truando, Colombia.

28. COPHIAS*.

Chalcides (*non Laur.*), *Bonnat. Encycl. Méth., Erp.* p. 67; *Daud. Rept.* iv. p. 367; *Wiegm. Herp. Mex.* p. 12.
Chalcis (*non Fabr.*), *Merrem, Tent. Syst. Amph.* p. 75; *Gray, Cat Liz.* p. 57.
Colobus (*non Illig.*), *Merr. l. c.*
Cophias, *Fitzing. N. Class. Rept.* p. 20; *Wiegm. l. c.*
Chalcis, part., *Wagler, Syst. Amph.* p. 196.
Chalcides, part., *Dum. & Bibr.* v. p. 450.
Microdactylus (*Tschudi*), *Gray, l. c.*
Bachia, *Gray, l. c.* p. 58.
Herpetochalcis, *Boettger, Ber. Offenb. Ver. Naturk.* 1883, p. 150.

Tongue moderately elongate, arrow-headed. Teeth conical. Head with large regular shields; frontonasal separating the nasals; præfrontals and frontoparietals absent; nostril pierced in the suture between the nasal and first labial. Eyes minute, eyelids developed. No ear-opening. Body and tail elongate, vermiform; limbs rudimentary, the anterior with clawless, tubercle-like digits, the posterior similar to the anterior or styliform, undivided. Scales smooth, forming regular annuli, and also longitudinal series on the belly; the dorsals narrow, either quadrangular and juxtaposed, or hexagonal and overlapping; the ventrals broader, quadrangular, juxtaposed. No collar fold. Males with præanal pores.

South America.

Synopsis of the Species.

I. Dorsal scales quadrangular, juxtaposed.

Hind limb tridactyle; an interparietal; three supraoculars .... 1. *tridactylus*, p. 418.
Hind limb didactyle; an interparietal; two supraoculars .... 2. *heteropus*, p. 418.
Hind limb undivided, styliform; no interparietal .... 3. *flavescens*, p. 418.

II. Dorsal scales hexagonal, overlapping.

Hind limb styliform, undivided .... 4. *dorbignyi*, p. 419.

* *Chalcides heteropus*, Licht. & Mart. Nomencl. Rept. Mus. Berol. p. 17.—Venezuela.

1. Cophias tridactylus.

Microdactylus schlegelii, *Gray, Cat.* p. 57.
Chalcides tridactylus, *Daud. Rept.* iv. p. 367, pl. lviii. fig. 3.
—— schlegelii, *Dum. & Bibr.* p. 457.

Both pairs of limbs with three clawed digital tubercles. Frontal pentagonal, longer than broad; interparietal triangular, oblong; three supraoculars; nasal large, oblong, quadrilateral; a loreal; six or seven temporals; six upper and four lower labials; chin-shields, one anterior and two pairs. Scales quadrangular, juxtaposed, narrow on the back and sides, broader on the belly; forty-six scales from occiput to base of tail, ten of which are anterior to the fore limbs; twenty-six scales round the body. Three elongate præanals. Tail scaled like the body. Upper surfaces olive-brown, lower rufous.

	millim.		millim.
Head and body	67	Fore limb	5
Head	8	Hind limb........	5

Hab. —— ?

2. Cophias heteropus.

Herpetochalcis heteropus, *Boettg. Ber. Offenb. Ver. Naturk.* 1883, p. 150.

Fore limb with three, hind limb with two clawed digital tubercles. Internasal triangular, truncate in front; a long, triangular interparietal; two supraoculars; six upper and four lower labials. Scales quadrangular, juxtaposed, thirty-four round the body. Six præanals. Four præanal pores. Yellowish, back darker; upper surface of tail with four light longitudinal bands.

From snout to vent 54 millim.

Central America?

3. Cophias flavescens.

Chalcis flavescens, *Gray, Cat.* p. 58.
? Chalcides flavescens, *Bonnat. Encycl. Méth., Erp.* p. 67, pl. xii. fig. 3.
Chamæsaura cophias, *Schneid. Hist. Amph.* p. 209.
Chalcides monodactylus, *Daud. Rept.* iv. p. 370.
Chalcis cophias, *Merr. Tent.* p. 75.
Colobus daudini, *Merr. l. c.* p. 76.
Chalcides cophias, *Dum. & Bibr.* p. 459; *Gravenh. Nova Acta Ac. Leop.-Carol.* xxiii. 1851, i. p. 306, pl. xxix.
—— trilineatus, *Peters, Mon. Berl. Ac.* 1872, p. 775.

Fore limb with three or four digital tubercles, hind limb undivided. Frontonasal trapezoid, nearly as broad as long; frontal large, pentagonal, longer than broad; parietals large; no interparietal; two supraoculars; nasal twice as broad as deep; loreal shorter, but a little deeper than the latter; two infraorbitals, anterior triangular and wedged in between the third and fourth labials; temporal shields five; five upper labials, fifth largest; four lower labials; chin-shields, one anterior and two pairs, the first pair forming a suture. Scales quadrangular, juxtaposed, narrow on the back and sides, broader on the belly; thirty scales round the middle of the body and fifty from occiput to base of tail. A pair of enlarged pectoral shields. Præanals elongate, three, the median sometimes divided into two. Male with two præanal pores on each side. Tail obtuse at the end, covered with scales like those of the body. Brown; back lighter, with two or three longitudinal darker lines.

	millim.		millim.
Total length	120	From end of snout to vent	63
Head	7	Fore limb	3
Width of head	4·5	Hind limb........	2·5
From end of snout to fore limb	14	Tail	57

Guianas; Venezuela.

a–b. ♂ ♀.	Cayenne.	
c. Hgr.	Caracas.	Paris Museum [P.].

4. Cophias dorbignyi.

Bachia dorbignii, *Gray, Cat.* p. 58.
Chalcides dorbignyi, *Dum. & Bibr.* p. 462; *A. Dum. Cat. Méth. Rept.* p. 146.

Fore limb with four (or three) digital tubercles; hind limb extremely minute, undivided. Rostral large, rather projecting; frontonasal rather small, trapezoid, as broad as long; frontal large, pentagonal, longer than broad; parietals large; no interparietal; two supraoculars; nasal more than twice as broad as high; loreal not half the size of latter; infraorbital triangular, wedged in between the third and fourth labials; temporal shields five; five upper and three lower labials; chin-shields, one anterior and two pairs, first pair forming a suture. Dorsal and lateral scales elongate hexagonal, overlapping; twenty-eight scales round the middle of the body, and fifty from occiput to base of tail. A pair of enlarged pectoral shields. Præanal shields three, median narrowest. Tail long, obtuse at the end, covered with scales similar to the dorsals. Back brownish white, sides and lower surfaces reddish brown; three reddish brown lines along the back.

	millim.		millim.
Total length	158	From end of snout to vent	71
Head	7	Fore limb	2·5
Width of head	4·5	Hind limb........	1·5
From end of snout to fore limb	14	Tail	87

Chili; Venezuela.

a–b. ♀?	Baranquilla.	Mr. Rippon [C.].

29. OPHIOGNOMON.

Ophiognomon, *Cope, Proc. Ac. Philad.* 1868, p. 100.
Hapalolepis, *Peters, Mon. Berl. Ac.* 1871, p. 399.
Propus, *Cope, Proc. Ac. Philad.* 1874, p. 70.

Tongue moderately elongate, arrow-headed. Teeth conical. Head with large regular shields; frontonasal, præfrontals, and fronto-parietals absent; nasals forming a suture; nostril pierced in the suture between the nasal and the first labial. Eyes minute, eyelids developed. No ear-opening. Body and tail elongate, vermiform; limbs rudimentary, the anterior with tubercle-like indications of digits, the posterior reduced to a minute tubercle or altogether absent. Scales smooth, forming regular annuli, and also longitudinal series on the belly; the dorsals narrow, hexagonal or quadrangular, juxtaposed or slightly overlapping; the ventrals broader, quadrangular, juxtaposed. A more or less distinct lateral fold. No collar fold. Males with præanal pores.

Upper Amazons.

Synopsis of the Species.

Hind limb present, half as long as præanal plates; 20 scales round the body ...... 1. *trisanale*, p. 420.
Hind limb present, reduced to a minute tubercle; 26 to 28 scales round the body............................ 2. *abendrothii*, p. 421.
Hind limb absent .... 3. *vermiforme*, p. 421.

1. Ophiognomon trisanale.

Ophiognomon trisanale, *Cope, l. c.*

Fore limb as long as three anterior labials, with three terminal tubercles; hind limb styliform, minute, half as long as anal plates. Snout obtuse, slightly projecting. Frontal large, hexagonal, posterior angle prolonged; parietals larger, obliquely hexagonal, truncate behind; no interparietal; two supraoculars; a square loreal; two very small infraorbitals; temporals three, anterior large; four upper labials, posterior largest; lower labials three; mental narrow; chin-shields, one anterior and three pairs, anterior pair extensively in contact, the posterior smaller and separated from temporals by four narrow plates, and from each other by four plates, the median pair of which are larger and join the anterior pair of chin-shields. A groove surrounds the throat behind the jaws, which is succeeded by five annuli of equal ovate scales; these are followed by a cross series of six more elongate, which precede a pair of large sternal plates. Dorsal scales hexagonal; twenty scales round the body and thirty-seven transverse rows between axilla and groin. Three elongate parallel anal plates. Caudal scales angulate, like dorsals. Below brown; sides with a brown blackish band, which is bounded above by a rather narrow yellow band, which commences on the supraciliary region; dorsal region between the latter yellowish grey, bounded exteriorly by a distinct blackish line, and divided medially by an indistinct blackish line; head brownish, paler below.

	millim.		millim.
Head	8	From end of snout to vent	67
Width of head	4·5		
From end of snout to fore limb	16		

Napo or Upper Marañon.

2. Ophiognomon abendrothii.

Chalcides (Hapalolepis) abendrothii, *Peters, l. c.* p. 399.
Ophiognomon trisanale, *O'Shaughn. Ann. & Mag. N. H.* (4) xii. 1873, p. 44.

The following characters disagree with those assigned by Cope to *O. trisanale*:—The fore limb is not as long as the three anterior labials, and the hind limb is reduced to an extremely minute tubercle which can hardly be distinguished with the naked eye. Five upper labials; second pair of chin-shields frequently forming a suture. Twenty-six to twenty-eight scales round the body; forty-nine to fifty-two transverse rows of scales from occiput to base of tail (forty to forty-two between axilla and groin). Males with two præanal pores on each side. Brown, sides darker; a light band, bordered on both sides by a dark line, along each side of the back, commencing from the supraciliary region; a median dark brown line.

	millim.		millim.
Total length	146	From end of snout to fore limb	12
Head	6·5	From end of snout to vent	59
Width of head	4·5		

North-eastern Peru.

a. ♂.	Sarayacu, Peru.	Berlin Museum [P.]. (One of the types.)
b–e. ♂ ♀.	Sarayacu, Peru.	Messrs. Veitch [P.].

3. Ophiognomon vermiforme.

Propus vermiformis, *Cope, l. c.*

Limbs one pair, the anterior only. Scales in twenty-six rows on the posterior and twenty rows on the anterior part (with closed lateral folds) of the body, and fifty-one transverse annuli between the nape and the vent; anteriorly they are imbricate on the dorsal surface, posteriorly they are truncate. Five upper labials; two pairs of chin-shields in contact. Two femoral pores on each side. Colour, a dark reddish brown; a dorsolateral series of pale spots on each side, separated by an average width of eight scales.

	millim.		millim.
Total length (extremity of tail lost)	122	Width of head	4
Head	6	From end of snout to vent	64
		Fore limb	2·7

Nauta, Peruvian Amazons.

30. HETERODACTYLUS.

Heterodactylus, *Spix, Spec. Nov. Lacert. Bras.* p. 25; *Dum. & Bibr.* v. p. 444; *Gray, Cat. Liz.* p. 59.
Chirocolus, *Wagler, Syst. Amph.* p. 157; *Wiegm. Herp. Mex.* p. 10.

Tongue moderately elongate, arrow-headed. Lateral teeth bicuspid. Head with large regular shields; a frontonasal separating the nasals; præfrontals and frontoparietals absent; nostril pierced in the suture between the nasal and the first labial. Eyelids developed, lower with a transparent disk. No ear-opening. Limbs short, with well developed digits; the inner finger rudimentary, tubercular, clawless. Dorsal and lateral scales equal, hexagonal-lanceolate, keeled, imbricate, forming transverse series; ventrals large, subquadrangular, rounded and overlapping posteriorly, smooth, arranged in regular longitudinal and transverse series. No collar fold. Tail cylindrical. Males with, females without, femoral pores.

Brazil.

1. Heterodactylus imbricatus.

Heterodactylus imbricatus, *Gray, Cat.* p. 76.
Heterodactylus imbricatus, *Spix, l. c.* pl. xxvii. fig. 1; *Dum. & Bibr.* v. p. 447; *Reinh. & Lütk. Vidensk. Meddel.* 1861, p. 217, pl. vi. fig. 11; *Peters, Abh. Berl. Ac.* 1862, p. 172; *Bouleng. Proc. Zool. Soc.* 1885, p. 1.

Head short, snout broad, truncate; body much elongate. Frontonasal large, hexagonal, broader than long; frontal narrower, pentagonal, a little longer than broad; a pair of hexagonal parietals, followed by two pairs of large transverse occipitals; interparietal very small or absent, seldom in contact with the frontal; three supraoculars; a narrow loreal; a row of small infraorbitals; temples shielded; six upper and four lower labials; chin-shields, one anterior and two large pairs, the first pair and sometimes also a short portion of the second forming a suture; median gulars transversely enlarged. A row of six elongate pectoral shields. Dorsal scales very narrow, lanceolate, ending in a long sharp point; thirty-three or thirty-five scales round the middle of the body, ventrals included; thirty-two to thirty-five scales from occiput to base of tail. Ventrals longer than broad, in six or eight longitudinal and twenty-five or twenty-six transverse series. Five elongate præanals, outer very narrow. Limbs with smooth shields. Male with four femoral and one or two præanal pores on each side. Tail extremely long and thick, tapering gradually at the end, covered with annuli of scales similar to the dorsals. Yellowish brown above, sides darker; a more or less indistinct, light, dark-edged, dorso-lateral band; lower surfaces yellowish.

	millim.		millim.
Total length	400	From end of snout to vent	110
Head	19	Fore limb	17
Width of head	14	Hind limb........	31
From end of snout to fore limb	30	Tail	290

Brazil.

a. ♂.	Montequeira Mountains.	G. L. Hunt, Esq. [P.].
b. ♀.	S. America.	Paris Museum.
c. ♀.	S. America.	Zoological Society.

2. Heterodactylus lundii.

Heterodactylus lundii, *Reinh. & Lütk. Vidensk. Meddel.* 1861, p. 214, pl. vi. fig. 10.

More slender, with shorter hind limbs and narrower head. Frontonasal as long as broad; parietals separated, the posterior angle of the frontal being truncate and forming a suture with a small interparietal. Scales on the upper and lateral surface of neck smaller; dorsal scales with shorter posterior point; thirty transverse series of ventrals. One præanal and one femoral pore on each side. Dorso-lateral light band narrower.

	millim.		millim.
Total length	160	From fore limb to vent	39
From end of snout to fore limb	13·5	Fore limb	7
		Hind limb..........	12

Serra da Piedade, Brazil.

31. PERODACTYLUS.

Perodactylus, *Reinh. & Lütk. Vidensk. Meddel.* 1861, p. 218; *Peters, Abh. Berl. Ac.* 1862, p. 188.

Head with large regular shields; frontonasal separating the nasals; præfrontals and frontoparietals present; nostril pierced in the middle of a single nasal. Eyelids developed, lower with a transparent disk. Ear exposed. Limbs well developed, the inner finger rudimentary, clawless. Dorsal scales hexagonal-lanceolate, keeled, imbricate, forming transverse series, similar to the caudals; lateral and ventral scales smooth, rounded, imbricate, the ventrals transversely dilated and forming four longitudinal rows. No collar fold. Tail cylindrical. Females without femoral pores.

Brazil.

1. Perodactylus modestus.

Perodactylus modestus, *Reinh. & Lütk. l. c.* pl. v. fig. 7.
Iphisa (Perodactylus) modesta, *Peters, l. c.*

Body elongate; snout truncate. Frontonasal large, broader than long; præfrontals and frontoparietals small; interparietal narrow, hexagonal; three supraoculars; a loreal; a row of small infra-orbitals; temporal shields irregular; two pairs of large chin-shields, in contact, followed by a double row of transversely enlarged gulars. Nuchal scales large, transverse, in two rows, smooth. Bronze-brown above, with two whitish longitudinal bands; lower surfaces white.

	millim.		millim.
Total length (tail injured)	112	From end of snout to vent	47
Head, to ear-opening	9	Fore limb	10
From end of snout to fore limb	14	Hind limb	16

Morro de Garza, Brazil.

32. IPHISA.

Iphisa, *Gray, Proc. Zool. Soc.* 1851, p. 39; *Peters. Mon. Berl. Ac.* 1862, p. 188.

Tongue rather elongate, arrow-headed. Lateral teeth tricuspid. Head with large regular shields; frontonasal separating the nasals; præfrontals and frontoparietals present; nostril pierced in the lower part of a single nasal. Eyelids developed, lower with a transparent disk. Ear exposed. Limbs well developed; inner finger short, clawless. All the scales roundish-hexagonal, imbricate, smooth, arranged quincuncially, two rows along the back and two along the belly much dilated transversely; caudal scales lanceolate, keeled. No collar fold. Tail cylindrical. Males with, females without, femoral pores.

Brazil, Guianas.

1. Iphisa elegans.

Iphisa elegans, *Gray, l. c.* pl. vi. fig. 3.

Head depressed, snout short, broad, truncate; body elongate; limbs short. Frontonasal much broader than long, sometimes longitudinally bisected; præfrontals forming a short suture; frontal hexagonal, narrower than the frontonasal; frontoparietals a little larger than the præfrontals, forming a short suture; parietals large; interparietal pentagonal, with straight lateral edges, as long as, but much narrower than, the parietals; occipitals not distinguishable from the nuchal shields; supraoculars three, anterior smallest, second largest; two superposed loreals; a row of infraorbitals; temples shielded; seven upper and six or seven lower labials; chin-shields, one anterior and two pairs, the first pair enormous, forming a very long suture, the second pair very small. Dorsal and nuchal scales forming a double alternating series of twenty-nine or thirty pairs; these enlarged scales rounded-hexagonal, three times as broad as long behind the occiput, becoming gradually narrower;

lateral scales large, elongate-cycloid, forming oblique series; gular and ventral scales like the dorsals, forming twenty-seven or twenty-eight pairs; three enlarged sternal shields, median triangular; twelve scales round the middle of the body. Three large præanals, median shortest and elongate-triangular. Limbs with smooth shields; hinder side of thighs granulate; upper surface of tibia with keeled scales. Male with ten femoral pores on each side. Tail long, tapering to a point; covered with hexagonal-lanceolate keeled scales forming regular imbricate annuli. Chestnut-brown above, with small darker marblings; a more or less distinct light lateral line, separating the lighter colour of the dorsal surface from the blackish brown of the sides of the head and body; lower surfaces yellowish white.

	millim.		millim.
Total length	144	From end of snout to vent	54
Head	11	Fore limb	10
Width of head	7	Hind limb	16
From end of snout to fore limb	19	Tail (broken)	90

N. Brazil; Guianas.

a. ♂.	Para	(Type.)
b. ♀.	Demerara Falls.	

33. TRETIOSCINCUS.

Tretioscincus, *Cope, Proc. Ac. Philad.* 1862, p. 184; *Bocourt, Miss. Sc. Mex., Rept.* p. 452.

Tongue rather elongate, arrow-headed. Head with large regular shields; frontonasal separating the nasals; præfrontals and fronto-parietals present; nostril pierced in the suture of a divided nasal. Eyelids developed, lower with a transparent disk. Ear exposed. Limbs well developed; inner finger rudimentary, clawless. All the scales roundish-hexagonal, imbricate, smooth, or feebly keeled on the back, arranged quincuncially. No collar fold. Tail cylindrical. Males with femoral pores.

Colombia; Central America.

1. Tretioscincus bifasciatus.

Heteropus bifasciatus, *A. Dum. Cat. Méth. Rept.* p. 182.
Tretioscincus castanicterus, *Cope, l. c.*
—— bifasciatus, *Cope, Proc. Ac. Philad.* 1864, p. 229; *Bocourt, l. c.* p. 453, pl. xxii. F. fig. 4.

Body moderately elongate. Frontonasal broad; frontal much narrower; parietals broad, short; interparietal subtriangular, as long as parietals; a loreal; a row of infraorbitals; eight upper labials; chin-shields, one anterior and three pairs, all forming a suture. Dorsal scales smooth on the nape, becoming gradually more and more distinctly keeled towards the tail; sixteen scales round the body; twenty-nine scales from parietals to base of tail. Præanal scales six. Five or six femoral pores on each side. Tail moderately long, tapering to a fine point; upper caudal scales strongly keeled. Brown above, with a yellowish longitudinal band on each side, beginning from the tip of the snout; another on the upper lip; lower surfaces yellowish, grey-spotted.

	millim.		millim.
Total length......	138	Fore limb	13
Head	12	Hind limb..........	20
Width of head	9	Tail	80
From end of snout to vent	52		

North-Eastern Colombia.

2. Tretioscincus lævicauda.

Tretioscincus lævicaudus, *Cope, Proc. Amer. Philos. Soc.* xi. 1870, p. 557.

Scales large, those of both body and tail entirely smooth; several behind the occipital plates larger than the others. Limbs short. Colour, a dark clear olive above, paler below; a light green band from the temporal region along the canthus rostralis to the end of the muzzle, and another parallel along the upper lip; no other markings. Size about that of *T. bifasciatus.*

W. Nicaragua.

34. MICRABLEPHARUS.

Gymnophthalmus, part., *Bocourt, Miss. Sc. Mex., Rept.* p. 466.
Micrablepharus, *Bœttger, Zeitschr. f. Naturw.* lviii. 1885, p. 217.

Tongue arrow-headed. Head with large regular shields; frontonasal separating the nasals; no præfrontals; frontoparietals present; Nostril pierced in the middle of a semidivided nasal. No eyelids. Ear exposed. Limbs well developed; inner finger absent. All the scales roundish-hexagonal, imbricate, smooth, arranged quincuncially. No collar fold. Tail cylindrical. Males with femoral pores.

Brazil, Paraguay.

1. Micrablepharus maximiliani.

Gymnophthalmus quadrilineatus (*non Linn.*), *Wied, Beitr. Nat. Bras.* i. p. 195, *and Abbild.*
—— maximiliani, *Reinh. & Lütk. Vidensk. Meddel.* 1861, p. 211, pl. v. fig. 6; *Bocourt, l. c.* p. 467, pl. xxii. H. fig. 5.
Micrablepharus glaucurus, *Bœttg. l. c.* p. 218.

Habit of *Gymnophthalmus quadrilineatus.* Frontonasal very large; frontal small; frontoparietals small; parietals and interparietals short, of equal length, latter pentagonal; two supraoculars, anterior largest and in contact with the frontonasal; two supraciliaries, anterior very large; chin-shields, one anterior and three alternating pairs forming a suture. Sixteen scales round the body. Four or five femoral pores on each side. Scales on the upper part of the posterior two thirds of the tail keeled. Yellowish brown above; a broad lateral dark brown band, commencing from the snout and passing through the eye and the fore limb, bordered above and below by a whitish streak; lower surfaces yellowish, throat and chin brown-dotted; tail blue.

	millim.
From end of snout to fore limb..........	13
From end of snout to vent	27
Tail	67

Brazil; Paraguay.

35. GYMNOPHTHALMUS.

Gymnophthalmus, *Merrem, Tent. Syst. Amph.* p. 74; *Wagler, Syst. Amph.* p. 157; *Wiegm. Herp. Mex.* p. 11; *Dum. & Bibr.* v. p. 819; *Gray, Cat. Liz.* p. 63; *Bocourt, Miss. Sc. Mex., Rept.* p. 466.
Epaphelus, *Cope, Journ. Ac. Philad.* (2) viii. 1876, p. 115.

Tongue moderately elongate, arrow-headed. Teeth conical. Head with large regular shields; frontonasal separating the nasals; præfrontals present; no frontoparietals; nostril pierced in the middle of a single nasal. No eyelids. Ear exposed. Limbs well developed; inner finger absent All the scales roundish-hexagonal, imbricate, smooth or feebly keeled, arranged quincuncially. No collar fold. Tail cylindrical. Males with, females without, femoral pores.

Tropical America.

Synopsis of the Species.

I. Dorsal scales perfectly smooth.

15 scales round the body .... 1. *quadrilineatus*, p. 427.
13 scales round the body .... 2. *sumichrasti*, p. 428.

II. Dorsal scales keeled.

15 scales round the body .... 3. *luetkenii*, p. 429.
17 scales round the body .... 4. *pleii*, p. 429.

1. Gymnophthalmus quadrilineatus.

Gymnophthalmus lineata, *Gray, Cat.* p. 63.
Lacerta quadrilineata, *Linn. S. N.* i. p. 371.
Seps lineatus, *Laur. Syn. Rept.* p. 60.
Scincus quadrilineatus, *Daud. Rept.* iv. p. 266.
Gymnophthalmus quadrilineatus, *Merr. Tent.* p. 74; *Gravenh. Nova Acta. Ac. Leop.-Carol.* xxiii. 1851, i. p. 354; *Bocourt, l. c.* p. 468, pl. xxii. H. fig. 4.
—— merremii, *Coct. Et. Scinc.*
—— quadrilineatus, part., *Dum. & Bibr.* v. p. 820.

Gymnophthalmus nitidus, *Reinh. & Lütk. Vidensk. Meddel.* 1862, p. 226; *Bocourt, l. c.* p. 470.

Body and limbs moderately elongate in the male, the body longer and the limbs weaker in the female; head depressed. Frontonasal large, broader than long, seven-sided; præfrontals forming a short suture; frontal very small, pentagonal; parietals rather small; interparietal large, subpentagonal, longer than parietals; a very large supraciliary, forming an oblique suture with the unique supra-ocular which forms a suture anteriorly with the smaller præfrontal; a loreal; seven or eight upper and six or seven lower labials; chin-shields, one anterior and two pairs forming a suture. Scales on the body perfectly smooth; fifteen round the body, and thirty-four (♂) to thirty-eight (♀) from occiput to base of tail. Four femoral pores on each side, in males. Tail cylindrical, tapering, covered with rounded scales which, except at the base, are feebly keeled; the portion of the tail which has been reproduced is no longer covered with rounded scales, but with overlapping whorls of lanceolate scales, exactly similar to those of *Heterodactylus* and allies. Dark olive above; a yellowish, black-edged band on each side, beginning from the tip of the snout; another, narrow, parallel line along the upper lip, through the ear-opening, and over the fore limb, to groin; tail pinkish posteriorly; throat and belly grey, black-dotted.

	♂. millim.	♀. millim.
Total length (tail injured)	83	85
Head	6·5	7
Width of head	4	5
From end of snout to fore limb	11	11·5
From end of snout to vent	31	41
Fore limb	7	7
Hind limb	12	12

Brazil; West Indies.

a–b, c–d. ♂ ♀. ——?

2. Gymnophthalmus sumichrasti.

Epaphelus sumichrasti, *Cope, Journ. Ac. Philad.* (2) viii. 1876, p. 115.

Gymnophthalmus sumichrasti, *Bocourt, Miss. Sc. Mex., Rept.* p. 471, pl. xx. H. fig. 2.

In most respects similar to the preceding; the difference of proportions between the sexes still more marked. Scales perfectly smooth, feebly keeled on the posterior part of the tail; thirteen scales round the body, and thirty-one (♂) or thirty-seven (♂) from occiput to base of tail. Male with three or four femoral pores on each side. Yellowish or olive-brown above, with metallic gloss; sides much darker, the two colours sharply defined, and limited by a more or less distinct light line; lower surfaces greyish, with more or less strong metallic gloss.

	♂. millim.	♀. millim.
Total length (tail broken)	63	—
Head	7	7·5
Width of head	4·5	5
From end of snout to fore limb.. .	10·5	12
From end of snout to vent	30	41
Fore limb	7	8·5
Hind limb	11	11·5

From Tehuantepec to Venezuela.

a. ♀.	Guatemala.	M. F. Bocourt, [P.].
b. ♂.	Venezuela.	

3. Gymnophthalmus luetkenii.

Gymnophthalmus lütkenii, *Bocourt, Miss. Sc. Mex., Rept.* p. 474.

Dorsal and upper caudal scales keeled; fifteen round the body and thirty-four from occiput to base of tail. Yellowish brown above, dotted and spotted with black; a dark lateral band, bordered above by a whitish line which begins on the tip of the snout; another parallel whitish line along the upper lip to the fore limb; greenish yellow inferiorly.

	millim.		millim.
Total length..........	83	From end of snout to vent	31
Head	7	Fore limb	9
Width of head	4	Hind limb	12

S. Lucia, West Indies.

4. Gymnophthalmus pleii.

Gymnophthalmus quadrilineatus, part., *Dum. & Bibr.* v. p. 820.
—— pleii, *Bocourt, Miss. Sc. Mex., Rept.* p. 473, pl. xxii. H. fig. 3.

Dorsal and upper caudal scales keeled; seventeen scales round the body, and thirty-two to thirty-four from occiput to base of tail. Bronze-green above, dotted with blackish, with dark and light longitudinal lines as in the preceding; lower surfaces greenish grey.

	millim.		millim.
Total length	129	From end of snout to vent	48
Head	9	Fore limb.............	11
Width of head	6	Hind limb........... ..	16

Martinique.

Fam. 15. AMPHISBÆNIDÆ.

Amphisbænidæ, *Gray, Ann. Phil.* (2) x. 1825.
Amphisbænoidea, *Fitzinger, Neue Classif. Rept.* 1826.
Angues, part., *Wagler, Syst. Amph.* 1830.
Amphisbænæ, *Wiegmann, Herp. Mex.* 1834.
Chalcidiens glyptodermes, *Duméril & Bibron, Erp. Gén.* v. 1839.
Trogonophidæ, Amphisbænidæ, Lepidosternidæ, Chirotidæ, *Gray, Cat. Tort.* &c. 1844.
Amphisbænidæ, *Cope, Proc. Ac. Philad.* 1864.
Amphisbænidæ, Trogonophidæ, *Cope, Proc. Amer. Assoc. Adv. Sc* xix. 1871.
Amphisbænidæ, *Boulenger, Ann. & Mag. N. H.* (5) xiv. 1884.

Tongue moderately elongate, arrow-headed, covered with imbricate scale-like papillæ, ending in two long, narrow, smooth points. Teeth large, few, anchylosed to the inner (*Prosphyodontes*) or to the upper (*Emphyodontes*) edge of the jaws; præmaxillary teeth usually in odd number; no pterygoid teeth. Skull thick, strongly ossified, without interorbital septum, nor columella cranii, nor postorbital and frontosquamosal arches; præmaxillary single; nasals two; frontals two; parietal single, very large; quadratum very oblique or nearly horizontal, owing to the shortness of the post-coronoid part of the mandible; occipital condyle frequently divided. Vertebræ very numerous, depressed, all except the foremost without spinose processes. Pectoral arch imperfectly developed in *Chirotes*, reduced to minute rudiments in the other, limbless, forms; pelvic arch reduced to minute rudiments. Worm-like, adapted to subterranean life; eyes concealed under the skin; mouth small, frequently inferior; no ear. Head covered with symmetrical plates; skin divided into soft squarish segments forming regular annuli. Tail short.

The shape of the skull varies considerably according to the genera; in this respect *Blanus* appears to be the least specialized, although the occipital condyle is divided, while it remains single (kidney-shaped when seen posteriorly) in the otherwise more specialized *Acrodonta*. The mandible especially undergoes the greatest modifications, as may be seen from Gervais * and Peters's † accounts of *Blanus, Amphisbæna, Lepidosternon, Monopeltis,* and *Agamodon.*

All the members of this family are burrowers, and many live in ants' nests. They bore narrow galleries in the earth, in which they are able to progress backwards as well as forwards. On the ground they progress on a straight line, by slight vertical undulations, not by lateral movements, as in other limbless reptiles; the tail of many species appears to be more or less prehensile. The food of these Lizards consists of small insects and worms. Little has been published on ther habits, and all that is known on their mode of parturition is that *Anops kingii* is oviparous, and deposits its ova in ants' nests.

* Ann. Sc. Nat. xx. pls. xiv. & xv.
† Reise n. Mossamb. iii. pl. xiii. A; Sitzb. Berl. Ac. 1882, pl. x.

As many as sixty-five species are characterized in this account; thirty-nine are American, out of which only two (*Chirotes* and *Rhineura*) occur north of the Tropic of Cancer, and four (*Amphisbæna*) in the West Indies. The twenty-six other species inhabit Africa, with the exception of the four species of the genera *Blanus* and *Trogonophis*, which belong to the Mediterranean district.

Synopsis of the Genera.

A. Prosphyodontes.

I. Fore limbs present .... 1. **Chirotes**, p. 432.

II. No limbs.

A. Segments of the pectoral region not differentiated.

Nostril pierced in the first labial .... 2. **Blanus**, p. 433.
Nostril in a separate nasal; head without cutting-edge .... 3. **Amphisbæna**, p. 435.
Head strongly compressed, the rostral shield enormous, with an arched cutting-edge .... 4. **Anops**, p. 451.

B. Segments of the pectoral region more or less enlarged, or forming angular series.

Snout compressed .... 5. **Geocalamus**, p. 453.
Snout depressed; nostril in a small separate nasal; tail cylindrical.... 6. **Monopeltis**, p. 453.
Tail depressed, above with rows of round tubercles .... 7. **Rhineura**, p. 459.
Nostril pierced in the rostral ...... 8. **Lepidosternon**, p. 460.

B. Emphyodontes.

No præanal pores .... 9. **Trogonophis**, p. 470.
Præanal pores; three large upper head-shields besides the rostral .... 10. **Pachycalamus**, p. 471.
Præanal pores; a single upper head-shield besides the rostral .... 11. **Agamodon**, p. 471.

A. PROSPHYODONTES.

Prosphyodontes, *Wiegm. Herp. Mex.* p. 20.

1. CHIROTES.

Chirotes, *Cuv. R. A.* ii. p. 57; *Merrem, Tent. Syst. Amph.* p. 161; *Fitzing. Neue Classif. Rept.* p. 24; *Wagl. Syst. Amph.* p. 196; *Wiegm. l. c.*; *Dum. & Bibr.* v. p. 472; *Gray, Cat. Tort.* 1844, p. 74, *and Cat. Sh. Rept.* ii. p. 34: *Bocourt, Miss. Sc. Mex., Rept.* p. 486.
Bimanus, *Oppel, Ordn. Rept.* p. 45.

Fore limbs present, with four clawed digits. Snout rounded; nostril lateral, pierced in a separate nasal. No enlarged pectoral segments. Lateral line distinct. Tail cylindrical, obtuse. Præanal pores.

Mexico and California; Platte River.

1. Chirotes canaliculatus.

Chirotes lumbricoides, *Gray, Cat. Tort.* p. 74, *and Sh. Rept.* ii. p. 34.
Lacép. Quadr. Ovip. i. p. 613, pl. xli.
Bipes canaliculatus, *Bonnat. Encycl. Méth., Erp.* p. 68.
Lacerta lumbricoides, *Shaw, Nat. Misc.* vi. pl. 212.
—— mexicana, *Donnd. Zool. Beitr.* iii. p. 135.
Chamæsaura propus, *Schneid. Hist. Amph.* ii. p. 211.
Lacerta sulcata, *Suckow, Anfangsgr. Naturg. Thiere*, iii. p. 147.
Chalcides propus, *Daud. Rept.* iv. p. 372, pl. lviii. fig. 4.
Bimanus propus, *Oppel, Ordn. Rept.* p. 46.
Chirotes canaliculatus, *Merr. Tent.* p. 161; *Guér. Icon. R. A., Rept.* pl. xvi. fig. 3; *Müll. Zeitschr. f. Physiol.* iv. 1831, p. 257, pl. xxii. figs. 6, 7; *Dum. & Bibr.* v. p. 474; *Bocourt, Miss. Sc. Mex., Rept.* p. 487, pl. xxi. c, fig. 8; *Strauch, Mél. Biol. Ac. St. Petersb.* xi. p. 374.
—— lumbricoides, *Flem. Phil. Zool.* ii. p. 278; *Long, Exped. Rocky Mount.* i. p. 484.

Snout broad, rounded, very convex, not projecting beyond the lower jaw. Limbs depressed, about as long as the skull, with four well developed clawed digits, the outer shortest. Rostral pentagonal, forming a suture with the frontal, which is very large, and broader than long; nostril in the anterior portion of a large nasal; the other head-shields are: three labials on each side, an ocular, through which the eye is distinguishable, a præocular, an infraocular, a supraocular, and a pair of small postfrontals. Two hundred and twenty or two hundred and twenty-four annuli on the body (counting from the postfrontals) and thirty-six on the tail; [from Bibron's and Strauch's accounts the number of rings on the body would vary between 209 and 260]; the divisions of the annuli longer than broad, of equal width on the dorsal and ventral surfaces; on the latter the length is greater, four ventral annuli corresponding to five dorsals. Six or eight narrow præanal shields. Præanal pores six. Brownish flesh-coloured.

	millim.
Length to vent	128
Tail	20
Diameter of body	5·5

Reaches a length of 200 millim.
Mexico; California; Platte River.

a–b. Hgr. Mexico.

2. BLANUS.

Blanus, *Wagler, Syst. Amph.* p. 197; *Gray, Cat. Tort.* 1844, p. 72, *and Proc. Zool. Soc.* 1865, p. 446.
Amphisbæna, part., *Dum. & Bibr.* v. p. 476.

Nostril pierced in the first labial; a large frontal, forming a suture with the rostral. No enlarged pectoral shields. A well-marked lateral line; a curved postgular fold. Tail pointed. Præanal pores.

Borders of the Mediterranean.

The skull shows a lesser degree of specialization than in the other Amphisbænoids: there is an indication of a postorbital process, and the lower jaw is relatively long.

Synopsis of the Species.

Snout not projecting beyond the lower jaw; four lower labials, first and fourth smallest 1. *cinereus*, p. 433.
Snout projecting beyond the lower jaw; four lower labials, first and fourth smallest.... 2. *strauchii*, p. 434.
Snout projecting beyond the lower jaw; three lower labials, first and second large ...... 3. *bedriagæ*, p. 434.

1. Blanus cinereus.

Blanus cinereus, part., *Gray, Cat. Tort.* p. 72, *and Sh. Rept.* ii. p. 34.
Amphisbæna cinerea, *Vandelli, Mem. Ac. Sc. Lisb.* i. 1797, p. 69; *Gerv. Mag. Zool.* 1836, p. 2, pl. x.; *Dum. & Bibr.* v. p. 500; *Strauch, Mél. Biol. Ac. St. Petersb.* xi. 1881, p. 416; *Bedriaga, Arch. f. Nat.* 1884, p. 24, pl. iv.
—— oxyura, *Wagl. in Spix, Serp. Bras.* p. 72, pl. xxxv. fig. 1.
? Amphisbæna rufa, *Hempr. Verh. Ges. Nat. Fr. Berl.* i. 1829, p. 130.
Blanus cinereus, *Wagl. Syst. Amph.* p. 197; *Schreib. Herp. Eur.* p. 334.
? Blanus rufus, *Wiegm. Arch. f. Nat.* 1836, p. 157.

Præmaxillary teeth 7; maxillaries 4—4; mandibulars 7—7. Snout rounded, not at all projecting beyond the lower jaw. Rostral moderate, trapezoid, forming a suture with the large frontal, which is as broad as long or a little broader than long; behind the frontal three pairs of slightly enlarged square shields; ocular small, eye rather indistinct; four upper labials, first largest, pierced by the nostril, fourth smallest; second labial forming a long suture with the ocular and a very short one with the frontal. Mental trapezoid, followed by a chin-shield of nearly equal size; four lower labials, the first and fourth small. A distinct transverse fold behind the occiput. One hundred and ten to one hundred and twenty-five annuli on the body, and twenty to twenty-two on the tail; an annulus on the middle of the body contains thirty to thirty-four segments, viz. fourteen or sixteen ventral and sixteen or eighteen dorsal. Lateral line well marked; a more or less distinct vertebral line. Four or six anal segments, median pair largest. Præanal pores six (exceptionally four). Brownish flesh-colour, the segments more or less dark brown.

	millim.
Length to vent	215
Tail	27
Diameter of body	7·5

Spain, Portugal, Morocco, Algiers.

a. Ad. (Annuli 118+21; pores 6).	Portugal.	Prof. B. du Bocage [P.].
b. Ad. (A. 120+22; p. 6).	Oporto.	E. Allen, Esq. [P.].
c. Ad. (A. 123+20; p. 6).	S. Europe.	P. B. Webb, Esq. [P.].
d–e. Ad. (A. 125+21, 125+22; p. 6).	Tangiers.	Mr. Fraser [C.].
f. Hgr. (A. 110+22; p. 6).	N. Africa.	

2. Blanus strauchii.

Amphisbæna strauchi, *Bedriaga, Arch. f. Nat.* 1884, p. 35, pl. iv.

Præmaxillary teeth 7; max. 3—3 or 4—4; mand. 8—8. Snout projecting beyond the lower jaw. Frontal a little longer than broad. No occipital fold. One hundred and eight to one hundred and nineteen annuli on the body and twenty to twenty-two on the tail; an annulus contains sixteen or eighteen dorsal and eighteen or twenty ventral segments. Præanal pores eight (less frequently four or six). In other respects like the preceding.

	millim.
Length to vent	170
Tail	21
Diameter of body	7

Asia Minor, Syria.

a. Ad. (A. 108+20; p. 8).	Smyrna.	Dr. J. von Bedriaga [E.]. (One of the types.)
b. Ad. (A. 112+19; p. 8).	Smyrna.	
c–d. Ad. (A. 119+20, 119+22; p. 8).	Valley of the Meinder.	R. McAndrew, Esq. [P.].
e. Ad. (A. 115+7; p. 8).	Constantinople.	

3. Blanus bedriagæ. (Plate XXIII. fig. 1.)

Blanus bedriagæ, *Bouleng. Ann. & Mag. N. H.* (5) xiii. 1884, p. 396.

Præmaxillary teeth 7; max. 3—3; mand. 7—7. Snout projecting beyond the lower jaw. Frontal as broad as long; the suture between the frontal and the second labial nearly as long as that between the latter and the ocular. Three lower labials, the first and second large, the third small. No occipital fold. One hundred and one to one hundred and fourteen annuli on the body and nineteen to twenty-one on the tail; an annulus contains sixteen to eighteen dorsal and eighteen to twenty ventral segments. Præanal pores ten. In other respects like *B. cinereus.*

	millim.
Length to vent	175
Tail	21
Diameter of body	7

Asia Minor.

a. Several spec.; ad. & hgr. (A. 101 to 114+7 to 21).	Xanthus.	Sir C. Fellows [P.].	(Types.)
b. Ad. (A. 112+19).	Xanthus.		(Types.)
c. Ad. (A. 110+20).	——?		

3. AMPHISBÆNA.

Amphisbæna, *Linn. Syst. Nat.* i. p. 392, *et omn. auct.*
Sarea, *Gray, Cat. Tort.* 1844, p. 71, *and Proc. Zool. Soc.* 1865, p. 449.
Cynisca, *Gray, ll. cc.* pp. 71, 448.
Cadea, *Gray, ll. cc.* pp. 71, 449.
Diphalus, *Cope, Proc. Ac. Philad.* 1861, p. 76.
Bronia, *Gray, Proc. Zool. Soc.* 1865, p. 448.
Ophioproctes, *Bouleng. Bull. Soc. Zool. France,* 1878, p. 300.
Aporarchus, *Cope, Proc. Am. Philos. Soc.* xxii. 1885, p. 187.

Nostril lateral, pierced in a separate nasal. Snout rounded or feebly compressed. No enlarged pectoral segments. Tail cylindrical, obtuse. Præanal pores.

Tropical America; Tropical Africa.

Synopsis of the Species.

I. Nasals forming a suture on the snout.

A. More than four præanal pores.

1. The segments of the two median ventral rows not twice as broad as long.

a. Nasal suture as long as præfrontal.

An annulus on the middle of the body contains less than 60 segments; the median ventral segments not longer than broad .... 1. *fuliginosa,* p. 437.

An annulus on the middle of the body contains at least 60 segments; the median ventral segments longer than broad .... 2. *alba,* p. 438.

b. Nasal suture much shorter than præfrontal.

An infraocular; no præocular; occipitals small .... 3. *beniensis,* p. 439.

An infraocular; a præocular; occipitals small .... 4. *subocularis*, p. 440.
Ocular forming a suture with second and third labials; occipitals small ...... 5. *pretrii*, p. 440.
Ocular forming a suture with second and third labials; occipitals nearly as large as frontals .... 6. *leucocephala*, p. 441.
Occipitals twice as large as frontals .... 7. *mertensii*, p. 441.

2. The segments of the two median ventral rows more than twice as broad as long.

a. Præfrontals distinct; three frontals .... 20. *kraussi*, p. 447.

b. Præfrontals united with the rostral.

First upper labial distinct; a supraocular 21. *leucura*, p. 447.
First upper labial united with the nasal; a small median frontal .... 22. *muelleri*, p. 448.
First and second upper labials united with the nasal; only two anal segments .. 24. *liberiensis*, p. 449.

B. Not more than four præanal pores.

1. Two præfrontals, distinct from the nasals and labials.

a. Suture between the nasals at least one third as long as that between the præfrontals.

α. Temporals small; 28 to 38 segments in an annulus on the middle of the body.

* Ocular in contact with the second and third upper labials.

216 or more annuli on the body; suture between the frontals longer than that between the præfrontals .... 8. *vermicularis*, p. 441.
214 or more annuli on the body; suture between the frontals a little shorter than that between the præfrontals .. 9. *darwinii*, p. 442.

** Oculars in contact with the second labial only.

Four upper labials, two large and two small .... 11. *angustifrons*, p. 443.
Two upper labials .... 13. *gracilis*, p. 444.

β. Temporals small; 46 segments in an annulus on the middle of the body.

No occipitals .... 12. *plumbea*, p. 443.

γ. A large anterior temporal.

Head compressed and arched; 256 annuli on the body .... 14. *steindachneri*, p. 444.
180 annuli on the body.............. 18. *violacea*, p. 446.

b. Suture between the nasals not one third as long as that between the præfrontals.

Ocular forming a suture with the second labial; a præocular .... 15. *mildei*, p. 445.
Ocular forming a suture with the second and third labials; no præocular .... 16. *cæca*, p. 445.
Ocular united with the second labial .. 17. *cubana*, p. 446.

2. Four præfrontals.

The two pairs of præfrontals in front of each other .... 10. *albocingulata*, p. 443.
The two pairs on a transverse line .... 19. *quadrifrons*, p. 447.

3. Præfrontals, nasal and anterior labial on each side united into a single shield .... 23. *leonina*, p. 448.

II. Nasals small, separated from each other.

A pair of præfrontals; rostral narrow superiorly; snout pointed.......... 25. *fenestrata*, p. 449.
A single præfrontal .... 26. *punctata*, p. 450.
A pair of præfrontals; rostral large; snout compressed .... 27. *brasiliana*, p. 450.

1. Amphisbæna fuliginosa.

Amphisbæna americana, *Gray, Cat. Tort.* p. 70, *and Sh. Rept.* ii. p. 35.
Amphisbæna fuliginosa, *Linn. S. N.* i. p. 392; *Daud. Rept.* vii. p. 406, pl. xci. fig. 2; *Guér. Icon. R. A., Rept.* pl. xviii. fig. 1; *Dum. & Bibr.* v. p. 480; *Strauch, Mél. Biol. Ac. St. Petersb.* xi. 1881, p. 376; *Bocourt, Miss. Sc. Mex., Rept.* p. 489, pl. xxi. f. 5.
—— vulgaris, *Laur. Syn. Rept.* p. 66.
—— varia, *Laur. l. c.*
—— magnifica, *Laur. l. c.*
—— flava, *Laur. l. c.* p. 67.

Usually: præmaxillary teeth 5; maxillaries 5—5; mandibulars 8—8. Snout broad, rounded, slightly prominent. Tail not so thick as the body, constricted at the base. Rostral moderate, subtriangular; nasals very large, the largest shields of the head forming a suture; a pair of large præfrontals and a pair of smaller frontals, followed by two or three pairs of square shields; eye distinct through the ocular; one or two infraoculars, the anterior, if present, formed by division of the second labial; upper labials three or four, posterior small; mental quadrangular, followed by a median subpentagonal chin-shield; on each side a pair of large chin-shields; three or four lower labials, first large, the others small. One hundred and ninety-three to two hundred and seventeen annuli on the body, and twenty-four to thirty-two on the tail; the segments of these annuli longer than broad on the back and sides, equilateral or broader than long on the belly; on the middle of the body, each annulus contains forty to fifty-two segments, twenty to twenty-six on the dorsal, and as many on the ventral side. Lateral line distinct. Eight or ten anal shields. Præanal pores seven to ten (6 to 12, *Strauch*), eight being the usual number. Black and white varied, either the one or the other colour predominating.

	millim.
Length to vent	390
Tail	55
Diameter of body	18

Tropical South America; West Indies.

a. Ad. (A. 193+27; p. 8).	W. Indies.	Miss Saul [P.].
b–d. Ad. & hgr. (A. 198+9, 201 +8, 204+30; p. 8, 8, 10).	British Guiana.	
e. Ad. (A. 199+9; p. 8).	Berbice.	Lady Essex [P.].
f. Ad. (199+29; p. 8).	Berbice.	
g. Hgr. (A. 203+28; p. 7).	Demerara.	Capt. Sabine [P.].
h–i. Ad. & hgr. (A. 201+29, 217 +28; p. 8, 9).	Upper Amazon.	
k. Ad. (A. 204+30; p. 8).	Moyobamba.	Mr. A. H. Roff [C.].
l. Ad. (A. 209+9; p. 8).	Cashiboya, N.E. Peru.	Messrs. Veitch [P.].
m–n. Ad. & hgr. (A. 207+30, 212+31; p. 8, 9).	Sarayacu. N.E. Peru.	Messrs. Veitch [P.].
o. Ad. (A. 211+32; p. 9).	Guayaquil.	Mr. Fraser [C].
p. Ad. (A. 207+27; p. 8).	S. America.	
q. Ad. (204+11; p. 8).	S. America.	Bristol Museum [P.].
r. Ad. (202+29; p. 8).	S. America.	
s. Ad. (A. 210+27; p. 7).	S. America.	
t. Ad., stuffed.	S. America.	
u. Hgr., skeleton.	S. America.	

2. Amphisbæna alba.

Amphisbæna alba, *Gray, Cat. Tort.* p. 70, *and Sh. Rept.* p. 35.

Amphisbæna alba, *Linn. S. N.* i. p. 393; *Daud. Rept.* vii. p. 401, pl. xci. fig. 1; *Dum. & Bibr.* v. p. 484; *Strauch, Mél. Biol. Ac. St. Petersb.* xi. p. 378.

—— rosea, *Shaw, Nat. Miscell.* iii. pl. 86.

—— pachyura, *Wolf, Abbild. u. Beschr. merkw. nat. Gegenst.* ii. p. 61, pl. xvii.

—— flavescens, *Wied, Abbild. Nat. Bras.*; *Wagl. Icon. Descr. Amph.* pl. xvi. fig. 1.

Præmaxillary teeth 5; maxillaries 5—5; mandibulars 8—8. Snout broad, rounded, slightly prominent. Tail as thick as the body, of equal width throughout. Rostral moderate, subtriangular; nasals and præfrontals very large; frontals smaller; postocular large, sometimes divided; no infraocular, the ocular, through which the eye is distinguishable, forming a suture with the second and third labials; four upper labials, fourth small; mental trapezoid, followed by a heptagonal or pentagonal chin-shield; three or four lower labials, second largest, first and second forming a suture with the median chin-shield; a large chin-shield on each side behind the second labial. Two hundred and nineteen to two hundred and thirty-three annuli on the body (205 to 242, *Strauch*) and seventeen to twenty-one on the tail; the segments of the annuli all much longer than broad, and scarcely broader on the belly than on the back; on the middle of the body each annulus contains sixty to eighty segments, twenty-eight to thirty-eight on the dorsal and thirty-two to forty-two on the ventral side. Lateral line distinct. Eight or ten anal segments. Præanal pores six to ten. Brownish above, whitish beneath, uniform.

	millim.
Length to vent	520
Tail	50
Diameter of body	23

Tropical South America; West Indies.

a. Ad. (A. 230+19; p. 9).	Brazil.	
b. Ad. (A. 226+18; p. 7).	Brazil.	
c–d. Ad. (A. 232+18, 233+18; p. 8).	Pernambuco.	W. A. Forbes, Esq. [P.].
e–g. Ad. (A. 219+19, 220+21, 222+19; p. 8).	Moyobamba.	Mr. A. H. Roff [C.].
h–i. Ad. (A. 232+17, 225+17; p. 7, 6).	S. America.	
k. Ad. (A. 225+20; p. 10).	S. America.	
l. Ad. (A. 231+19; p. 8).	S. America.	
m. Ad. (A. 222+20; p. 9).	S. America.	
n. Ad., stuffed.	S. America.	
o, p. Ad., skeletons.	S. America.	

3. Amphisbæna beniensis.

Amphisbæna beniensis, *Cope, Proc. Am. Phil. Soc.* xxii. 1885, pp. 184, 188, pl. —. fig. 2.

Snout moderately elongate, obtuse; rostral plate a little visible from above; nasals forming a suture, which is shorter than that between the præfrontals or the frontals; a pair of præfrontals and a pair of smaller frontals; occipitals small; eye visible through the ocular; a postocular and an infraocular; no præocular; four upper labials, third and fourth small, second in contact with the ocular; mental subquadrate, followed by a seven-sided chin-shield; three lower labials. Tail with sixteen annuli. Eight or ten anal segments. Six præanal pores. Colour above light reddish brown; below yellow.

Length to vent 310 millim.; tail 35.

Upper Beni River, Bolivia.

4. Amphisbæna subocularis.

Amphisbæna pretrei (*non D. & B.*), *Guichen. in Casteln. Amér. Mér. Rept.* pl. viii.; *Strauch, Mél. Biol. Ac. St. Petersb.* xi. p. 381.
Amphisbæna subocularis, *Peters, Mon. Berl. Ac.* 1878, p. 779, pl. —. fig. 2; *Strauch, l. c.* p. 380.

Præmaxillary teeth 7; maxillaries 4—4; mandibulars 6—6. Snout rounded, prominent. Tail cylindrical, as thick as the body. Rostral moderate, triangular; nasals forming a short suture; a pair of large præfrontals followed by a pair of smaller frontals, the suture between the former being scarcely longer than that between the latter; occipitals small; eye visible through the ocular; a postocular, one or two infraoculars, a præocular; four upper labials, fourth smallest, first either in contact with the præfrontal, or separated by a small shield; mental trapezoid, followed by a seven-sided chin-shield; three lower labials, second largest. Two hundred and forty to two hundred and fifty-three annuli on the body, and twenty-six to twenty-nine on the tail; the segments of the annuli longer than broad on the back and sides, a little broader than long on the two median ventral rows; on the middle of the body, an annulus contains forty-six or forty-eight segments, twenty-two or twenty-four on the dorsal, and twenty-four on the ventral side. Lateral line distinct. Ten anal shields. Six to eight præanal pores. Brownish above, whitish inferiorly.

	millim.
Length to vent	315
Tail	37
Diameter of body	13

Brazil (Pernambuco and Bahia).

a. Ad. (A. 240+26, p. 7).	Bahia.	St. Petersburg Mus. [E.].
b. Hgr. (A. 245+26, p. 8).	Brazil.	Prof. C. Machado [P.].

5. Amphisbæna pretrii.

Amphisbæna petrei, *Gray, Cat. Tort.* p. 70, *and Sh. Rept.* ii. p. 35.
Amphisbæna pretrei, *Dum. & Bibr.* v. p. 486.

Præmaxillary teeth 7; maxillaries 5—5; mandibulars 8—8. Head subconical, a little depressed; snout rounded. Rostral small, triangular; nasals forming a suture; a pair of præfrontals and a pair of frontals; occipitals small; eye hardly distinct; no præ- or infraocular; three upper labials; mental anvil-shaped, followed by a larger chin-shield, the posterior border of which forms an angle; three lower labials, posterior very small, second largest, forming a suture with a chin-shield posteriorly. Two hundred and thirty-five to two hundred and thirty-eight annuli on the body, and twenty-six to twenty-eight on the tail; on the back and sides the segments of the annuli are square or hardly longer than broad, on the belly they are distinctly dilated. Lateral line distinct. Ten or twelve anal segments. Eight præanal pores. White, the upper scales reddish brown.

Length to vent 280 millim.; tail 34 millim.

Brazil.

6. Amphisbæna leucocephala.

Amphisbæna leucocephala, *Peters, Mon. Berl. Ac.* 1878, p. 778, pl. —. fig. 1; *Strauch, Mél. Biol. Ac. St. Petersb.* xi. p. 384.

A pair of occipitals, nearly as large as the frontals; a small fourth upper labial; ten or twelve præanal pores. Otherwise I cannot see any difference, from the descriptions, between this and the preceding species. Two hundred and thirty-two or two hundred and thirty-three annuli on the body, twenty-nine or thirty on the tail.

Bahia.

7. Amphisbæna mertensii.

Amphisbæna mertensii, *Strauch, Mél. Biol. Ac. St. Petersb.* xi. p. 385.

The head-shields differ from those of the preceding species chiefly in the presence of a pair of very large fronto-parietals, or occipitals, each of which is as large as the two frontals taken together; no infraocular; three upper and as many lower labials. Two hundred and twenty-nine annuli on the body, and thirty-two on the tail; the segments of the annuli narrow, much longer than broad on the dorsal surface, the two median ventral rows nearly twice as broad as long; an annulus contains forty-four segments, twenty-four above and twenty below the lateral lines. Latter distinct. Eight anal segments. Seven præanal pores. Yellowish brown, lighter inferiorly.

Length to vent 290 millim.; tail 40 millim.

S. America.

8. Amphisbæna vermicularis.

Amphisbæna vermicularis, *Gray, Cat. Tort.* p. 71, *and Sh. Rept.* ii. p. 35.

Amphisbæna vermicularis, *Wagl. in Spix, Serp. Bras.* p. 73, pl. xxv. fig. 2; *Dum. & Bibr.* v. p. 489; *Rhein. & Lütk. Vidensk. Meddel.* 1861, p. 205; *Strauch, Mél. Biol. Ac. St. Petersb.* xi. p. 395.

Præmaxillary teeth 7; maxillaries 5—5; mandibulars 8—8. Snout rounded, prominent. Tail cylindrical, as thick as the body. Rostral moderate, subtriangular; nasals forming a suture; a pair of large præfrontals, a pair of smaller frontals, having together a subrhomboidal shape, and a pair of small occipitals; the longest suture is that between the frontals, then come those of the præfrontals, nasals, and occipitals; the length of the nasal suture equals only half that of the præfrontal; eye visible through the ocular; a postocular, no infraocular, the ocular forming a suture with the second and third labials; four upper labials, fourth small; mental trapezoid, followed by a seven-sided chin-shield; three lower labials; second largest, first two forming a suture with the median chin-shield; a pair of chin-shields on each side, in contact with the second and third labials. Two hundred and thirteen to two hundred and forty-seven annuli on the body (*Strauch*), and sixteen or seventeen on the tail (28–37, *Strauch*); the segments of the annuli longer than broad on the back and sides, a little broader than long on the two median ventral rows; on the middle of the body, each annulus contains thirty segments, fourteen on the dorsal, and sixteen on the ventral side. Lateral line distinct. Six anal segments. Præanal pores two or four, frequently (♀?) very indistinct. Light brown above, yellowish inferiorly.

	millim.
Length to vent	270
Tail	21
Diameter of body	8

Northern and Central Brazil.

a. Ad. (A. 221+17; p. 2).	Brazil.	Dr. Gardner [P.].
b. Ad. (A. 216+16; p. 4).	Porto Bello (?).	Capt. Austin [P.].

The determination of the latter specimen remains uncertain, the skin of the head having been entirely removed for investigation of the skull. However, the scaling of the body agrees well with that of *A. vermicularis*, and I am inclined to regard the identification as correct. It is not by any means identical with the species described below as *A. cæca*, as suggested by Strauch (*l. c.* p. 406), the body being much more elongate, and the scales less numerous below the lateral lines (sixteen instead of twenty-four).

9. Amphisbæna darwinii.

Amphisbæna darwinii, *Gray, Cat. Tort.* p. 71, *and Sh. Rept.* ii. p. 36.
—— camura, *Gray, l. c.* p. 35.
—— heterozonata, *Gray, l. c.*
Amphisbæna darwinii, *Dum. & Bibr.* v. p. 490; *Peters, Mon. Berl. Ac.* 1878, p. 781, pl. —. fig. 6; *Strauch, Mél. Biol. Ac. St. Petersb.* xi. p. 403; *Bouleng. Ann. & Mag. N. H.* (5), xvi. 1885, p. 296.
—— camura, *Cope, Proc. Ac. Philad.* 1862, p. 350.
—— vermicularis (*non Wagl.*), *Hensel, Arch. f. Nat.* 1868, p. 339.
—— heterozonata, *Burmeist. Reise La Plata,* ii. p. 527.
—— trachura, *Cope, Proc. Am. Philos. Soc.* xxii. 1885, p. 187, pl. —. fig. 5.
Aporarchus prunicolor, *Cope, l. c.* p. 189, pl. —. fig. 5.

This more southern form is so closely allied to the preceding that but a few slight differences distinguish them:—The suture between the præfrontals is slightly longer than that between the frontals; the fourth upper labial is generally missing; the annuli on the body are less numerous, one hundred and eighty-two to two hundred (177 to 214, *Strauch*). Caudal annuli, nineteen to twenty-five (17 to 25, *Strauch*). The number of segments in one annulus of the middle of the body varies between twenty-eight and thirty-eight, twelve to eighteen above and sixteen to twenty-two below the lateral lines. Anal segments six or eight. Occipitals frequently broken up, not distinguishable from the surrounding plates. All the specimens in the collection have four præanal pores, but they are frequently hardly visible. Colour either uniform brown above and yellowish beneath, or checked with dark brown on yellowish ground, in which case a light collar may be apparent.

	millim.
Length to vent	360
Tail (injured)................	23
Diameter of body	16

South Brazil, Paraguay, Uruguay, Buenos Ayres.

a–d, e. Ad. & yg. (A. 184–200+11–25).	Rio Grande do Sul.	Dr. v. Ihering [C.].
f–l. Ad. & hgr. (A. 182–188+9–27).	S. Lorenzo, Rio Grande do Sul.	Dr. v. Ihering [C.].
m. Ad. (A. 190+23).	Montevideo.	Zoolog. Soc.
n–q. Ad. (A. 183–190+9–25).	Uruguay.	
r. Ad. skeleton.	Uruguay.	
s. Ad. (A. 200+19).	Buenos Ayres.	E. W. White, Esq. [C.].

10. Amphisbæna albocingulata.

Amphisbæna albocingulata, *Boettg. Zeitschr. f. Naturw.* lviii. 1885, p. 215.

Distinguished from *A. darwinii* by the presence of four præfrontals, a smaller anterior and a larger posterior pair, and by the presence of a more or less distinct præocular. Annuli 190+27.

Paraguay.

11. Amphisbæna angustifrons.

Amphisbæna angustifrons, *Cope, Proc. Ac. Philad.* 1861, p. 70, *and Proc. Am. Philos. Soc.* 1885, xxii. p. 188, pl. —. fig. 4.

Differs from *A. darwinii* in the presence of an infraocular, separating the ocular from the third and fourth labials, which are minute.

Buenos Ayres.

[*Amphisbæna occidentalis*, Cope, Journ. Ac. Philad. (2), viii, 1876, p. 176, and Proc. Am. Philos. Soc. xxii. pl. —. fig. 3, from the Valley of Jequetepec, Peru, is another close ally of *A. vermicularis* and *darwinii*, but is not sufficiently characterized in the description to be incorporated in the system.]

12. Amphisbæna plumbea. (Plate XXIII. fig. 2.)

Amphisbæna plumbea, *Gray, Cat. Sh. Rept.* ii. p. 36.

Allied to *A. darwinii*, but well distinguished in the following points:—The nasals much larger, their median suture a little longer than that between the frontals, which are not followed by any occipitals; ocular smaller; labials much higher, the head being much more compressed; fourth upper labial very small; lower labials and chin-shields as in *A. vermicularis*. The number of annuli on the body given by Gray (270) is no longer controllable, on account of the bad state of the specimen; but I should feel inclined to regard it as nearly correct. Caudal annuli twenty-two. The segments of the annuli are much narrower than in the two preceding species; the two median ventral rows equilateral; on the middle of the body an annulus contains forty-six scales, twenty-two above and twenty-four below the lateral lines. Anal segments six. Præanal pores four.

Mendoza.

a. Bad state.	Mendoza.	(Type.)

13. Amphisbæna gracilis.

Amphisbæna gracilis, *Strauch, Mél. Biol. Ac. St. Petersb.* xi. p. 391.

Snout obtusely pointed. Rostral small, triangular; nasals forming a suture; a pair of præfrontals, larger than the nasals; frontals smaller; two occipitals; eye distinct through the ocular, which is situated between the præfrontal and the second labial; two upper labials; mental quadrangular, followed by a very large hexagonal chin-shield; three lower labials, third very small. Two hundred and twenty-four annuli on the body and twenty-eight on the tail; an annulus contains about sixteen dorsal and sixteen ventral segments, of which the two median ventrals are a little broader than the others. Lateral line very distinct; a slight vertebral line. Six anal segments. Four præanal pores. Uniform dark purplish brown.

Length to vent 170 millim.; tail 20 millim.

South America?

14. Amphisbæna steindachneri.

Amphisbæna steindachneri, *Strauch, Mél. Biol. Ac. St. Petersb.* xi. p. 407.

Head compressed and arched, separated from the body by a more or less distinct groove. Tail suddenly compressed at the end. Rostral moderate, triangular, entirely under the snout; nasals rather large, forming a suture; præfrontals two, very large; frontals either smaller or larger than the præfrontals; eye distinct through the ocular, which forms a suture with the second and third upper labials; latter three, subequal; a large anterior temporal; mental large, quadrangular, followed by a larger quadrangular chin-shield; three lower labials, first and third very small; a large chin-shield on each side. Two hundred and fifty-six annuli on the body and nineteen or twenty on the tail; the annuli on the body comprise sixteen dorsal and twelve ventral segments, the former elongate and narrow, the latter, especially the two median, broader. Lateral line very distinct; a vertebral and a ventral line. Six anal segments; four præanal pores. Yellowish brown.

Caiçara and Mattogrosso, Brazil.

15. **Amphisbæna mildei.**

Amphisbæna mildei, *Peters, Mon. Berl. Ac.* 1878, p. 779, pl. —. fig. 3; *Strauch, Mél. Biol. Ac. St. Petersb.* xi. p. 395.

Rostral moderate, pentagonal, much broader than high; nasals forming a short suture; præfrontals twice as long as latter; a pair of smaller frontals; a præocular, between the first and second labials, and an infraocular, between the second and third; three upper labials, first largest, second in contact with the ocular; three lower labials, second largest. One hundred and ninety-eight annuli on the body, and twenty-four on the tail. A lateral line. Six anal segments; four præanal pores. Purplish brown above, whitish inferiorly.

	millim.
Length to vent	293
Tail	37
Diameter of body	11

Porto Alegre.

16. **Amphisbæna cæca.**

Sarea cæca, *Gray, Cat. Tort.* p. 71.
—— ——, part., *Gray, Cat. Sh. Rept.* ii. p. 37.
—— innocens, *Gray, l. c.*

Amphisbæna cæca, *Cuv. R. A.* 2nd ed. ii. p. 73; *Dum. & Bibr.* v. p. 492; *Peters, Mon. Berl. Ac.* 1878, p. 781, pl. —. fig. 7; *Strauch, Mél. Biol. Ac. St. Petersb.* xi. p. 405.
—— innocens, *Weinl. Abh. Senck. Ges.* iv. 1862, p. 137, pl. v. fig. 2.

Præmaxillary teeth 7, maxillaries 5–5, mandibulars 8–8. Snout rather pointed, slightly prominent. Tail thinner than the body. Rostral small, triangular; nasals forming a short suture; a pair of very large præfrontals, followed by a pair of much smaller frontals; eye hardly distinguishable through the ocular; a postocular, no infraocular; three large upper labials, and an indistinct fourth, the second and third forming a suture with the ocular; mental quadrangular, followed by a large seven-sided chin-shield which is much longer than broad; a triangular chin-shield on each side, behind the second labial; lower labials three, second very large. One hundred and eighty-nine (to 247, *Strauch*) annuli on the body and nineteen on the tail; the divisions of the annuli longer than broad, nearly equilateral on the middle of the belly, but nowhere broader than long; on the middle of the body an annulus contains forty-two divisions, eighteen above and twenty-four below the lateral lines. Anal shields eight [or six]. Præanal pores four. Brown above, lighter beneath.

	millim.
Length to vent	210
Tail	21
Diameter of body	9

West Indies.

a. Ad. Porto Bello, W. Indies. Capt. Austin [P.].

17. Amphisbæna cubana.

Amphisbæna cubana, *Peters, Mon. Berl. Ac.* 1878, p. 780, pl. —. fig. 4; *Strauch, Mél. Biol. Ac. St. Petersb.* xi. p. 411.

Judging from the figure and description, I feel inclined to regard this so-called species as an *A. cæca* in which the ocular has fused with the second labial; although two specimens are known to present the same scutellation, it is not impossible that this is only a case of anomaly. Another character put forward for the separation of the two forms, viz. the presence of eight anal segments in *A. cubana*, instead of six, as in *A. cæca*, does not hold good, since the specimen described above, and in which the ocular is distinct from the labial, has eight.

Cuba.

18. Amphisbæna violacea.

Cynisca? violacea, *Gray, Cat. Sh. Rept.* ii. p. 36.
Amphisbæna violacea, *Peters, Mon. Berl. Ac.* 1854, p. 620, *and* 1862, p. 26, *and Reise n. Mossamb.* iii. p. 85, pl. xiii. fig. 2; *Strauch, Mél. Biol. Ac. St. Petersb.* xi. p. 411.

Præmaxillary teeth 7, maxillaries 4–4, mandibulars 7–7. Snout rounded, prominent. Rostral moderate, triangular; nasals forming a suture behind the rostral; a pair of præfrontals, a pair of frontals and a pair of narrow occipitals; the suture between the frontals is the longest, then come, in respect to length, the præfrontal suture, the parietal, and the nasal; a large temporal, between the frontal and the fourth labial; four upper labials, anterior smallest, posterior largest; ocular rather large, eye distinct; no infraocular, the ocular forming a suture with the second and third labials; mental trapezoid with concave lateral borders, followed by a pentagonal or heptagonal chin-shield; three lower labials, second very large; two large chin-shields on each side. One hundred and eighty to one hundred and eighty-eight annuli on the body and fifty-six to fifty-nine on the tail; an annulus contains eighteen or twenty dorsal, and sixteen or eighteen ventral segments; the two median ventral segments much broader than the others. Lateral line distinct. Six anal segments and four præanal pores. Dark purple above, lighter beneath.

	millim.
Length to vent	153
Tail	45
Diameter of body	5

Inhambane, S.E. Africa.

19. Amphisbæna quadrifrons.

Cynisca quadrifrons, *Gray, Cat. Sh. Rept.* ii. p. 36.
Amphisbæna quadrifrons, *Peters, Mon. Berl. Ac.* 1862, p. 25, *and* 1879, p. 277, pl. —. fig. 4; *Strauch, Mél. Biol. Ac. St. Petersb.* xi. p. 412.

Snout rounded, prominent. Rostral triangular, its point visible from above; nasals forming a suture, four elongate præfrontals, on a curved transverse line; a pair of frontals followed by a pair of small occipitals; eye distinct through the ocular, which forms a suture with the second and third labials; two large temporals; four upper labials, third largest; mental trapezoid, followed by a larger seven-sided chin-shield; four lower labials, second largest, fourth very small; a large chin-shield behind the second lower labial. Two hundred and twenty-six to two hundred and thirty-seven annuli on the body, forty-eight on the tail. Four anal segments. Four præanal pores.

Hereroland, S.W. Africa.

20. Amphisbæna kraussi.

Amphisbæna kraussi, *Peters, Sitzb. Ges. Nat. Fr.* 1878, p. 192, *and Mon. Berl. Ac.* 1878, p. 731, pl.—. fig. 5.

Allied to *A. leucura*. Rostral higher than broad, pointed above; nasals half as long as præfrontals; a very small median frontal; a pair of frontals, forming a point posteriorly, wedged in between two large postfrontals, which are followed by a pair of occipitals; a præocular, separating the ocular from the nasal; three upper labials, third largest, followed by a very large temporal, which is the largest of all the head-shields; an elongate mental and two lower labials, the anterior of which is very large, the second very small. Six anal segments; eight præanal pores.

West Africa.

21. Amphisbæna leucura.

Cynisca leucura, *Gray, Cat. Tort.* p. 71, *and Sh. Rept.* ii. p. 36.
Amphisbæna leucura, *Dum. & Bibr.* v. p. 498; *Peters, Mon. Berl. Ac.* 1879, p. 277, pl. —. fig. 5; *Strauch, Mél. Biol. Ac. St. Petersb.* xi. p. 388.

Præmaxillary teeth 5, maxillaries 4–4, mandibulars 7–7. Snout rounded, prominent. Rostral small, triangular; nasal and præfrontal on each side united, forming a long suture on the upper surface of the snout; a median pentagonal frontal, between the supraoculars; a pair of postfrontals followed by a pair of occipitals; three large temporals; ocular a little larger than supraocular; eye slightly distinct; a præocular, broader than deep; three upper labials, third largest; mental elongate, narrow, trapezoid; two lower labials, first extremely large, second small; a small chin-shield on each side, forming a suture with the two lower labials. Two hundred and thirty-four annuli on the body (206, *Dum. & Bibr.*, 217 to 233, *Strauch*), twenty-nine on the tail; an annulus in the middle of the body contains thirty-four segments, twenty above and fourteen below the lateral lines; the dorsal and lateral segments much longer than broad, the two median ventrals about twice and a half as broad as long. Lateral line distinct; a slight vertebral line. Anal segments eight; præanal pores eight [to ten]. Brownish, a little lighter inferiorly.

	millim.
Length to vent	195
Tail	27
Diameter of body	5·5

West Africa.

a. Ad. Calabar.

22. Amphisbæna muelleri.

Cynisca, sp., *F. Müller, Verh. Naturf. Ges. Basel*, vi. 1878, p. 704, pl. ii. fig. *c*.
Amphisbæna mülleri, *Strauch, Mél. Biol. Ac. St. Petersb.* xi. p. 389.

Allied to the preceding; differs in the following points:—Nasal, præfrontal and anterior labial united into a single shield; frontal small, triangular; no præocular, no supraocular. Two hundred and twenty-nine to two hundred and forty annuli on the body, and twenty-five to twenty-seven on the tail; an annulus contains twenty-four or twenty-six segments, twelve on the dorsal side and ten or twelve on the ventral. Six anal segments; twelve præanal pores.

Length to vent 175 millim., tail 20 millim.

Gold Coast; Sierra Leone.

23. Amphisbæna leonina.

Amphisbæna leonina, *F. Müller, Verh. Naturf. Ges. Basel*, vii. 1885, p. 700, pl. ix. figs. *a–e*.

Also allied to the two preceding species. Nasal, præfrontal, and first and second labials united into a single shield; frontal small, triangular, followed by a pair of large postfrontals and a pair of occipitals; no supraocular, no præocular. Two hundred and forty annuli on the body, twenty on the tail; an annulus contains twenty-four segments, fourteen on the dorsal side, and ten on the ventral; latter, two median rows very broad. Six anal segments; three (four?) præanal pores. Brown; lower surface of body whitish.

	millim.
Length to vent	155
Tail	15
Diameter of body	4

Tumbo Island, Sierra Leone.

24. Amphisbæna liberiensis.

Ophioproctes liberiensis, *Bouleng. Bull. Soc. Zool. France,* 1878, p. 301.
Amphisbæna liberiensis, *Strauch, Mél. Biol. Ac. St. Petersb.* xi. p. 390.

Snout rounded, prominent. Rostral small, triangular; nasals, præfrontals and anterior labials united into a large shield forming a long suture with its fellow and covering the whole of the snout; ocular small, eye hardly distinguishable; a pair of large frontals; a pair of small parietals; a large labial, posterior to the vertical of the eye, and two or three temporals, upper largest; mental quadrangular, twice as long as broad; three lower labials, anterior large, the two others small. Two hundred annuli on the body and twenty-eight on the tail; an annulus contains twenty-four segments, the dorsals longer than broad, the two median ventrals nearly three times as broad as long. Lateral line distinct. Anal segments two, very large. Eight præanal pores. Brown, rather lighter inferiorly.

Length to vent 135 millim., tail 18 millim.

Liberia.

25. Amphisbæna fenestrata.

Sarea cæca, part., *Gray, Cat. Sh. Rept.* ii. p. 38.
Sarea ? fenestrata, *Gray, l. c.*
Diphalus fenestratus, *Cope, Proc. Ac. Philad.* 1861, p. 76, *and Proc. Amer. Phil. Soc.* xi. 1869, p. 164.
Amphisbæna antillensis, *Reinh. & Lütk. Vidensk. Meddel.* 1862, p. 224.
—— fenestrata, *Strauch, Mél. Biol. Ac. St. Petersb.* xi. p. 415.

Præmaxillary teeth 5; maxillaries 4—4; mandibulars 7—7. Snout prominent, pointed. Rostral moderately large, prolonged posteriorly, separating the nasals, which are not larger than the ocular; a pair of very large præfrontals, twice as long as broad, followed by a pair of smaller posterior frontals; eye distinct through the ocular, which forms a suture with the second and third labials; no præocular; two temporals, upper largest; three upper labials, first smallest, second largest; mental quadrangular, followed by a larger octagonal chin-shield; three lower labials, first very small, second very large; a small chin-shield on each side behind the second lower labial. Two hundred and forty-two to two hundred and forty-seven annuli on the body and fifteen or sixteen on the tail; on the middle of the body, an annulus contains twenty-six or twenty-eight segments, twelve or fourteen above and fourteen below the lateral lines; the dorsal and lateral segments equilateral or slightly longer than broad, the two median ventrals distinctly broader than long. Lateral line distinct; a feeble vertebral line. Anal segments four or six; præanal pores four. Flesh-coloured, each segment above and on the sides marked with a central rectangular brown spot.

	millim.
Length to vent	240
Tail	18
Diameter of body	8·5

Virgin Islands, West Indies.

a. Ad. (A. 242+15).	St. Jean.	Copenhagen Museum [P.]. (As typical of *A. antillensis*.)
b–c. Ad. (A. 244+16, 247+15).	St. Thomas.	Hr. Riise [C.].

26. **Amphisbæna punctata.** (Plate XXIII. fig. 3.)

Cadea punctata, *Gray, Cat. Tort.* p. 72, *and Sh. Rept.* ii. p. 38.
Amphisbæna punctata, *Bell, Zool. Journ.* iii. 1828, p. 236; *Dum. & Bibr.* v. p. 494; *Coct. in R. de la Sagra, Hist. Cuba*, p. 195, pl. xxi.; *Strauch, Mél. Biol. Ac. St. Petersb.* xi. p. 417.

Præmaxillary teeth 5; maxillaries 5—5; mandibulars 8—8. Snout pointed, slightly prominent. Rostral moderately large, quadrangular, forming a broad suture with the anterior border of the præfrontal; latter large, single, covering the upper surface of the snout; longer than broad; a pair of narrow, elongate frontals followed by a pair of small square occipitals; eye slightly distinct through the ocular; an elongate supraocular produced forwards to the nasal with which it forms a suture, the two shields separating the labials from the frontal; three upper labials, first smallest, second largest; mental subquadrangular, followed by an elongate chin-shield; three lower labials, first small, second and third very large. Two hundred and nineteen annuli on the dorsal side, one hundred and seventy-eight on the ventral; fifteen annuli on the tail; an annulus on the middle of the body contains thirty segments; the dorsal segments oval, the ventrals square and distinctly broader than long. No lateral line. Anal segments eight; præanal pores four. Flesh-coloured above, with small brown spots.

	millim.
Length to vent	180
Tail	13
Diameter of body	7

Cuba.

a. Ad.	Cuba.	W. S. Mac Leay, Esq. [P.]. (Type.)

27. **Amphisbæna brasiliana.**

Bronia brasiliana, *Gray, Cat. Sh. Rept.* ii. p. 37.
Bronia brasiliana, *Gray, Proc. Zool. Soc.* 1865, p. 448, fig.
Amphisbæna brasiliana, *Strauch, Mél. Biol. Ac. St. Petersb.* xi. p. 413.

Præmaxillary teeth 7; maxillaries 5—5; mandibulars 8—8.

Head rather compressed, snout rounded, very prominent. Rostral large, forming a broad suture with the præfrontals ; nasal small, on the side of the rostral (in the two specimens in the Collection, this plate has fused with the rostral, but this should be considered rather as anomalous, since four other specimens examined by Strauch are said to have the nasal completely separated from the rostral); a pair of large præfrontals, followed by a pair of smaller frontals forming together a heart-shaped figure ; eye not distinct through the ocular; three upper labials, anterior smallest; mental subquadrangular, followed by a pentagonal chin-shield; three lower labials, anterior very small, second largest; a large chin-shield on each side, behind the second labial. Two hundred and twenty-two or two hundred and twenty-six annuli on the body, and sixteen or seventeen on the tail; on the middle of the body an annulus contains thirty-eight segments, twenty above and eighteen below the lateral lines; the dorsal segments longer than broad, the two median ventrals a little broader than long. Lateral lines distinct. Anal segments eight or ten ; præanal pores four. Brownish white, each dorsal segment with a small central brown spot.

	millim.
Length to vent	260
Tail	19
Diameter of body	10

Northern Brazil.

a. Ad. (Ann. 222+17). Santarem. H.W. Bates, Esq. [C.]. (Type.)
b. Ad. (A. 226+16). River Cuparia.

4. ANOPS.

Anops, *Bell, Proc. Zool. Soc.* 1883, p. 99 ; *Gray, Cat. Tort.* 1844, p. 72, *and Proc. Zool. Soc.* 1865, p. 450.
Amphisbæna, part., *Dum. & Bibr.* v. p. 476.
Baikia, *Gray, Proc. Zool. Soc.* 1865, p. 450.

Head strongly compressed, the rostral shield enormous, forming an arch with cutting-edge ; nasals small, lateral. No enlarged pectoral segments. Præanal pores. Tail cylindrical, obtuse.

South-eastern South America, West Africa.

1. Anops kingii.

Anops kingii, *Gray, Cat. Tort.* p. 72, *and Sh. Rept.* ii. p. 38.
Anops kingii, *Bell, l. c.* 1833, p. 99, *and Zool. Journ.* v. 1834, p. 391, pl. xvi. fig. 1.
Amphisbæna kingii, *Dum. & Bibr.* v. p. 476 ; *Hensel, Arch. f. Nat.* 1868, p. 343 ; *Strauch, Mél. Biol. Ac. St. Petersb.* xi. p. 418.

Head-shields rather variable ; rostral reaching the vertical of the anterior angle or of the centre of the ocular ; latter moderately large, eye more or less distinct; nostril pierced in the anterior portion of a small nasal situated in a notch of the rostral; an elongate præocular, about the size of the nasal; three upper labials; mental quadrangular, followed by a subpentagonal chin-shield of nearly equal size; a larger lateral chin-shield; three lower labials, third small. Two hundred and seventeen or two hundred and twenty-four annuli on the body, and twenty or twenty-one on the tail (214 to 240+20 to 24, *Strauch*); on the middle of the body an annulus comprises thirty-two or thirty-eight segments, sixteen or eighteen above, and sixteen to twenty below the lateral lines; the dorsal segments longer than broad, the ventrals rectangular or a little broader than long. Lateral line distinct. Six anal segments. Four præanal pores. Purplish brown above, whitish inferiorly.

	millim.
Length to vent	210
Tail..............................	22
Diameter of body	7

Southern Brazil, Uruguay, Argentine Republic.

a. Hgr. (A. 224+21).	Rio Grande do Sul.	Dr. v. Ihering [C.].
b–f. Ova & fœti.	Rio Grande do Sul.	Dr. v. Ihering [C.].
g. Yg. (A. 217+22).	Cosquin, Cordova.	E. W. White, Esq. [C.].

2. **Anops africanus.** (Plate XXIII, fig. 4.)

Baikia africana, *Gray, Cat. Sh. Rept.* ii. p. 39.
Baikia africana, *Gray, Proc. Zool. Soc.* 1865, p. 451, fig.
Amphisbæna africana, *Strauch, Mél. Biol. Ac. St. Petersb.* xi. p. 421.

Head separated from the body by a circular fold. Nasal confluent with rostral anteriorly (this may be an individual anomaly); two shields behind the rostral on each side, forming a short suture, the anterior, or præfrontal, narrow, band-like, descending to the second labial and the ocular; ocular extremely small, triangular at the point of intersection of the præfrontal, the frontal, the temporal, and the second and third labials; eye distinguishable, not under the ocular, but in the supero-anterior corner of the third labial; upper labials three, first very small, second largest; mental quadrangular, followed by an elongate quadrangular chin-shield; two lower labials, first very large, second very small; a large chin-shield on each side, behind the first labial. One hundred and forty-eight annuli on the body and twenty-five on the tail; an annulus contains thirty-eight segments, twenty dorsal and eighteen ventral; latter, two median a little broader than long; dorsal segments longer than broad. Lateral line distinct. Ten anal segments; four præanal pores. Flesh-coloured.

	millim.
Length to vent	180
Tail..............................	22
Diameter of body	5

a.	W. Africa.	(Type.)

5. GEOCALAMUS.

Geocalamus, *Günth. Ann. & Mag. N. H.* (5) vi. 1880, p. 234.

Head slightly compressed, snout rounded; rostral large; nasals small, lateral. Pectoral segments slightly enlarged, forming angular series. Tail cylindrical, obtuse. A lateral line; a gular fold. Præanal pores.

East Africa.

1. Geocalamus modestus. (Plate XXIII. fig. 5.)

Geocalamus modestus, *Günth. l. c.* fig.
Amphisbæna modesta, *Strauch, Mél. Biol. Ac. St. Petersb.* xi. p. 412.

Præmaxillary teeth 3; maxillaries 3—3; mandibulars 7—7. A strong circular fold separates the head from the body. Rostral forming an angle posteriorly, wedged in between a pair of large præfrontals, forming a short suture; a square median frontal, in one specimen partly united with the præfrontals; a pair of frontals: nasal small, smaller than the ocular; eye slightly distinct; three upper labials, first very small, three temporals, superior largest; mental trapezoid, followed by a rather smaller heptagonal chin-shield; three lower labials, first smallest; a lateral chin-shield. Two hundred and thirty-eight to two hundred and forty-one annuli on the body, and twenty-nine on the tail; an annulus on the middle of the body consists of thirty-four to thirty-eight segments, of which sixteen or eighteen are above, and eighteen or twenty below the lateral line; all the segments longer than broad, the median ventrals, however, nearly equilateral. The pectoral segments only feebly differentiated from the others, more elongate, forming angular series, pointing forwards. A distinct lateral line. Six anal segments, median pair largest. Four præanal pores. Dorsal and lateral segments brown; the interspaces between them, as well as the lower surfaces, whitish.

	millim.
Length to vent	205
Tail	32
Diameter of body	5·5

East Africa.

a–c.	Ad.	Mpwapwa, 200 miles inland of the coast opposite Zanzibar.	H. W. Bates, Esq. [C.]. (Types.)

6. MONOPELTIS *.

Monopeltis, *Smith, Ill. Zool. S. Afr., Rept.*
Phractogonus, *Hallow. Proc. Ac. Philad.* 1852, p. 62.

* The number of rings on the dorsal and ventral series do not correspond, as in most other Amphisbænians, not only on account of the pectoral shields, but also because very frequently the anterior ventral rings are larger than the dorsals. I have therefore constantly counted the annuli above the lateral line, commencing from the head.

Monotrophis, *Gray, Proc. Zool. Soc.* 1865, p. 454.
Dalophia, *Gray, l. c.*

Nostril pierced in a small nasal on the lower surface of the snout. Head depressed, with broad projecting snout, with cutting-edge. A strong gular fold. Pectoral segments enlarged. Præanal pores present or absent. Tail cylindrical, obtuse.

Africa.

Synopsis of the Species.

I. A single large shield on the head.

A. No præocular.

42 to 48 segments in an annulus of the body; three lower labials ..	1. *capensis*, p. 455.
32 or 34 segments in an annulus of the body; two lower labials	2. *sphenorhynchus*, p. 455.
34 segments in an annulus of the body; three lower labials; no præanal pores	3. *welwitschii*, p. 456.
B. A præocular	4. *guentheri*, p. 456.

II. Two large shields on the head.

A. Anterior head-shield not longer than the posterior; a præanal pore on each side.

1. Two pairs of elongate pectoral shields.

Nasals in contact with each other . ..	5. *galeata*, p. 457.
Nasals separated by the rostral, which is longitudinally bisected	6. *dumerilii*, p. 457.

2. Three pairs of elongate pectoral shields.

30 segments in an annulus of the body	7. *scalper*, p. 457.
18 segments in an annulus	8. *magnipartita*, p. 458.

B. Anterior head-shields longer than the posterior; no præanal pores.

Pectoral shields large, elongate......	9. *anchietæ*, p. 458.
Pectoral shields small, very numerous; no præocular	10. *jugularis*, p. 459.
Pectoral shields small, very numerous; a præocular	11. *koppenfelsii*, p. 459.

1. Monopeltis capensis. (Plate XXIV. fig. 1.)

Monotrophis capensis, *Gray, Cat. Sh. Rept.* ii. p. 41.
Monopeltis capensis, *Smith, Ill. S. Afr., Rept.* pl. lxvii.
Monotrophis capensis, *Gray, Proc. Zool. Soc.* 1865, p. 454.
Lepidosternon capense, *Strauch, Mél. Biol. Ac. St. Petersb.* xi. p. 462.

Præmaxillary tooth 1; maxillaries 3—3; mandibulars 7—7. Rostral quadrangular, twice as broad as high, separating the nasals; a single plate covering the head, bordered posteriorly by a pair of small band-like occipitals; ocular minute, eye not distinct; three upper labials, third large; mental small, followed by three or four chin-shields; three lower labials, first two small, third very large. Two hundred and four or two hundred and twelve annuli on the body and ten or twelve on the tail; an annulus on the middle of the body includes forty-two to forty-eight segments, twenty-four to twenty-eight dorsals and eighteen or twenty ventrals; the dorsal segments are very narrow; the ventrals much larger, the two median series but very slightly broader than long, except quite anteriorly. Pectoral shields six, large, very elongate, the median pair as long as the distance between the posterior edge of the occipitals and the tip of the snout. Lateral line distinct. Anal segments four, median pair largest. A præanal pore on each side. Rose during life, head-shield yellow.

	millim.
Length to vent	220
Tail	13
Diameter of body	7

South Africa.

a–b. Ad.	24° lat. S.	Sir A. Smith [P.].	(Types.)

2. Monopeltis sphenorhynchus.

Monopeltis sphenorhynchus, *Peters, Mon. Berl. Ac.* 1879, p. 275, *and Reise n. Mossamb.* iii. p. 87, pl. xiii. A. figs. 1–3.
Lepidosternon sphenorhynchum, *Strauch, Mél. Biol. Ac. St. Petersb.* xi. p. 465.

Well distinguished from the preceding in the following points:—Maxillary teeth two on each side. Two lower labials, first small, second very large. The segments of the annuli larger, thirty-two or thirty-four, of which eighteen or twenty are on the dorsal and fourteen on the ventral side; the two median ventral segments throughout the length of the body, much broader than long.

	millim.
Length to vent	245
Tail	13
Diameter of body	8

South-East Africa.

a. Ad. (Ann. 204+12).	Zambezi.	Sir J. Kirk [P.].
b. Ad. (A. 198+11).	Shire Valley, Zambezi.	

3. Monopeltis welwitschii. (Plate XXIV. fig. 2.)

Dalophia welwitschii, *Gray, Cat. Sh. Rept.* ii. p. 41.
Dalophia welwitschii, *Gray, Proc. Zool. Soc.* 1865, p. 454, fig.

Rostral very small, triangular, with truncate anterior angle, the inner angles of the nasals shortly separated; a single plate covering the head, with a short suture in front of the ocular, which is small: eye not distinguishable; a pair of band-like occipitals; three upper labials, third large; mental small, quadrangular, followed by a larger pentagonal chin-shield; three lower labials, first and second small, third very large. Two hundred and seventy-one or two hundred and seventy-five annuli on the body and twenty-two or twenty-three on the tail; on the middle of the body an annulus comprises thirty-four segments, twenty above and fourteen below the lateral lines; the dorsal segments much longer than broad, the two median ventrals much broader than long in the posterior part of the body. Pectoral shields six, large, very elongate, the median pair widening a little posteriorly, forming an angle anteriorly; its length equals the distance between the posterior edge of the occipitals and the end of the snout. Lateral line distinct. Anal segments six. No præanal pores. Colourless, head-shields yellow.

	millim.
Length to vent	265
Tail.........................	24
Diameter of body	6

Angola.

a–b. Ad.	Pungo Andongo.	Mr. Welwitsch [C.].	(Types.)

4. Monopeltis guentheri. (Plate XXIV. fig. 3.)

Præmaxillary tooth 1; maxillaries 2—2; mandibulars 6—6. Snout acuminate. Rostral pentagonal, widely separating the nasals; a single shield covering the head, with a short suture from above a præocular, which is larger than the small ocular; eye scarcely distinguishable; a pair of band-like occipitals; three upper labials, first smallest, second longest, third deepest; mental small, quadrangular, followed by a larger, heptagonal chin-shield; three lower labials, first and second small, third very large. Two hundred and fifty to two hundred and fifty-four annuli on the body; an annulus on the middle of the body comprises twenty-eight or thirty-two segments, sixteen or eighteen above and twelve or fourteen below the lateral lines; the dorsal segments much longer than broad, the two median ventrals much broader than long, especially in the posterior part of the body, where they are nearly or quite twice as broad as long. Pectoral shields six, large, very elongate, of equal width throughout their length, the inner pair slightly longer than the outer, and forming an angle anteriorly; its length equals the distance between the posterior border of the occipitals and the end of the snout. Lateral line distinct. Anal segments six. Three præanal pores on each side. Colourless, head-shields yellowish or brownish.

	millim.
Length to vent	250
Tail	32
Diameter of body	6

Congo.

a–e. Ad. & yg. Congo.

5. Monopeltis galeata.

Phractogonus galeatus, *Hallow. Proc. Ac. Philad.* 1852, p. 62, fig., *and* 1857, p. 50.

Lepidosternon galeatum, *Strauch, Mél. Biol. Ac. St. Petersb.* xi. p. 465.

Præmaxillary teeth 2; maxillaries 4—4; mandibulars 5—5 (?). Nasals in contact; two large shields covering the head, the anterior shorter, but broader, than the posterior; a pair of small occipitals; ocular rather large, eye not distinguishable; five upper labials; three lower labials, the posterior very large. Two hundred and fourteen annuli on the body and eighteen on the tail. No longitudinal sulci upon the body. Pectoral shields four, the two middle ones by their junction forming a triangle superiorly and preceded by two small equilateral shields. Six anal segments. One or two præanal pores on each side. Obscure white.

Length to vent 466 millim.; tail 35 millim.

Liberia.

6. Monopeltis dumerilii.

Phractogonus galeatus (*non Hallow.* ?), *A. Dum. Rev. et Mag. Zool.* (2) viii. p. 424, *and Arch. Mus.* x. p. 184.

Lepidosternon dumerilii, *Strauch, Mél. Biol. Ac. St. Petersb.* xi. p. 467.

Præmaxillary teeth 7; maxillaries 3—3; mandibulars 6—6 (*Dum.*). Nasals entirely separated by the small rostral, which is longitudinally bisected; a band-like narrow transverse shield between the two large shields of the head; three lower labials. Two hundred and twenty-four or two hundred and twenty-five annuli on the body, and twenty-one on the tail. Otherwise as in *M. galeata*.

Gaboon.

7. Monopeltis scalper. (Plate XXIV. fig. 4.)

Phractogonus scalper, *Günth. Proc. Zool. Soc.* 1876, p. 678, fig.

Lepidosternon scalprum, *Strauch, Mél. Biol. Ac. St. Petersb.* xi. p. 469.

Præmaxillary tooth 1; maxillaries 3—3; mandibulars 6—6. Rostral small, trapezoid, separating widely the nasals; two large shields covering the head, the anterior a little shorter, but broader, than the posterior; a pair of small transverse occipitals; ocular small, eye distinguishable; three upper labials, third largest; mental small, followed by a larger chin-shield; three lower labials, first two very small, third very large. Two hundred and thirty-four annuli on the body and eighteen on the tail; on the middle of the body an annulus comprises thirty segments, sixteen above and fourteen below the lateral line; the dorsal segments longer than broad; the ventrals larger, the two median series much broader than long throughout the length of the body. Three pairs of pectoral shields, of which the median is broadest and preceded by a pair of rhombic shields. Lateral line distinct. Two large pairs of anal segments, with a tendency to unite into one pair; on each side a minute additional segment. A præanal pore on each side. Brownish above, whitish beneath; cephalic scutes yellow.

	millim.
Length to vent	290
Tail........................	29
Diameter of body	9

Angola.

a. Ad. Angola. Capt. Cameron [P.]. (Type.)

8. Monopeltis magnipartita.

Monopeltis (Phractogonus) magnipartitus, *Peters, Mon. Berl. Ac.* 1879, p. 276.

Lepidosternon magnipartitum, *Strauch, Mél. Biol. Ac. St. Petersb.* xi. p. 469.

Agrees with *M. scalper* in the pholidosis of the head and the pectoral shields, but differs in the much larger segments of the body, of which there are only ten dorsal and eight ventral rows. Two hundred and twenty-four annuli on the body and twenty-one on the tail. Six anal segments. A præanal pore on each side.

Length to vent 455 millim.; tail 40 millim.

Gaboon.

9. Monopeltis anchietæ.

Lepidosternon (Phractogonus) anchietæ, *Bocage, Jorn. Sc. Lisb.* iv. 1873, p. 247, fig.

A small subtrapezoid rostral separating the nasals; two large shields on the head, the posterior narrower and not half as long as the anterior; a pair of small occipitals; ocular very small, eye invisible; three upper labials, third very large; mental small, trapezoid, followed by a much larger hexagonal chin-shield bordered on each side by a narrower one; three lower labials, third very large. One hundred and ninety-nine annuli on the body and ten on the tail. Pectoral shields large, six, the median pair longest, narrowed anteriorly, the second pair L-shaped. Lateral line distinct. Two large anal shields; no pores. Brown above, whitish inferiorly; anterior head-shield yellow.

	millim.
Length to vent	264
Tail	16
Diameter of body	10

Mossamedes, S.W. Africa.

10. Monopeltis jugularis.

Monopeltis (Phractogonus) jugularis, *Peters, Mon. Berl. Ac.* 1880, p. 219, pl. —. fig. 1.
Lepidosternon jugulare, *Strauch, Mél. Biol. Ac. St. Petersb.* xi. p. 469.

A small rostral separates the nasals; two large shields covering the head, the posterior narrower and a little shorter than the anterior; a pair of band-like occipitals; ocular small, eye distinct; three upper labials, third largest; mental roundish, followed by a large subcordiform chin-shield; three lower labials, third very large. Two hundred and six annuli on the body and thirteen on the tail; an annulus contains thirty to thirty-five segments, sixteen to nineteen dorsals and fourteen to sixteen ventrals; latter, two median rows broadest. Pectoral shields small, thirty-six, commencing immediately behind the gular fold, forming U-shaped rows. Six anal segments, the median pair very large; no pores. Yellowish, each segment with a brown spot, which is smaller and darker on the dorsal segments than on the ventrals.

	millim.
Length to vent	510
Tail	29
Diameter of body	22

West Africa.

11. Monopeltis koppenfelsii.

Lepidosternon koppenfelsii, *Strauch, Mél. Biol. Ac. St. Petersb.* xi. p. 469.

Allied to *M. jugularis*; distinguished by the presence of a præocular, and the sternal shields not beginning immediately after the gular fold. Two hundred and eight annuli on the body and thirteen on the tail; an annulus on the body comprises thirty-two or thirty-four segments, sixteen to eighteen on the dorsal and sixteen on the ventral side; the two median ventral segments much broader than long.

Length to vent 338 millim.; tail 22 millim.

Gaboon.

7. RHINEURA.

Rhineura, *Cope, Proc. Ac. Philad.* 1861, p. 75.

Nostril pierced in a small nasal on the lower surface of the snout. Head depressed, with broad projecting snout with angular edge

Pectoral segments enlarged. A gular fold; a vertebral line. No præanal pores. Tail rounded, depressed, above with regular transverse rows of round tubercles separated by granules.

Florida.

1. Rhineura floridana.

Lepidosternon floridanum, *Baird, Proc. Ac. Philad.* 1858, p. 255; *Strauch, Mél. Biol. Ac. St. Petersb.* xi. p. 460.

Rhineura floridana, *Cope, Proc. Ac. Philad.* 1861, p. 75; *Bocourt, Miss. Sc. Mex., Rept.* p. 491, pl. xxix. I. fig. 1.

Rostral small, hexagonal, separating the nasals which rest on the first and second labials; a broad plate forming the edge of the snout; behind the latter a præfrontal, a larger frontal, and a pair of occipitals; ocular small, the eye not visible; a large præocular, a small infraocular, and a small and a large loreal; four upper labials, fourth very large; mental small, pentagonal, narrowed anteriorly; four chin-shields on a row; five lower labials. Two hundred and forty-four annuli on the body and eighteen on the tail (246—250 + 17—18, *Strauch*); an annulus in the middle of the body consists of twenty-eight segments; dorsal segments much longer than broad; the two median ventrals broader than long. A vertebral line, no lateral line. Pectoral shields not much enlarged, forming four transverse series. Eight anal segments, median pair largest. The tubercles on the tail in thirteen transverse series. Colourless.

	millim.
Length to vent	265
Tail	17
Diameter of body	6·5

Florida.

a. Ad.	Florida.	E. D. Cope, Esq. [E.].

8. LEPIDOSTERNON.*

Leposternon, *Wagler, in Spix, Spec. Nov. Serp. Bras.* p. 70.

Lepidosternon, *Wagl. Syst. Amph.* p. 197; *Dum. & Bibr.* v. p. 503; *Gray, Cat. Tort.* 1844, p. 73, *and Proc. Zool. Soc.* 1865, p. 451.

Cephalopeltis, *Müller, in Teidem. u. Trevir. Zeitschr. f. Physiol.* iv. 1831, p. 257; *Gray, Cat. Tort.* p. 73, *and Proc. Zool. Soc.* 1865, p. 453.

Nostril inferior, pierced in the rostral. Head depressed, with projecting snout. Pectoral segments enlarged. A strong gular fold. Tail cylindrical, obtuse. No præanal pores.

South America.

---

* For the mode of counting the annuli, see footnote under *Monopeltis*, p. 453.

Synopsis of the Species.

I. Rostral separated from the frontal by the præfrontals.

A. First lower labial much larger than second.

1. Two upper labials, first larger than second.

Frontal as long as broad, or broader than long .... 1. *microcephalum*, p. 462.
Frontal longer than broad; supraocular distinct .... 2. *phocæna*, p. 463.
Frontal longer than broad; supraocular united with the præfrontal .... 3. *strauchii*, p. 463.

2. Three upper labials, first smaller than second.

Ocular separated from the third labial by a subocular .... 4. *infraorbitale*, p. 463.
No subocular .... 5. *rostratum*, p. 464.

B. First lower labial very small, second large.

Frontal broader than long.......... 6. *petersii*, p. 464.
Frontal small, longer than broad .... 7. *polystegum*, p. 464.

II. Rostral in contact with the frontal.

A. Frontal not, or but slightly, extending posteriorly beyond the ocular.

1. Frontal a little broader than long; parietals and occipitals present; head very uneven with longitudinal and transverse grooves .... 8. *crassum*, p. 465.

2. Frontal longer than broad.

a. No occipitals; nine or ten large elongate pectoral shields on a transverse line ...... 9. *wuchereri*, p. 466.

b. Occipitals distinct.

α. Supraocular distinct.

Frontal octagonal; occipitals twice as broad as long; three upper labials . 10. *guentheri*, 466.
Frontal octagonal; occipitals twice as broad as long; two upper labials .. 11. *boettgeri*, p. 466.
Frontal hexagonal; occipitals thrice as broad as long; two upper labials .. 12. *affine*, p. 467.

β. Supraocular fused with the præfrontal.

Only the point of the frontal in contact with the rostral; first upper labial twice as long as second .... 13. *boulengeri*, p. 467.

Frontal forming a broad suture with the rostral; first upper labial scarcely larger than the second .... 14. *onychocephalum*, p. 468.

B. Frontal very large, covering the occiput.

Præfrontals present, narrow, widely separated .... 15. *octostegum*, p. 468.
No præfrontals .... 16. *scutigerum*, p. 469.

1. Lepidosternon microcephalum.

Lepidosternon microcephalum, *Gray, Cat. Tort.* p. 73, *and Sh. Rept.* ii. p. 39.
Leposternon microcephalus, *Wagl. in Spix, Serp. Bras.* p. 70, pl. xxvi. fig. 2.
Amphisbæna punctata (*non Bell*), *Wied, Abbild. Nat. Bras.*
Lepidosternon microcephalus, *Wagl. Icon. Descr. Amph.* pl. xvi. fig. 1.
—— maximiliani, *Wiegm. Arch. f. Nat.* 1836, p. 157, pl. iii. fig. 3.
—— microcephalum, *Dum. & Bibr.* v. p. 505; *Strauch, Mél. Biol. Ac. St. Petersb.* xi. p. 424.

Præmaxillary teeth 1 or 2; maxillaries 4—4; mandibulars 5—5. Snout with three more or less marked longitudinal grooves; rostral large, but widely separated from the ocular; a pair of large præfrontals, forming a suture, separating the rostral from the frontal, which is hexagonal and as broad as long or a little broader than long; a pair of parietals and a pair of small occipitals; eye slightly distinct through the good-sized ocular; a large supraocular, followed by a rather larger upper temporal; two upper labials, first large, second smaller than and under the ocular; mental small, four-sided, broader than long, followed by a large median chin-shield; two lateral chin-shields on each side, in contact with the anterior infralabial; two lower labials, first very large, second very small. Two hundred and seventeen to two hundred and sixty-three annuli on the body, and thirteen to sixteen on the tail; an annulus on the middle of the body contains forty-six or forty-eight segments, twenty-two or twenty-four dorsal and as many ventral; the ventral segments a little larger than the dorsals. Pectoral shields numerous, a median pair longest, with two or three elongate ones on each side, the others small. Præanal segments six, median and outer pairs largest. Lateral, vertebral, and ventral lines distinct. Yellowish in spirits, the dorsal segments with a brown spot in the centre.

	millim.
Length to vent	470
Tail	28
Diameter of body	20

Brazil.

a. Ad. (Annuli 218+13).	Rio Janeiro.	Mrs. Fry [P.].
b. Ad. (A. 234+15).	Brazil.	
c. Hgr. (A. 217+13).	Brazil.	
d. Hgr. (A. 263+16).	Brazil.	Derby Museum.

2. Lepidosternon phocæna.

Lepidosternon phocæna, *Gray, Cat. Tort.* p. 73, *and Sh. Rept.* ii. p. 40.

Lepidosternon phocæna, *Dum. & Bibr.* v. p. 507; *D'Orb. Voy. Amér. Mér., Rept.* p. 9, pl. vi. figs. 7–11; *Strauch, Mél. Biol. Ac. St. Petersb.* xi. p. 427.

Differs from *L. microcephalum* chiefly in the following points:—Frontal much longer than broad, as long as its distance from the end of the snout; supraoculars large, broader than long; occipitals large, nearly as large as the parietals. Sternal plates all small, lozenge-shaped, forming oblique series, without any longitudinally dilated ones.

Argentine Republic.

a. Ad., stuffed. (Annuli 217+13.) S. America. Mr. Bridges [C.].

3. Lepidosternon strauchii.

Lepidosternum strauchi, *Boettg. Zeitschr. f. Naturw.* lviii. 1885, p. 221.

Allied to the preceding. Supraoculars united with the præfrontals. Two hundred and sixty-five annuli on the body and eighteen on the tail; an annulus contains twenty-six dorsal and twenty-four ventral segments.

Paraguay.

4. Lepidosternon infraorbitale.

Lepidosternon infraorbitale, *Berth. Gött. Nachr.* 1859, p. 179; *Strauch, Mél. Biol. Ac. St. Petersb.* xi. p. 429.

Snout very thick, little depressed, not very prominent, with the longitudinal grooves feebly marked; rostral rather small, widely separated from the ocular; a pair of irregular, suboval præfrontals, placed very obliquely and forming together a very short suture; a regularly hexagonal frontal, a little longer than broad; a pair of very irregular parietals, each about half as large as the frontal; occipitals considerably smaller than the parietals; eye rather distinct through the ocular, which is a little larger than the supraocular, and forms a suture with a subocular and a very small portion of the second labial; subocular trapezoid, wedged in between the second and third labials; temporal shields very small; three upper labials, first and third very small, second very large; mental small, broad, quadrangular, followed by a large median chin-shield, on each side of which is an elongate, narrow, lateral one; three lower labials, first large, second and third very small. Two hundred and sixty-four annuli on the body and seventeen on the tail; on the body an annulus contains fifty-four segments, twenty-eight dorsal and twenty-six ventral. Pectoral shields numerons, small, irregular, mostly roundish, forming oblique series. No enlarged anal shields.

Lateral line much stronger than the vertebral and ventral. Light yellowish-brown, the segments each with a darker spot.

	millim.
Length to vent	384
Tail	26

Bahia.

5. Lepidosternon rostratum.

Lepidosternon rostratum, *Strauch, Mél. Biol. Ac. St. Petersb.* xi. p. 433.

Differs from *L. infraorbitale* by the much more pointed, prominent, and strongly depressed snout, the horizontal edge of which is almost sharp, and by the absence of infraocular. Two hundred and forty to two hundred and forty-five annuli on the body, and fourteen to sixteen on the tail. The spots on the dorsal segments very indistinct.

Length to vent, 585 millim.; tail 35 millim.

Bahia.

6. Lepidosternon petersii.

Lepidosternon petersii, *Strauch, Mél. Biol. Ac. St. Petersb.* xi. p. 438.

Snout rather pointed, not turned upwards; rostral followed by a pair of short but very broad præfrontals; behind the latter, a hexagonal frontal, broader than long, followed by two pentagonal parietals which are but little smaller than the frontal; a pair of small occipitals; a rather small supraocular; eye not distinguishable; three temporals, upper largest; two upper labials, anterior very large; mental short and broad, followed by a median large chin-shield; a small lateral chin-shield on each side; three lower labials, first very small, second large. Two hundred and nine annuli on the body and sixteen on the tail. Pectoral shields numerous, small, polygonal, with three median pairs (on a transverse line) strongly dilated longitudinally. Six anal shields. Lateral line very distinct, vertebral and ventral lines less so. Light brown above.

Brazil.

7. Lepidosternon polystegum.

Lepidosternon grayii, *Gray, Cat. Sh. Rept.* ii. p. 40.

Lepidosternon polystegum, *A. Dum. Cat. Méth. Rept.* p. 149; *Peters, Mon. Berl. Ac.* 1879, p. 277, pl. —. fig. 3; *Strauch, Mél. Biol. Ac. St. Petersb.* xi. p. 441.

—— grayii (*Smith*), *Gray, Proc. Zool. Soc.* 1865, p. 452, fig.

Præmaxillary tooth 1; maxillaries 2—2; mandibulars 5—5. Snout depressed, prominent, turned upwards, broadly rounded, with angular horizontal edge; rostral large, widely separated from the ocular, with a pair of short grooves posteriorly; it is separated from the frontal by a pair of narrow, band-like, transverse præfrontals; frontal small, hexagonal, with very open anterior angle and acute posterior angle wedged in between a pair of large elongate pentagonal parietals; occipitals very small; eye distinct through the ocular; an infraocular, a præocular, and a supraocular; latter large, elongate; four temporals, upper very large; three upper labials, first small, second large, third very small; mental very small, four-sided, as long as, or longer than broad, followed by a large median chin-shield, which is separated from the labials by a pair of narrow lateral chin-shields on each side; another lateral chin-shield on each side, on a line with the third infralabial; infralabials three, first very small, second large. Two hundred and eighty-six to two hundred and ninety-five annuli on the body and seventeen on the tail; an annulus on the middle of the body contains twenty-six to thirty dorsal and twenty-four or twenty-six ventral segments. Pectoral shields small, numerous, longer than broad, forming oblique series. Anal segments in two rows, slightly enlarged. Lateral, vertebral, and ventral lines distinct. Colourless or pale brownish above.

	millim.
Length to vent	340
Tail........................	17
Diameter of body	9

Brazil.

a. Ad. (Annuli 295+17). Recife, Pernambuco. W. A. Forbes, Esq. [P.].
b. Ad. (A. 286+17). S. America. Sir A. Smith [P.]. (Type of *L. grayi*.)
c. Ad., escaping from the side of the body on an *Elaps lemniscatus*. Specimen with anomalous head-shields. See P. Z. S. 1885, p. 327.

8. **Lepidosternon crassum.**

Lepidosternon crassum, *Strauch, Mél. Biol. Ac. St. Petersb.* xi. p. 443.

Head as in *L. microcephalum*, but shorter. Rostral rather large, in contact with the anterior angle of the frontal; præfrontals large, broader than long, separated by the frontal; latter large, pentagonal, a little broader than long, with a deep median groove, continued from the rostral; a pair of parietals, quadrangular, somewhat broader than long, followed by a pair of small broad occipitals; eye feebly distinct through the ocular; supraocular imperfectly separated from the frontal, rather small, square; three temporals, the upper nearly as large as the two others together; two upper labials, first very large, second small. Mental square, as long as broad, apparently soldered with the anterior chin-shield; a lateral chin-shield on each side; two lower labials, first very large, second very small. Body remarkably thick; two hundred and twelve annuli on the body and fifteen on the tail; an annulus contains twenty-two dorsal and twenty ventral segments, those on the middle of the belly a little broader than long; pectoral shields rather irregular, numbering about eight, elongate, some, however, divided transversely into two or three, others confluent with those on the sides; seven anal segments; vertebral and lateral lines strongly, ventral feebly marked. Light brownish white, the dorsal and lateral segments each with a roundish dark brown spot.

Length to vent 505 millim.; tail 35 millim.

Brazil.

9. Lepidosternon wuchereri.

Lepidosternon octostegum (*non A. Dum.*), *Steind. Novara, Rept.*, p. 53.

—— wuchereri, *Peters, Mon. Berl. Ac.* 1879, p. 276, pl. —. fig. 2; *Strauch, Mél. Biol. Ac. St. Petersb.* xi. p. 447.

Snout turned upwards; rostral large, in contact with the anterior angle of the frontal; præfrontals trapezoid, broader than long, separated by the frontal; latter hexagonal, a little longer than broad, followed by a pair of smaller parietals which are broader than long; no occipitals; a supraocular, a little smaller than the ocular; two temporals, upper largest; two upper labials, first twice as long as second. Mental very small, quadrangular, followed by a large chin-shield; a pair of small lateral chin-shields on each side; two lower labials, first much larger than second. Two hundred and seventy-eight annuli on the body and seventeen on the tail; nine or ten large, elongate, narrow pectoral shields on a transverse line; four anal segments.

Length to vent 340 millim.; tail 20 millim.

Brazil.

10. Lepidosternon guentheri.

Lepidosternon güntheri, *Strauch, Mél. Biol. Ac. St. Petersb.* xi. p. 449.

Closely allied to the preceding, but differing in the presence of a pair of large occipitals and the different pectoral scutellation, which consists of numerous, mostly small, shields, of which the median pair alone is much elongate. Frontal octagonal, forming an obtuse angle anteriorly and posteriorly; occipitals twice as broad as long; three upper labials, the first largest, the third very small. Two hundred and thirty-one annuli on the body and thirteen on the tail; an annulus of the body contains thirty-two dorsal and twenty-four ventral segments. Six anal segments.

Length to vent 312 millim.; tail 18 millim.

Habitat unknown.

11. Lepidosternon boettgeri. (Plate XXIV. fig. 5.)

Præmaxillary teeth 3; maxillaries 4—4; mandibulars 5—5. Snout more depressed than in *L. microcephalum*, with sharper rostral edge; rostral large, widely separated from the ocular; frontal much longer than broad, narrowed, and in contact with the rostral anteriorly, separating the præfrontals; latter a little longer than broad a pair of parietals, longer than broad; a pair of occipitals, smaller than the latter, and broader than long; eye very distinct through the ocular; a large supraocular, followed by a nearly equal-sized upper temporal; two upper labials, first large, second smaller than and under the ocular; mental small, four-sided, broader than long, followed by a large median chin-shield; lateral chin-shields very small; two lower labials, first very large, second very small. Two hundred and forty-nine annuli on the body and twelve on the tail; an annulus contains fifty-eight segments, thirty dorsal and twenty-eight ventral; the segments of the two median ventral series as broad as long or slightly broader. Pectoral shields numerous, small, forming oblique rows, the largest only a little longer than broad. Præanal segments six (rather irregular). Lateral line strongly, vertebral and ventral lines feebly marked. Yellowish inferiorly, pale brownish above, without spots.

	millim.
Length to vent	350
Tail	20
Diameter of body	16

Argentine Republic.

a. Ad. Corrientes.

12. Lepidosternon affine.

Lepidosternum affine, *Boettg. Zeitschr. f. Naturw.* lviii. 1885, p. 233.

Allied to *L. guentheri*, but differing in the following points:—Rostral apparently more pointed, as long as the first upper labial, with sharper edge; frontal hexagonal, forming an acute angle anteriorly and posteriorly; occipitals thrice as broad as long; two upper labials, first much larger than second. Two hundred and sixty-six annuli on the body and sixteen on the tail; an annulus contains thirty dorsal and twenty-four ventral segments. Eight anal segments.

Length to vent 294 millim.; tail 16 millim.

Paraguay.

13. Lepidosternon boulengeri.

Lepidosternum boulengeri, *Boettg. Zeitschr. f. Naturw.* lviii. 1885, p. 220.

Snout rather pointed; frontal longer than broad, pyriform, its narrowed anterior extremity in contact with the rostral, its lateral edge forming a suture with the præfrontal and supraocular, which are fused into a single shield, its posterior slightly convex border in contact with the parietals; a pair of occipitals, twice as broad as long; eye more or less distinct through the trapezoid ocular; two or three temporals, upper large; two upper labials, first twice as long as second; mental four-sided, followed by an elliptic anteriorly truncate chin-shield; two lower labials, first twice as long as second, on the side of which there is a larger suboval lateral chin-shield. Two hundred and forty-four to two hundred and fifty-five annuli on the body, and eleven to thirteen on the tail; an annulus contains twenty-eight dorsal and twenty-four ventral segments. Pectoral shields numerous, small, the medians but little larger than the laterals, little longer than broad. Præanal segments eight. Lateral, vertebral, and ventral lines feebly marked. Yellowish, the segments on the posterior third of the body greyish in the centre.

	millim.
Length to vent	330
Tail.........................	15
Diameter of body	13·5

Paraguay.

14. Lepidosternon onychocephalum.

Lepidosternum onychocephalum, *Boettg. Zeitschr. f. Naturw.* lviii. 1885, p. 224.

Rostral rather pointed, its posterior border truncate and forming a broad suture with the frontal; latter longer than broad, larger than any other head-shield except the rostral; præfrontal and supraocular united into a single shield on each side of the frontal; two parietals, the suture they form with the frontal narrower than that between the latter and the rostral; a pair of very small occipitals, broader than long; eye distinct through the relatively large pentagonal ocular; three or four temporals; two upper labials, first scarcely larger than second; two lower labials, first twice as large as second. Two hundred and fifty-five annuli on the body and sixteen on the tail; an annulus contains thirty dorsal and twenty-four ventral segments. The median pectoral shields considerably larger than the others and elongate. Præanal segments six. Yellowish, the segments of the posterior part of the body brownish-grey in the centre.

	millim.
Length to vent	260·5
Tail.......................	15·5
Diameter of body	9

Paraguay.

15. Lepidosternon octostegum.

Lepidosternon octostegum, *A. Dum. Cat. Méth. Rept.* p. 150; *Peters, Mon. Berl. Ac.* 1879, p. 277, pl. —. fig. 1; *Strauch, Mél. Biol. Ac. St. Petersb.* xi. p. 455.

Frontal very large, covering the greater part of the head, forming a broad suture with the rostral, separating the elongate, narrow, præfrontals; a pair of minute occipitals; eye distinct through the rather large ocular, which is as much in contact with the præfrontal as with the supraocular; latter rather large, evidently representing also the upper temporal; a small postocular or temporal; a single, very large, upper labial; mental very small, followed by a single rather small chin-shield; a single, very large, lower labial. Three hundred and seventy-seven annuli on the body and fourteen on the tail; pectoral shields large, elongate, three pairs on the median line, and two or three lateral pairs, largest. Four anal segments. Brownish white, with very small darker spots.

Length to vent 336 millim.; tail 14 millim.

Brazil.

16. Lepidosternon scutigerum.

Cephalopeltis scutigera, *Gray, Cat. Tort.* p. 73, *and Sh. Rept.* ii. p. 40.
Amphisbæna scutigera, *Hempr. Verh. Ges. Naturf. Fr.* i. 1829, p. 129.
Cephalopeltis cuvieri, *Müll. Zeitschr. f. Physiol.* iv. 1831, p. 257, pl. xxii. fig. 5.
—— hemprichii, *Wiegm. Arch. f. Nat.* 1836, p. 157, pl. iii. fig. 2.
Lepidosternon scutigerum, *Dum. & Bibr.* v. p. 509; *Strauch, Mél. Biol. Ac. St. Petersb.* xi. p. 456.
Cephalopeltis scutigera, *Girard, U.S. Explor. Exped., Herp.* p. 217.

Præmaxillary tooth 1, maxillaries 3—3, mandibulars 5—5. Snout prominent, with subangular horizontal edge; two shields cover the upper surface of the head, viz. the large rostral, the lateral angles of which extend to the ocular, and the enormous frontal, which forms an obtuse angle anteriorly; a pair of very small transverse occipitals; eye slightly distinct through the ocular; no supra-ocular; three upper labials, first large, second and third small and under the ocular; mental coalescent with the single elongate chin-shield; a single, very large, lower labial. Three hundred annuli on the body and sixteen on the tail; an annulus on the middle of the body contains forty-eight or fifty segments, twenty-two dorsal and twenty-six or twenty-eight ventral; pectoral shields large, two pairs on the median line and three shields on each side, on a transverse line, largest. Anal segments six, median pair largest. Vertebral, lateral, and ventral lines well marked. Yellowish, the dorsal segments with a brown spot.

	millim.
Length to vent	410
Tail	25
Diameter of body	13

Brazil.

a. Ad.	Near Rio Janeiro.	Lords of the Admiralty [P.]. Dr. Cunningham [C.]

B. EMPHYODONTES.

Emphyodontes, *Wiegm. Herp. Mex.* p. 21.

9. TROGONOPHIS.

Trogonophis, *Kaup, Isis*, 1830, p. 880; *Wiegm. Arch. f. Nat.* 1836, p. 158; *Dum. & Bibr.* v. p. 467; *Gray, Cat. Tort.* 1844, p. 68, *and Proc. Zool. Soc.* 1865, p. 445.

Nostril pierced in a large nasal; two pairs of upper head-shields. A vertebral line and a stronger lateral line. Tail conical. No præanal pores.

North-west Africa.

1. Trogonophis wiegmanni.

Trogonophis wiegmanni, *Gray, Cat. Tort.* p. 68, *and Sh. Rept.* ii. p. 33.
Trogonophis wiegmanni, *Kaup, l. c.* pl. viii. fig. *a*; *Dum. & Bibr.* v. p. 469; *Gerv. Ann. Sc. Nat.* (3) xx. 1853, p. 308, pl. xv. figs. 3, 4; *Strauch, Mél. Biol. Ac. St. Petersb.* xi. 1881, p. 373.
Amphisbæna elegans, *Gerv. Mag. Zool.* 1836, p. 3, pl. xi.
—— wiegmanni, *Schleg. in Wagn. Reis. Alg.* iii. p. 122, pl. vi.

Præmaxillary teeth 5; maxillaries 4—4; mandibulars 8—8. Head moderately depressed, with obtuse, slightly projecting snout. Rostral large, pentagonal, its posterior angle wedged in between the nasals, which are large and form a short suture; a pair of large frontals, followed by a pair of small postfrontals; eye distinct through the small ocular; a supraocular; two infraoculars; a loreal; five upper labials, second and fifth smallest. Mental trapezoid, in contact with a small pentagonal anterior chin-shield; three pairs of lateral chin-shields, anterior largest, with a small shield wedged in between them posteriorly; four lower labials, second largest. Body short. One hundred and thirty-six to one hundred and fifty-one annuli on the body and twelve to fourteen on the tail; on the middle of the body an annulus contains fifty to sixty-four segments, viz. twenty-two to thirty dorsal and twenty-eight to thirty-four ventral; lateral line strongly marked; a feebly marked vertebral line. Anal segments six or eight, median pair largest. Yellowish, checked with brown.

	millim.
Length to vent	240
Tail	19
Diameter of body	17

Algeria; Morocco.

a–c. Ad. (Ann. 142 to 144 +12).	Tangiers.	Mr. Fraser [C.].
d, e–f. Hgr. (A. 147+13, 148+13, 150+14).	Mogador.	Rev. R. T. Lowe [P.].
g–h. Hgr (A. 141, 145+12, 14).	Oran.	Rev. A. Levi [P.].
i. Hgr. (A. 144+13).	Algiers.	Prof. Duméril [P.].

k. Hgr. (151+14).	Algiers.	Mr. Kemp [P.].
l, m. Ad. & hgr. (A. 140+12, 145+13).	Algiers.	
n–o. Ad. (A. 136+15, 141+13).	Tamesmida, Algeria.	M. F. Lataste [P.].

10. PACHYCALAMUS.

Pachycalamus, *Günth. Proc. Zool. Soc.* 1881, p. 461; *Peters, Sitzb. Berl. Ac.* xxvi. 1882, p. 583.

Nostril inferior, between two small nasals, on the side of the large rostral; three large upper head-shields. A ventral line, no vertebral or lateral lines. Tail depressed, obtusely pointed. Præanal pores.

Socotra.

1. Pachycalamus brevis.

Pachycalamus brevis, *Günth. l. c.*

Præmaxillary teeth 3; maxillaries 3—3; mandibulars 6—6. Head depressed, with truncate projecting snout. Rostral large, trapezoid, its posterior border largest, straight, in contact with a pair of large præfrontals; a large frontal, angular anteriorly, nearly as long as broad, with more or less distinct traces of median division; eye slightly distinct through the ocular; a large præocular; an infraocular; five upper labials, first very small, fourth and fifth largest. Mental narrow, elongate, a little broader anteriorly; chin-shields very small, median hexagonal; three lower labials, second largest. Body short, though less so than in *Trogonophis*. One hundred and sixty-five to one hundred and seventy-three annuli on the body, and nineteen or twenty on the tail; on the middle of the body an annulus contains forty-eight or fifty segments. A well-marked ventral line, but no vertebral or lateral. Anal segments six or eight. Four præanal pores. Brown above; head and lower surfaces yellowish white.

	millim.
Length to vent	198
Tail	15
Diameter of body	9·5

Socotra.

a–f. Ad. & yg.	Socotra.	Prof. J. B. Balfour [C.]. (Types.)

11. AGAMODON.

Agamodon, *Peters, Sitzb. Berl. Ac.* xxvi. 1882, p. 580.

Nostril inferior, pierced in a nasal which unites with the first labial; rostral large, followed by a single very large shield. A ventral line, an obsolete vertebral line, no lateral line; median dorsal and ventral segments roundish, scale-like. Tail compressed, pointed. Præanal pores.

East Africa.

1. Agamodon anguliceps.

Agamodon anguliceps, *Peters*, *l. c.* pl. x.

Præmaxillary teeth 3; maxillaries 2—2; mandibulars 5—5. Head with projecting supero-lateral edges; snout truncate, projecting. Rostral large, trapezoid, its posterior border largest, straight, in contact with a large shield; latter covering the frontal and parietal region, longer than broad, subpentagonal, with obtuse posterior angle; a large præocular; an infraocular; three upper labials, first very small and united with the nasal, third very large. Mental narrow, elongate; chin-shields very small; three lower labials, third largest. Body very short. One hundred and thirty-two annuli on the body, and seventeen on the tail. No lateral line, but a feebly marked vertebral line and a stronger ventral one; the segments adjoining these grooves round and scale-like. Eight or ten anal segments. Four præanal pores. Yellowish white; back with irregular black spots, which may be confluent into an interrupted longitudinal line.

	millim.
Length to vent	153
Tail	19
Diameter of body	12

Barava, East Africa.

ADDENDUM.

Page 117. Add a new species :—

7. Enyalioides leechii.

In habit very similar to *Enyalius catenatus*, but the digits strongly keeled and pectinate inferiorly and distinctly denticulated laterally. Nostril below the canthus rostralis, much nearer the end of the snout than the orbit. Upper head-scales small, equal, strongly keeled; occipital a little enlarged; a few longitudinal rows of strongly keeled supraoculars, smaller than the scales on the vertex, surrounded by smaller granular ones; twelve or thirteen upper and eleven lower labials. Gular scales small, granular, obtusely keeled. Dorso-nuchal crest quite indistinct (anomalous?). Dorsal scales small, granular, keeled, laterals smaller still, obtusely keeled; ventrals much larger, squarish subrhomboidal, very strongly keeled. Limbs long, covered with equal strongly keeled scales, the hind side of the thighs finely granulate; the adpressed hind limb reaches beyond the end of the snout; the length of the tibia equals the distance between the end of the snout and the gular fold. Tail scarcely compressed, not crested, covered with subequal keeled scales, which are largest inferiorly; its length about once and two thirds that of head and body. Chestnut-brown above; a black line from the eye to the fore limb, passing on the lower border of the tympanum; a short yellowish streak on each side of the nape; a triangular yellowish spot, pointing towards its fellow, above each shoulder; limbs with rather indistinct darker and lighter cross bands; lower surfaces brownish; an oblique arrow-headed dark-brown marking on each side of the throat; tail with a series of large, elliptical, longitudinal, blackish spots above and inferiorly, some confluent and forming rings.

	millim.		millim.
Total length	292	Fore limb	53
Head	26	Hind limb	109
Width of head	20	Tail	185
Body	81		

Upper Amazons.

a. ♀.	Santarem.	J. H. Leech, Esq. [C.].

ALPHABETICAL INDEX.

LIST OF PLATES.

PLATE I.

Fig. 1. *Anolis lineatopus*, Gray, p. 39, male.
2. —— ——, female.
3. —— *citrinellus*, Cope, p. 35.
3 *a*. —— ——, upper view of head, ×2.
4. —— *homolechis*, Cope, p. 28, upper view of head, ×2.
5. —— *cybotes*, Cope, p. 34, upper view of head, ×2.

PLATE II.

Fig. 1. *Anolis gingivinus*, Cope, p. 28.
1 *a*. —— ——, upper view of head, ×2.
2. —— *damulus*, Cope, p. 47.
2 *a*. —— ——, upper view of head, ×2.
3. —— *impetigosus*, Cope, p. 55.
3 *a*. —— ——, upper view of head, ×2.
3 *b*. —— ——, lower view of head, ×2.

PLATE III.

Fig. 1. *Anolis fasciatus*, Blgr., p. 59.
1 *a*. —— ——, upper view of head, ×2.
2. —— *gemmosus*, O'Sh., p. 60.
2 *a*. —— ——, upper view of head, ×2.

PLATE IV.

Anolis fraseri, Gthr., p. 65; and upper view of head, × $1\frac{1}{2}$.

PLATE V.

Fig. 1. *Anolis semilineatus*, Cope, p. 68.
1 *a*. —— ——, upper view of head, × 2.
2. —— *bitectus*, Cope, p. 71.
2 *a*. —— ——, upper view of head, × 2.
3. —— *leptoscelis*, Blgr., p. 92.
3 *a*. —— ——, upper view of head, × 2.

PLATE VI.

Fig. 1. *Anolis lentiginosus*, O'Sh., p. 93.
1 *a*. —— ——, upper view of head, × 2.
2. *Tropidodactylus onca*, O'Sh., p. 97.
3 *a*. —— ——, upper view of head, × 2.

PLATE VII.

Enyalius iheringii, Blgr., p. 120.

PLATE VIII.

Fig. 1. *Stenocercus torquatus*, Blgr., p. 133.
2. —— *humeralis*, Gthr., p. 134.
3. —— *varius*, Blgr., p. 134.

PLATE IX.

Fig. 1. *Stenocercus cupreus*, Blgr., p. 135.
2. —— *mœstus*, Blgr., p. 136.
3. *Anisolepis iheringii*, Blgr., p. 122.

PLATE X.

Fig. 1. *Liolæmus gravenhorstii*, Gray, p. 142.
1 *a*. —— ——, upper view of head, $\times 2$.
2. —— *fuscus*, Blgr., p. 144.
2 *a*. —— ——, upper view of head, $\times 2$.
3. —— *occipitalis*, Blgr., p. 156.
3 *a*. —— ——, upper view of head, $\times 2$.

PLATE XI.

Liocephalus aculeatus, O'Sh., p. 167; and upper view of head.

PLATE XII.

Liocephalus ornatus, Gray, p. 168; and upper view of head, $\times 1\frac{1}{2}$.

PLATE XIII.

Liocephalus guentheri, Blgr., p. 169; and upper view of head.

PLATE XIV.

Fig. 1. *Platysaurus guttatus*, Smith, p. 262.
2. *Sauresia sepoides*, Gray, p. 295.
2 *a*. —— ——, upper view of head, $\times 2$.

PLATE XV.

Fig. 1. *Ophisaurus gracilis*, Gray, p. 283; upper view of head and anterior part of body.
1 *a*. —— ——, lateral view.
1 *b*. —— ——, upper view of head, $\times 2$.
2. —— *koellikeri*, Gthr., p. 283; upper view of head and anterior part of body.
2 *a*. —— ——, lateral view.
2 *b*. —— ——, upper view of head, $\times 2$.

Plate XVI.

Fig. 1. *Diploglossus striatus*, Gray, p. 289.
1 *a*. —— ——, upper view of head.
2. —— *microlepis*, Gray, p. 292, upper view of head.

Plate XVII.

Diploglossus hewardii, Gray, p. 291; and upper view of head.

Plate XVIII.

Varanus caudolineatus, Blgr., p. 324.

Plate XIX.

Cnemidophorus espeuti, Blgr., p. 362; and lower view of hind limbs and anal region.

Plate XX.

Fig. 1. *Alopoglossus copii*, Blgr., p. 383.
2. *Oreosaurus oculatus*, O'Sh., p. 410.
a. Lateral view of head and anterior part of body; *b*. Upper view of head; *c*. Lower view of head and neck; *d*. Lower view of hind limb and anal region. ×1½.

Plate XXI.

Fig. 1. *Prionodactylus oshaughnessyi*, Blgr., p. 392.
2. —— *vertebralis*, O'Sh., p. 394.
a, *b*, *c*, *d*, as in the preceding plate, ×2.

Plate XXII.

Fig. 1. *Proctoporus simoterus*, O'Sh., p. 414.
2. —— *meleagris*, Blgr., p. 415.
a, *b*, *c*, *d*, as in the preceding plates, ×1½.

PLATE XXIII.

Fig. 1. *Blanus bedriagæ*, Blgr., p. 434, ×2.
2. *Amphisbæna plumbea*, Gray, p. 443, ×2.
3. —— *punctata*, Gray, p. 450, ×2.
4. *Anops africanus*, Gray, p. 452, ×4.
5. *Geocalamus modestus*, Gthr., p. 453, ×2.
a. Upper view of head; *b*. Lateral view of head
c. Lower view of head; *d*. Anal region.

PLATE XXIV.

Fig. 1. *Monopeltis capensis*, Smith, p. 455, ×2.
2. —— *welwitschii*, Gray, p. 456, ×2.
3. —— *guentheri*, Blgr., p. 456, ×2.
4. —— *scalper*, Gthr., p. 457, ×2.
5. *Lepidosternon boettgeri*, Blgr., p. 466.
a. Upper view of head; *b*. Lateral view of head; *c*. Lower view of head and pectoral region; *d*. Anal region.

PRINTED BY TAYLOR AND FRANCIS, RED LION COURT, FLEET STREET.

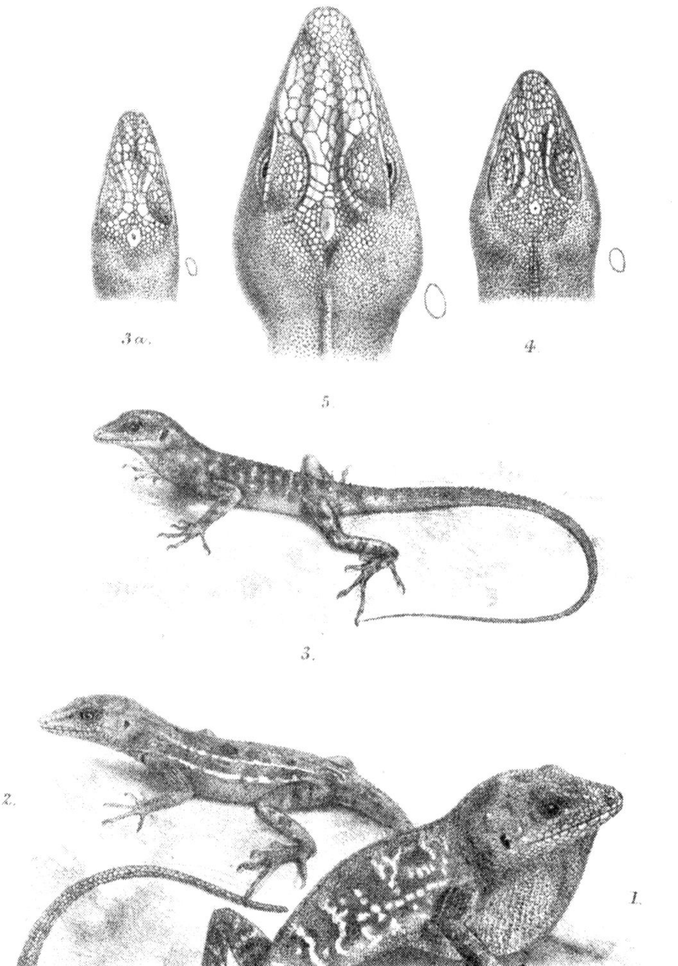

P. Smit del et lith. Mintern Bros. imp.

1.2. *Anolis lineatopus.* 3. *Anolis citrinellus.*
4. *Anolis homolechis.* 5. *Anolis cybotes.*

P. Smit del et lith. Mintern Bros. imp.

1. Anolis gingivinus. 2. Anolis damulus.
3. Anolis impetigosus

G. Mintern del. et lith. Mintern Bros. imp.

1. Anolis fasciatus. *2. Anolis gemmosus.*

BRIT. MUS. N. H.

Pl. IV.

P. Smit del et lith.

Mintern Bros. imp.

Anolis fraseri.

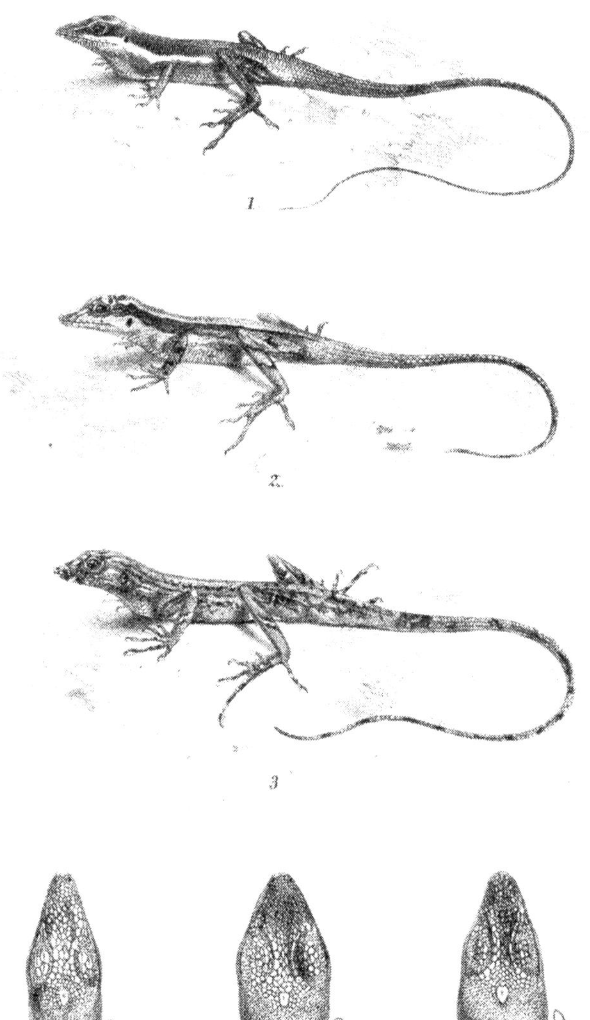

P. Smit del et lith. Mintern Bros. imp.

1. Anolis semilineatus. 2. Anolis bitectus. 3. Anolis leptoscelis.

BRIT. MUS. N. H. PL. VI.

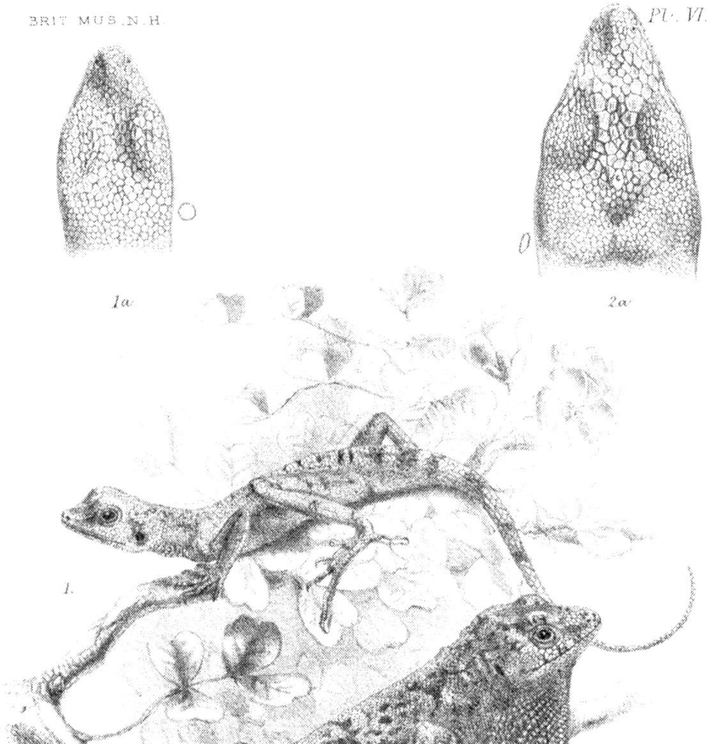

P. Smit del et lith. Mintern Bros. imp.

1. *Anolis lentiginosus.* 2. *Tropidodactylus onca.*

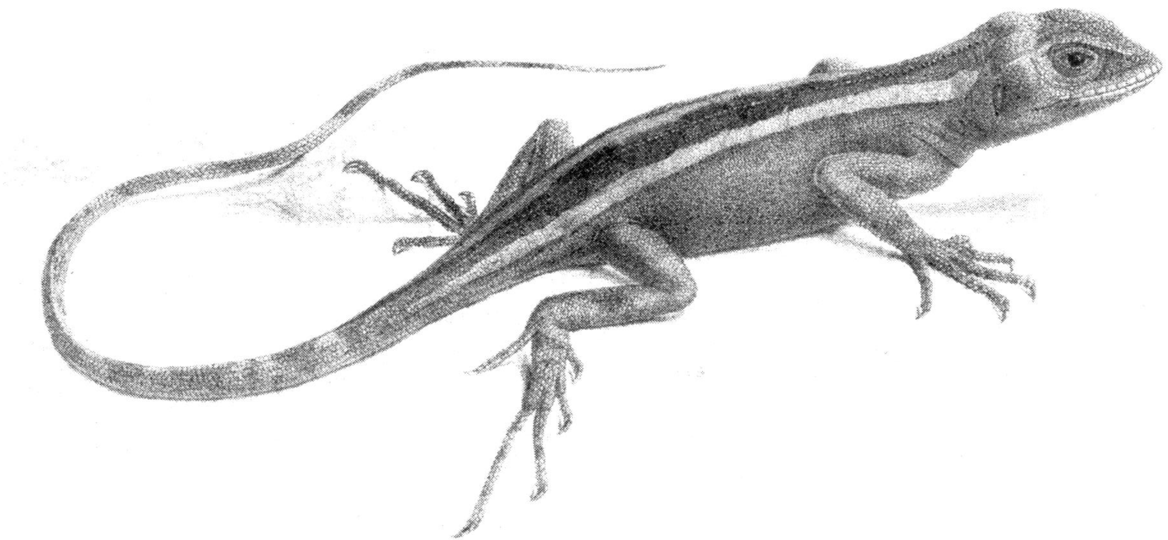

R. Mintern del. et lith. Mintern Bros. imp.

Enyalius iheringii

P. Smit del et lith. Mintern Bros. imp.

1. Stenocercus torquatus. 2. Stenocercus humeralis.
3 Stenocercus varius.

P. Smit del et lith. Mintern Bros. imp.

1. *Stenocercus cupreus.* 2. *Stenocercus moestus.*
3. *Anisolepis iheringii.*

P. Smit del. et lith. Mintern Bros. imp.

1. *Liolæmus gravenhorstii.* 2. *Liolæmus fuscus.*
3. *Liolæmus occipitalis.*

BRIT. MUS. N. H.

Pl. XI.

P. Smit del. et lith.

Mintern Bros. imp.

Liocephalus aculeatus.

P. Smit del et lith.

Minter Bros. imp.

Liocephalus ornatus.

BRIT. MUS. N. H.

Pl. XIII.

P. Smit del. et lith.

Mintern Bros. imp.

Liocephalus guentheri.

P. Smit del et lith. Mintern Bros. imp.

1. *Platysaurus guttatus.* 2. *Sauresia sepoides.*

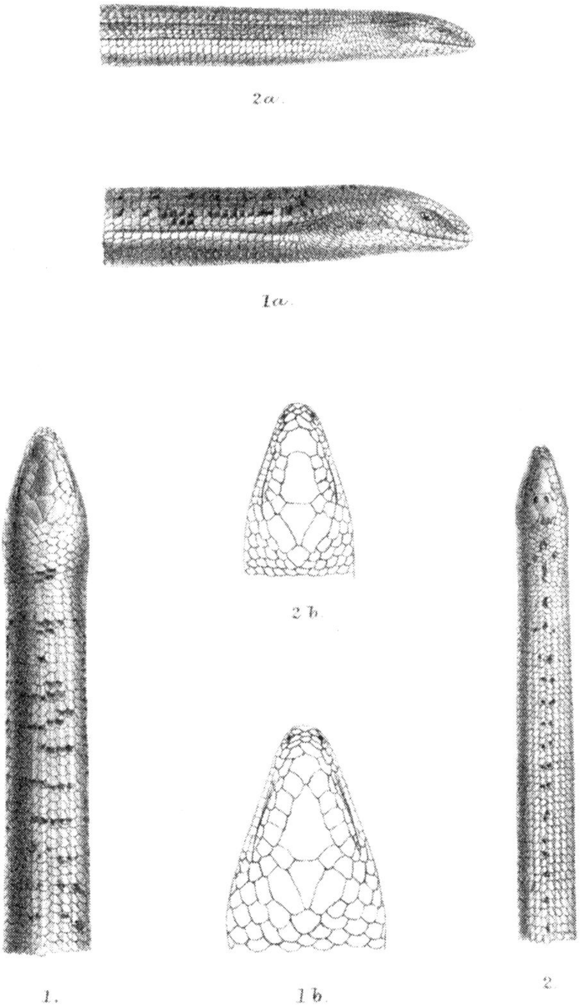

P. Smit. del et. lith. Mintern Bros. imp.

1. *Ophisaurus gracilis.* 2. *Ophisaurus kœllikeri.*

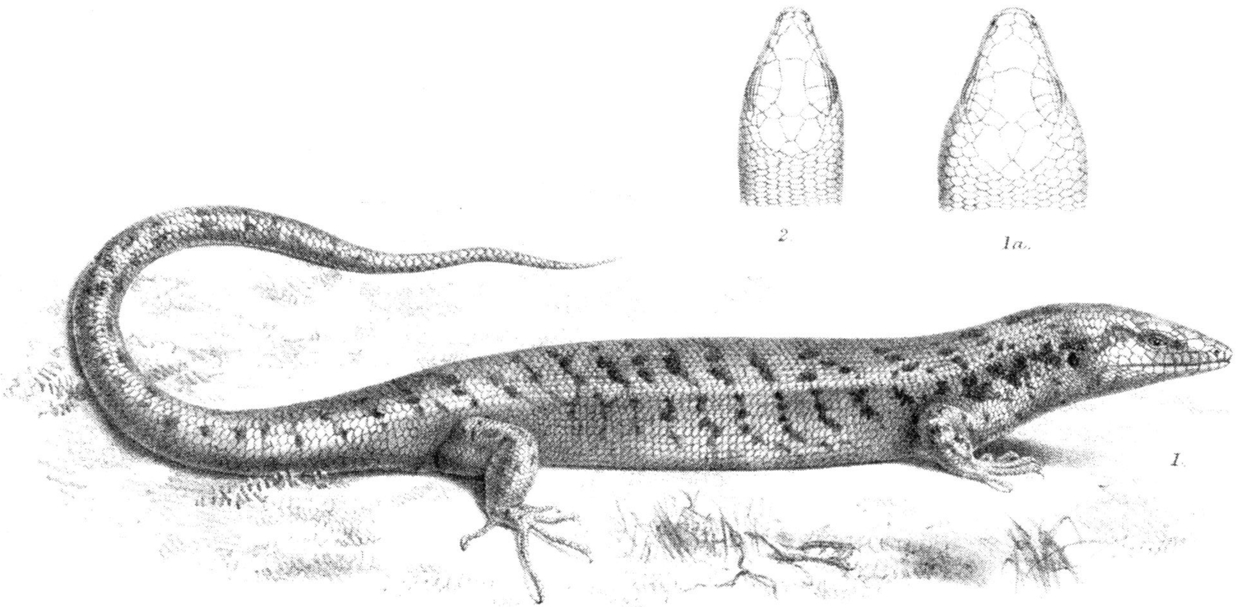

P. Smit del et lith.

Mintern Bros. imp.

1. Diploglossus striatus. 2. Diploglossus microlepis.

P. Smit del et lith.

Mintern Bros. imp.

Diploglossus hewardii.

R. Mintern del. et lith.   Mintern Bros. imp.

Varanus caudolineatus.

BRIT. MUS. N. H. PL. XIX.

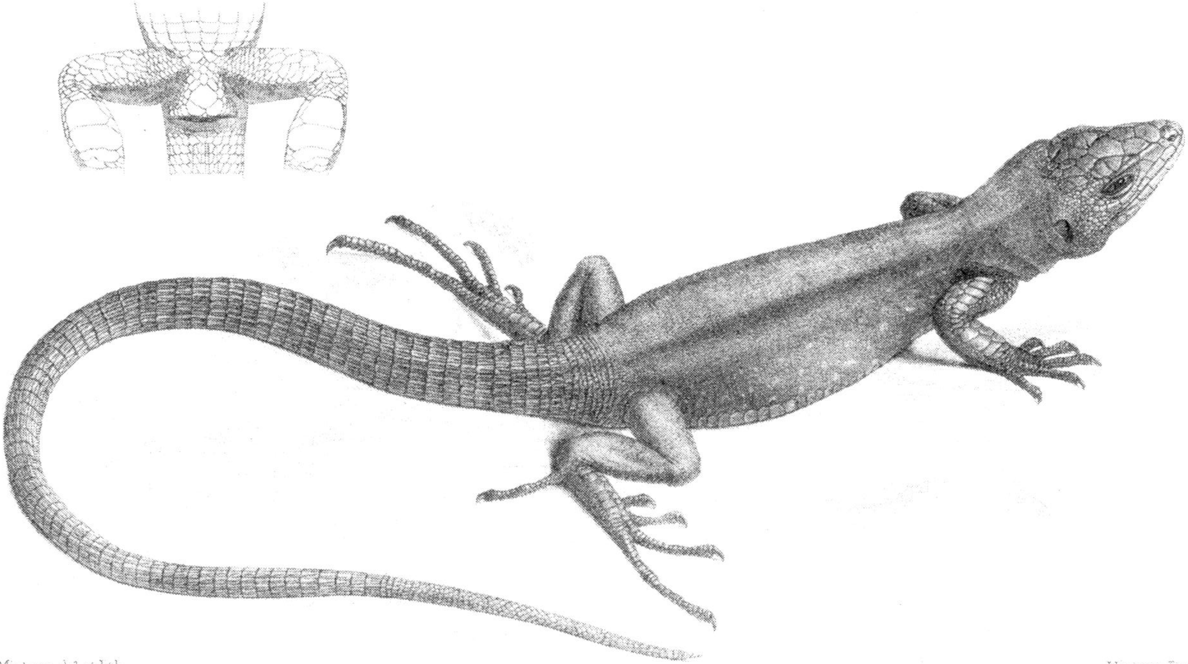

R. Mintern del. et lith. Mintern Bros. imp.

Cnemidophorus espeuti.

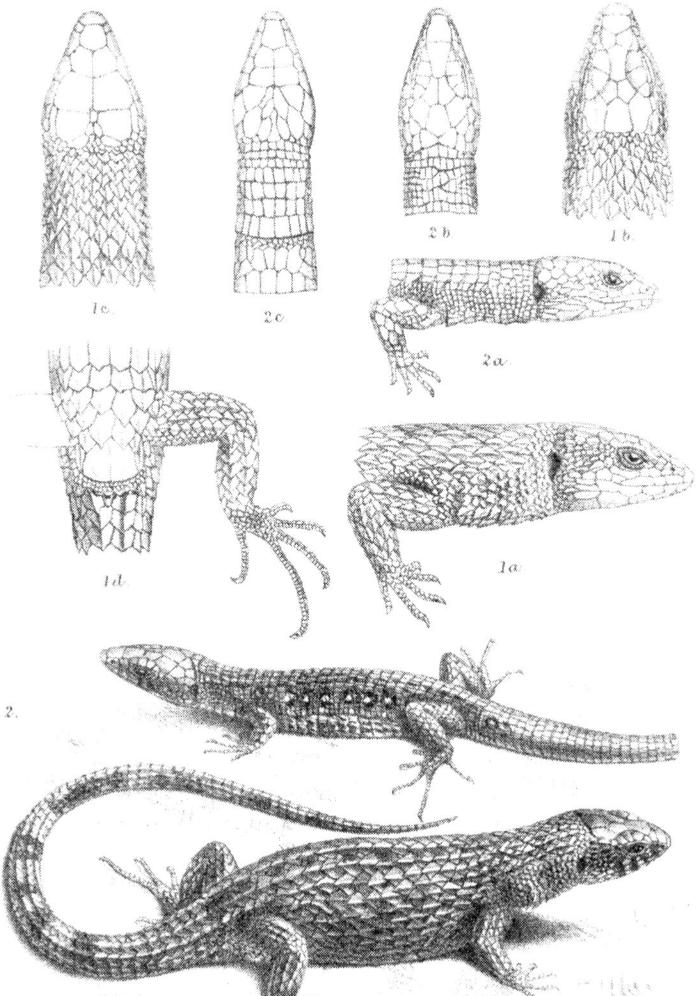

P. Smit del. et lith.

Mintern Bros. imp.

1. *Alopoglossus copii*. 2. *Oreosaurus oculatus*.

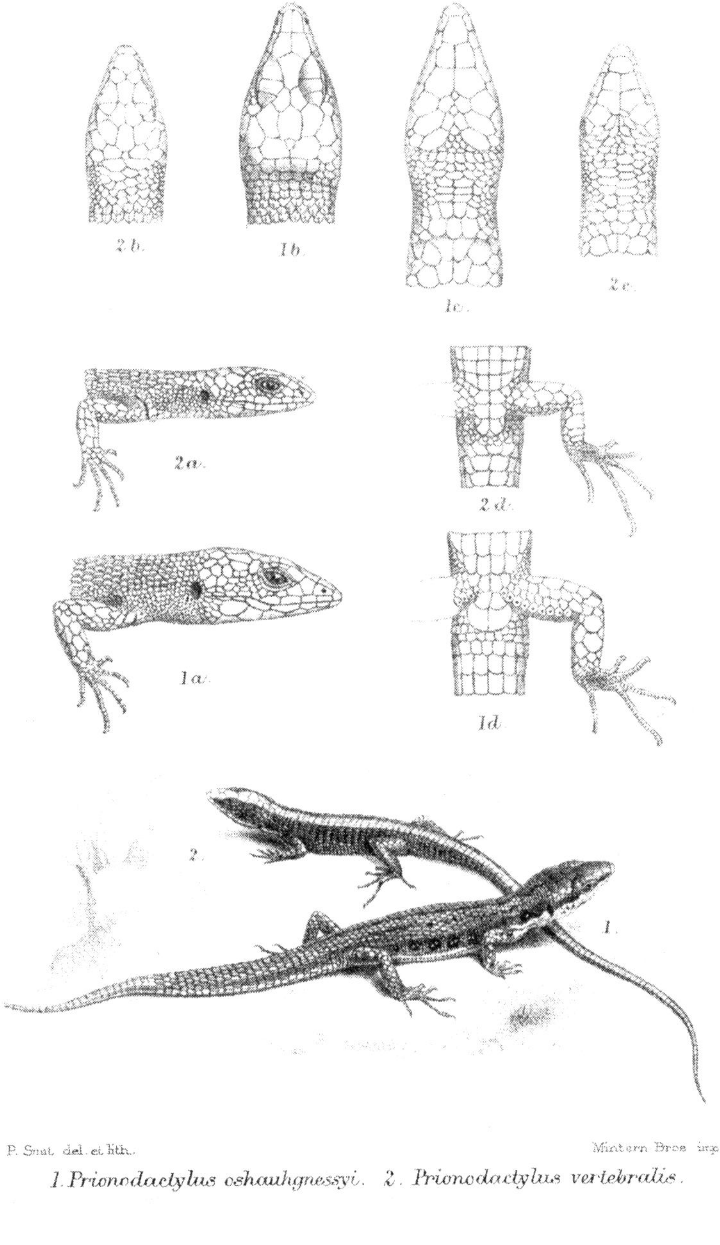

P. Smit. del. et lith. Mintern Bros. imp.

1. *Prionodactylus oshauhgnessyi.* 2. *Prionodactylus vertebralis.*

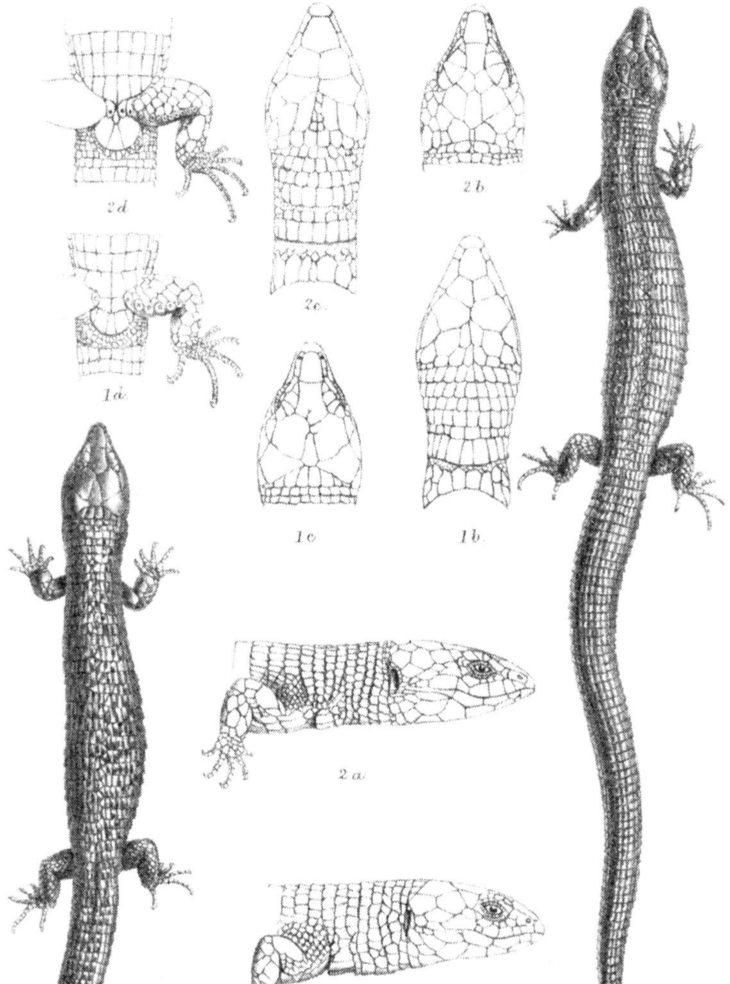

P. Smit del et lith. Mintern Bros. imp.

1. *Proctoporus simoterus.* 2. *Proctoporus meleagris.*

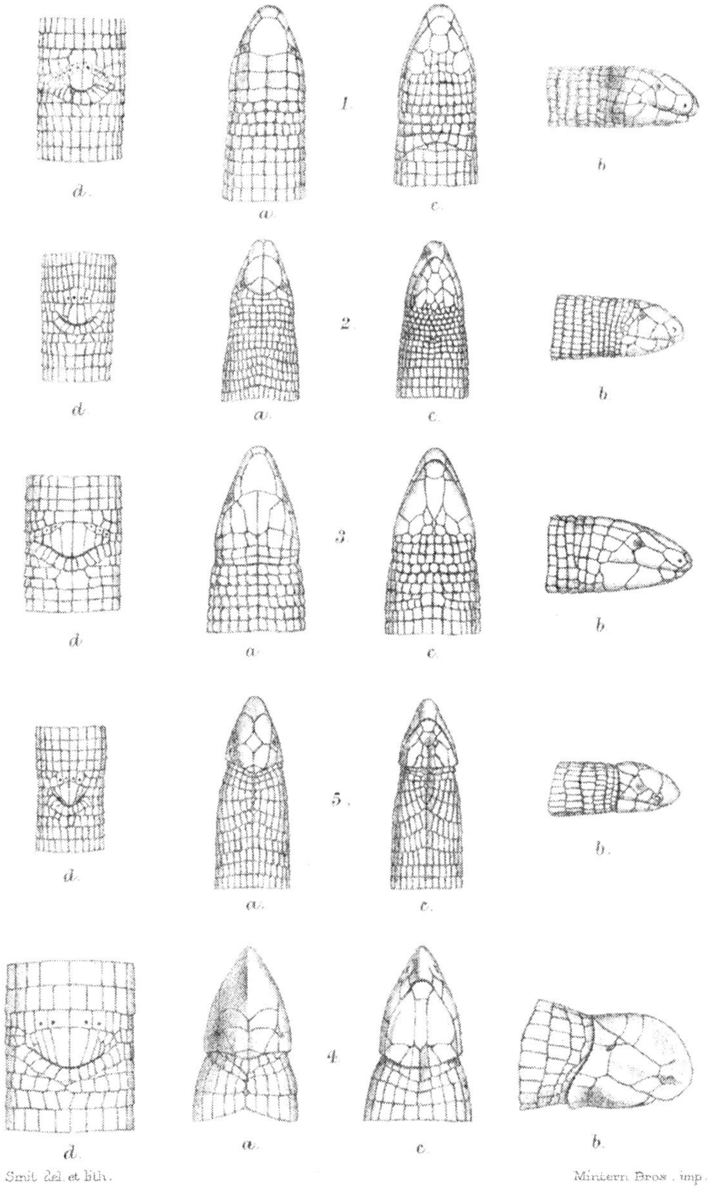

P. Smit del. et lith. Mintern Bros. imp.

1. Blanus bedriagæ. 2. Amphisbæna plumbea. 3. Amphisbæna punctata. 4. Anops africanus. 5. Geocalamus modestus.

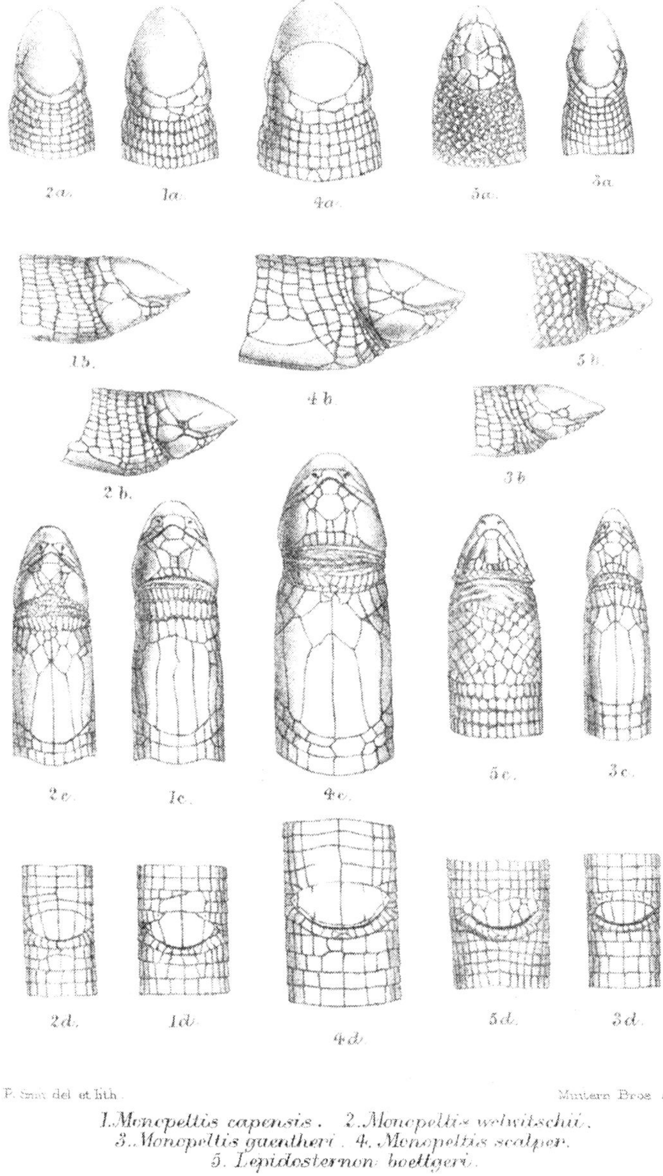

P. Smit del et lith. Mintern Bros imp.

1. Monopeltis capensis. 2. Monopeltis welwitschii. 3. Monopeltis guentheri. 4. Monopeltis scalper. 5. Lepidosternon boettgeri.

Printed in Dunstable, United Kingdom

91224064R00315